This work describes and analyzes the environmental history of the mountain areas of the Mediterranean world. It focuses on five sample areas, one in each of Turkey, Greece, Italy, Spain, and Morocco. Examining the peasant and pastoral economies, and the soils, forests, fields, and pastures on which they depended, it concludes that great changes in mountain landscapes occurred in the nineteenth and twentieth centuries, creating the often barren and depopulated countryside of today. It suggests that the social and political turbulence of modern times derives some of its impetus from the desperation of mountain folk whose forests and soils thinned with each passing generation.

STUDIES IN ENVIRONMENT AND HISTORY

The Mountains of the Mediterranean World

STUDIES IN ENVIRONMENT AND HISTORY

Other Books in the Series

Donald Worster *Nature's Economy: A History of Ecological Ideas*
Kenneth F. Kiple *The Caribbean Slave: A Biological History*
Alfred W. Crosby *Ecological Imperialism: The Biological Expansion of Europe, 900–1900*
Arthur F. McEvoy *The Fisherman's Problem: Ecology and Law in the California Fisheries, 1850–1980*
Robert Harms *Games Against Nature: An Eco-Cultural History of the Nunu of Equatorial Africa*
Warren Dean *Brazil and the Struggle for Rubber: A Study in Environmental History*
Samuel P. Hays *Beauty, Health, and Permanence: Environmental Politics in the United States, 1955–1985*
Donald Worster *The Ends of the Earth: Perspectives on Modern Environmental History*
Michael Williams *Americans and Their Forests: A Historical Geography*
Timothy Silver *A New Face on the Countryside: Indians, Colonists, and Slaves in the South Atlantic Forests, 1500–1800*
Theodore Steinberg *Nature Incorporated: Industrialization and the Waters of New England*

THE MOUNTAINS OF THE MEDITERRANEAN WORLD

AN ENVIRONMENTAL HISTORY

J. R. McNeill
Georgetown University

CAMBRIDGE
UNIVERSITY PRESS

PUBLISHED BY THE PRESS SYNDICATE OF THE UNIVERSITY OF CAMBRIDGE
The Pitt Building, Trumpington Street, Cambridge, United Kingdom

CAMBRIDGE UNIVERSITY PRESS
The Edinburgh Building, Cambridge CB2 2RU, UK
40 West 20th Street, New York NY 10011-4211, USA
477 Williamstown Road, Port Melbourne, VIC 3207, Australia
Ruiz de Alarcón 13, 28014 Madrid, Spain
Dock House, The Waterfront, Cape Town 8001, South Africa

http://www.cambridge.org

First published 1992
First paperback edition 2002

A catalogue record for this book is available from the British Library

Library of Congress Cataloguing in Publication data
McNeill, John Robert,
The mountains of the Mediterranean world: an environmental
history / J. R. McNeill.
p.   cm. – (Studies in environment and history)
Includes bibliographical references and index.
ISBN 0 521 33248 6
1. Man – Influence on nature – Mediterranean Region.
2. Deforestation – Mediterranean Region. 3. Soil erosion –
Mediterranean Region. 4. Landscape changes – Mediterranean Region.
5. Mountain ecology – Mediterranean Region. 6. Mediterranean Region –
Social conditions. I. Title. II. Series.
GF541.M35 1992
333.73'13'091822–dc20 91-44389 CIP

ISBN 0 521 33248 6 hardback
ISBN 0 521 52288 9 paperback

For J.A.B.

CONTENTS

List of Illustrations and Tables — *page* ix
Preface — xiii
Note on Transliteration — xv
List of Abbreviations — xvii

1 ***The Argument: Ecology, Economy, Shells, and Skeletons*** — 1
Population and Ecology: Overshoot — 2
Market Integration — 8

2 ***Mediterranean Mountain Environments*** — 12
The Mediterranean Zone — 12
Five Mountain Ranges — 19
A Few Villages — 49
Conclusion — 66

3 ***The Deep History of Mediterranean Landscapes*** — 68
Prehistoric Landscapes — 69
Ancient Landscape Change — 72
After the Ancients, A.D. 500–1000 — 84
The Slow Renaissance of the Mountains, 1000–1700 — 86
Five Mountain Landscapes, 500–1700 — 93
Conclusion — 102

4 ***Material Life in the Mountain Environment, 1700–1900*** — 104
The Land and the Seasons — 106
Vulnerabilities of Mountain Life — 114
Auxiliary Activities — 117
Basic Necessities — 126
Seasons of Life — 138
The Village and the Wider World — 142
Conclusion — 145

5 ***Population, Settlement, and Landscapes*** — 147
The Quality of the Data — 149
The Taurus — 152

The Pindus 162
The Lucanian Apennines 176
The Alpujarra 189
The Rif 202
Conclusion 218

6 *Political Economy and Mountain Landscapes* 221
The Loss of Auxiliary Activities 223
Economic Change and Environmental Decay 236
Politics and War 260
Conclusion 270

7 *The Changing Landscape since 1800* 272
Economy and Landscapes 272
Evidence of Landscape Change 284
Deforestation 286
Erosion 311
Consequences of Landscape Change 325
Interventions 346
Conclusion 349

8 *Conclusion* 351

Bibliography 359
Index 419

ILLUSTRATIONS AND TABLES

Photographs

2.1 Fields, houses, and ancient theater of Zerk (Altınkaya) with Bozburun Mountain in the background (July 1989). In the fourth century A.D. about twenty thousand people lived here. *page* 50

2.2 Bringing in fodder for the late summer dry season near Koçular, in the Yılanlı Plateau (July 1989). 52

2.3 Pindus villages still depend on the pastoral economy. High pasture between Vovousa and Perivoli toward the end of summer (1987). 55

2.4 Frangadhes is a village in central Zagori, typical in its architecture (September 1987). 58

2.5 Trevélez shrouded in snow (February 1988). Trevélez is the highest village in the Sierra Nevada. 61

2.6 Springtime (1988) in the *barranca* of Poqueira, with the villages of Capileira, Pampaneira, and Bubión. The tallest peaks of the Sierra Nevada lie behind this *barranca.* 62

2.7 The village and terraces of Mecina Bombarón, a typical Alpujarra landscape (February 1988). 63

2.8 Some of the rare terraces in the Rif, just south of Ketama. 65

6.1 The caravan system of the western Balkans required bridges like this one near Kipi in Zagori. This bridge was built in 1853, a busy time in Pindus commerce, although earlier bridges may well have existed on the same site. 233

7.1 Pindus goats in search of sustenance. Goats and fire are the chief agents through which humankind influences the spontaneous vegetation of Mediterranean mountain landscapes. 278

7.2 Mount Mitsikeli in October 1987. Two centuries ago this mountainside was well forested. 293
7.3 Mount Olytzika in October 1987. This slope lost most of its forest cover in the nineteenth century, but vegetation has recovered somewhat in recent decades. 296
7.4 Lucanian soils are especially prone to erosion. This slope, about 2 kilometers south of Pietrapertosa, is a typical example. 317
7.5 Tree crops offer minimal obstacles to accelerated erosion. Near Torviczón, in the Sierra Contraviesa, an almond grove loses ground to erosion and to encroaching maquis (February 1988). 321
7.6 A scene from the Nekor River valley where erosion rates are among the highest in the world. Hillslope erosion produces the braided river bed, but irrigation and hard work continue to produce good crops. 323
7.7 The barren slopes behind this *ksar* in the Nekor River valley were once cultivated. Faint traces of old terraces remain. 324
7.8 This *rambla*, or dry riverbed, serves as a highway except after a rain. Then its waters tumble into the Río Guadalfeo en route to the sea. A threshing floor is in the foreground. 341
7.9 Planted pines above this wheat field between Boured and Bab Sidi Kebat now slow soil loss. Severe gully erosion has inspired one of the few reforestation projects in the Rif. 348
7.10 Another reforestation project in the Rif, this one southwest of Bab Sidi Kebat. In this case there is no soil left to protect: the intervention came too late. 348
8.1 These ruins are the village of Campomaggiore Vecchio, destroyed by a mudslide in 1885. In Lucania, nineteenth-century deforestation increased the frequency of *frane*. 357

Maps

Mountains of the Mediterranean region xviii
2.1 The western Taurus mountains 23
2.2 The Pindus mountains 28
2.3 The Lucanian Apennines 33

2.4 The Sierra Nevada and Alpujarra 38
2.5 The Rif 45

Figures

1.1a Constant carrying capacity and periodic overshoot. 4
1.1b Carrying capacity rising in response to population pressure: overshoot is avoided (or forestalled). 5
1.1c Carrying capacity falling as a result of overshoot. This situation most closely approximates the modern environmental history of the mountains of the Mediterranean. 5

Tables

5.1 Population of Eğridir Kaza (District), 1475–1985 153
5.2 Population of Sütçüler District, 1831–1985 154
5.3 Population of selected western Taurus villages, 1935–1985 155
5.4 Antalya city population, 1530–1980 159
5.5 Population of selected Pindus villages, 1805–1981 164
5.6 Population of Pindus towns, 1735–1981 166
5.7 Population of Ioannina Kaza/Eparch, 1800–1981 168
5.8 Estimated Ioannina city population, 1731–1981 170
5.9 Mountain village households in Lucania, 1277–1669 177
5.10 Population of selected Lucanian mountain villages, 1735–1981 178
5.11 Basilicata population estimates, 1400–1791 180
5.12 Differential population history in southern Italy, 1793–1981 182
5.13 Population of selected Alpujarra villages, 1586–1986 192
5.14 Population of the Alpujarra, 1568–1981 194
5.15 Comparative population growth, 1845–1986: Alpujarra, Province of Granada, Spain 199
5.16 Population of selected Rif tribes, 1933–1982 210
5.17 Beni Bou Frah population, 1926–1979 (according to Pascon) 211
5.18 Beni Ouriarhel population, 1929–1960 (according to Hart) 212
5.19 Population of Morocco, Spanish Zone, and Fez, 1550–1988 214
5.20 Population of northern Morocco cities, c. 1500–1982 216

6.1	Imports to Corfu from the mainland, 1811	250
6.2	Imports to Corfu, 1827–1858	251
7.1	Forest Area of Basilicata, 1800–1962	301
7.2	Erosion rates in Rif watersheds	322
7.3	Price of fuelwood imported at Corfu, 1827–1859	326
7.4	Average yields in Potenza *circondario*, 1879–1894	332
7.5	Land use in Potenza *circondario*, 1879–1894	332
7.6	Rif cereal yields, 1934–1977	335

PREFACE

My interest in the Mediterranean world in a personal sense comes from my parents. They first met there, while serving with American forces in World War II. It was either 1944 in Egypt or 1945 in Greece: as so often in historical inquiry, available testimony is in conflict. My interest in the Mediterranean in a scholarly sense comes chiefly from brief exposure at age twenty to one of the titans of modern classical scholarship, a diminutive Oxford don named Russell Meiggs. His lectures at Swarthmore College on ancient Greek history made me want to go to Greece. I did go and, on my way there, stopped in Oxford and impudently (I was still twenty) called upon Meiggs, who with his wife fed and housed me for two days. He was too gracious to make it clear that he had better things to do with his time than entertain a hungry and grubby former student. In the course of my unannounced visit, he said many things which I still remember. One was that the Greek landscape I would soon see was a vastly different one than that inhabited by Leonidas and Alexander, and that I must try to imagine it with forests now vanished. In his last book – Meiggs died in 1989 – he suggested that a large part of the deforestation of the Mediterranean world came only recently. I suppose that one of the purposes of my book is to prove Meiggs right. I certainly owe him a good turn.

Without knowing it, I began working on this book soon after I left Meiggs and arrived in Greece. I had a job in Athens but took every opportunity to leave the city for the mountains of Attica, the Peloponnese, Evvia, and Macedonia. On weekends and holidays I walked in the mountains, enjoying the unsurpassed hospitality of Greek villagers. Since then I have walked in the mountains of Turkey, Italy, Spain, Morocco, and Tunisia. Some of this I did for fun, some, eventually, explicitly as research – which did not diminish the fun. Everywhere I met with peasants and shepherds who fed and housed me, and taught me useful things, from the names of trees to small lessons in the art of sheepherding. I owe a good turn to all those who shared with a stranger their food, water, shelter, and tiny fractions of their knowledge. Many of them are now dead, I expect, since most were old men when we met. They no less than Meiggs have inspired me to try to find out how their world got to be the way it is, and why its day has passed.

In addition I have debts to dozens of archivists, librarians, scholars, and friends (overlapping categories) who have helped this book along at one stage or another. For assistance in Rabat I thank Professor Omar Abboulabes, M. Tayaa, M. Chrayah, and Edward Thomas. In Madrid, Juan Pan-Montojo and Patricia Zahnisser helped out in many ways. Dr. Miguel Guirao directed me to manuscripts in Granada. Fernanda Serenella, Stefanie Ayazi, and Giulia Barrera guided my researches in Rome. In Athens, Epaminondas and Beata Panogopoulos offered hospitality, as did Spiros Zournatzis and Chip Ammerman. Dr. Niki Goulandri and Professor G. Mavrommatis answered many questions with admirable patience. In Thessaloniki I enjoyed the hospitality and wisdom of Bruce and Tad Lansdale, Bill McGrew, John Koliopoulos, V. Rotsios, and Anastasios Papastavrou. In Ioannina welcome help came from Eleni Kinga, Christos Raptis, and Michael Arapoglu. In the tiny village of Koukouli, Kostas Lazaradis explained local history and botany to me. In Paris, Florence Martin permitted me to make camp in her apartment for several weeks, and in London Jim Stoner discussed every aspect of my work with me. Professor Juan Martínez Alier of Barcelona asked a good question that focused my thinking. Scholars who have answered my questions include Professor Halil Inalcik, Professor Karl Butzer, Dr. M. Reille, Dr. Abdellatif Bencherifa, and Dr. Turhan Istanbullu. Pietro Turilli, Tina Zournatzi, Luisa Fonseca, and Pinar Gürler, former students of mine, have helped me procure or translate documents from their native lands. Eileen Scully typed the bibliography with dispatch and good humor. Alan Karras, Karl Butzer, Tommaso Astarita, Julie Billingsley, William McNeill, Scott Redford, Alfred Crosby, and Donald Worster read all or parts of the manuscript for me and have improved it markedly. Frank Smith, my editor at Cambridge University Press, showed faith in me at the start and patience ever after. Faith, patience, and financial support came from the Fulbright program and Georgetown University's School of Foreign Service. To all these people and institutions, and to many more, I owe great thanks. None of them, of course, bear any responsibility for what errors and misjudgments remain.

NOTE ON TRANSLITERATION

Many systems exist for the transliteration of modern Greek, none of them satisfactory. I have aimed for phonetic fidelity. Thus, for instance, the Greek letter gamma may come out as *y* or as *g*, depending on its sound. With respect to Arabic and Berber, I have transliterated place names in the French style, because this is the system used in official Moroccan publications. The only (intentional) exceptions are places that have well-known English names, such as Fez or Tangier. As for plurals, I have normally added an *s* to foreign words where that will be clearer to readers than would be preserving the original plural forms. Thus *yaylas* instead of *yaylalar*. In practice this affects Greek, Arabic, and Turkish.

ABBREVIATIONS

ACEM	Alta Comisaria de España en Marruecos (Tetouan)
ACSR	Archivio Centrale di Stato (Rome)
AHN	Archivo Histórico Nacional (Madrid)
AMAE-M	Archivo del Ministerio de Asuntos Extranjeros (Madrid)
AMAE-P	Archives du Ministère des Affaires Etrangères (Paris)
AMAPA	Archivo del Ministerio de Agricultura, Pesca y Alimentación (Madrid)
Arch. Parl.	Archivi Parlamentari (of the ACSR, Rome)
ASAEAG	Archivio Storico della Accademia Economico–Agrario dei Georgofili (Florence)
ASB-P	Archivi di Stato Basilicata (Potenza)
BL AM	British Library, Additional Manuscripts (London)
BNM	Biblioteca Nacional (Madrid)
BRAH	Biblioteca de la Real Academia de la Historia (Madrid)
BSA	British School of Archaeology (Athens)
CO	Colonial Office papers (in the PRO)
DERRO	Développement économique rural du Rif occidental (Morocco)
DGA	Direzione Generale dell'Agricoltura (section of the MAIC)
EHM	Society for Epirot Studies (Ioannina)
FAO	Food and Agriculture Organization of the United Nations (Rome)
FO	Foreign Office papers (in the PRO)
ICONA	Institución Nacional para la Conservación de la Naturaleza (Spain)
MAIC	Ministero dell'Agricoltura, Industria e Comercio (section of the ACSR)
PC	Presidenza del Consiglio (a section of the ACSR, Rome)
PRO	Public Record Office (London)
SHM	Servicio Histórico Militar (Madrid)

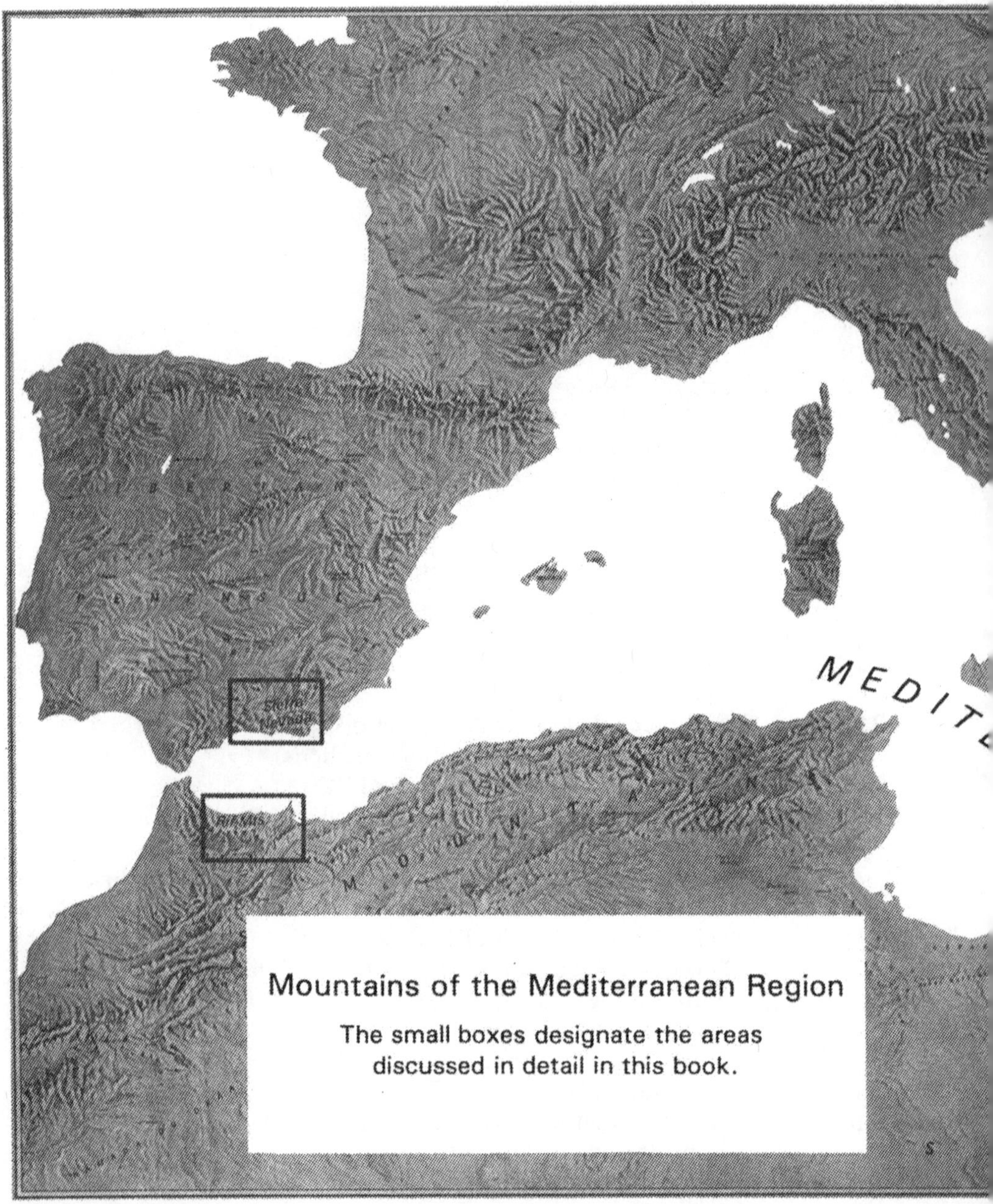

Mountains of the Mediterranean Region

The small boxes designate the areas discussed in detail in this book.

Reproduced courtesy of the National Geographic Society.

1

THE ARGUMENT: ECOLOGY, ECONOMY, SHELLS, AND SKELETONS

> I have loved the Mediterranean with a passion, no doubt because I am a northerner like so many others in whose footsteps I have followed.
>
> Fernand Braudel

The Mediterranean world has a special beauty to it. The stark light, the soaring mountains, the sparkling sea, and much else contribute to make landscapes that have long attracted people from afar. Dr. Johnson once said that the grand object of all travel is to see the shores of the Mediterranean. Apparently little has changed since the eighteenth century: today one-third of all international tourism in the world involves Mediterranean countries. But in fact a good deal has changed, including the landscapes themselves.

The beauty of the Mediterranean mountains is in a way a sad one. Skeletal mountains and shell villages dot the upland areas of the Mediterranean world, dominating the physical and social landscape. Rugged limestone ridges or smooth schist shoulders, bare of all but the scantiest vegetation, make the famous and apparently timeless vistas from Granada, Marrakesh, or (on a clear day) Athens. Between the ridges, usually situated so as to enjoy the Mediterranean sun, or perhaps a source of fresh water, lie quiet and moribund villages. They are shell villages, home only to the very old and sometimes the very young, but, perhaps with the brief exception of some summer weeks, home to no one in the prime of life. Both the mountains and the villages are usually picturesque. But their beauty is that of a still-life painting – *nature morte* as the French put it. They are dying villages and sterile mountains.

It has not always been so in the mountains of the Mediterranean, and indeed it is not yet so everywhere. Not very long ago, forests clothed the high slopes and hard-working peasants and shepherds, scraping an often difficult living from their surroundings, filled the villages. And in some corners, like the Rif, while the mountains are almost bare, the villages are by no means empty. But they may well be soon. A mountain way of life is coming to an end there, as it has throughout the Christian Mediterranean world. In this book I will try to explain something of this way of life, now dead or dying, what it depended on, and how and why it has ended – or is ending.

Skeletal mountains and shell villages are the legacy of two slow changes: one is essentially ecological and the other economic. The modern physical and social landscape of Mediterranean mountain areas is a product of these two classes of change. Generally speaking, ecological processes have asserted themselves on the local level whereas economic ones have operated on a far vaster scale. That does not mean, however, that economic processes have been more important than ecological ones. Often it has been the other way around. In each village and landscape the combination has been unique. The only safe general conclusion is that both classes of event were important: both the topsoil erosion caused by the sudden collapse of a peasant's terrace, and the gradual development of a world-scale wheat market contributed to the demise of the mountain way of life.

Ecological and economic processes often combined, sometimes simply – as in the demand for timber to build the Suez Canal that denuded parts of the southern flank of the Taurus – and sometimes in devilishly complicated ways. In either case, for most of the Mediterranean mountain world, the changes that ended a way of life and left behind skeletal mountains and shell villages are comparatively recent, in most settings less than two hundred years old. This is not long in the history of the Mediterranean. Naturally peasant terraces collapsed time and again over millennia, and large-scale economic systems shifted too. But in the mountains the changes of the past two hundred years – in some places the past one hundred – have normally been greater than the slower, more modest ecological and economic shifts of earlier times. The recent ones will, I suspect, also prove more decisive: not mere fluctuations within a broad and resilient equilibrium, but a sea change.

Population and Ecology: Overshoot

After about 1800, earlier in some locales and much later in others, many mountain communities began to experience what ecologists colorfully refer to as *overshoot.* At roughly the same time, they felt the impact of what economists, less colorfully, refer to as market integration. Overshoot is a concept derived from population ecology.[1] When

[1] Catton 1980; Hutchinson 1978.

a population has grown beyond the level that is sustainable with its given resources, it has overshot. In the Mediterranean world, both in the uplands and the lowlands, population began to grow in the late eighteenth century, as it did in many other parts of the world. This trend was no simple recovery from famine or epidemic, but a basic change in demographic regime. In the lowlands this population pressure led to several changes that eventually produced higher living standards, although this result came slowly and haltingly, with much suffering along the way. In the uplands, such beneficial changes – say, the use of machines in agriculture – proved impossible to bring about,[2] and rapid population growth meant that people were faced with the prospect of lower standards of living. Their numbers had grown beyond the point that was sustainable given their resources, and their resources proved difficult to expand through technical means. Indeed very often their resources shrank, as upland peasants desperately cleared and farmed slopes that, they well knew, could not hold soil. In getting one or two years' needed crop, they undermined prospects for the future.

Among animal populations this overshoot usually means demographic catastrophe, normally through starvation. This pattern can occur among human beings, and some ecologists explain recent Ethiopian and Sahelian famines this way. But human affairs are much complicated by our capacity to adapt culturally and individually to our circumstances. Overshoot among a human population generally produces a desperate search for ways in which to expand the resource base, to enlarge the niche of that population, so as to avoid disaster. Sometimes overshoot and impending disaster can lead to technical revolutions that quickly expand usable resources, or to slow but sustained improvements in agriculture that avoid (or at least forestall) crisis.[3] This is the story of rising population and rising agricultural productivity found in much of lowland Europe since 1750, in China, in Java, and in many places in between. It may be that those who made the earliest conversion from hunter and gatherer to cultivator and pastoralist did so in a desperate attempt to avoid the consequences of overshoot.[4] Sometimes overshoot can lead to war and attempted conquest, as one community seeks to acquire the resources of another. It can also lead to emigration, as individuals seeking to solve their own problems accidentally mitigate or postpone the problem of their communities (unless too many go at once). In complex economies overshoot can inspire people to try to make a living in new ways, to exploit neglected niches through trade

[2] Mignon (1981, 207–14) explains this for the Alpujarras.

[3] Boserup 1965; Geertz 1963. [4] M. Cohen 1989, 21–3.

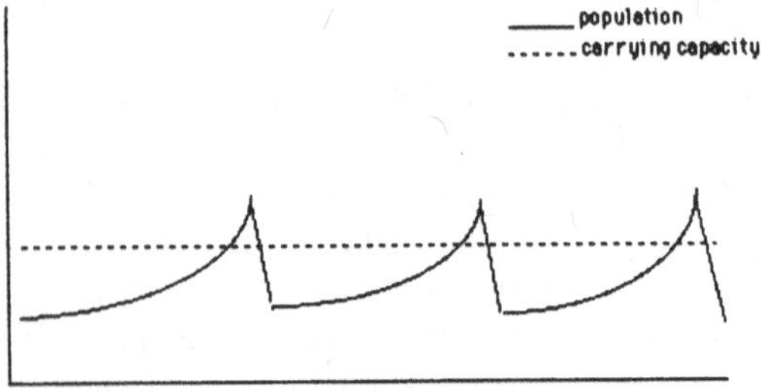

1.1a. Constant carrying capacity and periodic overshoot.

and manufacture. One way or another, overshoot must produce adaptations, changes in a group's way of life that either extend the resource base or permit more intensive exploitation of the existing one; otherwise, it will produce demographic disaster.

In the Mediterranean mountains overshoot promoted several adaptations, ranging from agricultural experimentation to specializations such as silk manufacture, military service, and brigandage – and to emigration. None of these adaptations were entirely new, of course. Overshoot did not produce them but merely drove far larger numbers to try them.

Perhaps reference to an abstract model (even in an introduction) will help clarify matters. When ecologists speak of carrying capacity they illustrate the concept with a simple graph that has two lines.[5] One is straight (and usually dotted) and refers to a maximum sustainable population. The other represents population history, and it curves up and down. When it rises above the dotted line, it soon falls far below it. This is overshoot (Fig. 1.1a). This model is useful in the abstract and among animal populations may be fine as a first approximation to actual population dynamics. But it is too simple to be of use in reference to human populations. The difficulty is that the dotted line has a history too.

Carrying capacity can rise or fall, sometimes suddenly. It may rise as a population acquires new skills or technology (Fig. 1.1b), bringing opportunity and growth, and forestalling the pangs of population pressure. These are happy times in a community's history. But the dotted line may curve cruelly downward as well (Fig. 1.1c), if skills are lost or resources depleted.

[5] Ellen (1982, 41–6) reviews the pitfalls of carrying capacity in the context of a fine discussion of the great questions of human geography.

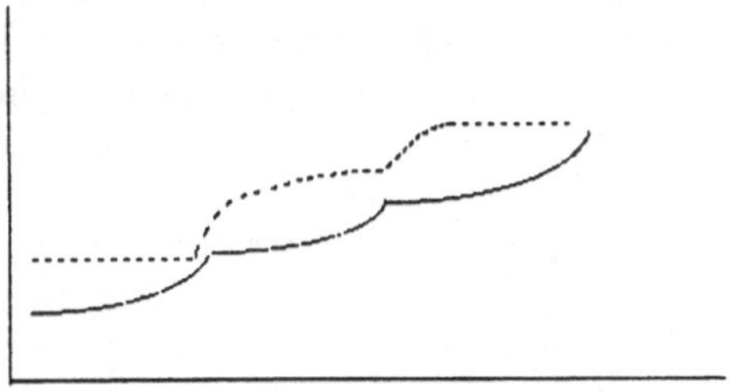

1.1b. Carrying capacity rising in response to population pressure: overshoot is avoided (or forestalled).

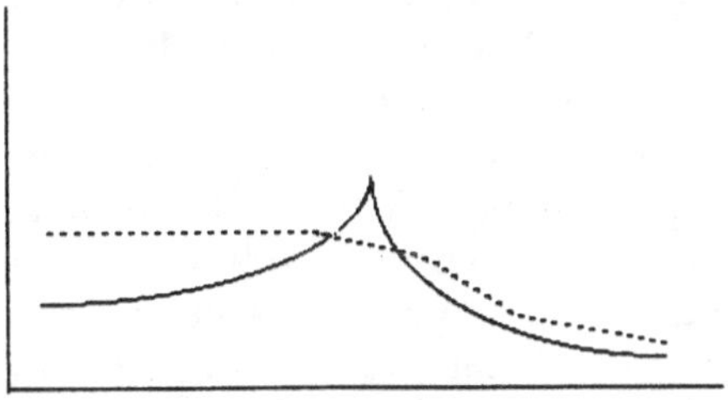

1.1c. Carrying capacity falling as a result of overshoot. This situation most closely approximates the modern environmental history of the mountains of the Mediterranean.

The mountain people of the Mediterranean world have experienced both expansions and contractions of their local niche. For example, after the sixteenth century, American food crops, notably maize and the potato, arrived in the region, making mountain survival much easier; the dotted line ascended. But more recently, as I will argue at some length, deforestation and soil erosion have reduced the productivity of mountain lands; the dotted line turned downward. That both carrying capacity and actual population can change over time does not invalidate the ideas of carrying capacity and overshoot. These notions, while harder to apply confidently, retain their importance. They are, I think, essential to understanding the pressures under which mountain people lived and live.

In gentle environments the dotted line, over the long haul, has climbed with each practical application of human ingenuity. But in difficult environments, history is less open-ended and constraints are far more important. As a rule, gentler environments have felt the pinch of population pressure less often, because the landscapes are less subject to degradation (a downturn of the dotted line) and they are more receptive to technical improvements (upturns in the dotted line). Human breeding behavior being what it is, many landscapes at many times have felt population pressure. But more marginal ones, less flexible ones, have felt it more often and have found fewer answers.

Furthermore, gentle environments can be more resilient than marginal ones. Human effect on ecosystems is of two sorts: temporary perturbations and permanent changes. A fire, a poisoning, an episode of deforestation – these will be temporary perturbations in many parts of the world, and after the event flora and fauna will slowly tend toward whatever condition current climate and soils encourage. Permanent changes normally require ongoing human action, say agriculture or city life. But in marginal environments temporary perturbations are often not so temporary. Recuperation is slow and scars last long.

The ecological fragility of the Mediterranean lands, and especially their mountains, adds yet another complication to the delightfully simple abstraction of carrying capacity and overshoot. It makes these lands especially vulnerable to deterioration from population pressure. But in addition, it has made them vulnerable to population fluctuation and even to underpopulation. This curious claim requires some explanation.

Like marginal environments everywhere, say the desert edge or the moist tropical forest, the Mediterranean mountains have a finite carrying capacity, a maximum threshold population, which changes with levels of technology and productivity. But they also have a minimum threshold, a population level beneath which labor shortage prevents agriculture in all but the garden spots. Mountain agriculture needs many hands and backs, because few labor-saving devices apply, and terraces and irrigation ditches require constant surveillance and maintenance. If the available supply of hands and backs falls beneath a certain minimum, terraces and irrigation cannot be maintained, agriculture becomes notably less productive, and the survival of one and all is imperiled: undershoot. This has often happened in Mediterranean Europe, and has combined with damaged community morale – the feeling that no one is going to stay so let's not be left behind – to provoke mass emigration. Emigration can be a bit like soil erosion: it might begin with a tolerable trickle, but can pass a threshold where morale (like the soil's integrity) is punctured, whereafter it accelerates uncontrollably.

Overpopulation normally led to desperate and reckless land clearing and accelerated soil erosion. Underpopulation normally led to terrace collapse, field abandonment, destruction of irrigation works. In *either* event landscape suffered, topsoil – sometimes all soil – was lost, and the potential for vegetation recovery reduced. Once human communities had established themselves in the mountains, either overpopulation or underpopulation could lead to landscape degradation, even desertification.

In gentler environments, such as northwestern Europe or eastern North America, such tight constraints have never existed. Land clearing for agriculture has led to only modest erosion, neither enough to prevent the return of forest to abandoned lands, nor enough to prevent the reestablishment of stable ecosystems that retain some utility for humankind. Underpopulation cannot exist in these lands, and overpopulation comes, if at all, only at a much higher threshold, because of the suitability of level landscapes to agricultural innovations. But the Mediterranean, like other fragile ecosystems, was fundamentally different: once villages existed, they could function in a sustainable way only within a fairly narrow band of population density. If human numbers rose too high or dipped too low, the landscape suffered, chiefly through soil erosion. When this happened, the narrow band tightened, as the impoverished soils could support slightly fewer people at the maximum, yet needed slightly more labor to be productive. The range of suitable population shrank, making fluctuation in one or another direction all the more dangerous.

Until quite recently population levels have *normally* fluctuated in every human community, in an irregular rhythm defined by famine, disease, and war. Of late, population almost everywhere seems only to rise, but this is a recent trend, not characteristic of much of human experience (and probably not destined to endure much longer). In level places, or rugged places without aggressive rainfall, population fluctuation has had limited durable impact upon landscapes. Vegetation might change, it might even come and go, but as long as soils did not wash away, vegetation could return, and farming might return if people needed it. But in marginal environments this resilience does not exist, and population fluctuation outside of a certain band has had a downward ratchet effect on the productive potential of landscapes, making overshoot – or undershoot – all the more likely. In effect, human communities in marginal environments live under a death sentence, which can be stayed only through careful husbandry of the land. In the mountains of the Mediterranean, other forces have helped to prevent an indefinite stay of execution.

Market Integration

Market integration is generally a good thing for people who have a comparative advantage in the production of one or many goods and services. It means that one can sell more of one's specialty. But it is very bad indeed for people who have no comparative advantage in anything except their labor power, and who find that whatever they can produce, others can produce better or more cheaply. For those who have few things to sell, and for whose goods demand is unpredictable, joining a market system adds to the uncontrollable elements to which one's fate is hostage: weather, war and politics, and now prices as well. For Mediterranean mountain peasants in the nineteenth and early twentieth centuries, as for African and South American peasants since 1950, this has often been a change for the worse, because added economic insecurity outweighs the importance of possible new wealth obtained through the market. In the case of Mediterranean mountain villagers, market integration proved a mixed blessing, but in the end, it was corrosive of the mountain way of life.

Most mountain villagers had never been entirely cut off from the currents of economic life on the plains or even on the seas. Many worked as seasonal laborers, in agriculture, construction, or warfare in the plains below. Women as well as men often walked for days to find seasonal work on the farms of Thessaly or the Basilicata coasts. This was greatly to their advantage, as it meant they knew something of the world beyond, and how to exploit its niches. It made them more adaptable than they could otherwise have been. So when their villages came in closer and steadier contact with the outside world, some of these mountain folk took advantage of it, changing their lives for the better. But all had to change their lives, for better or for worse.

Like overshoot, market integration was a gradual process. While most Mediterranean mountain villages were never fully isolated from the economies of the plains, neither are they yet entirely integrated into larger markets. A fair amount of subsistence production still goes on. But by and large a big change took place between 1800 and 1950, a slow evolutionary change with revolutionary consequences. The technology of transport and the organization of markets improved to the point where outside goods, even agricultural ones of high bulk and low value, could be imported into mountain villages. And mountain folk could sell whatever goods or services in which they might enjoy comparative advantage. Usually, however, these were not their principal products, most of which were foods that the plains – or distant but now economically closer America – produced better. So villagers had to adapt to offer things for which outsiders would pay: timber, silk, honey, and, most often, labor services.

Sometimes distant changes in demand opened up whole new opportunities for mountain villagers, although they often reaped only a small share of the benefits. When nineteenth-century urbanization, and in particular urban sanitation efforts, raised the market price of lead everywhere, villagers around the Sierra de Gádor in southeastern Spain found they could get work digging lead out of the mountains. They did so prodigiously, temporarily making theirs the world's largest lead mining region. Catalan and English companies made great sums from this, while tens of thousands of Spanish peasants made modest livings from it.

Lead mining in the Sierra de Gádor is a poor example of market integration and its effects on mountain villages in that it rose and fell rather quickly between 1830 and 1860. Most other cases proved more enduring. But it is an excellent example of an important feature of this integration because it was ecologically very destructive, deforesting thousands of hectares of Almería and Granada. This region now contains the starkest desert in Europe. (Parts of Almería so resemble the American southwest that Spanish spaghetti Westerns are filmed there.) In general, the economic processes in which mountain villages enjoyed comparative advantage – herding, logging, and charcoal making, for example – were potentially or necessarily destructive of vegetation and soils: they were hard to sustain indefinitely and could easily deepen the ecological difficulties villagers had already begun to encounter. As population growth threatened villages with overshoot, the new directions their economies were beginning to take often reduced the resource base just at a time when villagers needed to expand it.

Market integration meant that distant demand could be concentrated on a relatively small producing area. This can often have an ecological effect roughly analogous to the focusing of the sun's rays on a single point, as children do with a magnifying glass in order to set a leaf on fire. When, in the late nineteenth and early twentieth centuries, the world demand for ivory – for piano keys, billiard balls, and combs – focused on East Africa, where lived millions of elephants with the soft ivory tusk best for carving, the result for the elephants was catastrophic. Africans, Arabs, and eventually Europeans quickly and severely depleted a population, which, for various reasons, has never recovered. Much the same thing happened to the beavers of northern New England between 1620 and 1700 when beaver fur came into fashion in western Europe. The beaver here has recovered a bit better than East African elephants. Similarly, but on a far larger scale, in the sixteenth and seventeenth centuries the Andean silver mountain of Potosí was systematically connected to Europe and East Asia, and legions of miners soon exhausted the silver veins while their fellows stripped the surrounding vegetation for fuel. The silver, of course, has never recovered. In this sad case concentrated demand also rapidly depleted the surrounding human population, and it too has never recovered. The history of the Andes contains two other curious examples of ecological degradation produced by the impact of market integration. When European agriculture began to use large quantities of Peruvian guano in the nineteenth century, Andean highlanders could no longer afford their customary fertilizer. Their solution was to raise more sheep in their mountains. They now have the fertilizer they need for their fields, but at the cost of overgrazed slopes and hydrological disorder.[6] More recently the eastern face of the Peruvian Andes has responded to the vigorous North American demand for cocaine by deforesting tracts and vastly expanding the cultivation of coca. As cultivated today (some evidence suggests it once was grown only on terraces), coca permits very rapid erosion.[7]

Similar ecological disturbances resulted from distant demand concentrated into small zones in the Mediterranean mountains. Timber demand inspired a German company to deforest the Sila uplands in Calabria around the turn of the century. The presence of Russian, French, and British garrisons on the Ionian Islands from 1797 to 1863 helped cause a great expansion of flocks in Epirus, leading to overgrazing and deforestation. More recently, the strong demand for marijuana in western Europe has led to its widespread production in the Moroccan Rif, with unfortunate and complex ecological consequences detailed in Chapter 7. In other places, at other times, and in other ways concentrated demand from afar proved ecologically destructive in Mediterranean mountain areas. And where it did not, improved trade links often brought local mountain production into devastating competition from more efficient lowland producers, obliging mountain people to adapt or emigrate. Either course could deepen ecological problems. Market integration, as much as overshoot, has created the contemporary physical and social landscapes of the Mediterranean mountains.

A few of the millions of peasants affected, while forced to alter their inherited way of life, found very happy landings. Quite a few found satisfactory ones, and very, very few starved to death. So it is tempting to think that while overshoot and market integration may have meant

[6] Gulliet 1984.

[7] Dourojeanni 1990. In addition, transforming coca into cocaine requires chemicals, notably kerosene and sulfuric acid, that are turning up in toxic concentrations in the Peruvian Amazonian river system.

declining living standards in recent centuries in the Mediterranean mountains, the adaptations provoked and pursued, most notably emigration, have on balance made for a happy story. Certainly it is hard to find a mountain emigrant who prefers to return to his native or ancestral village. But many will say that mountain life is better than life in the plain or the city. It may be that while they prefer not to live in the shell villages of today, they think they would prefer the villages of yesterday, if not overcrowded. Perhaps they pine for a simpler past, one without anonymous buying and selling, without concrete and glass, and idealize it accordingly. In any event, the mountain life they remember (or have been told about) is gone, and whether the demise of the mountain way of life is an improvement, on balance, is impossible to say.

In the pages that follow I will argue that ecological and economic changes combined to force the abandonment of the mountain way of life in the Christian Mediterranean, and that the same forces are currently operating in similar ways in much of the Muslim Mediterranean. I will maintain that most of the deforestation and anthropic erosion in upland areas is comparatively recent. I will contend that peasants suffered from fuel shortages and land shortages, as wood and soil grew ever scarcer, while population grew more abundant. Further, I will claim that this difficult situation led to desperate attempts to preserve customary life, most of which merely exacerbated ecological troubles, as peasants in effect consumed their capital in an effort to avoid or delay starvation or a risky venture into the outside world. Finally, I will claim – with little to offer as proof – that the straitened circumstances of mountain folk in the nineteenth and early twentieth centuries created an enduring resentment, a sense of grievance, that made mountaineers ready followers (and occasionally leaders) of political movements that promised basic changes. That is, the mountains, which in previous centuries supplied so many loyal recruits for the princes and sultans, helping them maintain and expand their authority, now served as a reservoir of rebelliousness, promoting political instability where they had once done the opposite. The tumultuous political history of the modern states of the Mediterranean world has among its many tributaries a current of mountain ecosystem history.

2

MEDITERRANEAN MOUNTAIN ENVIRONMENTS

There is no species of animals on earth, nor birds on the wing, who are not part of a community, just as you are.

Koran, 6.38

The casual observer finds in the Mediterranean world a certain unity. In ancient times the Mediterranean exhibited considerable cultural unity, many traces of which still survive. Belief in the evil eye exists on all shores of the Mediterranean. The same gesture of the wrist is used in the Casbah in Algiers and Trastevere in Rome to mean "I don't know." At various times the Mediterranean has exhibited a certain economic unity, its phases of expansion and contraction felt throughout.[1] Whereas cultural and economic unity may have come and gone, ecological unity has not. Ecologists do not agree as to what constitutes an ecosystem – some prefer the scale of a fishpond while others feel Amazonia qualifies – but none fail to see the Mediterranean as an ecological unit. They may call it a single ecosystem or a series of parallel ecosystems, but they all recognize its distinctiveness just as easily as the casual observer feels it.

The Mediterranean Zone

The Mediterranean is not so much the sea between the lands, as its name asserts, but the sea among the mountains. Except for the long stretch of lowland desert between Tunisia and Sinai, one is almost never out of sight of mountains if within sight of the Mediterranean coast. African and European tectonic plates have been slowly and intermittently colliding for millions of years, forcing up mountains all along this geologic frontier. Most show their marine origins clearly: they are predominantly of limestone, and one can find fossil seashells above the timberline. Almost every range parallels the nearest coast. In these

[1] Sarton 1936, on cultural unity; Braudel 1976, on economic.

mountain environments on all sides of the Mediterranean, history, geography, and economics, and much else besides, share some of the same patterns and rhythms, distinct from the rest of Europe, Africa, or Asia. It is the mountains and the climate that make the Mediterranean different.[2]

Mountains

Beginning in the west, the Moroccan Atlas and the Spanish Baetic ranges mirror one another across the narrows of the western Mediterranean. The Atlas consists of three distinct chains, the High Atlas (Jebel Toubkal, 4,165 meters) in the southwest; the gentler Middle Atlas (Jebel Bou Naceur, 3,340 meters) to the northeast; and then the Little Atlas, more commonly known as the Rif (Jebel Tiddiquin, 2,448 meters), a lower but very rugged range whose ridges run down to the Mediterranean coast. Napoleon said that Africa begins at the Pyrenees, and Spanish imperialists a century later said it began at the High Atlas. But geologically it starts between the Rif and the Middle Atlas; standing there in the Taza gap one can detect a change in the color of the earth that corresponds to what are essentially African and European tectonic plates. The African Atlas continues throughout the length of the Maghreb to Tunisia, growing lower and less rugged toward the east. But all through the Atlas there are mountain villages striking much the same bargains with nature as their counterparts elsewhere around the Mediterranean.

The Baetic range is a set of cordilleras in southeastern Spain, one of several ranges that rib the Iberian Peninsula east to west. By far the tallest chain in the Baetic complex is the Sierra Nevada (Mulhacén, 3,481 meters), a gentle schist massif towering over the city and the plain of Granada. Low east–west ranges mark the Iberian Peninsula up to the Pyrenees. Southern France has a narrow stretch of coastal lowland, lying between the Cévennes (lower than 1,300 meters) and the sea, but toward the east the aptly named Alpes Maritimes rise to almost 3,300 meters, linking the Alps and the Italian Apennines. These latter (Mount Corno, 2,912 meters) run the length of Italy, to Sicily. The Alps also connect with the long Balkan system of the Dinaric Alps and the Pindus, making the Balkan Peninsula as mountainous as the Italian. The undersea mountains of the Aegean (the peaks of which are the Greek islands) link the Pindus with the mountain chains of Asia Minor. Turkey is almost all mountain and high plateau, with the Koroğlu (up

[2] *La mobilité* 1984, 11–276, for a recent summary of the geology and plate tectonics of the Mediterranean. Also: Ardos 1969; Bousquet 1974; and for current controversies, Stanley 1990.

to 2,378 meters) and the Kuzey Anadoğlu mountains along the Black Sea coast, and the snow-capped Taurus along the southern coast (Aladağları, 3,734 meters). A mountainous crescent skirts the shore of the eastern Mediterranean, from Syria to Sinai, reaching 3,088 meters in Lebanon. Even the islands of the Mediterranean are rugged, and some of the larger ones, like Crete and Corsica, are truly mountainous.

All of these chains are characteristically Mediterranean mountains, not merely by location but by dint of climate, ecology, and human geography as well. They are the wettest parts of a semiarid zone, support much the same vegetation, and once supported much the same economic activities – transhumant pastoralism, cereal cultivation, garden agriculture, timber cutting, and forest gathering.

Climate

Despite considerable local diversity there is a distinct Mediterranean climate, characterized by hot dry summers and cool wet winters. This climate is produced by special geographic conditions that exist in only five zones on earth. The other so-called mediterranean areas are all much smaller, generally no more than a coastal strip: southwest Australia; the western Cape in South Africa; south-central Chile; and southern California. All of these share a location on the shifting frontier between tropical climes dominated by easterly trade winds and temperate latitudes dominated by prevailing westerlies; they are all situated between roughly 32 and 39 degrees latitude. They all share westward exposure to an ocean. This combination is required to create the pattern of arid summer, when the desiccating trade winds slide toward the pole, and the wet winter, when the moisture-bearing westerlies slide toward the equator. As the wind belts shift with the season, mediterranean latitudes feel the alternating effect of trade winds and westerlies.[3]

The Mediterranean is the largest of all mediterraneans because no north-south mountain chain impedes the winter's wet winds. Whereas the Andes and the Californian coastal ranges confine Pacific

[3] The trade winds are dry for two reasons. Hot air rising in the tropics loses its moisture with its ascent (hence heavy tropical rains), and then descends from the upper atmosphere, warming with its descent. As air warms, its capacity to bear moisture increases – hence the dry air masses circulating in the subtropics, and the location of most of the world's deserts. Furthermore, the trade winds (in the Northern Hemisphere) blow from the northeast, warming as they go south, and thus, again, increasing their capacity to hold moisture. A recent analysis of Mediterranean climate is Perry 1981. See also Beckinsale and Beckinsale 1975, 23–36.

mediterraneans to a narrow littoral fringe, the north Atlantic version penetrates 4,000 kilometers inland from Gibraltar. Winter winds driven by the north Atlantic high pressure (usually anchored around the Azores in winter) whistle past the east–west mountain chains of southern Europe and North Africa, in corridors between the Massif Central and the Pyrenees, between the several ranges of Iberia, and between the Baetic ranges and the Atlas. All mediterranean zones depend for their rainfall on the irregular seasonal migration of the wet westerlies, whose latitude fluctuates yearly as well as seasonally, so the annual rainfall variations are often great. Local conditions, naturally, complicate this general picture.

Location and exposure help determine rainfall in Mediterranean mountain areas. As a rule, ranges further north and west get more rain, as prevailing winds bring moisture east from the Atlantic. So the Pyrenees get more rain than the anti-Lebanon. Local exposure, however, is usually more influential than location. The north flank of the Pyrenees is wetter than the south, and the eastern slope of the anti-Lebanon is dryer than the western. The north face of the High Atlas, looking down on Marrakesh, gets far more rain than does the south face, which is part of the pre-Sahara. The western flank of the Pindus, the Epirus side, gets much more rain than the eastern slope in Thessaly and Macedonia.

But most pronounced of all is the variation with the season. In the summer when the dry northeast trades blow across the Mediterranean, they join forces with northerly winds rushing to fill the low pressure created by the intense heat of the Sahara. In the eastern Mediterranean, where both Greeks and Turks call it *meltemi,* this wind generally comes from the Russian steppe, but whatever its origin it warms as it travels southward, increasing its capacity to absorb and retain moisture – hence, in part, the dry summers. On rare occasion the winds come from the south, from the Sahara. This is the famous *scirocco* that brings heat and dust but little moisture to southern Italy, Spain (where it is called *leveche*), and at times to Greece and Turkey.[4] The scirocco blows most often in the spring, when a hot and lengthy one can wither crops even before

4 Laborde (1809, 2:125, 148) says in Andalusia the scirocco is known as *solano.* He attributed to it considerable power: "It is a hot wind, and dries up the plants as soon as it touches them; it penetrates the body, and produces effervescence of the fluids, a sudden rarefaction of the humours, but above all it makes a dangerous impression on the head. It exalts, it kindles, and ferments the imagination, and drives many persons into a state, little better than madness: it is when this blows that murders and assassinations are most frequent in the country. Sometimes too, when it is violent, and lasts for a long time, women are seen in a state of extraordinary fury."

the long, dry summer. In Morocco the scirocco is a foehn wind, coming over the Atlas, and can wither man and beast. Mountains and plain alike are subject to the dry northerlies of summer, and some of the mountains, the south face of the Taurus for instance, can also feel the desiccating breath of the scirocco. Only rarely does a thunderstorm break the dry monotony of the Mediterranean mountain summer.

Winter is another matter, however. Mediterranean winter weather is shaped by low-pressure cells that drift eastward from the Atlantic, attracting winds from northern Europe (as in the spring the low pressure attracts the scirocco). These north winds can come in dry chilly blasts, like the Rhone's *mistral* or the *bora* of the Adriatic, but they can also bring storms. Every chain in the Mediterranean gets at least 1,000 millimeters of precipitation most every year, and almost all falls between November and April, much of it as snow in the taller ranges. Snow lasts all year round in the higher peaks of the Sierra Nevada and the Atlas, in the Pyrenees and the Alps, and in the fastnesses of eastern Turkey. Late autumn and winter rains are often torrential, having what soil erosion experts call a high aggressivity index. Gerald Brenan, a native of a wet climate, wrote that he had never seen a rain as strong as the one he encountered in Chechouan (Xauen), in the western Rif, in the winter of 1923, when it came down "like pencils."[5] As a rule, most rainfall comes in fall and winter in the eastern Mediterranean, and in winter and spring in the Maghreb and Iberia. But almost everywhere the consequence is that it is very often either too dry or too wet in the Mediterranean, inconvenient for agriculture and other human activities. Because winter is too cold (though not too wet) for agriculture and summer too dry, the brief fall and spring are the growing seasons.

One of the consequences of the Mediterranean's rainfall regime is high vulnerability to soil erosion. Other things being equal, concentrated rainfall leads to more rapid erosion than evenly distributed rainfall. Other things are of course never equal. Slope, soil composition, vegetation, and other factors also matter. Most Mediterranean mountains are rugged, with slopes of less than 15 degrees rather rare. Unstable clayey soils, especially in the Apennines and the Rif, lead to landslides as well as less spectacular erosion. Vegetation cover (or its absence) is crucial.

Vegetation

The Mediterranean climate has given rise to a distinct Mediterranean flora. It is highly diverse, despite millennia of human modification,

[5] Brenan 1974, 65. For rainfall aggressivity, see *La mobilité* 1984, 319–21.

because of the irregular relief of the Mediterranean littoral and because again it represents a frontier between tropical and temperate zones with elements of both. About ten thousand plant species are represented. In general Mediterranean flora are adapted to months of summer drought, to frequent fire, and to great fluctuation in temperature.[6]

Like climate, vegetation in the Mediterranean mountains varies somewhat with location and with local exposure (to sun, wind, and rain), but altitude is easily the most important factor. At least it is the most important factor in determining what *might* grow, because humankind (through the use of fire) and associated animals are the key factor now in what *can* grow. Spontaneous vegetation in Mediterranean uplands below 1,000 meters is primarily *maquis* and *garrigue.*[7] In most cases maquis and garrigue formations represent degraded forest perpetuated by fire and grazing, although in some lowland locales they are probably climax vegetation. Opinions differ.

Maquis is the French term for tall, thick scrub, which often grows taller than a man and can be quite impenetrable. It is found everywhere in the Mediterranean, consisting of dozens of woody and herbaceous plants, many of which are evergreens and all of which are supremely adapted to drought. Some have roots that tap water 8 meters below the surface. Junipers, holm and kermes oak, Spanish broom, lentisk, and cistus are characteristic, although maquis varies considerably in composition from place to place, depending on climate, soils, and the frequency of fire.

Garrigue is low scrub, usually no more than half a meter tall, with tufts separated by bare soil. It flourishes on calcareous (meaning derived from limestone) rather than silicacious (sandy or silica-based) soils. Hundreds of plants are found in garrigue, often aromatic ones like lavender, sage, rosemary, and thyme. Napoleon claimed that blindfolded he could recognize his native Corsica by its scent. But he would have found that Sardinia and Cyprus exude much the same sweet smell of garrigue. Only botanists, and perhaps goats, can distinguish varieties of

[6] Le Houérou 1974. Authoritative general floras include Baroni 1955; Caballero 1940; von Haláczy 1901–4; Jahandiez and Maire 1931–41; Turrill 1929; Mouterde 1970; Rikli 1943–8; Davis 1965–88.

[7] *Macchia* in Italian; *monte bajo* or *matorral* in Spain (*chaparral* in American Spanish; in Andalusia *Quercus ilex* is called *chaparra*); *longos* or *xerovoúni* in Greek; *al arachd* in Arabic, but *raba srira* in Rifian Arabic; *amayou* or *amayou imzi* in Rifian Berber. Garrigue (from the Catalan *garric*) is *tomillares* (*tomillo* = thyme) in Spanish, *phrygana* in Greek. A convenient glossary of botanical terms appears in Polunin and Huxley 1974, 15–22. Maquis and garrigue are discussed on pp. 9–12. Quézel 1982 is a general survey. See also Naveh and Kutiel 1990.

garrigue, and botanists must use their eyes. Garrigue also contains the bright flowers that give Mediterranean hillsides their brilliant but brief April burst of color. Garrigue is the delight of honeybees, as Mediterranean mountain honey is the delight of the honey cognoscenti. Hymettos honey earned fame in ancient times and is justly prized today. Honey from Sütçüler in the western Taurus enjoys local renown. Rif honey is so tasty that Generalissimo Franco allegedly had 2 liters a week flown to him in Madrid.[8] A small consolation for the deforestation of the mountains perhaps.

In many places, mostly between 800 and 2,000 meters, high forest survives. In general, scarcity of water and abundance of light discourage the growth of tall timber in the Mediterranean, but these conditions obtain more in the plains than in the mountains. Where it is found, tall forest normally consists of pines in the lower altitudes, with fir, oak and beech, or cedar higher up. In the Rif, Middle Atlas, the Taurus, and the Lebanon, cedars still crown some heights. The precise elevation at which these trees flourish varies with latitude and local conditions. As a rule, everything can grow a little higher up the further south and east one goes, but there are many exceptions. Beech does not grow naturally in North Africa, for instance, or cedar in Europe. Above the forest, in those ranges sufficiently tall, is alpine meadow, with tufted grasses and wildflowers; or else bare rock. Much of the area above the tree line (2,500 to 3,000 meters) is under snow for half of the year.

The classic Mediterranean species, found at almost all elevations, and especially common in maquis, are the holm oak and kermes oak, *Quercus ilex* and *Quercus coccifera*. They form either shrubs or trees, depending on the attention they get from goats. They are almost indifferent to soils, they can survive heat and cold and severe drought, and can regenerate from their deep roots after fires. Their leaves are spiny so that only goats will eat them (and only when nothing more succulent is available), but even determined goats cannot kill them. Holm oak can manage at sea level and up to 2,000 meters (2,800 in the Rif). Some hardy specimens live seven hundred years. These oaks are the most characteristic of Mediterranean plants, and their range, as well as or better than the olive's, defines the extent of the Mediterranean world.[9]

[8] *Isparta 1967 Il Yıllığı* 1967, 196; Woolman 1969, 24. The secret extravagance of the Spanish dictator suggests he had never tried Alpujarras honey, which stands comparison with any rival. Gerald Brenan ate 50 pounds (23 kilograms) of it per year when he lived in Yegen (Brenan [1957] 1988, 128).

[9] In Anatolia and the Levant *Quercus calliprinos*, a close cousin of *Q. coccifera*, flourishes in formations and at altitudes where to the west one finds *Q. coccifera.* Zohary 1973, 2:356–7; Daget 1980; Ruíz de le Torre 1979.

The physical relief and the climate of the Mediterranean have changed only in slight detail over the past ten thousand years. The preceding descriptions correspond equally well to the late Neolithic and to the late twentieth century. Coastlines have changed a little here and there. Earthquakes, landslides, geologic upthrust, and erosion have modified topography to a small degree. (The effect of erosion on soils is another matter.) Epochs of cooler and warmer, wetter and drier weather have come and gone, affecting agricultural productivity especially in marginal areas. But in the large context of ancient and modern history, they may almost be regarded as constants. The same is decidedly untrue for Mediterranean vegetation and soils. As far as they are concerned, the Mediterranean is now an anthropic (man-made) landscape, except at the highest altitudes. The key variable in Mediterranean environmental history is vegetation.

Every chain of mountains, indeed every mountain village, in the Mediterranean world has its own geology, climate, vegetation, and history. I will consider five mountain zones more closely, from the point of view of their biogeography, which emphasizes altitude and vegetation belts, and from the point of view of accessibility, which emphasizes relief, routes, and connections to the wider world. The key variables in mountain vegetation history are elevation and human access.

Five Mountain Ranges

The Taurus

The Taurus parallels the Mediterranean shore of Turkey, beginning in the west near Lake Eğridir and extending eastward as far as the Seyhan River. Beyond that is the long curve of the anti-Taurus, slicing into eastern Turkey. The Taurus is mostly porous (karstic) limestone, and its soils limestone-derived, although there are pockets of volcanic soils here and there.[10] In the central Taurus (Bolkar Dağları), several peaks exceed 3,000 meters. In the west, where the mountains tower over a series of lakes, most summits are between 2,000 and 2,500 meters. Only one, the Dedegöl (or Anamas) Dağ between Lake Eğridir and Lake Beyşehir, approaches 3,000 meters.

On the map the Taurus appears a long sinuous range, but on the ground it is a confusing array of massifs and peaks, arranged randomly

[10] Taurus geology: Ardos 1979; J. Dumont et al. 1979; Brinkmann 1976; Planhol 1956.

at all angles. The relief is rugged, often precipitous. At two points along the length of the Taurus, rivers cut through from the Anatolian Plain to the Mediterranean: at Mut, in the central Taurus, where the Göksu flows by, and at the Cilician Gates in the eastern Taurus, carved out by the Çakit River.

Smaller rivers, some merely seasonal watercourses, link the mountains and nearby plains. On a small scale the western Taurus and the coastal plain of Antalya combine in the same sort of happy union as the Sierra Nevada and the Vega de Granada, or the High Atlas and the plain of Marrakesh. Water from the mountains allows irrigation in the Antalya Plain, producing fine crops of cotton and maize. The same arrangement makes the plain of Adana (the Çukurova) productive. On an even smaller scale this is true of valleys and plateaus amid the large lakes of the western Taurus. Pastoralists too have taken advantage of the proximity of mountain and plain. Until the end of the nineteenth century, nomads and seminomads routinely moved between the Taurus and the coastal plains in seasonal transhumance (see Chapter 4).

Two belts of climate duel for sovereignty over the Taurus. It lies on the climatic frontier between a typically Mediterranean zone and one that is continental. This pattern makes for tremendous uncertainty in the weather, not so much day to day but year to year. The rainfall maximum comes in the winter, as is usual in the Mediterranean, but there is a second rainy season in May, associated with local thunderstorms caused by convection. The rains are heavy, favoring erosion, but the late spring rains make for more water in the springs and underground limestone passages in summer. Isparta, at 1,050 meters, the nearest weather station to the western Taurus, averages about ten days of snow cover per year, although it can remain blanketed for over a month. Every winter the temperature gets down to –10°C, and the January average minumum is just below freezing. Blasts of continental winter alternate with the soft breath of the Mediterranean through March, when the milder climate takes over. As a wry local proverb has it: "March 15 it's summer, March 16 it's winter."[11] The last days of frost come in April and do not return until November, or mid-October at the earliest. The summer heat (average maxima around 30°C) is moderated by cool nights (average minima in June and July around 15°C). The hottest it ever gets in Isparta is around 40°C.

The rainfall regime in the western Taurus also varies tremendously, again because it is on the frontier between Mediterranean and con-

[11] Planhol 1958, 37. The original: "martın onbeşinci günü yaz, onaltıncı günü kıstır." Further details on Taurus climate are in Zohary 1973, 2:30–3.

tinental climates. One year the typical continental pattern of humid summer and dry winter might prevail, whereas the next brings the dry summer and wet winter of the Mediterranean. In a disastrous year, such as 1932, a dry Mediterranean summer can be followed by a dry continental winter. The worst of all is an early continental winter without autumn rains followed by an early Mediterranean summer, preventing the May rains. If this happens, crops fail completely.[12] This variability adds to the hazards of agriculture: although a full drought is unlikely, one never knows with assurance in what season the heavy rains might come, whether one should plant winter wheat, spring wheat, or maize, or leave a field fallow. Adjacent to the mountain climate is the reliably Mediterranean climate of Antalya Plain, where January rains and sweltering summer heat are certainties.

The potential natural vegetation of the Taurus is high forest and has been for about seven to ten thousand years.[13] In remote corners it still survives in abundance. But generally, especially at the lower altitudes, it has given way to species more compatible with humankind, with grazing animals, and with fire. The lower elevations, especially below 500 meters, support typical Mediterranean maquis featuring junipers, *Pistacia*, and *Quercus calliprinos* (a subspecies of *Q. coccifera*). On the hillsides above 500 meters a few patches of pine (*Pinus brutia*, an invasive species that does well in disturbed settings) appear. Maquis becomes rarer above 600 meters, although it is found in places up to 1,200 meters. The most common forest tree of the western Taurus, the limestone-loving and drought-resistant black pine (*Pinus nigra*), reigns between 900 and 1,500 meters. Black pine does well wherever Mediterranean climate and continental climate overlap; it is abundant on the north, west, and south sides of the Anatolian Plateau. Extensive stands adorn the slopes of the mountains around the Köprü River, where *Pinus silvestris* is also found. Various junipers join the pines between 1,000 and 2,400 meters. The mighty cedar (*Cedrus libani*) occasionally appears in mixed stands on the Dedegöl Dağ between 1,500 and 2,300 meters.[14] Its preferred range is from 1,500 to 1,800 meters. Fir (*Abies*

[12] Planhol 1958, 37–8.

[13] Akman, Banbero, and Quézel 1978; Quézel and Pamukçuoğlu, 1973 and 1975; and Kürschner 1984 are the most detailed works on Taurus vegetation. See also Kasapliğil 1952; Zohary 1973, 1:157–9 and 2:550–65; P. Davis 1965–88, 1:18–21; Planhol 1958, 42–50; Louis 1939, 94ff.; Roberts 1982, 233–5.

[14] These upper limits come from Planhol 1958, 45; Zohary (1973, 2:396) gives 2,000 meters as the upper limit of *C. libani*. P. Davis (1965–88, 1:20) says the tree line is about 1,700 meters in the Taurus. I have seen plenty of trees above this height on Bozburun Dağ and elsewhere and consequently incline toward Planhol and Zohary in this case.

cilicica) likes slightly higher altitudes (1,600 to 1,900 meters), and is usually mixed with black pine and juniper in the western Taurus. In the central Taurus fine forests of mixed fir and cedar still exist between 1,400 and 1,800 meters. Oak (*Quercus peduncula* and *Quercus sessiflora*), the most common broadleaf, is rare in the western Taurus, least so around Sütçüler. It generally is found mixed with black pine. In the western Taurus today extensive mature forest, with trees aged over centuries, exists almost exclusively between the Aksu and Köprü Su. Larger stands dot the Bolkar Dağları. The woods are broken by frequent clearings and meadows; fires, goats, and fuelwood gatherers keep undergrowth to a minimum; lumber operations nibble away at the edges, wherever roads permit; but withal this remains some of the thickest and least disturbed forest in the Mediterranean. The western Taurus has long been only on the fringes of economic life, with happy consequences for vegetation.

The Taurus has always had connections with the sea to its south and the Anatolian Plain to its north. For millennia armies and caravans have passed through the eastern Taurus at the Cilician Gates. The route through the central Taurus at Mut, connecting the Konya Plain to the sea at Silifke, has seen almost as much traffic. At its western end the Taurus is less of an obstacle to north–south traffic than anywhere but the two gaps. Lake Eğridir is drained to the south, to the Mediterranean, by the Aksu, and many historic routes connected the Antalya Plain to the lake region and the Anatolian Plateau.[15] From at least the eleventh century a considerable caravan traffic crossed the length of Anatolia, growing in the thirteenth century with the establishment of the Pax Mongolica in Asia. Much of it passed immediately north of the Taurus, where mountain water made the grass a bit greener than further north. When political conditions were right, caravans also wended north–south across Anatolia, making the routes through and around the Taurus very important. When the Byzantines controlled the Bosphorus (and sometimes even when they did not), Seljuk and Ottoman caravans connected the Black Sea littoral and the Russian rivers to Egypt. A prosperous trade in slaves, among other items, passed this way, especially in early Ottoman times (c. 1400–1520). Antalya and Alanya were at times important ports. Burdur and Eğridir were major crossroads, visited by Xerxes and his Persian army, Barbarossa and his Christian crusaders, twice by the Moroccan world traveler Ibn Battuta, and often by hosts of less famous soldiers, merchants, and travelers. So the western Taurus, though not pierced by any major routes like the central

[15] Planhol 1958, 23–7. The Lake Region is ancient Pisidia and Ottoman Hamidili.

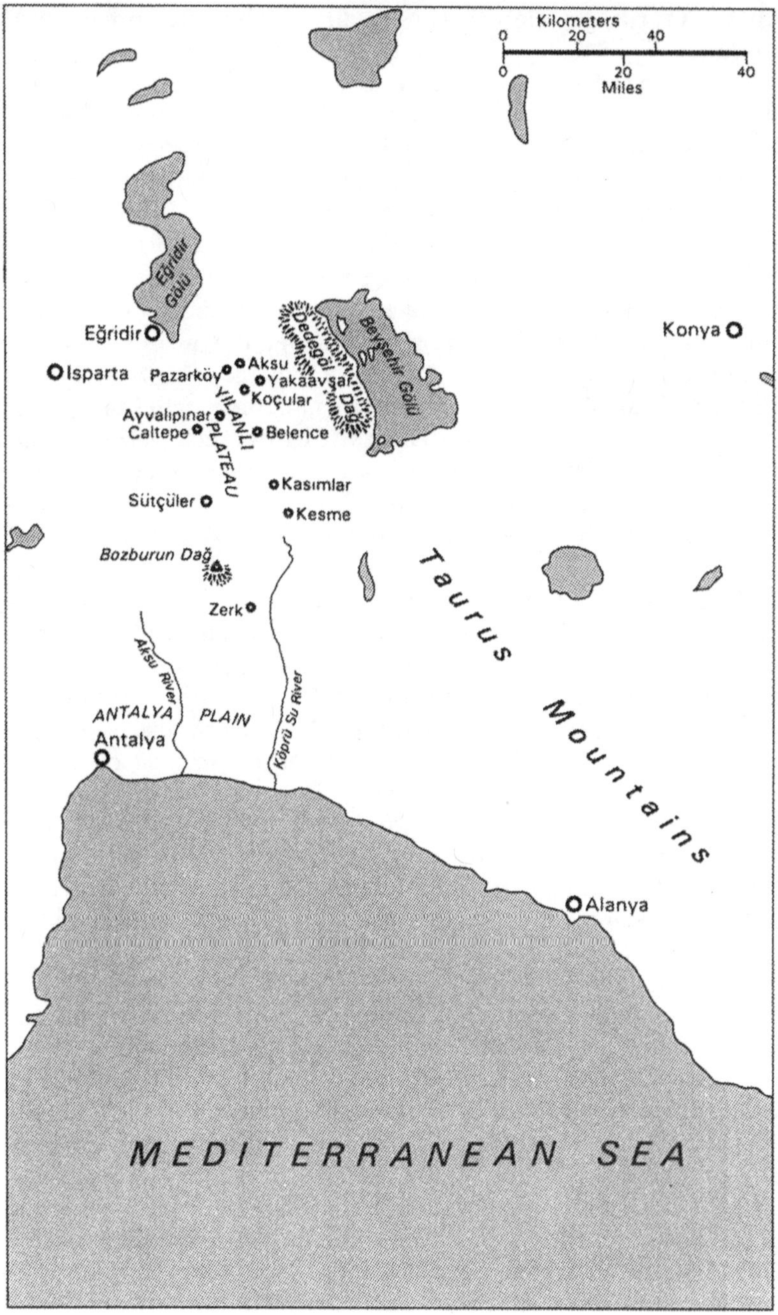

Map 2.1. The western Taurus mountains

and eastern Taurus, lay close by important commercial and military thoroughfares.

These routes faded in importance with the decline of Antalya in the seventeenth century and the slow demise of the caravan traffic generally. Only in the twentieth century have avenues of transport grown up again around the Taurus. Today the massif is surrounded by roads, although only three or four paved roads penetrate the mountains, and only two of these, at Mut and at the Cilician Gates, carry much traffic. In general roads only serve the periphery once served by caravans or coasting vessels. Many villages are at the end of a rough track, and some summer encampments (*yaylas*) are two days' walk from the nearest road. A railroad now traces the northern edge of the Taurus, going from Konya through the Cilician Gates to Adana, and a small one reaches to Isparta, near the western end of the range. Neither carries much traffic. So the situation today is similar to that of the past: the Taurus remains on the fringe of a world of easy travel and communication, its fringes caught up in interaction with the wider world but its heart still difficult of access.

Human settlement is sparse in the western Taurus and probably always has been. Reliable sources of water are few and soils are thin except in a few favored valleys and bowls. The one exception to this is the region southeast of Lake Eğridir, west of Lake Beyşehir, around the headwaters of the Köprü Su: the upland of Yılanlı. Here villages dot the countryside, and people are everywhere, giving it the look of a fully occupied landscape. But this is deceiving. Here, as elsewhere in the western Taurus, settlements sprout where the land is flat or slightly sloping, but nowhere else. Terracing, widely practiced in the Pindus and Sierra Nevada and in the mountains of Pontic Anatolia, is rare here. Every promising valley up to 1,500 meters has its village or villages, generally situated in piedmont, but the mountain slopes are virtually uninhabited. This may not always have been so, of course. Indeed abandoned terraces are visible here and there. But there are no abandoned or depopulated mountain villages (although many abandoned nomad encampments). In recent times, at least, the mountains proper have been the domain of summer pastoralists and their flocks, and peasant cultivators have stuck to the valleys and plateaus. Around the yaylas occasionally herders raise a summer crop of maize or barley. Probably cultivators have always avoided the mountains, preferring the piedmont, because of the harsh winters and short growing season, but one would require an archaeological survey to know for sure.

The economy of the western Taurus consists of farming, herding, and, to a small degree, logging. Agriculture is a small-scale business, based on family farms of a couple of hectares – perhaps 5 hectares in a prosperous family. Maize is the most important crop, raised on the flattest and best-watered land. Wheat and barley grow on gentle slopes. Garden vegetables – cucumbers, tomatoes, onions – surround most houses. Sunflowers, vines, and fruit trees, mostly apples, complete the picture. Herding is an essential aspect of life. Goats forage freely around every village and yayla. In most villages the better-off keep a cow or two. Sheep are rare, perhaps because the nearby Anatolian steppe and Antalya Plain have long supported large flocks very economically. Commercial timber cutting is also of some consequence. In Seljuk and early Ottoman times, Antalya and Alanya and other ports exported timber to Egypt. Now the trade is local, but huge lumberyards with tens of thousands of pine logs line the banks of the Köprü Su. The roads through the western Taurus, for example that between Manavgat and Beyşehir, carry little traffic besides timber trucks.

Ancient monuments and sunny beaches draw plenty of tourists to southern Turkey, but only rarely do they wander far from the coast. Consequently they have, as yet, no economic importance in the Taurus. Mining, important in other mountain districts, is and always has been negligible in the Taurus, especially the western part. Useful mineral deposits are few. No prehistoric mining existed, and an Austrian survey of the nineteenth century showed only a small lead mine at Bozkır, twelve hours south of Konya.[16]

Highly commercial agriculture dominates the borders of the western Taurus region. The Antalya Plain grows seemingly endless amounts of cotton with the help of mountain water. To the northeast, in the steppe of central Anatolia, big farms raise wheat, often with modern American machinery (although in some cases with ox-drawn plows and handcrafted sickles). Around Isparta the Turkish rose-growing industry prospers, supplying the nation's perfume manufacturers. It used to be a center of opium production.

The commercial connections of the western Taurus are chiefly with the city of Antalya and neighboring coastal towns such as Serik and Manavgat. Antalya is now the largest city in southwestern Turkey (population, c. 250,000), having grown prodigiously in recent decades. The famous religious center of Konya, the nearest city to the east, is too distant to exert much sway over the western Taurus. The modest towns of Isparta, a provincial capital of about 50,000, and Eğridir, a

[16] De Jesus 1980; Issawi 1980, 284.

delightfully sleepy burg of 15,000, are the only other urban centers of consequence. Today Antalya serves as the chief link to the outside world, and it has only a small port.

The Pindus

The Pindus, the backbone of continental Greece, is a series of parallel limestone ridges extending along a northwest–southeast axis from Albania to the Gulf of Corinth. It is the southern extension of the calcareous Dinaric Alps that run through Yugoslavia and Albania, hugging the Adriatic coast.[17] Its relief is often precipitous. The range is a chaotic one, with many distinct massifs (the highest of which, Smólikas, is 2,637 meters) and very few passes. The only pass of any significance is at Metsovo (called the Zygos, 1,450 meters), which links Thessaly and Epirus.

The western flank of the Pindus receives plenty of precipitation, most of it from November to April, often in spectacular thunderstorms, or in snow. One nineteenth-century traveler wrote that the Turks said there are only three things worth seeing in Ioannina: the lake, the tomb of Ali Pasha, and the rain.[18] Annual precipitation at Ioannina is about 1,100 millimeters, in the Zagori district about 1,300, and in the Arachthos watershed almost 1,400. December is usually the rainiest month. January temperatures in Ioannina average about 5°C and bitter cold is very rare. But winter has its cold spells in the mountains, and frosts are routine from October to May. Many winters bring deep snow above 1,500 meters, hindering transport in the mountains. Summer is dry and warm, but less so than in most of Greece. Average July maxima are about 25°C in Ioannina, and a few degrees less in the mountains. In typical years, rain falls on about five or six days even in the driest months, although summer drought is not unknown. Ioannina averages about 33 millimeters of rainfall even in July. As in most mountain ranges, the weather is capricious and the diurnal range high. Zagori peasants say the cold can kill an animal before noon and the heat will make it decompose before evening.[19]

The Pindus once supported a mantle of tall forest, pine, fir, oak, and beech. Remnants exist in eastern Zagori and on certain shoulders of

[17] Pindus geology: Desprairies 1979; Institut Français du Pétrole 1966.

[18] Chirol 1881, 176.

[19] This is a rough rendition of a proverb from Théophilou 1983, 116. Meteorological statistical data from Greece, Ethniki Statistiki Iperesia 1986, 6–9. Dimitriadis (1980, 152–67) has some climate data for various points in Ioannina province. Greece, Iperesia Syntonismou [C. A. Doxiadis] (1950, 22) has some data on the Arachthos basin.

the Smólikas and Timfi massifs. Lowland Epirus supports the usual Mediterranean maquis and garrigue associations. Above 500 meters the most common forest formations are of black pine, which grows up to 1,900 meters. *Pinus nigra* flourishes in degraded soils and prospers where more demanding species cannot. On the roof of the apse of a small Samarina church (where soil conditions are sure to be poor) a lonesome black pine keeps watch over the adjacent square. The Balkan pine (*Pinus heldrecheii*) grows at up to 2,400 meters. Two fir species grow in the northern Pindus, at up to 2,000 meters, the widespread Cephalonian fir (*Abies cephalonica*) and the silver fir (*Abies pectinata*). Several species of tall oak grow at altitudes from below 500 to above 2,000 meters (*Quercus cerris, Quercus aegilops, Quercus robur,* and *Quercus conferta* are the most common). *Quercus aegilops*, or valonea oak, has had considerable economic importance as a source of vegetable dye. *Quercus ilex* is found up to 1,700 meters, enjoying a primacy it owes to its indifference to drought and to soil conditions. European beech (*Fagus sylvatica*) stands adorn a few slopes between 1,500 and 1,700 meters as well, but beech does not do well on calcareous soils. Further south in the Pindus, on the ridges known as Mitsikeli, Peristeri, and Xirovouni, patches of tall pine cover remain here and there, but most vegetation is scant garrigue. Good pasture land is rare below 1,600 meters. Above the timberline is an alpine stratum of meadow grasses and small shrubs.[20]

The surviving vegetation owes its good fortune in part to the remoteness of the Pindus and its dissected relief. Humankind, sheep, and goats have probably covered every inch of the area but have not carried on regular economic activity everywhere. Historically only the route over the Zygos had any commercial or military importance. In the early nineteenth century the local tyrant Ali Pasha (d. 1822) built good roads, but they did not long survive him. Cobblestone tracks (*kalderimi*), suitable for mules but not for wheels, served as the only routes for most of the modern history of the Pindus. Wheeled traffic could not routinely cross the Zygos until the late 1930s. Only a few roads penetrate the Pindus today, and most of those are of recent construction. The great majority are unpaved. No railroad track comes closer than Kalabaka in Thessaly, and that one has never carried much. So the connection between the Pindus and the outside world – that is, the plains of Thessaly and Arta and the lands beyond the sea – has generally required slow going by mule or on foot. Indeed, there are a few villages that today (as of 1990) cannot be reached but on foot. Many more are cut off in

[20] Pindus vegetation: Quézel 1967; Quézel and Contandriopoulos 1965; Strid 1986; on ecological succession on Pindus forest soils, see S. Sakellaridis 1968.

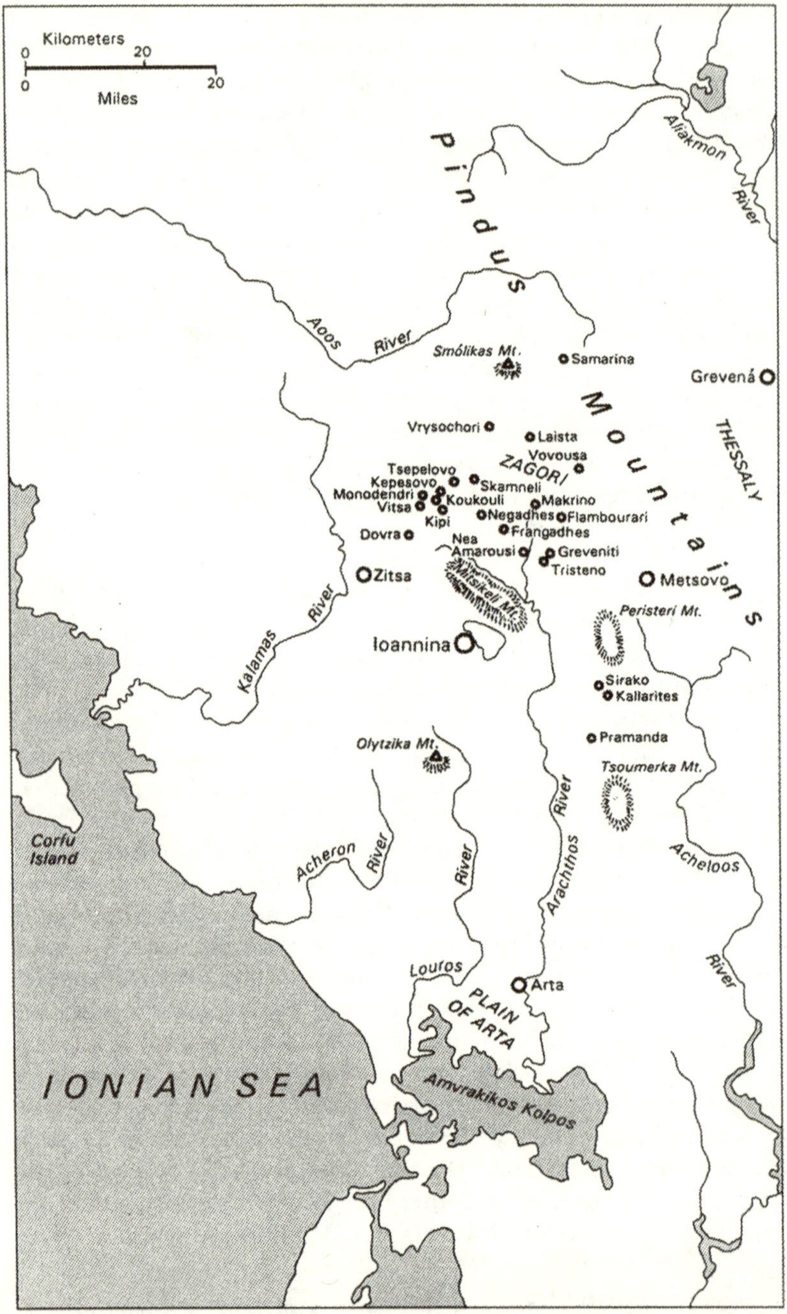

Map 2.2. The Pindus mountains

winter by deep snow and at other times by landslides. Pindus villagers have struggled against this natural isolation, with some success. But the survival of tall timber is an indication that their success has not been total.

Villages number perhaps two hundred now, although any count is an approximation because the Pindus has no clear boundaries. Most have always been small, with fewer than a thousand people, and are much smaller today, with a permanent population usually of less than two hundred. Much of the population is Vlach, a disappearing minority, whose language is akin to Rumanian. One can still hear Vlach spoken in villages such as Vovousa, Samarina, or Metsovo, but generally only by the very old. The young, who are few to begin with, have gone to school in Greek and prefer that language and identity.[21]

The economy of the Pindus rests on garden agriculture, usually of maize, beans, carrots, and tomatoes. The herding of sheep and especially goats remains important, although much less so than previously. In some areas, notably eastern Zagori where flysch soils encourage the tall beech stands, forest products also matter. Tourism has not lived up to hopes as yet. Historically many Pindus villagers have worked in the lowlands of Thessaly or Arta, or gone further afield as masons, tinkers, herders, or mercenaries (see Chapter 4). To some extent this pattern of migrant labor – or migrant entrepreneurship – persists to the present, although in the last forty years many have gone to the cities and never come back. Remittances from these emigrants, together with pensions, play something of the same role that migrant work did in earlier centuries.

The only urban center close to the Pindus is Ioannina. It is a city of fifty thousand today, situated across the lake of the same name from the foot of Mount Mitsikeli, and in an upland plain of considerable fertility. The locals uphold a tradition of wool- and metalworking, and once were known for widespread commerce and refined education.[22]

[21] Winnifrith 1987 is a recent study; Weigand 1894–5 a classic. Winnifrith suggests the language will disappear in fifty years (p. 3) and estimates the Greek Vlach population at 30,000, most of whom live in the Pindus. In 1853 an observer with long experience of European Turkey (M. Ubicini 1853, 2:177) estimated the Vlach population in Greece, Thessaly, and Macedonia at 200,000 to 220,000. Fifteen years later the British consul at Ioannina guessed that Vlachs in Epirus, Macedonia, and Thessaly numbered 45,000 (Stuart 1868, 316). Pouqueville (cited by Stuart 1868, 316) thought 74,000 as of about 1812. Clearly Vlachs are hard to count. In World War II the Italian occupying forces sponsored the Vlach tongue, which led to the suppression of schooling in Vlach after the war.

[22] Aravantinos 1856, 2:250–3, 272–84, on Ioannina crafts in the nineteenth century. Philippson (1897, 201–7) has a description of Ioannina from 1893.

The rhythms of economic life in Ioannina have had strong effect on the physical and social landscapes of the Pindus. Grevená, the chief center to the immediate east of the Pindus, and Kónitsa, a small town with a plain as fertile as Ioannina's, attracted people to their annual trade fairs. Metsovo, a Vlach center and now a very modest resort town, is the largest community truly in the heart of the Pindus. Thessaloniki and Athens are a long way away, a full day by bus, and before the bus, at least a ten-day walk. Ioannina has no rivals as a center for the people of the Pindus.

The Lucanian Apennines

The Roman province of Lucania acquired the name Basilicata some time around the tenth century, perhaps from the Byzantine Greek term for an administrator. The fascists changed the name back to Lucania in 1932, aiming to purify Italian place-name geography of foreign influences, but in 1947 the official name became Basilicata once more. Local people usually call it Lucania. By whatever name, it is an administrative region comprising the provinces of Potenza and Matera. It lies at the instep of the Italian boot, between flat Apulia to the northeast and rugged Calabria to the south. Basilicata is mountainous (47 percent of its 9,992 square kilometers is mountain, according to the official categories), with the highest peaks in the southern Apennines, but it is not rugged. With the exception of a few rocky outcroppings near Castelmezzano and Pietrapertosa, called the Lucanian Dolomites, the Basilicata mountains are gentle slopes and not at all forbidding.

The geology of the Lucanian Apennines is complex and controversial. The mountains are relatively young (Miocene), still growing in fact, by about a millimeter a year on average. Some settling still takes place in the form of earthquakes, to which the northern part of the region is especially prone. In 1857 one killed more than ten thousand people in Lucania alone. The last serious one came in 1980, and although its epicenter was in neighboring Campania, it damaged much of the north, especially the capital city of Potenza, which in 1988 was still rebuilding. Soils throughout most of the mountains up to 1,000 meters are argillaceous (clayey) schist and flysch, and when waterlogged often slide downhill in what are known as *frane.* Long scars testify to the prevalence of frane throughout Basilicata, especially the eastern slopes of the Apennines, those that drain into the Ionian Sea by the Sinni, Agri, and Basento rivers. No other part of Italy or the Mediterranean world suffers more from landslides than Basilicata. Indeed the Lucanian and Calabrian Apennines, together with the central and western Rif, are the most geologically unstable slopes in the Mediterranean world. But geology is only part of the problem.[23]

The other part is the heavy winter rains. Basilicata is much influenced by sea winds, and the mountains usually get well over 1,000 millimeters of rain a year, but it varies markedly year to year. The western slopes of Monte Sirino average 2,000 millimeters. The eastern slopes get less, and the lowlands around Metaponto, in the rain shadow of the Apennines, get as little as 300 or 500 millimeters. Winter downpours can be epic: in 1935 at Moliterno 100 millimeters fell in twelve hours, and in 1941 254 millimeters fell in one November day at Terranova di Pollino. Very little snow falls, except in the highest mountains.[24] The surrounding sea moderates temperatures, so that the Lucanian mountains have less pronounced annual variations than almost anywhere else in the Mediterranean mountains.

Rainfall distribution and altitude (and of course humanity) govern vegetation. Basilicata has more than 3,500 species and subspecies of plants. The lowland plains and foothills (up to 500 meters) contain the usual Mediterranean flora of shrub and Aleppo pine amid wheat fields and olive groves. Between 500 and 1,000 meters, fields and orchards are fewer; woods and pasture divide the landscape. In forests, oaks predominate, especially *Quercus cerris* and *Q. farnetto,* and chestnuts (*Castanea vesca*) are common. Esparto (*Lygeum spartum*) and broom (*Genista sagitallis*) dominate the scrub below 1,000 meters. Above 1,000 meters, where the clayey soils preferred by *Q. cerris* give way to calcareous ones, oaks and chestnut share space with beech and, above 1,500 meters, with associations of beech and fir (*Abies alba*). The oak forests normally have considerable shrub undergrowth (*Carpinus orientalis, Ostrya carpinofolia, Prunus spinosa*). Forests in southern Italy are generally poor, except for the Gargano, parts of the Sila Plateau in Calabria, and Pollino. On the Pollino Massif, as in Calabria to the south, one can see the flagship of local flora, the *pino loricato* (*Pinus heldrecheii*), usually standing alone. Above 1,800 meters, and there is not much of Basilicata in this category, a few beech and fir stands join the grasses, shrubs, and flowers of subalpine and alpine meadow. Asphodel is common.[25]

Humankind, sheep, and goats have drastically altered the vegetation

[23] On geology of Basilicata, Kayser 1961, 15–17; Puglisi 1977. On instability, *La mobilité* 1984, 305.

[24] Ranieri 1961, 95–103; Kayser 1961, 16; Kayser 1963, 93–100; Puglisi 1977, 92. A rainfall map covering 1921–70 appears in Versace 1986, 124.

[25] Di Stefano 1969, 26–7; Ranieri 1961, 103–15; Pratesi and Tassi 1979, 131–56; Bonin and Gamisans 1976; Tichy 1962, 28–34. For Monte Pollino vegetation, De Leo 1984, 237–42. Historical glances at Pollino vegetation are available in Rodolico 1963, 267–74; Gavioli

of the Lucanian mountains. Although these landscapes are among the more isolated in Italy, they have never been cut off. The five major rivers of the region, the Bradano, Basento, Cavone, Agri, and Sinni, form natural avenues of communication with their wide beds, and have often guided invading armies. In ancient and medieval times they were navigable, but today they have no commercial importance as they are so often either dry or in spate.[26] Roads (if often in disrepair) have crisscrossed the area since Roman times; people, animals, and goods have moved up and down the slopes continually. No one is more than two or three days' walk from the sea. Today paved roads reach every village. The Lucanian Apennines are the least isolated of the five mountain groups considered in this book.[27]

The mountains are thinly populated. The region of Basilicata as a whole has about sixty people per square kilometer, but this includes the fertile volcanic plain of Melfi and other favored districts. Villages are often large, with two or three thousand people, but they are far apart, often 10 or even 15 kilometers from their nearest neighbor.[28] As in the Pindus, an ethnic minority inhabits some of the remoter villages: Albanians live in a handful of communities on the north slope of Pollino, and in several communities in Calabria (and in Sicily). They are descendants of immigrants from the fifteenth and sixteenth centuries. Albanian is still spoken in some villages, although like Vlach in the Pindus it will soon die out, as have already the more colorful Albanian styles of dress and the Orthodox rite. The rest of Basilicata speaks local dialect, laced with Spanish, Arabic, and Greek influence. Outsiders can find it quite unintelligible.[29]

The mountaineers, whatever their language, are hardy folk. For many centuries, and until quite recently, they lived in grinding poverty and

1932, 1934, and 1936. The most exhaustive study is Gavioli 1947. A roster of trees and shrubs in the *circondari* of Potenza and Lagonegro, with Latin, Italian, and local names, appears in ACSR, MAIC, DGA, verso I, busta 350, f. 413, "Indice alfabetico degli alberi, arboscelli e frutici . . . 1872."

[26] Kayser 1961, 116.

[27] Di Gianfrancesco (1979, 34–64) details the history of roads in southern Italy from 1790 to 1860.

[28] On average, villages are 7 kilometers from their nearest neighbor in Basilicata, the largest such average in Italy. Galasso 1982, 30.

[29] By 1842 twenty-five of the forty villages settled by Balkan Christians had given up Orthodoxy; See Filangieri 1980, 237. On Basilicata speech: Serra 1983. My own experience is that standard Italian (even my version of it) is universally understood, but that most country people cannot, or prefer not, to speak anything but local dialect. This is truer, as everywhere, of the old than the young.

Map 2.3. The Lucanian Apennines

routinely bore great physical hardships. Lucanians recruited into Murat's army distinguished themselves during Napoleon's retreat from Russia by their astonishing resistance and fortitude. They attributed their toughness to living outdoors with few clothes, walking barefoot over frozen ground, and wading cold rivers. A gentler life in recent decades has bred mountain Lucanians little softer than those of 1812.[30]

The productive economy of the Lucanian mountains depends on agriculture and grazing. Farming in Basilicata gives low yields by the standards of southern Italy, and the lowest are in the mountains. Farms are generally very small, unlike much of southern Italy where latifundia are common – in lowland Calabria, Apulia, and the Campania, for example. Machinery and chemical fertilizer, while rare, are far more common than twenty years ago, and far more common than elsewhere in the Mediterranean mountains. Many farms now use a tractor that rolls on tank treads, well suited to the mud and slippery slopes of the plowing season. Mountain agriculture features potatoes, especially in the Lagonegro district, maize, and garden vegetables, but wheat is king. Wheat covers much of the rolling uplands up to 1,200 meters or so, and accounts for about half the arable in the province. Wheat yields are slightly better than half of the national average. But the cultivated area in the mountains is small; most utilized land is in pasture. Pastoralism has declined greatly in recent centuries, but sheep and goats together still outnumber human beings in Basilicata by about 10 or 20 percent. Generally they are grazed in small flocks, rarely more than one hundred together, by self-employed shepherds. Every mountain village has its pastures and flocks, but the shepherds are generally old men and one wonders who will tend the flocks in twenty years.

Two urban centers, neither in Lucania, compete unequally for influence over the Lucanian Apennines, as they do over all of southern Italy: Naples and Bari. Both are large cities, Naples especially, and attract people and goods from great distances. Naples has historically been the more important, certainly for the high mountains, which are in the western part of Basilicata and closer to Naples. The new autostrada, opened in the late 1960s, reinforces the importance of Naples, as it links western Basilicata with the Campania. Bari still gets its share of mountain emigrants because in recent years its economy has grown faster than has Naples's. Smaller centers, such as Potenza, while the largest in Basilicata, are easily overshadowed by Bari and Naples.

[30] Lenormant 1883, 1:245.

The Sierra Nevada and Alpujarra

The Sierra Nevada, as its name suggests, is snow-covered. After the eastern Taurus, it is the highest of Mediterranean ranges (unless one counts as Mediterranean the Alps or the High Atlas). Like the High Atlas, the Caucasus, the Elburz, the Tian Shan, and many other ranges, the Sierra Nevada provides year-round water to an area that would otherwise be too dry for agriculture. These tall ranges, with their snow-melt streams, make possible the rich plains of Marrakesh, Granada, Teheran, and the lush valleys of the Alpujarras, Soviet Georgia, the Fergana Valley. In effect they create desert oases at their feet and in their foothills. These high ranges, seen differently, are vertical Niles: whereas Egypt is the gift of the Nile, which, by virtue of its length, brings moisture from the rainy season in the Ethiopian highlands through the Egyptian desert, these mountain streams bring water from the moisture banks of the snows to arid lowlands. Granada and Teheran are the gift of the snow-capped mountains, as are other mini-Egypts all around the Mediterranean and indeed all along the arid belt to Xinjiang. Such tall mountain systems in the arid latitudes permit some of the *loci classici* of Islamic civilization, even if, as is the case with the Sierra Nevada, they have been Christianized in recent centuries. The Alpujarra, at least historically, is a fine example of an anthropic landscape painstakingly arranged to exploit this particular geography.[31]

Mountain communities exist on both the north and south slopes of the Sierra Nevada. The north flank (or most of it) is called the Marquesado del Cenete, where cereals, arboriculture, and iron mining form the basis of life. I will only occasionally make reference to the Marquesado and will focus instead on the Alpujarra.[32] The Alpujarra forms the southern slope of the Sierra Nevada, drained by the Río Guadalfeo. It runs east–west, cut off from the sea by the Sierra de Gádor and the Sierra de Contraviesa, and separated from Granada and its rich *vega* by the tall Sierra Nevada. If one counts together the Almería Alpujarra and the Granadine Alpujarra, the region covers about 1,400 square kilometers and has a population density (1975) of twenty-eight

[31] Similar geography underlay pre-Columbian American civilizations in the Andes and the southern Rockies.

[32] The Marquesado is discussed thoroughly in A. Cohen 1987 and Arias Abellán 1984. The Alpujarra Valley is sometimes referred to in the plural, Alpujarras, to mean either the Granadine and Almerian sections together, or the whole region between the Sierra and the sea. In the singular I shall use it to mean the Granadine segment, from Orgiva to Ujíjar.

per square kilometer. It is among the remotest places in all of Spain and has been, over the centuries, a refuge for Christian resistance to the Emirate of Córdoba, of Muslim resistance to the Castilian reconquista, of protonationalist resistance to the Napoleonic invasion, and of antinationalist resistance to the Francoist state. It is also among the most delightful places in Spain.

The Sierra Nevada and upper Alpujarra are mostly crystalline slate, gneisses, and schist.[33] This range is often smooth-surfaced, except for the narrow *barrancas* (ravines) cut by cascading streams. Unlike porous limestone massifs, the Sierra Nevada has little subterranean water and few springs. The Sierra Nevada has no accommodating glaciated U-shaped valleys (unlike the Alps and Pyrenees), and consequently no interior population. It is ringed round by villages, but there are none in the heart of the massif.[34] Most of the Sierra has slopes between 15 and 30 degrees, but the Alpujarra is generally a little less steep.[35] The youth of the rock and the pattern of rainfall make the Sierra very prone to erosion. Specialists see signs of considerable geologic erosion, and anyone can see signs of massive anthropic erosion in recent decades.

The Alpujarra climate is decidedly mediterranean. Summers are bone dry and beastly hot in the valley bottom (c. 400–500 meters), scarcely less dry but somewhat less hot in the higher villages (c. 900–1,100 meters). There, where most Alpujarreños live, summer temperatures vary around 19–25°C. As in the Mediterranean generally, it is a little hotter and drier to the east. A village near the western end of the Alpujarra, Pampaneira, gets an average of 865 millimeters of precipitation annually, whereas Trevélez, further east, gets only 568, and villages in the Almerian Alpujarra get less still.[36] Rainfall is concentrated in the late fall and early spring, and often comes in torrential downpours, especially at the higher altitudes. A storm in October 1973 dropped 222 millimeters of rain on Mecina Bombarón in twenty-four hours.[37] Winter brings snow, which can fall in the upper villages by October and can linger until April. It lasts through the summer on the peaks. Trevélez, at 1,460 meters, has snow cover about 125 days per year. It is perfectly

[33] Sierra Nevada geology: Aldaya 1968; Carandell 1935; Thornes 1976, 6–9.

[34] Jean Sermet is the originator of this conception of the Sierra Nevada. See his 1942, 739–46. Sermet wrote a fourteen-volume thèse d'état on southern Spain, finished in 1969. I believe it is available only at the Université de Toulouse. Regrettably, I have not seen it.

[35] Pezzi Cerreto et al. 1982, 189–90.

[36] Data from Navarro Alcala-Zamora 1979, 35–6; Thornes 1976, 8–11; and Ortega Alba 1981, 158. Ortega Alba has contradictory data on p. 54. See also Castillo Requeña 1981.

[37] Thornes 1976, 46.

ordinary on a hot day in Orgiva or Ujíjar to see cool snow up above. Alpujarreños say they have their heads in the arctic and their feet in the tropics.[38] Granadines often sunbathe and ski at the same time. Exposure makes a crucial difference too, as everywhere in mountain zones. In the Alpujarra it is possible to dry figs on a sunny slope while conserving potatoes under the snow on the shady side.

Altitude governs the flora of the Sierra Nevada and Alpujarra, as elsewhere in the Mediterranean mountains.[39] In the valley bottom fruit trees, vines, and gardens compete with maquis species and cork oak (*Quercus suber*, widespread in the western Mediterranean and absent in the eastern). Maquis and garrigue dominate uncultivated hillsides up to 1,100 meters, where taller forest emerges in spots. Chestnut grows to 1,500 and oak to 1,800 meters, although both are represented now by only scattered stands. Oak is the potential climax formation from 1,500 to 1,700 meters, and *Pinus silvestris* from 1,800 to 2,000 meters.[40] Pines (*P. silvestris* and *P. larisis*) are more common now between 1,000 and 1,800 meters, as the chosen species of ICONA (Instituto Nacional para la Conservación de la Naturaleza), the agency in charge of reforestation in Spain. Trees of any sort are very rare above 1,800 meters, where garrigue (broom, holm oak, juniper, gorse) flourishes up to about 2,200 meters. Above that is the final vegetation belt, largely meadow, grassy where there is enough water, and small shrubs where there is not. The heights of Mulhacén and Veleta are always blanketed in snow.

The Sierra Nevada has long felt the ungentle touch of human action. Although the massif has been an effective barrier to transport, except for brief periods in the past, and it has no villages higher than Trevélez, it has long been surrounded by sizable village populations. From time to time economic booms – silk, wine, mining – have attracted people to zones around the Sierra. Roads only penetrated the upper Alpujarra in the 1930s, but mule traffic between the villages and the outside world has often been heavy. The Alpujarra is a little more isolated than the Marquesado but for centuries has had contact with both Granada and the coast. The valley floor, which produces fruits, olives, and once produced quantities of silk, has long had market connections with Almería, Málaga, and Granada. Since Moorish times traffic between the coast and the Marquesado used passes such as the Puerta de la Ragua.

[38] "Tiene la cabeza en el polo y los pies en el trópico."

[39] Molero Mesa and Pérez Raya 1987; Molero Mesa and García Martínez 1981; Losa Quintana, Molera Mesa, and Casares Porcel 1984; Prieto Fernández 1975 and 1982; Prieto 1971; Rodríguez Martínez 1985, 31–50; Ruíz de la Torre 1979; Navarro Alcala-Zamora 1979, 36–7.

[40] Rivas Goday and Mayor López 1965, 360–80.

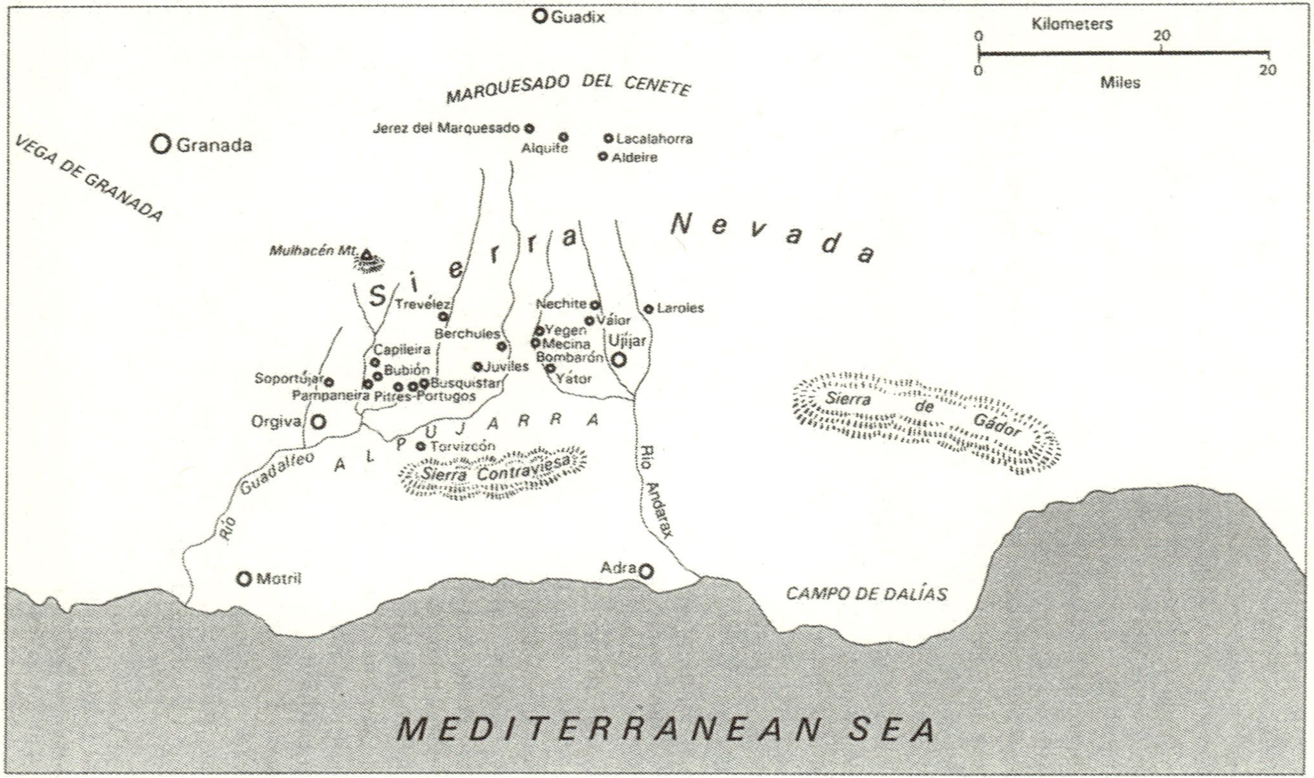

Map 2.4. The Sierra Nevada and Alpujarra

Mule trains thus passed through the Alpujarra, and linked it to a trading system involving salted fish, fruits, wine, and olive oil from the coast and flour from the Marquesado.[41] Now a railway connects the iron mines of Aldeire, in the Marquesado, with Almería. This change has served to isolate the Alpujarra rather than tie it to the coast, because the products of the Marquesado now travel to the east, wide of the mountains, to Almería, whereas before the railway they went by mule over the passes. The terrain and the nature of the surrounding economies has made the Alpujarra, and the Sierra Nevada in general, more accessible than the high Pindus or Taurus, if less so than the Lucanian Apennine villages. Today good paved roads and daily bus service link all the villages. If one has a motor vehicle, or bus fare, it is easy to get to Granada or the coast from anywhere in the Alpujarra in half a day or less.

The villages in the Alpujarra number about forty. Large ones like Orgiva and Ujíjar lie in the valley bottom, next to the Guadalfeo. Most however are around 800 to 1,100 meters, sheltered from the weather in one of the barrancas, located so as to get maximum sunshine. They lie a short distance from a stream, safe from flood yet near to the lifeblood of agriculture.[42] This elevation allows villagers to exploit agricultural possibilities from 500 to 2,000 meters, in effect giving them access to ecological zones from the subtropical to the subarctic. This verticality allows them to raise both Mediterranean crops and mountain crops: olives, vines, citrus, figs, and wheat below the villages, and barley, wheat, maize, apples, chestnuts, potatoes, kidney beans, walnuts, and cherries near or above the villages. In the past they cultivated barley and potato patches amid the grass up to the line of permanent snow, although this is no longer economical.[43]

This elevation, 800 to 1,100 meters, is also ideal for exploiting the irrigation potential of the Sierra. Much above 1,000 meters the waters

[41] BNM, Sección de Mapas, M8V, "Sierra Nevada y Alpujarra, 1913," shows the major tracks over the Sierra from the Alpujarra to the Marquesado. Soler y Pérez (1906, 25) discusses the routes. Brenan ([1957] 1988, 156–7) says he once (c. 1921) walked from Granada over the Sierra to Yegen, about sixty miles, in nineteen hours. He often met brigands when crossing the mountains. I walked over the Sierra at Puerta del Lobo in 1988, from Ujíjar almost to the Marquesado and back (less than forty miles) in a very long day, and happily met only ibex.

[42] Sermet (1942, 745–6) says that villages are located near springs. This is true, but there are many springs higher up that have no villages. I believe altitude is the key to village location. The distribution of population in 1887 by elevation in the Alpujarra (from Orgiva to Ujíjar) is given in Pino Artacho 1978, 87: 250–500 meters, 13.8%; 501–1,000 meters, 38.4%; 1,000–1,500 meters, 41.0%; and above 1,500 meters, 6.8%.

[43] Bosque Maurel 1971, 86. Sorre 1932, 301–8.

are often too diffuse, so that irrigation would require thousands of tiny channels. Much below 1,000 meters the waters are too concentrated, having flowed together into only half a dozen major tributaries of the Guadalfeo. Because keeping irrigation channels clear and functioning requires an enormous amount of work, doing it at the altitude where the natural network of streams most resembles a useful grid makes compelling sense. Muslims, who lived here before 1571, pioneered the irrigation system, showing the same delicate regard for water that imbues the Alhambra and Generalife in Granada. Their villages often lay a little lower than those now in the upper Alpujarra, because mountain crops like the potato were not available to them. Some of the higher villages date only from the Christian resettlement of the late sixteenth century, although most are of Moorish origins. The division of land rights between villages follows the old Moorish *tahas*: each village has lands at every altitude in a strip going up the mountainside. (Taha was the Arabic word for the transverse valleys of the Alpujarra now called barrancas.)

This principle of verticality, permitting each village access to several ecological zones, maximized the diversity of products and provided some insurance against bad harvests of one or another crop. It is a sensible cultural adaptation to mountain landscape, practiced by the Incas in the Andes and others elsewhere, and can make the villages involved self-sufficient. Indeed it can make them comparatively rich in a world of self-sufficient villages, as indeed the Alpujarra once was. But villages so well provided for self-sufficiency proved, in fact, poorly endowed for survival in a world of specialization and trade. We shall see (Chapter 6) how Alpujarreño life was well adapted to a world of slow communications and low trade, but not highly adaptable as the world changed.

The economy made possible by the peculiar geography of the Sierra Nevada rested on agriculture. Mining, stock raising, and forest industries have rarely amounted to much, and amount to almost nothing now. Iron mining still goes on in the Marquesado, and did until recently in Berchules in the western Alpujarra. Goats and sheep graze on the upper slopes in small numbers. A hydroelectric plant, the Central Poqueira, provides power (but little employment) in the westernmost barranca. Tourism adds something to the economy, especially since the highest road in Europe opened over the top of the Sierra, uniting Granada and its ski slopes to picturesque Capileira. With the Costa del Sol so nearby, Alpujarreños (and the Spanish national tourism organization) have sought to capitalize on the beauty of the region, but so far have met with little success.[44] Alpujarreños have made much more money going to the tourists and working in hotels and restaurants along the coast than they have by luring tourists to the Alpujarra.

Initiatives designed to change the basis of life in the Alpujarra have come to little so far, and agriculture, the gift of the snows, remains king, although the size of the kingdom is much reduced in the past thirty years. The villagers seem divided about the prospects for changing the basis of life in the Alpujarra. Many seem especially attached to local customs, proud of the Alpujarra's distinctions, although this has not kept thousands from emigrating. Outsiders have always considered the Alpujarra backward and clannish. One sixteenth-century visitor found Alpujarreños "ignorant" and their language a "monstrous mixture of Arabic and Spanish."[45] To most Spaniards, "Alpujarreño" means hayseed or hick, although in Granada Alpujarreños also have a reputation for hard work and shrewdness.[46]

The Rif

The Rif is as vague a term as the American Middle West. No two people agree on its boundaries, not the local tribesmen, Moroccans in general, or foreign scholars. I will use it to mean the area between Oued Laou and Oued Nekor, the mountainous terrain curving along the Mediterranean coast of northern Morocco and inland to a distance of about 100 kilometers. This arbitrary definition bears no relation to the political divisions of contemporary Morocco, or colonial Morocco before it. The Rif is an inhospitable land, more so than the Pindus and far more so than the Apennines or the Alpujarra, composed of chaotic ridges, slopes, and valleys. From the air it looks like so much crumpled-up brown paper.[47] The coast is equally inhospitable, with no good anchorages between Oued Laou and al-Hoceima. The Rif is a tough place to make a living, and has bred a population of very, very tough men and women.[48]

[44] Calatrava Requeña 1984.

[45] Cited in Braudel 1976, 1:35, 615.

[46] See Luna 1984, 20–1; Navarro Alcalá-Zamora 1979, 31; and Carrascosa 1960, 16. Probably the greatest influence on the image of the Alpujarra is the book of the famous Granadine novelist, Pedro Antonio de Alarcón, *La Alpujarra* (Madrid: Miguel Guijarro, 1874). To the extent that the region has a reputation outside of Spain, it is based on the marvelous book of Gerald Brenan ([1957] 1988). This has also earned a small audience within Spain.

[47] Ernest Gellner's phrase, cited in Hart 1976a, 28.

[48] According to Woolman (1969, 30), a North African proverb has it that "The Tunisian is a woman, the Algerian a man, the Moroccan a lion." To this it must be added that

The Rif range, geologically distinct from and much younger than the Atlas, is about 350 kilometers long and 80 to 100 kilometers wide. Its soils are derived chiefly from schist and marne and are heavy and clayey, although in most locales it is an exaggeration to speak of soils.[49] The upland plain of Targuist divides the range in two. The mountains are taller in the west, in the districts known as Gomara and Senhaja, with many peaks over 2,000 meters, whereas to the east gentler and lower slopes prevail. Short torrential streams drain the Rif to the north, although they carry water all the way to the Mediterranean only in the rainy season. The south flank, which on average is half as steep as the north, drains through a series of *oueds* into the Oued Ouerrha, itself a tributary of the Sebou, which debouches into the Atlantic. Drainage on both north and south flanks is fast and prone to floods, because the soils are generally impermeable and there is little subterranean water. In the western half of the Rif there are only about two hundred springs, one per 17 square kilometers or so, and this is an area of considerable rainfall.[50]

The watercourses are seasonal because rainfall is seasonal. Summers are parched and uncomfortably hot in the lowlands, while dry and much cooler at Ketama (27°C average July temperature) and other upland points. The precipitation comes in the winter and spring, often in downpours. In 1962, Chechouan (Xauen) got 279 millimeters of rain in a single day. Next year, at Bab Taza, 1,120 millimeters fell over a twenty-day span. Throughout the western Rif, week-long rainy spells that drop 300 to 400 millimeters are routine in the wet season. Winter brings snow, occasionally lots of it. Spanish troops floundered in 3-meter snow drifts during the Rif War in 1926 in Zerket, and in Senhaja the Spanish (after winning the war) set up ski troops. The high mountains are under snow from November to May. Ketama gets about seventy days of frost per year. Rifians are quite accustomed to snow and can walk barefoot in it for hours on end.[51] Precipitation trails off sharply as one goes east in the Rif, from over 1,000 millimeters in places (perhaps 2,000 on the peaks) to less than 400 millimeters: the influence of the Atlantic

Moroccans regard Rifians as especially hardy. (This is probably a Moroccan proverb rather than a North African one.) Leo Africanus, a Moroccan whose father owned property in the Rif, found Rifians courageous but "rude and uncivil" (1896, 516) – in the French translation (1956, 273), "brutes ignorantes."

49 Rif geology: Barathon 1989; Maurer 1968a, 40–212; Gharbaoui 1986; Morocco, Ministère de l'Agriculture 1960, 13–14. On soils generally, see Jiménez Salas 1946, 1949.

50 Morocco, Ministère de l'Agriculture 1960, 15–17. Russo 1928, 688, on low water retention of eastern Rif soils.

51 Coon 1931, 10.

diminishes to the east, and the taller mountains of the west intercept much of the moisture. The northern side of the mountains generally gets a bit less rain than the southern slopes, which are more open to the westerly Atlantic winds. Consequently, the southern Rif is far more densely settled than the north. Ketama, in the west, gets more rainy days (seventy-four per year on average) than anywhere along the Mediterranean coast of Africa. Most of the western Rif gets as much rain as Ketama (about 1,500 millimeters per year on average).[52] Annual fluctuation is considerable, and killing drought affects the eastern end of the Rif from time to time (most recently in 1944–5).

The distribution of rainfall in combination with altitude shapes the spontaneous vegetation of the Rif mountains. As everywhere in the Mediterranean, today's flora shows the consequences (often unintended) of human settlement. At least 80 percent of spontaneous vegetation today is scrub. The lower slopes, where not cultivated, support garrigue and maquis, a bit taller and thicker in the west than in the east. Many patches of the eastern Rif carry no vegetation to speak of. Lowland Rif maquis consists of the usual scrub oaks (and the economically important cork oak as well) and lentisk, but also of palmetto (*Chamaerops humilis*) and camel thorn (*Zizyphus lotus*). Like the kermes oak, palmetto can grow to 5 or 10 meters when left alone, as it is in some sacred sites in the Rif. But palmetto has edible buds and its fiber has several uses, so most plants get a lot of attention and thus stay small. Palmetto is virtually indestructible, invades disturbed landscapes quickly, and spreads everywhere fields are abandoned or forests burned. Above 500 meters holm oak dominates the maquis. Drought-resistant Aleppo pine (*Pinus halepensis*) grows up to 1,000 meters here and there, and *Pinus pinaster* appears up to 1,500 meters. Deciduous oaks cover a few patches around 1,000 meters in the western end of the chain, while fir and cedar are found further up. Firs appear between 1,500 and 2,000 meters in a couple of locations, interspersed with oak and pine on drier slopes. Neither fir nor pine is extensive. Cedar (*Cedrus atlanticus*) is the most important timber tree of the Rif, as in the Atlas ranges. In 1980 it covered about 15,000 hectares in the Rif.[53] The best stands are on silicaceous soils above 1,600 meters, but its range extends from 1,300 to 2,400 meters. Where undisturbed, cedar forests form a closed canopy with little undergrowth – cedar is sensitive to wind and flourishes only as closed forest.

[52] Mikesell (1961, 16) has a good rainfall map. See also Fay 1984, 6; Fay 1979, 83; G. Maurer 1979, 49; and G. Maurer 1968a:16–29 and 34–9, on Rif climate.

[53] M'Hirit 1982, 11. Cedar covers 116,000 hectares in the Middle Atlas (p. 17). This is the standard work on Moroccan cedar. See also P. Boudy 1948–54, 2:532–83, and Deil 1988 on Rif cedar.

Mature trees reach 40 meters, and the largest are 60 meters tall and 12 to 15 meters in circumference. Such trees may be six to seven hundred years old. But few patches of undisturbed cedar remain, and generally stands are now mixed with up- and, coming oak often *Q. faginea.* No part of the Rif is above the tree line, and there is no real subalpine or alpine stratum. Almost no tall trees grow east of the Senhaja mountains. Now soils are too degraded and rainfall too slight – both of which may be a consequence of deforestation – for forest to return to the eastern Rif. Except for sacred groves the east is all scrub, and the domain of lentisk and thuya (*Collitris articulata*) as tough as the palmetto and more drought-resistant.

The botany of the Rif is very complex, subject to extreme local variations on account of moisture and soil conditions. It is slightly more distinct than that of the Taurus, Pindus, Apennines, or Sierra Nevada, on account of cork oak, palmetto, and cedar, common species in the Rif but rare to nonexistent in the European chains. The historical botany of the Rif, about which more later, is obscure in detail but a clear disaster in its main lines. It reflects the impact of man and goat more than the other Mediterranean ranges, because eroded soils and summer drought make regeneration more difficult here than elsewhere, and thus human impact more enduring.[54]

The Rifians are great walkers and they have to be.[55] Only a few paved roads penetrate the chain, the best a sinuous highway running from Tetouan to Chechouan to al-Hoceima. Legend claims that the stretch from Tetouan to Chechouan was built after letting a donkey loose and building the road wherever it wandered.[56] Most Rifians walk or ride donkeys wherever they go, although the kif merchants prefer a Mercedes-Benz (kif, derived from *Cannabis sativa,* is the local term for marijuana). Roads carry little traffic: on a cool sunny day people lie down in the roads to enjoy the heat of the asphalt. The rough terrain has always made communication and transport difficult, and the proximity of the sea has helped but little because of its rugged coast. Isolation can

[54] Mikesell 1961, 18–31; Hart 1976a, 25–8; Grohmann-Kerouach 1971, 209; G. Maurer 1968a, 29–34; Emberger 1929; Ruíz de la Torre 1955. A curious feature of Rif botany is the great extent of prickly pear cactus, a native of North America. It surrounds every house in the central and eastern Rif. It must have returned from Mexico with sixteenth-century Spaniards, established itself in southern Spain and then in Morocco. It can be found in many locations in the Mediterranean, including the approaches to the Acropolis.

[55] Windus (1725, 25–6) noted this some 270 years ago and many twentieth-century observers agree. Some Rifians are great runners too: the 1928 Olympic marathon winner came from the Rif.

[56] Gaudio 1981, 89. Traveling the road does nothing to discredit the legend.

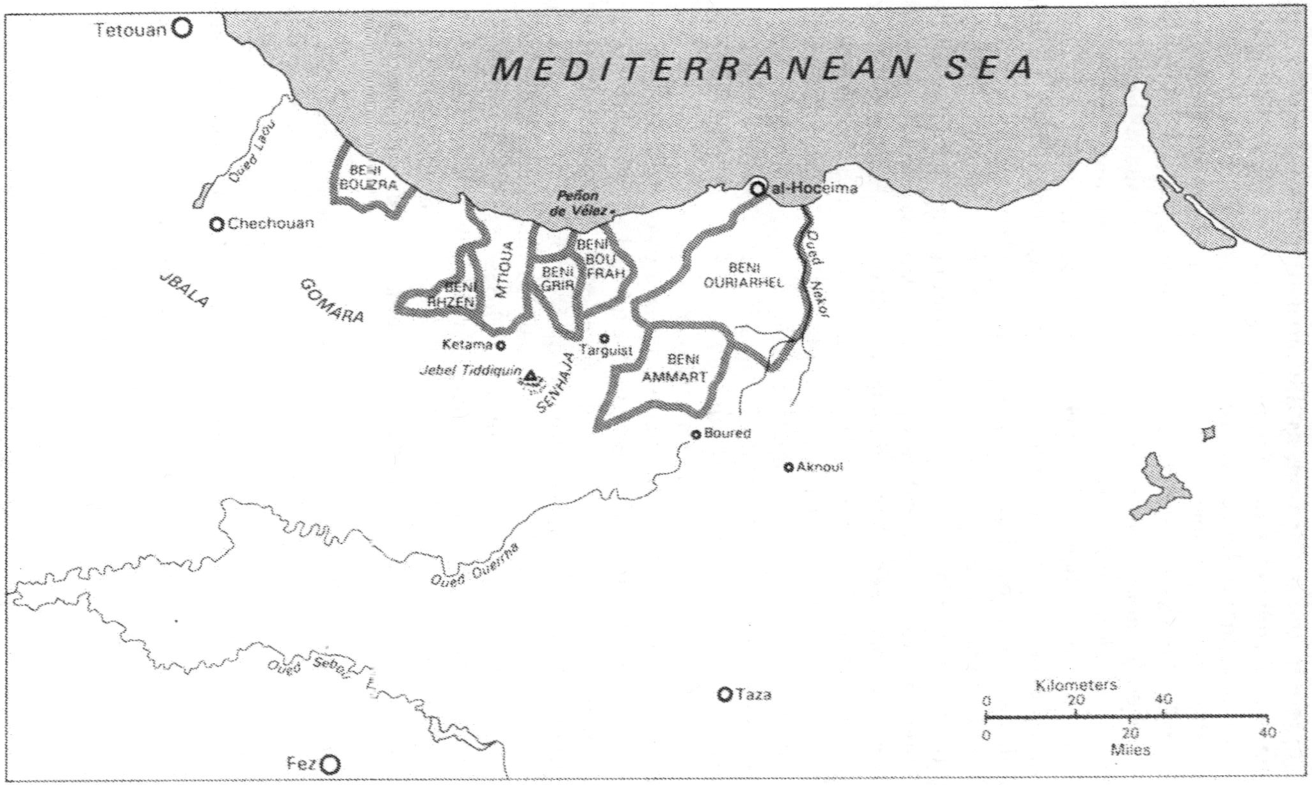

Map 2.5. The Rif

be complete in the wet season, as the oueds flood, mud makes trails impassable, or snow cuts off communities entirely. These mountains have been marginal in Moroccan and Mediterranean history generally, rarely visited, rarely conquered, rarely governed except internally, and rarely involved in extensive trade. In softer environs this isolation would have resulted in more luxurious surviving vegetation, but in the Rif the effects of overshoot alone have produced skeletal mountains and will, in all probability, suffice to produce shell villages too.

Villages are small outside Gomara and Senhaja. Everywhere smaller ones of fewer than two hundred people predominate. They generally contain members of an extended family. In the past they fissioned when they grew too large; today this is becoming impossible as land is too scarce. Villages are located well up on slopes, surveying all approaches. Defense needs, lowland malaria, and water supply determined settlement location.[57] Formerly settlement avoided the frontier lines between tribes. In the east settlement is often dispersed and houses are normally 100 meters or more apart, surrounded by a wall and a rampart of cactus.[58] Population density in the Rif as a whole as of 1982 was 111 per square kilometer, and rising.[59] This is by far the highest figure of any mountain area in the Mediterranean.

The Rifians are tribal people, acutely conscious of their kinship ties and purported ancestry. They recognize, and feel allegiance to, many social units. These begin with the family (generally an extended family) and proceed to the *douar*, the lineage or fraction, the tribe, and, in some cases, a confederation of tribes.[60] Beyond this, many Rifians feel an identity as Berbers, even some of those who now speak Arabic. Many of the sixty or so tribes speak Rifian Berber dialects, but others, especially in the west, speak Arabic. The Berber dialects can be mutually unintelligible. Some are dying out, and in the long run Berber in the Rif is

57 Fay 1972, 1:145, 165–6. An inspector general of primary education complained that he could never visit a school unannounced in the Rif: his approach was always visible from afar.

Recently settlement has returned to low-lying areas as fear of malaria has vanished. See Pascon and Wusten 1983, 91–102, for morphology of settlement among the Beni Bou Frah.

58 The reasons for these different settlement patterns in different parts of the Rif remain unclear. Dispersed settlement is rare among mountain people of the Islamic world. The Berbers of the Middle and High Atlas, the Druze, the Kurds all live in villages, although the Pathans of Afghanistan and northern Pakistan live spread out like the people of the central and eastern Rif.

59 M'Rabet 1986, 1.

60 *Douar* is the Arabic word used to denote a small hamlet, which often consists of related people.

likely to disappear, like Vlach speech in the Pindus, and so many other local languages the world around. Most men throughout the mountains understand Arabic now, although often women and children do not. Berbers are keenly aware of their minority status within Morocco, and few feel any allegiance to the nation, the state, or the throne. The mountain folk of the Taurus, Pindus, Apennines, and Alpujarra normally have much weaker senses of ancestry and broad kinship, and stronger senses of membership in a nation. These allegiances, however, shift over time, and great differences exist among individuals, classes, and regions. All things considered, however, the Rifians' social world is distinct.[61]

Rifians are almost all Muslims. The Arabs brought Islam to Morocco in the early eighth century, but it did not filter into the Rif mountains until the eleventh and twelfth centuries. In adopting Islam it seems the Rifians only slowly renounced alcohol. Leo Africanus, writing in the sixteenth century, said they had a penchant for drunkenness, whereas a twentieth-century Spanish observer noted their sobriety.[62] Today Rifian Islam has a few distinct features, such as a preference for pilgrimage to local shrines and an alleged casual attitude toward prayer.[63] During the Spanish occupation (which on offshore islands began in the sixteenth century, but in a substantial way began in 1927 with the end of the Rif War), a small Spanish Christian population lived in the towns, including mountain centers like Chechouan. They all left after 1956, when the Spanish Protectorate ended. Until the 1950s substantial Jewish quarters existed in towns like Chechouan and Tetouan, inhabited by descendants of refugees from Andalusia in the era of the Spanish reconquest. The Sephardim have mostly emigrated to Israel, and the traces of Andalusian influence are few, confined largely to architecture.

The economy of the Rif mountains rests on smallholder agriculture. Until very recently, Rifians have farmed as if labor were scarce, with no terraces and a heavy reliance on a form of shifting cultivation. Barley and wheat are the principal cereals, and almost every family grows some. Where soils and climate permit, citrus fruits, almonds, grapes for raisins, and figs complement the cereals. Higher up, walnuts are more important. These tree crops dominate the southern slope of the Rif and are much rarer on the north. The higher elevations are used as

[61] Hart 1976a is the most detailed social anthropology of a Rif tribe. See more generally G. Maurer 1968b, 16–18.

[62] Leo Africanus 1896, 2:516; Fernández de Castro y Pechera 1945, 47. Liger (1963, 104–5) says Rifian Berbers had a special dispensation from Almohad and Almoravid rulers allowing them to drink wine.

[63] Mikesell 1961, 52.

pasture or for temporary fields. Every year Rifians burn over many mountainsides to plant a cereal crop. After a year or two they abandon these fields to spontaneous vegetation, in which their goats may graze for twenty or so years. Then they burn the same area again. This is a system of long-fallow shifting agriculture, practiced widely (although illegal since 1956) and probably for a long time in the Rif. It gives better yields than settled cultivation, except where irrigation applies. In the late twentieth century, various pressures have made the system impossible to sustain (Chapter 7).

At lower elevations, or where irrigation is practical, vegetable gardens prosper. Where irrigation is possible, Rifian peasants can produce a variety of crops and manage to be self-sufficient in food. Naturally irrigable zones, like the lowland Oued Laou or Oued Nekor, support dense populations. Upland irrigation is possible in parts of Gomara and Senhaja, where Rifians raise the American trinity of maize, beans, and squash in the summer. Without irrigation, and most of the Rif is without it, Rifians are hostage to the unreliable rains. They must grow a narrower range of crops while also running a higher risk of crop failure.[64]

Weekly markets all over the Rif allow for the exchange of some goods, permitting a degree of specialization among farmers. But the only cash crop of major importance is kif, grown especially in Senhaja around Ketama, and not sold, at least not openly, in the markets. It is a summer crop and requires intensive irrigation. Formerly its utility lay in the fiber, used for rope, mats, and baskets. Now its value has skyrocketed with the demand for marijuana in western Europe.

Stock raising is of only secondary importance in the Rif mountains. In the well-watered west, richer families own a cow or two. Most every family has a goat or six, but flocks are small. Goats predominate, because the vegetation is often too poor for fussier creatures. They roam everywhere, in garrigue, in fallow fields, along roadsides. Herders do not practice long-distance transhumance, as in other Mediterranean mountains, perhaps because of the history of intertribal conflict. Local migrations between river bottom and high slopes still go on to a small degree. The beast of burden in the mountains is a donkey or mule, although women carry many a heavy load.

Hunting occupies men especially during winter months when there is little to do in the fields, but now the activity is as much ritual as it is economic. Along the coast fishermen cast their nets into the eastbound stream of cool Atlantic water, much richer in fish than most of the

[64] In 1960 only 1.4% of the Rif was irrigated. Morocco, Ministère de l'Agriculture 1960, 18.

Mediterranean, but the contribution to Rifian diet and economy is meager. Leo Africanus indicates that in the sixteenth century fishing played a major part in the local economy near Oued Laou, with exports perhaps as far as Fez.[65]

Economic life in the Rif is more active but perhaps – despite kif – less commercial than elsewhere in the Mediterranean mountains. No productive land goes unused, although plenty of able-bodied people, especially males, are underemployed. Weekly open-air markets (*suqs*) attract large crowds, but they are social events as much as economic ones (often segregated by sex so that there are men's markets and women's markets). Sundays once brought everyone in rural America to church, but that did not mean rural Americans were especially pious: services were a time of gathering and gossiping, and the same is true of the suq in Morocco. Suq attendance does not mean Rifians are especially market-oriented. Different villages hold suqs on different days in the Rif, so that the determined can attend a market every day of the week with a little traveling. Locals exchange nuts and fruits at the suq (generally not cereals) and might buy a few goods from a distance, such as batteries, salt, or tea. Specialists also sell services, ranging from shoe repair to letter writing. Indeed most of the functions that cities normally provide for a society are performed in the suq. In effect Rifians have temporal rather than spatial towns. In a landscape in which people travel much more easily than do goods, mobile market towns of this sort may represent an adaptation to environment.[66] This market system notwithstanding, production and consumption are mostly localized, often staying within the same family. Rif mountaineers consume a greater percentage of their own product than mountain folk anywhere in Europe, and probably the Taurus. The only fully market-oriented producers are the kif growers.

A Few Villages

Two Taurus Villages

Zerk (also known as Altınkaya) is a small village hidden away in the western Taurus. It lies at about 950 meters in a depression surrounded

[65] Leo Africanus 1896, 2:517–20; Grohmann-Kerouach 1971 treats fishing at al-Hoceima.

[66] Ethiopian kings used to move around their mountainous country with their royal retinues, rather than preside from a single capital city. It made sense for their capital city to move rather than to try to bring food to it. On Moroccan suqs, see Fogg 1939 and 1940; Mikesell 1961, 90–1; Hart 1976a, 69–88. An excellent map of Rif suqs appears in Grohmann-Kerouach 1971, 200. The classic work is Troin 1975.

2.1. Fields, houses, and ancient theater of Zerk (Altınkaya) with Bozburun Mountain in the background (July 1989). In the fourth century A.D. about twenty thousand people lived here.

by mountains, most notably Bozburun Dağ, which looms to the north. It is in the watershed of the Köprü Su. Zerk nestles against the acropolis of the ancient Greek city of Selge, which according to Gibbon had a population of twenty thousand.[67] Locals usually cite the same figure, or add a few thousand. Zerk's summer population is about eight hundred, less in winter. Little children abound, although teenagers are a bit rare. There is a school and one mosque.

The people of Zerk are almost all peasant farmers. Maize is their basic crop, giving 3,000 kilograms per hectare without irrigation (although here they speak of tenths of hectares, not hectares: 300 kilograms per deka). Soils are good in the bottomlands – the surrounding slopes are not farmed – and only shortage of water prevents still higher yields. Villagers also raise sunflowers, vines, walnuts, plums, and tomatoes. A few cattle and goats wander around, grazing on the stubble of harvested fields in late summer, while chickens patrol the grounds of most every house. The economy is for the most part subsistence. There are no shops, and villagers do not buy much from the outside, except occa-

[67] Gibbon 1909, 3:388.

sionally rice from the coast, a radio, or a lightbulb. Cooking is done with firewood, but every house now has electricity.

A notable exception to the subsistence economy is a couple of households whose members can speak some German or English: they earn a few thousand Turkish lira (about 2,200 to the U.S. dollar in 1989) every day in the summer by showing a jeepload or two of (normally German) tourists around the extensive ruins of Selge. Selge is hard to reach, and only the theater is recognizable to the casual observer, so it has nothing like the tourist trade of some of the ancient sites on the coast. But if a meal is added into the bargain, and the Germans are generous, the transaction might yield the equivalent of several days' wages for a day laborer, all accomplished in a few hours. As children can do the guiding, and women invariably do the cooking, this business is prized among the men of Zerk. But only those households who have a linguist can participate.

No one goes hungry in Zerk, at least not in normal times. The yields are adequate to feed the small population, and most families get some help from a son working in Antalya or Serik. Houses are of stone, solidly built, usually two storeys tall, with the ground floor often reserved for animals. Many families can afford to educate their children, or at least some of their children, beyond primary school, which means boarding them in some other town and forgoing the income from their labor. A few families are building new, larger houses. If Zerk had the aqueduct system that ancient Selge must have had, it would be prosperous indeed.

To the north of Zerk, up the Köprü Su Valley beyond its deep canyons, is the most fruitful land in the western Taurus, the Yılanlı Plateau. Koçular is representative of its cluster of villages, all situated between 1,050 and 1,400 meters. Koçular is richer than Zerk. Here, in the foothills of the Dedegöl Dağ a system of irrigation permits a flourishing agriculture. The government has recently built concrete conduits, troughs open to the air, to bring the mountain water to the village fields. The effect is to raise yields and living standards, and at least temporarily to reduce the drive to emigrate. Irrigation existed before, of course, but dirt channels required more labor to maintain and lost more water to infiltration. The pitch of the land is often modest hereabouts, and so the use of machinery is possible. Giant Massey-Fergusons, the tractor of choice, toil on one field while on the next a family wields scythe and sickle. Pickup trucks do not yet exist here, however, so transport proceeds at animal speeds. Oxen, mules, and donkeys trudge along every road, often pulling carts piled high with hay or watermelons or people.

2.2. Bringing in fodder for the late summer dry season near Koçular, in the Yılanlı Plateau (July 1989).

Fields here are larger than in truly mountain areas, although few are larger than a single hectare. Stone walls and scrub, or the occasional row of poplars, demarcate their boundaries. The steeper ones produce wheat or barley, whereas the flatter ones yield beans, maize, and sunflowers. Grape vines and pear and apple trees occupy the remaining cropland.[68]

Strictly speaking, terraces do not exist around Koçular. Steeper slopes now out of cultivation often have stone barriers built to slow soil loss. In the not-too-distant past, perhaps as recently as twenty years ago, these fields were in use. But with better irrigation, and in some places population decline, villagers now farm only the valley bottoms and lowest slopes. The abandoned fields scarcely have any soil left, but much of what has been lost is now useful in the maize fields below. So far, erosion has provoked only minute and inefficient efforts at soil conservation.

There is no mature forest to be seen anywhere around Koçular or its

[68] Planhol 1958, 152, 160, on the crops of Yılanlı. Maize seems to have become more important since he wrote, and beans less so.

neighbors. An occasional evergreen oak stands in the middle of a field as a shade tree, and planted groves of poplars exist here and there. A few aged plane trees decorate the banks of some rivulets. But all else is almost treeless, with scarcely any of the pines that clothe the mountains flanking the Köprü Su further south. Winter is long and cold, and firewood supply a serious concern. It comes from afar, or from the poplars. In late summer and autumn, the roads are alive with mobile brush piles, creeping toward the villages. Women and children ransack the thin undergrowth for enough fodder to get their animals through the lean season, which might come in September in a dry year, as well as in the cold winter.

All ages are represented in Koçular, as in Zerk. There is no sign of the skewed age distribution typical of shell villages. Emigration remains modest enough that its effect is hard to detect, camouflaged in part, no doubt, by a high birth rate. One emigrant who had worked many years in France (Angoulêmes) as a woodcutter had returned to Koçular and thought that many of those who had left the surrounding villages had come home. Since 1973 regulations have made it difficult to get work in Germany, leaving Turkish cities and the Persian Gulf as the most plausible options for would-be emigrants.[69] This may have helped keep people home.

All ages work, at least in the summer harvests. Men monopolize the farm machinery, but women do most of the scythe work – indeed, most of whatever work requires bending double for hours. Women also carry the fodder and firewood, if no animals are available. Men and boys wear Western dress and often a cloth cap, always dark and often tattered. Women wear headscarves, blouses, baggy trousers sometimes covered by a long skirt, all their vestments brightly colored. No one exposes much skin to the sun, no matter what the temperature. As in Zerk, the houses are of stone, often two storeys but sometimes three, solidly built, with red tile roofs. A few are coated with stucco, presenting a smooth face to the world, a sure sign that many families are doing well. If crop yields continue to climb, or population does not, they should continue to do so.

Five Pindus Villages

Samarina is the highest village in Greece, at about 1,650 meters, nestled against the east flank of the Smólikas Massif in the heart of the Pindus.

[69] I wrote these words before the Persian Gulf War. This source of employment may be more open to Turks in the future, as Iraqis, Jordanians, and Palestinians will likely not be welcome in Kuwait and Saudi Arabia.

Its communications have always been much greater with Macedonia and Thessaly than with Epirus. The forbidding mountains to its west make it in effect a cul-de-sac. It acquired a road and a telephone before 1940, but electricity came only in 1956. Its many charms have attracted a fair share of visitors to this remote spot, including two teams of scholars who have left portraits of the village as it was in 1911–13 and in 1967.[70]

To judge by its churches, it was founded in the eighteenth century. In 1987 it had three thousand inhabitants, according to a sign posted on the outskirts, and seven according to the latest census. The discrepancy arises because Samarina in the summer months is a bustling place, one of the largest communities in the Pindus, but in winter is deserted. Most of the people, perhaps 90 percent, prefer to speak Greek, although many of the adults can speak Vlach. Historically it has been home to transhumant shepherds, and although pastoralism is in decline, the seasonal ebb and flow of population still characterizes Samarina. There are many new houses, far more than is usual in the Pindus, as well as several in ruins. Some of the ruins are victims of erosion, which threatens the village as well as the surrounding pastures. Retaining walls and other devices have not prevented deep gullies through and around Samarina.

Shops, especially coffee shops, are surviving nicely on the summer trade. There is no agriculture in Samarina, and very little forest industry. Its sawmills have been out of business for over thirty years. It lives off its flocks, off money made and remitted by its natives elsewhere, and to a small degree off summer tourism. There are two small hotels and a large hostel, which cater to urban Greeks who come to enjoy the mountain air. Most, of course, have Samarinot ancestors. Pasture and sheep are everywhere in the summer months, but goats are rare. The sheep number forty thousand (1987), not a patch on what they used to be, according to two blue-eyed shepherds old enough to remember Turkish times (before 1912). But in the 1930s villagers gave Nicholas Hammond the same figure of forty thousand.[71] The sheep winter in Thessaly, but the people winter everywhere: Grevená, Larissa, Thessaloniki, and further afield. Almost no one stays up in Samarina through the winter.

South and west of Samarina but north and east of Ioannina lie the forty-six villages of Zagori, a region with its own identity and traditions in everything from architecture to feasting.[72] Eastern Zagori recalls the

[70] Wace and Thompson [1913] 1972; Sivignon 1968. Wace and Thompson were English classicists who lived and traveled with Samarina natives, on and off for three years. Sivignon headed a Greco-French team of geographers and sociologists.

[71] Hammond 1967, 266.

[72] On Zagori generally: Théophilou 1983; Campbell 1964; Lambridis 1870; Nikolaidis 1939.

2.3. Pindus villages still depend on the pastoral economy. High pasture between Vovousa and Perivoli toward the end of summer (1987).

Alps, with high pasture and tall forest. Vlach villages such as Vovousa and Perivoli live off timber and flocks, with small patches of farming on bottomlands along the Aoos River. Vovousa was burned by the Germans in 1944, so now a new village exists, centered on a bridge across the Aoos. Mountains clothed in second-growth pine overlook the village, and in summer keep it in shade for a third of the day. Vovousa owns about 100,000 *stremmata* (about 10,000 hectares) of official forest, much of which is actually maquis or garrigue. There is a sawmill just south of the village in an alluvial meadow, but production is now uneconomic what with Greek participation in the European Community. Better-quality pine from northern Europe sells in Ioannina for half the price.

About 150 people live here through the winter, and perhaps 240 spend the summer, making Vovousa a sizable village by contemporary Zagori standards. Roads built within the past twenty years afford easy communication with Ioannina so on a Saturday night in the summer those working afar can come back to the village for a variation on venerable social traditions. A lamb on a spit in one of three public establishments in the village attracts forty people, several of them grandparents, but with many little children and even a contingent or two of teenagers and those in their twenties, who arrive on motorcycle. The children and the middle-aged watch Greek soap-opera drama on a video cassette recorder (television signals do not reach Vovousa). The youths chatter boisterously among themselves, and the very old, whose eyes are not good enough to watch the television, sit quietly. Around midnight the television goes off and an old American jukebox is called into service, playing Epirot songs, featuring mournful clarinet melodies. The music is loud enough to wake the dead, so even the aged with weak ears can hear just fine. Old and young, and everyone in between, dance, tap a foot, sing along, hum, tightening ties of family and community that are constantly threatened by emigration and cultural change that divides one generation from another. Even the priest stops by, a man of thirty-five or so (of whom it is said that he likes beer more than he does God), to smoke and drink and chat with his contemporaries. On Sunday, and on every other day of the week but Saturday, Vovousa is a quiet, almost desolate place.

South of Zagori, beyond Metsovo and the Zygos, is the other historic heart of Vlach culture in the Pindus, a cluster of villages, Sirako, Kallarites, and Pramanda notable among them. Sirako is the northernmost of these, a community of two or three hundred in the summer, but home to only one family after the snows come in October. This is the birthplace of Kolettis, the French-educated private physician to Ali Pasha and the first prime minister of independent Greece. There are many ruined houses, some of two and three storeys, with fine nineteenth-century iron grille work on the windows, and spacious rooms inside. A century and more ago Sirako was a prosperous place. But now there are only two coffee shops, where old men play backgammon, and slowly sip coffee or an occasional ouzo. Children are rare and the elders refer to Sirako as a village of the dead. It is a classic shell village. The Germans burned much of it in 1944, but unlike Vovousa it has not been rebuilt, and whole neighborhoods have been empty since the war. There is no agriculture, no forest to speak of, and the pasture is thin. In 1987 the sheep numbered six or seven thousand according to locals, whereas before the war Sirako had over fifty thousand.[73]

[73] Sheep population numbers are rough guides at best. Official statistics (concerning all animals) are now often inflated so as to take advantage of subsidies provided by the European Community. The numbers here are those told me by locals and are unconfirmed. Sirako in the mid-nineteenth century had 42,000 sheep according to Lambridis 1887–90, 5:50. In the 1960s it had 25,000 according to Syndesmos Koinotitos Syrrakou 1969 (unpaginated). A description of Sirako in 1893 appears in Philippson 1897, 243–4. On the district as a whole: Papakostas 1967, 93–103.

In one typical house, a two-storey stone structure built in 1860, live a father aged sixty-six, a mother, daughter, and the father's nephew. None live there for more than two months of the year. Most spend the other months in Ioannina. None in the house are under thirty years of age (other children are elsewhere, in Greece and America). The rooms are large and airy, furnished in the Turkish style with divans around the perimeter, and the hospitality is warm, even by the high standards of rural Greece. The father's experience is typical of many in Sirako, and in surrounding villages. As a young man before the war he had sixteen hundred sheep. After the conclusion of the Greek Civil War (1949) he sold his sheep and became a bus driver in Ioannina and Athens. He had six sons, none of whom he wanted to become shepherds, and all of whom agreed with him. They have scattered, living in Ioannina and Preveza, where one is a teacher. The sons rarely return to Sirako.

Sirako has a sister village, Kallarites, to which it is connected by a footpath, once a mule track, that winds through the small canyon of the Kallaritikós River. The path has stairs painstakingly cut into rock, and once was a thoroughfare of some consequence. A century and a half ago Kallarites had a population of five thousand. It stood as the southernmost Vlach village in the Pindus. Now fifteen or twenty people are resident through the winter, while a couple of hundred are in evidence in summer months. Vlach is seldom heard. Kallarites once had considerable agriculture, but today the terraces are all abandoned, and crumble with every rain and passing hoof. Most are given over to sheep pasture, or else have been colonized by shrubs and a few hardy trees. Springs and even aqueducts once provided and controlled the flow of water. Today the springs are often dry and overgrown. Sheep raising is the only source of income now. The villagers now graze about fifteen thousand sheep in the surrounding mountains, a fraction of the total in former times. The main square (*plateia*) is large, dominated by a plane tree that must be three hundred years old. Several large coffee shops, most of which are dark and quiet, surround the square. They are patronized only by a few old men. A few young people summer in Kallarites, but they live elsewhere most of the time, go (or went) to school elsewhere, and look upon Kallarites as a summer refuge and the home of their ancestors. Most would prefer to be in Larissa (a town of no conspicuous charm). Most people in Kallarites have relatives in America, and those they speak of seem to have done well. The young would like to follow, and some will. The government is building a road that approaches Kallarites from the south, progressing slowly through

2.4. Frangadhes is a village in central Zagori, typical in its architecture (September 1987).

difficult terrain. By the time it is finished there may well be no one left.[74]

Pramanda is a few kilometers south of Kallarites, accessible by a winding road. It is a much bigger village than either Sirako or Kallarites is today, with fifteen hundred people, several restaurants, grocers, bakers, and even two shoe stores. Signs of abandoned agriculture are everywhere in the surrounding mountains, sometimes almost up to the summits. Little cultivation goes on today, just enough "for the houses." Nowadays sheep are the core of the economy, although some timber cutting goes on in the pine forest above the village – although there is precious little forest and much erosion, both in the pastures and in the heart of the village. Many houses will be undermined in the next few years unless drastic measures are taken, of which as yet there is no sign. In all probability the threatened houses will be abandoned instead.

These Pindus villages are typical of scores more in their depopulation, the age structure and morale of their remaining populations, the

[74] An 1893 description of Kallarites is in Philippson 1897, 321–2.

abandonment of agriculture, and their environmental degradation. They are shell villages, soon to be ghost towns, as the elder generation dies off. Their children are less likely to return to the Pindus for their golden years, and so a further rapid population decrease is almost certain. Everyone is aware of this. Some depopulated Greek villages are near enough to urban centers that they can expect prosperous vacationers and weekend people to buy the houses and maintain some sort of presence, and perhaps even a sense of community, if not much sense of continuity with the past. But in the Pindus this will not happen – the villages are too far from the big city.

Two Lucanian Mountain Villages

Among all the mountain villages of Basilicata two merit brief introductions, Calvello and Terranova di Pollino. Calvello sits on a steep hillside at 730 meters. Most villages in this area are on hilltops, or at least well up the slopes. Slightly over three thousand people live in the village. It is surrounded by rather good pasture and not so good farmland. Peasants now drive to their fields in cars, commuting distances their parents took hours to walk. The coat of arms of Calvello very appropriately features a sheep. Pastoralism has declined greatly of late but is still important. Thin woods, mostly oaks and chestnuts, crown a few of the ridges around the village. Much of the forest is in coppice, and in the fall and winter donkeys (and occasionally women) carry giant loads of firewood into Calvello. For centuries Calvello has produced fine tiles and ceramics. Although these traditions are in decline, they have not vanished and may perhaps survive indefinitely. Old people predominate in Calvello but not to the degree that they do in Pindus villages.

Terranova di Pollino is among the highest villages in southern Italy, at 926 meters, home to slightly more than two thousand people. It is on the north flank of the Pollino Massif (Serra Dolcedorme, 2,271 meters), in a flysch zone amid pasture and second-growth forest, none of which appears to be more than forty years old. It gets 1,085 millimeters of rain per year on average, mostly in the winter as at Calvello. The higher reaches of Pollino are limestone, through which waters percolate year round, to the advantage of the villagers below. Terranova di Pollino fits the classic European conception of a mountain village: it has an Alpine feel to it. Firewood is stacked everywhere, the smell of woodsmoke is in the air except for a few months out of the year. Men wear corduroy hats shaped like the Alpine peasants' felt ones; women wear colorful scarves. Fields are few, and most agriculture is confined to garden plots, although some sloping fields are in use below the village. To the northwest, above Terranova di Pollino, are row after row of abandoned terraces and signs of serious erosion. Fine pasture dots the mountainside, which is steep only near the summit. But flocks are now few, and more and more land is reverting to forest, generally to pine. Immediately to the west of the village, no more than 5 kilometers over a rough road is the tiny hamlet of Casa di Conte, beyond which is a fine tall beech forest where pigs once foraged. Now it is disturbed only by the occasional chain saw, by four-wheel-drive vehicles belonging to the Bari bourgeoisie, and, on the first Sunday in July, by colorfully caparisoned pilgrims on their way to a festival at the sanctuary of the Madonna di Pollino.

The mountain villages of the southern Apennines remain more viable than those of other mountain areas in southern Europe (though less vigorous than those of the Taurus). Their age structure is not as distorted, their depopulation not as severe. There are probably many reasons for this, but an obvious one is that the relief in Basilicata is mild compared with that of the Pindus, the Sierra Nevada, and other ranges, meaning that agricultural machinery has not been completely impractical. Even at Terranova di Pollino peasants use small bulldozers to pull broad disk plows. The machines mean that fewer hands are needed. As we shall see, agricultural labor shortage, paradoxically, has contributed to the depopulation of mountain villages elsewhere. The Lucanian Apennines, while terribly backward by lowland standards, have nonetheless made the transition to modern mechanized agriculture, something probably impossible in other Mediterranean mountain areas.[75]

Two Alpujarra Villages

From a distance the villages of the Alpujarra look much alike, their whitewashed houses on the mountain slopes resembling a cascade of sugar cubes against a green or brown background. But each has its distinctions. Trevélez's (population 877 in 1986) special claim is that it is the highest village in Spain, although this is occasionally disputed by partisans of the Pyrenees. It is certainly the highest in the Sierra Nevada, well up in a long barranca, nicely protected from the winter winds and open to the southern sun. But it gets only about eight and one-half hours of direct sun in summer and six in winter, which retards plant growth. Statistically, the climate resembles that of Stockholm far more than it does nearby Málaga. The village is divided into three

[75] An intimate portrait of the social world of an anonymous Lucanian mountain village c. 1959 is Cornelisen 1969.

2.5. Trevélez shrouded in snow (February 1988). Trevélez is the highest village in the Sierra Nevada.

parts, contiguous yet distinct: lower, middle, and upper. These correspond in inverted order to the social rank of the inhabitants. The poorest live in the highest section, originally home to shepherd families, whereas the richest live in the lower part of town, served by the road. But no one is very rich here, or anywhere in the Alpujarra, statistically among the poorest sections of Andalusia, the poorest region of Spain. Social distinctions do matter however, and patterns of paternalism and deference common in rural Spain persist here, even if the material gap between classes is narrow. Bars outnumber churches eight to two, a typical ratio in these parts. There is one school: education levels in the Alpujarra are as low as anywhere in Spain, if much improved in the past twenty years. A century ago illiteracy was 85 percent and curious superstitions held sway: In several villages, according to the Granada physician Oloriz, who made a tour of the Alpujarra in 1894, curative powers were attributed to old shoes of men named Juan.[76] Agriculture

[76] Pino Artacho 1978, 195–217. Oloriz also tells us that crime was rare in 1894, property crime extremely so. Most crimes were those against persons. Drunkenness was also rare, but Oloriz says this might be a matter of supply: phylloxera had destroyed the vines. In 1882, as today, Trevélez had two churches. Willkomm 1882, 169.

2.6. Springtime (1988) in the *barranca* of Poqueira, with the villages of Capileira, Pampaneira, and Bubión. The tallest peaks of the Sierra Nevada lie behind this *barranca.*

is still the business of Trevélez. Fields extend about 1 kilometer above the village, along the banks of the Río Trevélez and the *acequías* (irrigation ditches) that it feeds. In 1977 about 80 percent of cropland was irrigated.[77] Peas, beans, onions, potatoes, peppers, and squash are grown in and just below the village. Wheat, barley, and maize occupy the higher fields, giving modest yields, about 10 to 12 metric quintals per hectare. Meadows and chestnut groves go up to about 3 kilometers above the village. Acequías begin about 2 or 3 kilometers further up, where abandoned terraces indicate that agriculture was practiced perhaps thirty years ago. It is no longer economic to exploit the huge labor investment of previous generations more than about 1 kilometer outside of the village. Peasants use hoes in Trevélez because plots are tiny. In the bottom of the valley plows are routine, and a few are pulled by tractor.

Even in summer there is little grazing above Trevélez, because there is no road. Elsewhere in the Sierra Nevada the forest service has

[77] Liszewski and Suliborski 1977, 70.

2.7. The village and terraces of Mecina Bombarón, a typical Alpujarra landscape (February 1988).

built roads, which shepherds use to carry their flocks to high pasture. A rough carriage road, built by a mapping commission intent on taking measurements from the top of Mulhacén, existed above Trevélez a century ago.[78] Its ruin is visible in places, but unless the forest service decides to rebuild it, Trevélez will not have much livestock. Hogs used to be important in the local economy, and Trevélez ham still has a high reputation in Spain. Every bar and restaurant in the Alpujarra has hams hanging from the ceiling, all of which are called Trevélez ham, no matter where they might come from.

[78] Marín 1896, 192.

Trevélez may not cure many hams these days, but it still bakes its own bread, made from its own flour, ground and threshed from its own wheat and barley. The baker, as in every Alpujarra village, has a mountain of firewood outside his premises. Most every house has a supply stacked on the roof. Firewood is the principal fuel, and its shortage or abundance has meant a lot in the lives of the locals. At the moment (1988) firewood is not so hard to come by if you know whom to ask, as the baker does. Solar heaters are sold here, as they are in northern Greece, but while they may have a future at present the energy base of the village is not so far removed from its state five hundred years ago: wood supplies most of the fuel for heating and some for cooking; gas is used only in cooking; electricity powers only the lights, and the rare household appliance.

Capileira is one of the high villages settled in the sixteenth century by, among others, immigrants from Galicia in northern Spain (hence the Gallego name). It is in the first large barranca near the western end of the Alpujarra, and is the highest village on the highest road in Europe. The road means that Capileira gets a fair tourist trade, and it now has a large inn run by twin brothers. On a weekend evening its two restaurants are packed with locals and Granadines. No other village in the Alpujarra has made as much of tourism.[79] Tourism has moderated the decline of Capileira, although it may have accelerated the abandonment of agriculture, as it offers peasant youths easier livings. From the west edge of the village, looking across the barranca one can see six threshing floors that are now overgrown and out of use. More than half of the terraces on the same slope are unused now. Sheep and goats graze on the abandoned plots. Capileira is one of the few places where there are young shepherds, although I suspect they do not intend to remain shepherds for long.

Two Rifian Towns

Ketama is the center of the kif trade, and a strange place indeed. It lies on a well-watered upland plain at the intersection of important roads, amid a much-thinned cedar forest, and is the most important suq in Senhaja. In the days of the Spanish Protectorate it was a resort of sorts, and a fine hotel remains, now patronized chiefly by buyers and sellers of the local specialty. Animals graze in the grassiest meadows in the Rif

[79] Except perhaps Lanjarón, at the western mouth of the valley, really outside the valley many would say, where mineral waters used to bring thousands in search of cures. There are still twenty or thirty hotels in Lanjarón, although now it generally bottles its water for export rather than attracting visitors to its waters.

2.8. Some of the rare terraces in the Rif, just south of Ketama.

mountains, between stumps of ax-hewn oak and cedar. Irrigation supports horticulture, but any man with ambition can make better profits in kif than in vegetables. It is illegal to sell kif, but this scarcely depresses trade. Any outsider in Ketama is assumed to be there for the sole purpose of buying, and competition for his attention is stiff. Late-model Mercedes-Benzes follow a stranger everywhere, as far as 50 kilometers from Ketama, trying to strike a deal. This occasionally goes as far as running recalcitrant customers off the road. Ketama businessmen do not easily take no for an answer. In an earlier day, according to Coon, it was the center of a trade in young boys, but he did not actually see this.[80] Any visitor can see, indeed cannot help but see, the kif market in action.

Chechouan, perched on slopes at the western edge of the Rif mountains, is a whitewashed town of perhaps thirty thousand, recalling mountain villages in Andalusia, whence in fact most of its original settlers came.[81] It gets its name from the "horns" of the two-peaked mountain to its north, and its water too. Chechouan has extensive irrigated

[80] Coon 1931, 110–11.

[81] The town is also spelled Chaouen, Chefchaouene, Chauen, Sheshwan, Shawan, Shawen, and Xauen. A quest for originality seems to inform transliteration from Rifian Berber (or *thamazigth*) and from Rifian Arabic.

vegetable gardens and fruit trees all around the town, even in it. The streets are narrow and winding, often leading to dead ends, like a smaller version of the Fez or Marrakesh medina. People crowd the larger streets, and children are everywhere. Almost all young women carry a baby on their backs. Men generally wear the long cloak of the Arab world, the *jellaba*, but the younger among them often wear Western clothes. Few women wear the veil; most have blue tattoo on their faces and henna dye on their hands and feet. Little children are dressed as they are in Europe, if a bit more raggedly. All have shoes. In Chechouan most people live by agriculture and stock raising, or by the tourist and kif trade. It is on an easy route from Tangier and Ceuta and so attracts a share of the European tourists brought by the car ferries from Spain. It is also a haven for those Europeans attracted to the kif, and to the Moroccans who supply their wants.

Rifian towns and villages differ from mountain communities on the north side of the Mediterranean in that they are poorer (except for the kif dealers), dirtier, and far more crowded and alive: they are anything but shell villages. Rather, they are bursting with people, more heavily populated than ever before. They are, it seems, in the early phase of overshoot. They have a far more interesting – and tempestuous – future ahead of them than do the mountains of Greece, Italy, or Spain, although it may resemble their past. The consequences of overshoot may prove different in the Rif; but if Morocco is fortunate, Rifian experience will in its main lines follow that of the Pindus and Alpujarras.

Conclusion

Despite considerable variation from place to place, for reasons of climate and relief, all Mediterranean mountain areas are ecologically fragile. Their slopes and rainfall regimes invite accelerated erosion if vegetation cover permits. Drought discourages forest recovery in the wake of deforestation. Hence the impact of man and goat is more durable, more consequential here than in gentler environments. A denuded hillside in rural New England will normally recover in fifty years; a denuded hillside in Greece or Morocco will more likely uncover bare rock. Furthermore, the ordinary fluctuations in patterns of land use degrade Mediterranean mountains – and other ecologically fragile zones as well – much faster and more fully than in more blessed zones. Until quite recently population expanded and contracted in irregular rhythms defined by food supply, disease exposure, and war, and land was cleared or abandoned accordingly. In the Mediterranean mountains, these fluctuations reduced the productivity of the land, reduced its carrying capacity (except where technological improvements applied). Similar fluctuations inflicted far less severe costs in flatter or more temperate zones. Although it is not true that vegetation and soils never recover in the Pindus or the Alpujarra, they do so more slowly and less often. The ratcheting toward degraded and useless landscape is not quite predetermined, but the ratchet rarely slips.

Whereas ecological degradation and economic marginalization characterize all these five mountain zones, and almost all of the villages in them, the current social situation varies considerably. In the Taurus and the Rif the villages are full, teeming with activity. In the Rif the countryside is full, and it is hard to imagine further population growth.[82] In Lucania the villages are like the philosopher's glass, half empty or half full, depending on how one looks at it. In comparison to the Rif, or to their own recent past, they are half empty. In comparison to the villages of the Pindus or the Alpujarra, where mountain depopulation is furthest advanced, they are half full. Nature and history have combined to chart a different course in each setting, even if the general direction around the Mediterranean is much the same. Every landscape tells a slightly different story. These stories begin in Chapter 3.

[82] Although G. Maurer (1968b) thought the same a generation ago, when population was about 70% of today's levels.

3

THE DEEP HISTORY OF MEDITERRANEAN LANDSCAPES

> My Italy. . . . Words are in vain, for the mortal wounds, which I see so numerous in your beautiful body.
>
> Petrarch, *Italia Mia*

Millennia of human activity have weighed heavily upon Mediterranean lands, but far less so in the mountains than on the plains and low hills. The reasons for this are simple: the economic activities that have underlain Mediterranean civilizations rested on lowland crops and trade between coastal zones. The vine and the olive do best below 500 meters, and wheat, the third member of the Mediterranean agricultural trinity, is planted at higher altitudes only when all fertile lowland is used or used up. The timber trade, though an important business in ancient times, scarcely reached the high mountains. Until rather recently, only grazing animals and fires set by their keepers affected mountain vegetation and soils in an important way. Even grazing seems to have expanded greatly in Mediterranean areas in the past thousand years, so that its previous impact may well have been slighter than its recent effect. In general, where human populations have been great, so has landscape degradation. And where human populations have fluctuated greatly, degradation has been greater still. Only very rarely in ancient and medieval times could small numbers of people inspired by the demand of distant markets affect landscapes fundamentally. The modern period, as we shall see, was quite different in this respect.

The extreme landscape degradation of many parts of the Mediterranean world tempts one to think that the deterioration must be very old indeed. But this is not necessarily so. The mediterranean environments of Chile and southern California are often as severely eroded as anywhere in the Mediterranean, yet strong human influence there has been comparatively recent. Most, if not all, of the arroyos in Arizona and California developed since 1850. The Atlantic island of Madeira, whose name means wood or timber, has a mediterranean climate. It was uninhabited until 1425, but after a few short years of human occupation its vegetation resembled that of the western Mediterranean. After a few short centuries several Caribbean islands (admittedly not a mediterranean zone) have lost almost all their forests and much of their soils. The vulnerability of mediterranean ecosystems is such that when deprived of forest cover, erosion can follow quickly, if slopes and soils are propitious. Local conditions will determine subsequent ecological succession, but in most instances vegetation will be dominated by fire-resistant (and goat-resistant wherever there are goats) maquis. The landscape degradation of the Mediterranean *might* be very old, but its thoroughness is no proof.[1]

Prehistoric Landscapes

Prehuman Landscapes

Until the Neolithic revolution, the landscapes of the Mediterranean evolved very slowly. Human beings played almost no role. An endless contest between upthrusts and geologic (as opposed to anthropic) erosion defined the landforms of the earth's surface. These moves and countermoves – still going on – make glaciers seem quick by comparison. Climate change governed vegetation patterns, moving to a rhythm tapped out by the advance and retreat of glaciers across Europe. Investigation of pollen accumulated in bogs or lake sediments (palynology) shows that forests have come and gone. Species appeared and disappeared, dominated landscapes or became marginal, all without the slightest help from humankind. In glacial epochs, Mediterranean lands had little forest and less maquis; instead, steppe vegetation predominated. The higher mountains developed ice packs, although they had forest cover during interglacials. These shifts took tens of thousands of years to work themselves out, and in fact we know very little for certain about how landscapes looked then. We now live toward the end of an interglacial period (these tend to be brief compared to glacial periods), at the beginning of which the present flora and fauna of the Mediterranean took shape.[2]

Since the last ice age, about twelve thousand years ago, climate change has proceeded modestly, with only minor effect on vegetation. What we consider the "natural" vegetation of the Mediterranean world is that

[1] Aschmann 1973, 363; Crosby 1986, 75–9, for the early settlement history and ecology of Madeira. On the Caribbean, Richardson 1983, 3–31; on American arroyos, Cooke and Reeves 1976.

[2] Bintliff and van Zeist 1982; Roberts 1982. Mediterranean climate and vegetation seem to have made their first appearance two to three million years ago: Suc 1984.

which came with our current interglacial. As the Mediterranean basin warmed, forest species recolonized widely, marching out of their ice age refugia. Cold-climate trees such as birch led the way, but by roughly seven thousand years ago they had moved on to northern Europe and in the Mediterranean had given way to oaks and pines. After the last ice age but before our species had a noticeable impact on vegetation, almost all Mediterranean areas, lowland as well as upland, supported forest.[3]

Throughout these long millennia of slow changes, humankind played only a small role in shaping Mediterranean environments, and a smaller one in the mountains than in the plains. Hunters and gatherers lived in the Mediterranean basin at least 500,000 years ago. They were few in number, because their way of life can only support sparse populations, and it took them hundreds of thousands of years to make their way throughout the region. They hunted forest animals, primarily deer, and burned woodland as part of the hunt; but they had no goats to keep forest from regenerating, and, more important, their slender numbers limited their effect. Their activities scarcely affected vegetation complexes already selected for capacity to regenerate after fire. Their fires did, however, begin the long history of anthropic erosion in the Mediterranean basin. Mediterranean slopes erode many times faster when denuded by fire (in the Alpes Maritimes thirteen times as fast). But quick regeneration of scrub species reduced the scale of this Paleolithic erosion, and its extent was certainly small compared with what was to come later.[4] Furthermore, hunters and gatherers prudently preferred the flatter lowlands.

Breakthrough to Agriculture

When humankind learned to manipulate plant communities so as to favor those that provide usable energy (agriculture), and to tame edible and otherwise useful animals (animal husbandry), our species became the most important shaper of landscapes. This happened about twelve thousand years ago, not long after the climate change and sea level rise associated with the end of the last ice age. Geologic and climatic conditions continued to shift, of course, at their own speeds, but anthropic

[3] Expert opinion is summarized in Seigue 1985, 54–7. See also Thirgood 1981. Useful palynological studies include Bottema 1974; van Zeist and Bottema 1982; van Zeist, Woldring, and Stapert 1975, 55–143; Roberts 1982; Reille 1977. An excellent summary of palynological evidence is Pons and Reille 1986.

[4] Orengo and Rossi 1973.

change now accelerated – wherever agriculture and stock raising were practiced. These practices began first (it is almost certain) in southwest Asia and spread, rather slowly at first, in many directions. It took several thousand years before all the shores of the Mediterranean supported farmers and herders, and more time still until these populations spread beyond the most favorable places. By four thousand years ago, cereal cultivation had diffused widely, even to upland plateaus in the Taurus and the Rif, but its imprint remained thinly scattered.[5]

Early Neolithic farmers probably had only a modest impact on their landscapes. In the southern Argolid (Greece), a lowland area where extensive excavations have recently been carried out, the conclusion is that "little land needed to be cleared for agriculture, and we should not expect, nor do we see, any evidence of soil erosion."[6] After about 2500 B.C. farming seems to have expanded and hillside soils began to erode, although not enough to jeopardize settlement.[7] Grazing in the Argolid may have begun the history of forest degradation there well before 2500 B.C., but the total impact of Neolithic culture in this comparatively inviting environment remained modest. In the mountains it was probably negligible. Paleobotanists pretty well agree that four thousand years ago Mediterranean climate encouraged forest and humankind had not yet managed to destroy forest on any scale, so that almost all the Mediterranean world supported "continuous forest cover."[8]

By about three thousand years ago, through long trial and error, the food crops best suited to the rhythm of the Mediterranean seasons were established and spreading in the eastern part of the basin. Cultivation destabilized soils of the hillsides and rolling plains. The plains of northern Greece, for instance, lost much of their forest and began to suffer accelerated erosion as early as the Bronze Age.[9] Plains elsewhere around the inland sea shared the same destiny. The absence of tides and summer storms, together with strong shore winds, good harbors, and countless islands, made the Mediterranean very easy to sail, and so nets of exchange quickly grew up, encouraging agricultural zones to specialize in either wheat or olives and vines. This division of labor permitted great economic and demographic growth in the affected regions, and served as the basis for ancient Greek and Roman civilization. Nothing alters ecology quite like civilization.

[5] Pons 1984, 435.

[6] Van Andel and Runnels 1987, 67.

[7] Van Andel and Runnels 1987, 92.

[8] Van Zeist and Bottema 1982, 319; Pons and Quézel 1985, 35.

[9] Davidson 1980, 155–7. Pons and Quézel 1985, 36–7. Rackham (1982) thinks this is so for Greece generally (on the basis of observations of Crete and Boeotia).

Ancient Landscape Change

The Ancients and the Natural Environment

Assessing the environmental impact of Greek and Roman civilization is a popular and venerable exercise. The ancient writers provide a fair amount of evidence about vegetation, the uses of timber, agricultural practices, and so forth. Modern scholars have done their best to judge what it all meant. The most careful and informed analyses are those of Meiggs (1982) and Hughes (1982). The following general assessment draws heavily from them.[10]

The ancient Greeks and Romans energetically transformed the landscape around the Mediterranean, installing groves and fields where they could, often through seaborne colonization. They developed mining and smelting operations here and there, which voraciously devoured forest (unless fed by managed coppice). They built navies and merchant marines out of wood. A great deal of their other construction required wood. And they used wood – or charcoal – for over 90 percent of their fuel needs. Without a doubt a substantial measure of Mediterranean deforestation and consequent erosion happened in classical times, say between 500 B.C. and A.D. 500.[11]

It happened mostly in the low plains and hills, where Mediterranean agriculture flourished. As Lucretius put it: "Continuously men compelled the forests to withdraw higher up the mountains and to leave the lower slopes to cultivation."[12] The famous passages in Plato about deforestation and its effects apply to Attica (at least most of Attica), but not generally. He wrote that after nine thousand years of floods and erosion: "The result is that Athens is now like one of the small islands, the skeleton of a sick body with barely any flesh on it. In those early days the land was unspoilt; there was soil high upon the mountains. . . . There was abundant timber on the mountains and of this you can still see some evidence. Some of our mountains can now only support bees."[13] Plato apparently had Hymettus in mind in this passage (Hymettus then as now was covered by garrigue), because other mountains near Athens probably retained their forest through the fifth century B.C. Some, like

[10] But see also Trotta Treyden 1916; Semple 1919; Thirgood 1981; Hughes 1975 and 1988; Heichelheim [in Thomas 1956]; Darby 1956; Marsh [1864] 1965; Semple 1931; Winters 1974, 160–85; and Pons and Quézel 1985 among general treatments.

[11] Rackham (1982) disagrees, arguing for widespread preclassical deforestation and erosion. His is a maverick opinion.

[12] Lucretius 5.1370f, quoted in Meiggs 1982, 372–3.

[13] Plato, *Critias* 111B–C, cited in Meiggs 1982, 188. See also *Phaedo* 110E.

Parnes in northern Attica, had wild boars (and hence forest) in the days of Pausanias (late second century A.D.), and an "inexhaustible supply of timber" in the early nineteenth century.[14] In fact Parnes still has some pine forest, now protected by the Greek forest service. Attica surely felt the imprint of the ancients as deeply as anywhere, what with high population density, silver mining at Lavrion, and an imperial city-state enamored of monumental buildings and dependent on its wooden fleet. To be sure, Athens in the time of the Peloponnesian War desperately needed Macedonian timber. But only a few other areas shared the fate that Plato properly bemoaned.

The ancients transformed most lowland zones, even if they did not strip them quite so thoroughly as the demands of Athens did Attica. By the time the Roman Empire began to totter (third century A.D.) it is likely that no extensive forest remained in the plains or low hills surrounding the Mediterranean. Agriculture and timber cutting spearheaded the assault, although in mining areas such as southeastern Spain the demands of the smelter would have outstripped them. Wherever timber could easily float down to the sea, the ancients cut it. Wherever it stood close to cities, it could survive only if protected – as it often was, for religious or other reasons. But timber cutters would not venture into the remote and wild mountains, difficult of access, since they could not transport and sell the fruits of their labor. And agriculture did not beckon, as long as trade united the fertile littorals of the Mediterranean and the Black Seas. In Hughes's judgment, forest cover remained in mountain districts of Macedonia, Corsica, the Atlas, Cyprus, the Black Sea coast, the Taurus, and the Lebanon. In Meiggs's view, good forest remained a bit more widely. He adds Crete, Sicily, most of France, parts of the Apennines, the Dalmatian coast and hinterland (although I suspect that he is generous here), parts of the Anatolian Plateau, and the Aurès in Algeria.[15]

Ancient civilization certainly made its imprint on the Mediterranean basin. Low-lying areas were cleared for crops, stimulating considerable erosion on the low hillsides. Timber demand and in places smelting added to deforestation. Where goats followed in the wake of fire or the ax, the consequences to soils and vegetation were long-standing. Hydrological regimes were upset: springs dried up in some places (Plato's Attica for instance), floods became more common in others (the

[14] Pausanias 1.32.1; Leake 1835, 2:420; Meiggs 1982, 188–91. Taygetos, the rocky mountain spine near Sparta, was covered in forest in the fourth century B.C. according to Plato's student Theophrastus. Semple 1919, 28.

[15] Hughes 1982, 66–7; Meiggs 1982, 371–403; Thirgood (1981, 40–6) gives a similar judgment.

Tiber). Microclimate probably changed as a result of deforestation (Theophrastus records one such case near Philippi).[16] Silt deposits created deltas and marshes, preparing the ground for malaria, which haunted Mediterranean coasts from about the fourth century B.C. until quite recently.[17] It may well be true that one of the reasons for the decline of ancient civilization is environmental deterioration and consequent falling economic productivity.[18] But the mountains remained marginal, touched here and there by the environmental changes of ancient times, but not transformed.

Only a few scraps of information exist regarding landscape, population, and resources in ancient times in the western Taurus, northern Pindus, southern Apennines, Sierra Nevada, or Rif. None of them lay close to mining centers, to naval yards, or to great cities. But nonetheless in some cases landscapes changed considerably and permanently. Population fluctuations and war account for the changes.

The Taurus in Ancient Times

Coastal Pamphylia (now Antalya Province) and upland Pisidia (the Lake district at the western edge of the Taurus) flourished in Roman times. The coastal cities built theaters that held as many as thirty thousand people. Recent estimates, probably too high, put the population of Pamphylia at its height at six million.[19] In upland Pisidia and the western Taurus, settlement was much sparser, except around the site of today's village of Zerk.

At Zerk, where eight hundred people now live, a metropolis (Selge) of twenty thousand Hellenized Anatolians toiled away. The acropolis alone covers several hectares, and the theater could seat perhaps two thousand. Selge struck coins as early as the fifth century B.C., and was a local power in Alexander's day. It occasionally competed with his successors, the Seleucids and Attalids, for control over the Pamphylian plain (third and second centuries B.C.). Strabo reports that Selge had vines, olives, abundant pasture and rich forests. (The olives and vines must have been at some distance below the city, because they could not do well at 900 meters.) Selge produced and exported perfumes, sold as incense to the superstitious according to Strabo, but most of the population must have lived from agriculture.[20] Rough terrain prohibited

[16] Theophrastus 5.14.5. [17] Bruce-Chwatt and Zulueta 1980; Grmek 1989, 275–83.

[18] Hughes (1988) makes this case strongly. [19] *Antalya 1967. Il Yıllığı* 1969, 273.

[20] Pliny, *Natural History* 12.125, cited in Planhol 1958, 79. On ancient Selge: Strabo 12:7; Fellows 1838, 172; Bean 1963, 137–42.

importing grain from any distance. The soils at and around Selge are fertile; only water shortage constrained cultivation. Selgeans solved this problem with a system of aqueducts, the remains of which are visible today. Centuries of prosperity flowed from this adaptation to circumstances, preserved by the remoteness of Selge from the wars of Hellenistic times, by sage diplomacy (Selge welcomed Alexander and the Romans), by renowned martial prowess that kept warfare on their enemies' terrain, and eventually by the Pax Romana. Selge was never sacked or conquered. But in the late fifth century A.D. disaster struck. The city shrank to the size of a small town, never to recover.[21]

No known military event accounts for this fate. Perhaps the demographic collapse represents overshoot within the small confines of Selge; perhaps unrecorded political violence destroyed the irrigation works; perhaps the decline of Pamphylia (beginning in the third century A.D.) reduced the opportunities for trade, although this could only be a contributing cause to a decline as sharp as Selge's. Or perhaps epidemic disease is more likely. At any rate, catastrophic depopulation of this sort was not rare in Asia Minor after the third century.[22]

Archaeology reveals little about the ancient history, landscapes, and vegetation of the western Taurus beyond Selge. Of mountain communities in the general vicinity, only Selge's rival, Termessos, has left much behind. It lay to the west, in the Güllük Dağ, not in the Taurus. Exports of timber from the seaport of Side attest to lumber operations, presumably in the Taurus. Literary sources imply that Pisidia acquired considerable population in the second and third centuries A.D., partly though the settlement of Roman veterans. Isparta and other Turkish towns in old Pisidia certainly had ancient predecessors (some founded by colonists from Sparta). It is quite likely, but quite unknown, that these towns and villages also suffered catastrophic depopulation when brought into contact with lethal infections with which they had no prior experience. Such a fate is normal among populations isolated from broader currents for any length of time. If so, the neglect of fields and terraces might have led to local erosion. But it is impossible to know.

Palynological evidence seems to indicate alternating ages of deforestation and forest regeneration in Anatolian history. Forest returned to the Taurus gradually after the ice age (during which the mountains supported steppe grasses), and colonized slopes up to 2,000 meters by 4000 B.C.[23] In the Taurus it appears that Mediterranean forest

[21] Gibbon 1909, 3:388n. [22] Wagstaff 1985, 141.
[23] Van Zeist, Woldring, and Stapert 1975, 141.

established itself at the same time as Neolithic settlement; indeed perhaps the more varied habitat and thinner forest of southern Anatolia explain why Neolithic colonists avoided northern Anatolia, where thicker forest stood. At any rate, pollen records suggest population growth and land clearing in the Taurus in the unstable times between the fall of the Hittite Empire (c. 1200 B.C.) and the establishment of Seleucid control (third century B.C.). Thenceforth, through Roman times, the pollen record shows no agricultural activity in the Taurus.[24] Perhaps mountain population fluctuated in a rhythm defined by the interplay between mountain and plain. Although considerable linguistic, ethnic, and economic differences separated mountain folk from lowlanders, their interplay, whether peaceful or violent, would have played a strong role in determining settlement patterns and population densities in the mountains. When peace prevailed in the lowlands, and irrigation was possible, some mountain folk may have descended to try their hands in the more fertile plains – certainly in more recent times mountain people have done this in many settings.[25] But when rural insecurity plagued the lowlands of Konya, Antalya, and Adana, mountaineers would have found scant reason to abandon the security of their fastnesses. Indeed, people of the plain may well have sought refuge in the mountains in such times – although if mountain folk were numerous enough this would have been difficult indeed.

The Pax Romana came to southwestern Anatolia in the first century B.C. Great cities soon developed along the southern coast. Lowland agriculture spread widely with peaceful conditions; pollen records from the southwestern Konya Plain show increasing forest clearance. But the mountains, it seems, became peripheral and almost certainly lost population.[26]

Recent research on the Taurus in Roman times implies very low population in the mountains.[27] Despite their strong presence in the coastal lowlands of Cilicia and Pamphylia, neither the Romans nor their Byzantine successors could control the population of highland Isauria (the Taurus). The mountains belonged to a society ethnically and linguistically distinct from the lowland populations. The language of Isauria remained Luvian long after Greek had spread throughout the lowlands (although Christianity and later Islam did reach the mountain folk).

[24] The fertile Tol Ovası, a plain on the middle Köprü Su, was unsettled and forested in ancient times. Planhol 1958, 77.

[25] Roberts 1982. See also Todd 1980.

[26] Planhol 1958, 68–80.

[27] Shaw 1990. Shaw's article treats that part of the Taurus known as Isauria. Just where it began and ended is unclear.

Violent bands dominated mountain society. The Romans called these men brigands or bandits when strong enough to fight them, but called them barons when they paid them off instead. These mountain bands raided the plains regularly, often carrying off women. This implies a small population, unable to sustain itself biologically, and unable therefore to maintain a balanced sex ratio. Probably then, as more recently, mountain bands consisted predominantly of young men, to whom death often came early. Until at least the seventh century A.D., and often enough thereafter, the Taurus remained a "land of dissidence," in permament resistance to the plains and centralized states.[28] Such circumstances limited cultivation and herding in the mountains, and spontaneous vegetation, including tall forests of cedar, fir, oak, and pine, covered large areas. Ecologically, as in other respects, the Taurus had a very different experience from the neighboring plains.

Clues elsewhere in Anatolia support this idea. Geological evidence from the riverbeds of western Anatolia suggests considerable deforestation and erosion in Roman times, especially from A.D. 100 to 300. This is probably connected to the rapid expansion of Greek agriculture in the Meander (Büyük Menderes) and other river valleys well suited to the vine and olive, and not to depopulation, which came in the fifth century. Climate change is not a plausible explanation. Analysis of the sediment shows that this erosion affected lowlands, cleared and plowed by the Greeks. Mountain areas generally, and some patches of the Anatolian Plateau, remained heavily forested beyond antiquity.[29]

The Pindus in Ancient Times

Prehistoric remains in the Pindus suggest that regular human use of the land began during the last ice age. Seasonal transhumant exploitation followed for some millennia. Crops appeared in the lowlands about the seventh millennium B.C., but were few and scattered. The palynological record shows no changes in vegetation patterns until the

[28] Shaw 1990, 258–70. Shaw regards the struggle between mountain and plain almost as a constant and even refers to it as "trench warfare" (p. 263). He says the situation remained little changed for four thousand years. For life among early twentieth-century Taurus brigands, see the novel of Yashar Kemal, *Memed My Hawk.*

[29] Eisma 1964, 1061; Eisma 1978; Rowton 1967. Ramsay (1890, 387–415) discusses locations of ancient sites in Pisidia. As we shall see in Lucania, Greek colonization indirectly led to heavy erosion by pushing indigenous populations up onto unstable soils. This may have been the case in western Anatolia too, but Eisma does not say so: he attributes erosion directly to Greek settlement.

Bronze Age, in the second millennium. Little more than that can be claimed with confidence about man and environment in prehistory.[30]

A good deal more is known about the ancient history of the Pindus; everything that was known as of 1967 is collected in Hammond's encyclopedic study of Epirus.[31] In the 1930s he scoured the country for archaeological remains, village by village. He concluded that after 1700 B.C. Epirus was inhabited by shepherd peoples who eventually became Hellenized and attained considerable population in the era of the Epirot League, especially 240–167 B.C. The population was centered on Molossi – the high Pindus and its flanks. The Molossians based their economy on sheep and must have practiced the same transhumance that Vlachs more recently have pursued. Molossian sheepdogs enjoyed a reputation for relentless ferocity, which their descendants still deserve.[32] When the Romans under Aemilius Paullus conquered Epirus in 167 B.C., they sacked seventy settlements. They must have burned crops and forest, for it was their intention that central Epirus should never recover, and the Romans were thorough in these matters. They carted off 150,000 Epirotes into slavery – everyone they could find except those, like the Thesprotians, who had sided with Rome. Hammond says that most of those enslaved must have been from Molossi, and that another 150,000 must have escaped, so that the high Pindus was thenceforth depopulated. At any rate he postulates a huge population of mountain folk before the Roman conquest and almost none afterward.

If Hammond's interpretation of Polybius and Strabo is correct, the Pindus certainly suffered a catastrophe at least as great as the one that *might* have befallen the western Taurus about six centuries later. Terraces collapsed as agriculture was abandoned (for such a population could not live by sheep alone). The rivers must have run with silt, inundating lowlands and braiding channels. A suddenly depopulated mountain landscape is an environmental disaster, and this must have been one of the first order – if Hammond is correct.[33]

[30] Higgs 1978, 46–7; Exarchos 1982, 201.

[31] Hammond 1967. See also Oost [1954] 1975, 83–8, for the Roman impact.

[32] PRO FO 195/791, General Report on Epirus, 1 December 1863: "The Molossian dog is still as large and fierce as in the days of Virgil." I do not know about Virgil's day, but Pindus sheepdogs today fit this general description nicely.

[33] Spencer (1851, 2:191) came to this conclusion: "Previous to the invasion of the Romans history tells us that Epirus abounded in splendid forests; these were burnt down, consequently the rocky mountains denuded of their foliage, and the rains of centuries having washed away the soil, and for the want of moisture dried up the rivers, the climate of the country and its character for fertility has been entirely changed." Spencer goes on to accuse the Ottoman Turks (less justly I believe) of the same ecological crimes.

In Roman times only the coast and northern Epirus (now in Albania) prospered. Secondary Roman roads secured the coast and its immediate hinterland, and the Via Egnatia crossed well to the north, making the Pindus an isolated backland.[34] Nikopolis (near today's Preveza) developed into a Roman shipbuilding center, which presumably made heavy demands on the timber in the watersheds of the Louros and Arachthos rivers, exacerbating the ecological degradation that the Roman depopulation had brought on.[35] This logging, however, could scarcely have affected the high Pindus. The Romans had access to the timber of Dalmatia and the Black Sea coasts, and would not have found it rewarding to log the Pindus. Here, where soils (and livestock) permitted, forest may eventually have recolonized in the wake to Aemilius Paulus. Not for two millennia would the Pindus support as dense a population as in the days of ancient Molossi.

Unfortunately the palynological evidence to support the implications of Hammond's assessment does not exist. The pollen record deposited in the sediments of Lake Ioannina fails to show clear changes such as could be expected from a major settlement of the mountains in the centuries preceding 167 B.C. Nor does it show a rise in opportunistic species invading fields and pastures after 167 B.C. But technical difficulties with the cores collected here obscure the vegetation history during what may have been a crucial time.[36] At the end of antiquity, it seems, the Pindus (and perhaps all of Epirus) carried little population, and tall forest (perhaps second-growth) prevailed widely throughout.[37]

The Lucanian Apennines in Ancient Times

Palynology is no help in Lucania, where no suitable sediments have been found.[38] Geologic and literary evidence, however, points clearly to a considerable population in lowland areas in ancient times, especially

[34] Hammond 1967, 657–705. Hammond does not recognize that this depopulation affected the landscape: on pp. 267–8 he says in the high Pindus conditions have changed very little since the Neolithic and that pastures there are natural and not the product of deforestation. This is inconsistent with the idea of hundreds of thousands of people living in these mountains; inconsistent with the testimony of Pouqueville (see Chapter 7), who saw shepherds burning high forest; and inconsistent with the palynological evidence as presented by Bottema (1974), which of course was not available when Hammond wrote.

[35] Exarchos 1982, 202. [36] Bottema 1974, 138–40.

[37] Letsas 1953, 59; Exarchos 1982, 202.

[38] Neboit 1980. A University of Texas team is currently working on a site in Calabria, which may one day illuminate vegetation history in southern Italy.

after Greek colonization. Paleolithic and Neolithic remains center around the Materano (the area around Matera) and the Melfi-Venosa plain. Iron-wielding immigrants called Lyki arrived about 1200 B.C., followed by the first Greek colonists in the eighth century. The Greeks pushed the Lyki out of the most-favored locations along the coasts, establishing cities that flourished for centuries. They greatly expanded the scope of vine and olive culture, because they traded their wine and oil far and wide. As colonists faced with abundance in a new land, the Greeks no doubt showed much the same indifference to natural resource preservation as American, Russian, and Brazilian colonists would in their day. The Lyki had to move up into the hills, but gradually acquired some of the skills of their enemies. By the fourth century B.C. the people of the interior (now called Lucanians) could fight the Greeks on equal terms and did so, plunging the area into long years of warfare, as on the ethnic frontier between the mountain Taurus and lowland plains. The Romans eventually became involved in these struggles in Lucania and emerged as masters, as they never could in the Taurus.

War intensified when late in the third century B.C. the struggle between Rome and Carthage came to Lucania. When many Lucanians sided with Hannibal against Rome, the Romans punished this disloyalty by razing settlements to the ground and slaughtering and enslaving the survivors of the wars. Hannibal showed no less energy in destroying settlements as he retreated into Calabria. The Second Punic War raged in Lucania and Calabria from 217 to 203 B.C., during which both Hannibal and Fabius pursued a scorched-earth policy. After Hannibal's departure, the Romans punished the locals for about three more years. The fury and length of this war, and the fact that it visited slaughter and rapine upon the native tribes and was not just an affair between armies, led to widespread depopulation and devegetation in Lucania and Calabria. Roman postwar policy ensured further environmental damage by transforming the surviving peasantry into squatters with insecure tenure on large estates, and hence without interest in preserving newly exposed soils. Annexation of more than half of Lucania and the establishment of latifundia delayed forest recovery by encouraging transhumant pastoralism in sheep and goats. Huge tracts of soil must have been opened to erosion, crippling the potential of the region indefinitely.[39]

[39] Toynbee 1965, 2:11–35. After explaining where revivals took place Toynbee summarizes on p. 35: "Taking South-East Italy as a whole, it would be true to say that the peak of populousness and prosperity which had been attained there in the sixth century B.C. was never again approached until after the Second World War.... At the time of

Geological evidence shows heavy silt deposition through Greek and Roman times, followed by stream incision. Probably this is sad testament to the ecological cost of generations of war and fluctuating population. No one can know for certain that this erosion was anthropic or climatic, but the leading student of Lucanian coastal geology discounts climate change as a factor.[40]

At roughly the same time as the Roman absorption of Lucania and Calabria, the great cities of the coast, Metaponto, Heraclea, and Sibari, declined and were largely abandoned; the roads connecting the Ionian and Tyrrhenian coasts fell into disuse. All this is probably evidence of the establishment of malaria in Lucania and Calabria. Presumably the combination of forest clearing by the Greek colonies, the upland agriculture practiced on unstable soils by the displaced Lucanians, and then the destructive warfare and massacres of the fourth and third centuries so upset the hydrology of the Lucanian rivers that siltation and swamps resulted, inviting malaria. Certainly the river basins and coastlands of southern Italy have hosted malaria from Roman times until the 1940s.[41]

The deforestation and accelerated erosion of ancient times still left considerable areas of forest in the Lucanian mountains. The interfluvial ridges between the Bradano, Basento, and Cavone suffered the greatest damage. But the Pollino massif remained unaffected by the Greeks and Lucanians and had no importance at all in Roman and later Byzantine times.[42] It remained sufficiently remote that its foothills were suitable for monks seeking solitude, which led to the establishment of Byzantine monasteries. Pollino, and all Lucania, abounded in forest animals as well as in monks: in Rome gladiators fought Lucanian bears, and in the fifth century A.D. Roman meat came from "large droves of wild hogs" fattened on Lucanian acorns.[43]

The survival of substantial forest cover in Lucania did not sharply

writing in A.D. 1962, the marks of dirus Hannibal's presence in South-East Italy during the fifteen years 217–203 B.C. were still discernible."

On postwar agrarian policy, ibid., 117–21, 239–47, 288–9. See also Brunt 1971, 269–81, who puts less emphasis on war and more on Roman postwar policy in explaining the decline of Lucania and Calabria after the third century B.C.

40 Neboit 1980 and 1983. See also Brückner 1982. Vita-Finzi (1969) is the leading proponent of climate-induced accelerated erosion.

41 Bruce-Chwatt and Zulueta 1980, 89–105. Brunt (1971, 608–24) reviews the role of malaria in ancient Italy as a whole.

42 De Leo et al. 1984, 17–24, 119–22.

43 On monks: Rugolo 1988, 325–6. Italy, Ministero dell'Agricoltura e delle Foreste 1976; 116, on bears. On hogs: Gibbon 1909, 3:320. Hogs still ran wild in the forests of northern Calabria in the late eighteenth century. Grimaldi 1770, 122.

distinguish it from Italy as a whole. Here is Dionysius of Halicarnassus writing in about 20 B.C.:

> Italy is full, to overflowing, with practically everything that makes for both aesthetic and economic satisfaction. . . . [praise for olives, vines, pastures] Most marvellous of all, however, are the oakwoods on the crags and in the glens and on the hills that are left uncultivated. These woods teem with abundant ship-building timber of the first quality, as well as with timber that is excellent for other industrial purposes. Moreover, none of these natural resources are difficult to get or are inconveniently located for human use.[44]

Dionysius elsewhere rhapsodizes about the extent of forest in Calabria's Sila Plateau, where Rome from time to time found tall timbers for its navy and Spartacus and his rebel ex-slaves found temporary refuge from Rome's army. Forests probably covered about half of Calabria through Roman times.[45] Other Roman writers noted extensive forest in Etruria, Tuscany, the Tiber headwaters, and elsewhere, of which small patches remain today.[46]

Considerable forest certainly remained in Lucania in Roman times. Indeed, depopulation there might have permitted partial recovery of spontaneous vegetation. It was a land without terraces, more exposed to erosion when under annual crops than when under almost any other vegetation regime, so maquis and garrigue checked erosion better than did cultivated fields. Transhumance, which expanded notably in Lucania after the Hannibalic war, would have limited but not prevented vegetation recovery through Roman times, perhaps favoring garrigue and maquis in most places. This may have helped form the late classical landscape into a fairly stable combination of lowland maquis and pasture and upland forest and pasture. In short, the economic and political events of the eighth through the third centuries B.C. permitted what in retrospect was a brief florescence of Greek civilization, but led to an enduring environmental degradation in the lowlands and especially the hill country. Partial recovery would take many centuries. The mountain forests, however, largely withstood the Lyki, Greeks, Romans, Carthaginians, and their struggles.

[44] Book 1, chap. 37, quoted in Toynbee 1965, 2:594.

[45] Rugolo 1988, 323.

[46] Toynbee 1965, 2:596–9; Mosso 1906, 369–400; Ahlmann 1926, 85. Spartacus's revolt ended in Lucania after two years of guerrilla campaigns (73–71 B.C.) that did their part in damaging that landscape. Blunt 1971, 288.

The Sierra Nevada and Alpujarra in Ancient Times

The impression the ancients made upon the Sierra Nevada and Alpujarra came to little. Before humanity came on the scene, steppe vegetation alternated with assemblages dominated by oak and pine. After the retreat of the ice age, deciduous oak colonized the slopes of the Sierra, dominating the vegetation by 6000 B.C.[47] Prehistoric remains indicate settlements on the coast, and many paleolithic sites exist north of the mountains in the Guadix tableland. Bronze Age agriculturalists employed irrigation in lowland Almería, although it seems they left much of it in forest.[48] Whatever the impact of prehistoric societies on the landscapes of Guadix or the coast, they probably had none at all on the Alpujarra or Sierra Nevada. The area has no known history until the arrival of the Arabs in the eighth century.[49] Strabo's "mountain chain covered with dense woods and thick trees, which separates the coast from the interior," is usually interpreted to refer to the Sierra Nevada.[50] (But it might be the smaller coastal range, the Sierra Contraviesa: Strabo himself never saw Spain but relied instead on Polybius.) Neither Roman agriculture nor Roman mining touched the Alpujarra, although Greek, Carthaginian, and Roman remains litter the coastline at Almería. Berja, and its low plain just behind the coast, also attracted settlers from afar in ancient times. So did the Guadix tableland. But in the Alpujarra no one has ever found a Roman coin or inscription. It flourished only with the coming of the Arabs.[51]

The Rif in Ancient Times

Very little is known about the Rif before the arrival of the Arabs in A.D. 640. The prehistoric remains are very slight, and it is likely that the area supported only sparse population.[52] Palynological evidence suggests a very modest human presence in a forested landscape dominated, after

[47] Pons and Reille 1988. Cedar had grown in the Sierra Nevada during some of the earlier interglacials, the only instance I know of cedar in Europe. Florschutz, Menéndez Amor, and Wijmstra 1971. Menéndez Amor and Florschutz 1962.

[48] Chapman 1978, 264–5.

[49] Luna 1984:27; Trump 1980; Pareja López 1981c, 215–360, for the archaeological record.

[50] Quoted in Bosque Maurel 1969, 169. Bauer 1980, 110; Luna 1984, 26; Semple 1919, 33, all agree that Strabo meant the Sierra Nevada. Semple's translation: "thickly wooded with gigantic trees."

[51] Tapia 1965, 317; Pareja López 1981c, 327. On Roman mining, see Rickard 1928. On Almería's archaeology: Brenan [1957] 1988, 214–29.

[52] Trump 1980, 155.

the end of the ice age about twelve thousand years ago, by oaks.[53] Some settlements apparently existed when Phoenicians first established coastal trading posts after the twelfth century B.C. Carthaginians built more such posts after the ninth century, the locations of which remain unclear, and perhaps even a city (Rusaddir) at the site of today's Melilla. They must have traded with someone, although their Moroccan outposts probably specialized in entrepôt trade, with only modest commerce to the interior. No written sources shed any significant light on the ancient inhabitants of the Rif. It is likely that Carthaginian culture and trade affected the Rif less than any other part of North Africa.

After the conquest of Carthage (146 B.C.) the Romans created a colony in northern Morocco, to the west of the Rif in the Atlantic side of the Tingitane Peninsula (Mauretania Tingitania), and another to the east of the river Muluya (Mauretania Caesaria). They too had outposts on the Rif coast where sailing vessels might put in for water and supplies, and their armies marched through the Taza corridor to the south of the Rif more than once. But they left the interior alone. Strabo in the first century A.D. reported that the Rif was wooded right down to the sea; Herodotus and Pliny (who wrote that elephants could hide among the tall trees in the Muluya Valley) have equivalent passages.[54] In all likelihood none of them ever saw the Rif, but their remarks fit with modern palynological investigations and the history of Roman (salutary) neglect of the Rif. In prehistoric and ancient times no shore of the Mediterranean was as isolated and uninvolved in the broader currents of Mediterranean life as the Rif. It is a safe guess that here natural vegetation survived less disturbed than anywhere else so close to the sea.

After the Ancients, A.D. 500–1000

Almost everywhere that the ancients practiced settled agriculture, the end of ancient civilization led to landscape deterioration. Population decline of considerable proportions, beginning as early as the end of

[53] Reille 1977; Ballouche et al. 1986.

[54] Strabo 17.3.4–6 quoted in Semple 1919, 33. Gozalbes Cravioto, 1982, 12–27; Pliny is cited in Sangroniz 1921, 113–21. There are no elephants and very few trees anywhere near the Muluya today; nor have there been for a very long time. Trolard (1891, 57–59) says (on the strength of mentioned passages in Herodotus, Pliny, and Strabo) that before the Arabs one might stroll from Tripoli to Morocco among tall trees. This is unlikely given Roman agriculture in Tunisia and eastern Algeria; Trolard quite likely sought to implicate the Arabs in desertification. But high forest probably covered the better part of the Maghreb into Roman times and perhaps beyond. See Shaw 1985, 391–2.

the second century A.D., bled the countrysides of labor. Cities shrank, undermining commercial agriculture especially in Italy and North Africa. Between A.D. 350 and 600 Mediterranean population probably fell by a third, chiefly as a result of new infections inadvertently imported along caravan routes from the east. Land was abandoned, terraces fell apart, and the standard of soil conservation, an art clearly appreciated by Roman agronomic writers, declined generally.

Some geological evidence points to increased soil erosion in late antiquity all around the Mediterranean. Experts disagree as to what this might mean.[55] I think the best explanation is that low slopes cleared for vines and olives no longer received the care and maintenance of earlier times, for want of labor, and perhaps for want of knowledge lost in the collapse of Roman civilization. Classical civilization had intensified deforestation and erosion wherever it took root, in Lucania in the age of Greek colonization, in Anatolia in Hellenistic times. Maquis and garrigue spread to appropriate soils as pastoralism expanded, especially in Epirus and Lucania. Where classical civilization scarcely mattered, as in the Alpujarra and the Rif, the natural vegetation of the postglacial, and the soils that underpinned it, remained in place.

The establishment of classical civilization in any landscape proved destabilizing. Of course, the same is true of any civilization: the Greeks and Romans did what the Chinese and Incas did in other contexts as they expanded their way of life. And like the Chinese and Incas, the Greeks and the Romans, where their civilization endured, created a new equilibrium – or perhaps one might better say a long era of only slow landscape change. The replacement of one vegetal pattern by another took its toll on soils (as geologic records show), but once those vulnerable soils disappeared a new stability emerged, with cultivation as one element. A second element was wild plant associations that could flourish on thin and often dry soils – maquis and garrigue. A third was surviving high forest in the remoter districts, especially the mountains.

With the demise of classical civilization, this equilibrium altered once again. It made scant difference in the far western regions of the Mediterranean where the classical imprint had been lightest. In the Alpujarra and the Rif it made none, as Visigoths and Vandals avoided these zones as completely as had the Romans. But in Italy and Greece, and Anatolia to a lesser degree, after the depopulation of late antiquity and the decline of the commerical agriculture on which all rested, land-use

[55] Vita-Finzi (1969) has argued for climatic change as the key reason behind the erosion of late antiquity. Butzer (1974) is the champion of purely anthropic causes. Other authors have adjudicated this dispute: Wagstaff 1981; Bintliff 1982; Davidson 1980; Eisma 1978. As far as I can tell, the weight of expert opinion is with anthropic origin.

patterns shifted fundamentally. Malaria, now firmly ensconced, combined with a labor shortage to discourage agriculture in the lowlands, opening more land to sheep and goats. Presumably they kept forest from returning to abandoned lowland, keeping evergreen oaks and other aggressive colonizers down to the size of shrubs. In some landscapes, where grazing animals did not enjoy free rein, maquis probably grew into tall forest and garrigue turned to steppe grasses.[56] Palynological evidence suggests that the domain of tall forest expanded where sufficient soil and moisture permitted, as in Sicily, for example.[57] In the ensuing centuries the mountains scarcely mattered in human affairs, except as summer pasture. With shrinking population, more land at lower altitudes came open, and the logic of living high up diminished. The ideal niche, in terms of elevation, lay above the malarial stratum but low enough for tree crops: in effect where temperatures proved inhospitable to the anopheles mosquito, yet amenable to olives and citrus, as well as to the less demanding cereal and vegetable crops. The mountains disappeared from Mediterranean history.

The Slow Renaissance of the Mountains, 1000–1700

Eventually the mountains came to play a basic part in a new system of human ecology in the Mediterranean. This came at different times in different places, depending on the timing of two major changes: peasants moving up to the mountains and pastoralism expanding everywhere, including the mountains.

The motors behind these two changes were many. The important ones were ecological and geopolitical shifts: the expansion of the domain of malaria, a slow agricultural involution throughout the Mediterranean, and the political chaos that followed upon the dismantling of the Roman and Byzantine empires. These all worked together to make mountain life more attractive and the lowlands more dangerous.

Slowly, and at different rates in different places, mountain population grew. Malaria no doubt played a role in Lucania and Epirus, indeed throughout southern Italy and all of Greece. Its history remains enigmatic, but it became endemic in ancient times and never ceased to plague lowland populations. Its incidence fluctuated over the centuries,

[56] Rackham (1982) discusses ecological succession in Mediterranean lands when grazing is eliminated. See also Kürschner 1984 for the Bolkar Dağları section of the Taurus.

[57] Pons 1984, 436, for the pollen evidence. On Sicily: Laure 1953; Pecorano 1974, 37, 70–1. The Romans deforested much of Sicily, but after the fifth century a good deal recovered, so that between the sixth and eighth centuries much of the island was in forest. In many ways Sicily's forest history was typical of the Mediterranean as a whole.

making the plains and coasts alternately more and less attractive. But its general effect was to render higher altitudes preferable to lower ones. It is not clear how important malaria became in North Africa and in Spain after classical times. It may well have spared al-Andalus, although it afflicted much of Spain and the Rif by the twentieth century. Perhaps the absence of significant silt deposition on the coasts of Spain (until the sixteenth century) and Morocco meant that the mosquito's breeding grounds did not expand and malaria's influence correspondingly stagnated – if indeed it had any. The connection between malaria and landscape change was much firmer in Italy, Greece, and Anatolia than in Spain and the Maghreb.[58] In Anatolia, for example, the Mediterranean coastal plains, so populated and prosperous in the first century A.D. and again after 1945, were all but deserted in the intervening centuries. Malaria is almost certainly to blame. The Ottoman geographer Evliya Çelebi remarked that Silifke remained a small settlement (he visited in the early 1670s) on account of malaria.[59] Spontaneous vegetation, sometimes even tall forest, returned to lowland landscapes from which malaria barred human beings. My suspicion is that malaria in effect increased mountain population and settlement, with attendant effects on landscapes. But all this is clouded by uncertainty.

A second ecological factor improved the viability and attractiveness of the mountains: a slow agricultural involution. No uncertainty surrounds its importance. It had three components. The first was the delicate techniques of irrigated garden agriculture imported to the Mediterranean by the Arabs. The second, also conveyed by the Arabs, was a series of subtropical crops that flourished in coastal and lowland zones. The third component was American food crops, chiefly maize, well suited to altitude and prodigiously responsive to irrigation. These three factors combined to make the mountains attractive refuges from malaria and the insecurity (soon to be discussed) of the lowlands. This agricultural involution happened slowly, beginning in the tenth century in some places, but ending only in the sixteenth or seventeenth century.

The Arabs brought to the low coastlands intensive garden agriculture,

[58] Bruce-Chwatt and Zulueta 1980, 33–42, on malaria in Greece; 123–30, on Spain and Portugal. For twentieth-century malaria incidence rates in Epirus, Livadas and Sphangos 1940–1 1:63–5 and 2:174–85. On Greek coasts, malaria worsened notably in the first and nineteenth centuries A.D. The latter event, at least, is connected to landscape change: the expansion of rice, deforestation and siltation, the creation of marshes. Malaria in Anatolia was especially severe in Antalya and Cilicia in the early twentieth century, but affected Eğridir and the Konya Plain too. League of Nations 1925, 5–7. On the Rif: Soria Marco 1948, 16; Spain, Alta Comisaria en Marruecos, Delegación de Asuntos Indigenas 1929.

[59] E. Çelebi, *Seyahatnamesi* (Istanbul, 1896–1938), 9:320, cited in Faroqhi 1990, 149.

based on complex irrigation systems.[60] Experience of the desert and oasis agriculture, and the sophisticated waterworks of Mesopotamia and Egypt, equipped the Arabs to make good use of the Spanish coasts and similar environments in Morocco, Sicily, and elsewhere. They developed snowmelt irrigation to a high art, building networks of channels and tunnels to bring water to the sun-scorched plains in summer. They did not, by and large, apply their sophisticated horticulture far up the mountains from whence the water came (although Berbers did in the Atlas)[61] but concentrated on the plains and above all the coasts. From southwest Asia they brought suitable crops – sugar cane, cotton, rice, and citrus fruits among them – which prospered in the hot summers and survived the mild winters of the coastlands. The chronology of this diffusion is unclear, but probably most of these plants took root in the Mediterranean lands between the tenth and thirteenth centuries. Regular Seljuk communications with Egypt helped establish the new crops in suitable coastal plains of Anatolia, such as Antalya and Adana. The cultural and commercial cohesion of North Africa and Spain after the eighth century smoothed their path to the west. These subtropical crops made coastal agriculture far more profitable than ever before, and they spread widely, wherever the Arabs introduced them, and even a few places the Arabs never reached. These new crops flourished wherever malaria permitted the survival of sufficient labor, and whenever peace permitted sufficient irrigation. Where and when these conditions were met, these subtropical crops from the east transformed the coasts just as fundamentally as maize from the west would transform the mountains a few centuries later.

But this new system of coastal horticulture had its own direct effect on the mountains: it required seasonal bursts of labor for the harvests and promoted labor migration from the mountains to the coasts. Many coastal areas hosted malaria seasonally and pirates occasionally, so they carried a minimal permanent population, incapable of coping with a harvest season. So an opportunity developed for peasants from nearby areas whose soils provided insufficient food, or whose agricultural calendar permitted absence for a month or more. Seasonal work along

[60] Watson, 1983. K Butzer, Mateu, E. Butzer, and Kraus (1985) argue, based on Valencia, that the extension of irrigation in medieval times was a revival of Roman technique more than a transplant from the Arab world.

[61] The Berbers terraced the Atlas and the Algerian Aurès, allowing them to practice Arab-style horticulture. The Romans had never built terraces in North Africa. Nor did the Berbers build them in the Tell, the Kabylie, or (until recently) in the Rif. Despois 1961, 228–9; Despois 1956.

the coasts made survival in the mountains a bit easier, a theme I will detail in the next chapter.

Maize came on the scene much later than Arab horticulture. It originated in the Americas and arrived in the Old World only after 1492. In the sixteenth century it began to make its mark in the western Mediterranean, quite early in Morocco (1523) and the kingdom of Granada. It "abounded" in all the Alpujarra by 1800.[62] By the early seventeenth century it had appeared in Croatia. It probably came to the Balkans via Egypt (where it appeared in the sixteenth century) and Anatolia (seventeenth). The Turks called it Egyptian grain, and in the Balkans and Italy it often went by the name of Turkish grain. By the eighteenth century it had firmly established itself in the Balkans, although by no means throughout. Epirus seems to have been an early conquest, to judge by the reports of inhabitants of the Morea to the French consul there in the 1780s. Arta exported maize to Italy in the 1770s. In the village of Chouliarades it was the principle food by 1800. Maize became important in Italy in the seventeenth century, although it may have migrated to the south only in the eighteenth.[63] Wherever it went in the Mediterranean, maize caught on quickly – for good reasons.

Maize does well in the dry Mediterranean summers and can mature in the short mountain growing season. In the Rif today farmers manage two crops per summer; from planting to harvest takes only forty-five days.[64] Although its upper elevation limit is lower than wheat's or barley's, maize provides notably higher yields, in some cases eight to ten times higher, at similar altitudes. It will grow at 1,400 meters in Cosenza,[65] and higher in the Taurus. Perhaps by virtue of its novelty in North Africa and Europe, maize did not yet (in the seventeenth and eighteenth centuries) suffer from pests and infections that often jeopardized other cereals. Although it is nutritionally inferior to wheat, being deficient in protein and niacin (so that those who wholly depend upon it

[62] On Morocco: Magalhaes Godinho 1963 and Rosenberger 1980, 483–9. Maize did not become widespread in Morocco until the nineteenth century. In Granada it appeared around 1560, but became important only in the eighteenth century. Vincent 1982b, 399; Sanz Sampelayo 1981, 502. Sermet (1965) puts the arrival of maize at Granada in the seventeenth century, but Vincent is more trustworthy here. Garzón Pareja (1984, 571–2) says maize arrived on the Málaga coast only in 1690. On the Alpujarra: Martínez Ruíz 1977, 27. In the Basque country maize had appeared by the 1570s and dominated agriculture by 1800, leading to considerable forest clearance. Bilbao 1984.

[63] Stoianovich 1962 and 1966. Bertagnolli 1888, 11, and Filangieri 1980, 106, on Italy. On Chouliarades see Benekos 1974, 153–4.

[64] Coon 1931, 50. [65] Braudel 1976, 1:42.

suffer from pellagra), maize has three advantages over its rival grains besides the matter of yield. First, its response to irrigation is better than that of wheat or barley, making maize as good in a garden plot as in the field. In the Balkans at least it spread initially in irrigated gardens. Second, it can grow in association with vegetable crops, providing shade for those that do not like the Mediterranean sun. Throughout the mountain zones, maize was often raised in irrigated patches with garden vegetables. It is customarily grown with beans, which use the maize to climb on, and which in turn fix nitrogen in the soil, slowing its depletion. Third, maize stalks have many uses, from fuel to fodder, and the cobs make good charcoal. Where fuel as well as food is important, as in a mountain winter, maize recommended itself strongly.[66] It had a revolutionary effect on the mountains.

With maize came the potato, which spread throughout northern Europe as a field crop, but in southern Europe and North Africa often became a garden crop. It does well in the cold but needs irrigation. Even in irrigation zones the potato, for all its nutritional charms – one can live off the potato alone if necessary – has not become a crop of the first importance. It seems to have made its strongest imprint in Iberia, especially the well-watered northwest, and its slightest where irrigation water is scarce: Morocco and Anatolia. It was raised in quantity in the Pindus and the Apennines only on slopes facing west, where more rain falls – in Lagonegro, for example.[67] In the rainiest part of the Rif the potato had also become an important crop by the time the Spanish arrived in 1928.[68] In the Alpujarra, where irrigation is a fine art, the potato became something of a staple, but only in the twentieth century. The villagers of Mecina Bombarón allegedly eat potatoes every day.[69] Aside from the modern Alpujarra, the potato, although a moun-

[66] Stoianovich (1966, 1038) points out that maize did not interest landlords and tax collectors, who did not consider it fit for their consumption. In the plains this would have recommended it strongly to peasants who had the right to choose what to plant. In the mountains landlords were rare.

[67] The potato came to northern Spain, where irrigation was unnecessary, in the eighteenth century and soon appeared in Andalusia. In southern Italy it seems to have arrived with the French armies around 1800. Kapodistrias allegedly introduced it to Greece in 1830, but in Epirus "though everywhere suited to the soil, [it was] in no request for local use" in 1863 (R. Stuart, "General Report on Epirus," December 1, 1863, PRO FO 195/751). It came to southern Anatolia later than to Greece: in the 1830s Fellows was told it did not grow anywhere near Side. Jiménez Blanco 1986, 2:788–96; Morano 1981, 525; Filangieri 1980, 106; Benekos 1974, 154; Fellows 1838, 201.

[68] SHM, Monografías de Kabilas, 1–16, Beni-Jaled, p. 16. Potatoes may have existed for some time before 1928 here, but there is no evidence one way or another.

[69] Navarro Alcalá-Zamora 1981, 103.

tain crop in its land of origin (the Andes), never acquired the importance of maize in the mountains of the Mediterranean.

The tomato, also an introduction from tropical America, caught on well in parts of Italy and Spain as a garden crop, and can do well at altitudes of a 1,000 meters or more. But it never had the importance of maize, nor (unlike the potato) could it have.

When Mediterranean peasants adopted these American crops and adapted their cultivation (and that of traditional Mediterranean vegetables) to Arab irrigation techniques, they opened new niches in formerly inhospitable mountain zones. And they had good reasons to seek the austere hospitality of the mountains.

The great upheavals and conquests of history, as well as the migrations of maize and the peregrinations of the potato, encouraged mountain settlement. The downfall of the Roman and Byzantine empires, and their replacement in many areas by peoples of nomadic origin, rearranged matters everywhere in the Mediterranean. Lowland peoples found themselves harried by violent politics as well as by virulent diseases. Difficulties in public order, new landlords, and insistent tax collectors all drove people to seek refuge up in the mountains. The Arab conquest of the Maghreb sent Berbers in search of remote and safe districts, which they found in the Atlas and the Rif. The Castilian reconquest of Andalusia concentrated Muslims around the Sierra Nevada from 1200 to 1571, although some had lived there for centuries previously. The Ottoman conquest of Greece and the Balkans generally encouraged Christians (and some non-Turkish Muslims) to retreat up into the isolated fastnesses of the western Balkans, although not perhaps until maize had made its entry. Continual warfare and widespread brigandage in Anatolia accompanied the slow retreat of the Byzantine Empire from the eleventh century until the establishment of the Ottoman Empire in the fifteenth, inspiring Armenians, Türkmens, and others to find security in the high mountains of the Black Sea chains, of eastern Anatolia, and the Taurus. In short, conditions of insecurity that followed the breakdown of Roman control in the western Mediterranean and Byzantine authority in the east made self-defense a necessity and hence the mountains more attractive.

Political factors fundamentally influenced land use and human ecology in other ways too. The success of the Arabs, the Seljuks, and the Castilians extended the domain of pastoralism much further than ever before in the Mediterranean. In all likelihood pastoralism had expanded in late antiquity as labor grew short with depopulation, and the seasonal incidence of malaria made transhumance especially appealing. But a far greater expansion came with the irruption into cultivated lands of the

Muslim herdsmen.[70] The Arabs, Seljuks, and Ottomans had nomadic origins, and considered stock raising a preferable way of life to settled farming. This affected most strongly North Africa and Anatolia, but also Spain, Sicily, and, with the Ottoman expansion of the fifteenth century, the Balkans as well.

In Spain the Christians built an economy around the merino sheep, an Arab import from North Africa, establishing legal privileges for stock raisers organized as the *mesta.* Like the Arabs and Turks, the Castilians promoted extensive land-use patterns at the expense of more productive systems of cultivation, in this case Moorish, and had the military means to drive their enemies up into marginal lands. Often these were forest lands, where the horses and sheep of the conquerors could not prosper.

Emergent markets for wool, as well as the success of herdsmen in war, helped to spread pastoralism in the Mediterranean lands. In southern Italy, as in Spain, large-scale patterns of transhumance developed in late medieval times, connecting more firmly mountain and plain in a net of seasonal migrations, organized by the sheep raisers guild, the *dogana.* Monarchs in the western Mediterranean favored the expansion of sheep raising because wool exports were easy to tax. A growing wool trade to northern Italy and northwest Europe, and perhaps a general labor shortage after the arrival of the plague in the mid-fourteenth century, helped spread pastoralism at the expense of arable and, inevitably, of forest.

The expansion of pastoralism meant more intensive exploitation and burning of upland areas, probably significant deforestation in some, and certainly more erosion wherever the hoofs passed by. The sudden shift to pastoral dominance probably promoted erosion in Anatolia. It certainly extended the steppe. The ancient port of Miletus (Balat in Turkish) remained a port until the fourteenth century, but by the seventeenth it was 2 kilometers inland. Silt deposition by the Büyük Menderes accounts for the shifting shoreline.[71] Thus the rise of pastoralism gradually promoted deterioration of vegetation on mountain slopes. But initially, at least, more sheep and goats meant the mountains became easier to inhabit.

Political factors may also have helped mountain populations to get by. The Mediterranean world reverberated with the sounds of war in the fifteenth and sixteenth centuries, and seasonal work in the armies of sultans, kings, and princes beckoned to mountain folk (to males at

[70] Brice 1978, 145; Lindner 1983, on nomads and Ottomans in Anatolia.

[71] Erinç 1978, 97, 103–4. Balat is now about 5 kilometers inland.

any rate) who could not otherwise make ends meet. In ancient times the violent centuries had sent people scurrying away from fertile lowlands to the security of the mountains but had also tempted the adventurous and the desperate back down to serve as warriors. Much the same happened in early modern times, a golden age of mercenary warfare.

Five Mountain Landscapes, 500–1700

This gradual resurgence of settlement in the mountains had its impact on landscapes. Evidence is slender, of course, for these centuries, but the general impression it gives is of a small reduction in forest area, an expansion of pasture and meadowlands, and a tiny amount of land cleared for agriculture.

The Western Taurus Landscape

The western Taurus scarcely felt these changes. The Seljuk struggle with the declining Byzantine Empire may have driven some up into the hills, but if so the migrants have left no archaeological or palynological record. With the rise of Ottoman power, Christians fled Pisidia but not for the mountains.[72] Almost certainly the new nomad populations, Seljuk and Ottoman, created new summer pastures here and there in the mountains, but in the open forest of the Taurus they probably found plenty of good grazing without needing to alter mountain vegetation. In any event, they pastured their animals above the tree line and spent rather little time on migration within the forest belt (see Chapter 7).

In Seljuk and early Ottoman times the ports of southern Turkey – Alanya, Antalya, and Finike notably – did brisk business in lumber with Egypt. The *tahtacı*, an Alevi sect that specialized in timber cutting, felled trees in the foothills of the Taurus for this trade, but the higher elevations remained unaffected, leaving widespread tall timber. Furthermore, the tahtacı found religious significance in trees, which translated into self-imposed restrictions on their cutting.[73] The trade with Egypt fell off after 1522 when the Ottomans secured the entire sea route from the Black Sea to Egypt, and better timber from accessible coasts temporarily undermined the business of the southern ports. When Evliya Çelebi visited Alanya in 1671–2 he found it sunk into decline. Antalya became

[72] Vacalopoulos 1976, 251.

[73] Cahen 1968, 146; Roux 1970, 181–212; Pryor 1988, 143; Andrews 1989, 68–71, on the tahtacı.

an "insignificant local port" by the seventeenth century.[74] The timber trade revived only in the nineteenth century. The imprint of nomads upon the western Taurus landscape, modest enough when nomad numbers were at their height, grew fainter after the fifteenth century when Ottoman policy removed yörüks from the Taurus for resettlement in newly conquered lands in southeastern Europe. With minimal peasant settlement after the fifth century, a long depression in trade after the sixteenth century, and a reduction in nomad population in the fifteenth century, the pressures on western Taurus landscapes gradually diminished from levels that, except around Selge, had never been high in the first place.

Elsewhere in mountain Anatolia humankind also had a comparatively mild effect on landscapes in these centuries. Pollen evidence from highlands in eastern Turkey suggests that human impact upon vegetation remained very slender until five or six hundred years ago, when grasses seem to have spread at the expense of oak. This probably represents the extension of pasture at the hands of Seljuk or early Ottoman nomads. But a sharp decline in the presence of oak came much later, between one and two hundred years ago, when pine and walnut became prominent. This strongly implies forest burning, settlement, and cultivation in the nineteenth century but not before.[75] Roughly the same probably held true for the western Taurus – indeed, for parts of the central Anatolian plains as well: Tamerlane managed to hide elephants in forest near Ankara in 1402.[76] The population history of the western Taurus (see Chapter 5) seems consistent with these inferences.

The Pindus Landscape

The history of the Pindus environment after ancient times is obscure. It seems that the forests of northern Greece were used only for firewood and grazing and that population was so slight that forests expanded.[77] Population did not recover significantly from the Roman conquest until the medieval Despotate of Epirus (1267–1479), a semiautonomous entity under the Byzantine Empire. Unfortunately, virtually nothing is known about economic life during the Despotate.[78] A chronicle from the early

[74] Lloyd and Rice 1958, 7. Inalcik 1960; Inalcik 1973, 128.

[75] Van Zeist and Woldring 1978, 270–1 and 257–62.

[76] Thirgood 1981, 52. Thirgood cites Kasapliğil as saying that much of the Anatolian Plateau was deforested only in and after the nineteenth century. This strikes me as unlikely, but I have not investigated the question.

[77] Lazos 1935, 20. [78] Nicol 1984, 223–32.

fifteenth century describes Ioannina as rich and lush.[79] Surviving records indicate trade between Arta and the rest of the Adriatic world, chiefly in grain and salt, items produced locally in the plain and Gulf of Arta. Ioannina had only very local commercial importance. No evidence exists of a timber trade in the Pindus between ancient times and the eighteenth century. A local trade must surely have existed of course. But even in lowland areas tall timber existed in the sixteenth century. When the Ottoman traveler Pîrî Reis sailed along the Epirot coast in 1520, he noted that Batina, a harbor about 3 kilometers north of modern Preveza, was "surrounded entirely by forest." He also remarked that the inhabitants of Parga specialized in carpentry, suggesting that the town had access to a ready supply of wood.[80] In all likelihood, logging and other market-related activities in Epirus came to little, and what forest destruction took place in Byzantine and early Ottoman times was done for fuelwood.[81] If so, little happened in the mountains until the surge in mountain settlement between the sixteenth and eighteenth centuries. In the absence of information to the contrary, it is tempting to conclude that the landscape history of the Pindus was a dull and uneventful one between the ecological aftermath of the Roman depopulation of 167 B.C. and the arrival of the Ottomans and maize between the fifteenth and seventeenth centuries.[82] But one cannot know for sure.

The Landscape of the Lucanian Mountains

In the southern Apennines after ancient times cultivation receded and pastoralism advanced while plenty of tall forest remained. Pastoralism and transhumance in southern Italy are very old. They probably date from prehistoric times and both attracted the attention of several ancient writers.[83] With the collapse of ancient civilization in southern Italy, pastoralism expanded as the ideal adaptation to conditions of abundant land and scarce labor. In medieval times a local transhumance developed between the Metaponto coast and the high country of Lagonegro and Monte Pollino. It disappeared (at least from Pollino) in the fourteenth century, perhaps a result of the Black Death, but returned in the sixteenth century, coming into conflict with newly settled Albanian villages. This local transhumance formed a small part of a

[79] Schirò 1975, 379. [80] Reis 1988, 2:699.

[81] Grispos (1973, 150) makes this case for Greek lands generally.

[82] Kontos (1929, 135) suggests that forests expanded everywhere in Greece, Epirus included, during later Byzantine times, but this appears to be a hypothesis based solely on the fact of general Byzantine decline.

[83] See Franciosa 1951, 51–2.

much larger system that ranged over much of southern Italy and involved perhaps five to ten million sheep in the seventeenth century.[84]

Until the sixteenth century the economy of the southern Italian highlands rested on a system of sylvo-pastoralism. Villages were few and small, especially in the higher reaches like Pollino. Natural vegetation had comparatively few enemies, aside from shepherds who no doubt set fires to improve grazing for their flocks. Tall forest – harder to burn than maquis or young second-growth forest – dominated most mountain landscapes. It existed even in accessible areas in the seventh through ninth centuries, when southern Italy provided naval timber to Egypt and Tunisia. As late as the 1570s, a traveling Venetian ambassador, who (being an ambassador) probably did not seek out the remoter districts but who (being a Venetian) probably did have a keen eye for ship timber, noted that Basilicata had great stretches of forest with fine naval and construction timber.[85] Venice never managed to exploit these forests, despite the ambassador's suggestions.

It seems the mountain ecosystems of Lucania felt little stress in these centuries. Even though transhumance was part of a larger market system that supplied wool to northern Italian cities, sylvo-pastoralism in combination with low population permitted stability in mountain vegetation and soils, and perhaps a continuing recovery of forest from the depletions of ancient times. Even the fluctuations in population, which affected southern Italy after the arrival of plague in the fourteenth century, would have meant little for mountain landscapes because cultivation played such a small role in the highland economy. After the sixteenth century, as Chapter 5 will show, population began to grow and another era of rising pressure on mountain ecosystems would begin.

The Alpujarran Landscape

Archaeological evidence indicates the Alpujarra had acquired permanent inhabitants by at least A.D. 800. The evidence, fortifications at Busquistar, implies that life was not peaceful. Indeed the first historical mention of the Alpujarra is a ninth-century rebellion led by one Omar ibn-Hafsun against Moorish authority. But irrigation technique from North Africa soon made the Alpujarra a prosperous valley, and by the tenth century the political power of the Umayyad Caliphate made it peaceful. The subsequent struggles among Umayyads, Almoravids,

[84] Marino 1988; Franciosa 1951, 54–67; on Pollino's medieval history see P. De Leo 1984, 29–34.

[85] Lombard 1959, maps; Alberi 1858, 2:270–1.

Almohads, and Christians seem to have passed the Alpujarra by until the final denouement of Muslim civilization in Spain in 1571. So from the tenth to the sixteenth centuries the Muslim Alpujarra toiled away, free from the destruction that repeatedly ravaged much of the rest of Andalusia.[86]

In these centuries the Alpujarra developed a system of subsistence garden agriculture and commercial silk culture. Most of the existing villages were founded during these centuries. But the villagers had no maize, no potato, and no apparent interest in the higher slopes, leaving vegetation there in its natural condition.[87] Pastoralism involved only local transhumance between the high Alpujarra and Berja, Dalías, and the coast; even after the Castilian *reconquista*, the Alpujarra did not participate in the large-scale commercial sheep raising of Aragon, Catalonia, and other parts of Spain. Inaccessibility must have limited timber exploitation. Although the Moors did cut some timber in the valleys of the Sierra Nevada, it as a whole remained "oak country."[88] Throughout southeastern Spain, forests were far more extensive from the twelfth to the fourteenth centuries than now.[89]

The Alpujarran landscape as created by the Moors represented a sensible adaptation to existing conditions. It required great amounts of labor, as it depended on terraces and irrigation, needing ceaseless vigilance and maintenance. Thus the stability of the Moorish Alpujarra landscape was vulnerable to labor shortage. Although there is no record of it, the Black Death may have temporarily reduced population beneath the threshold necessary to maintain such landscapes, thus permitting rapid soil erosion. But whatever disruption the Black Death entailed came to nothing compared with the environmental disaster of the 1570s.

War and expulsion led to a general depopulation (Chapter 5) and deprived the Alpujarra agro-ecosystems of the skilled agricultural labor on which they depended. Every successful Spanish siege of an Alpujarran village was followed by a legion of two thousand men whose job it was to cut every tree and shrub, burn all crops, houses, and fields.[90] Terraces crumbled and collapsed, and the Guadalfeo and Andarax rivers ran thick with silt. With the Moors gone, the Castilian crown settled strangers in the Alpujarra, Christians from all over Spain. None of them had

86 Pareja López 1981a, 2:401–2; Bosque Maurel 1969, 169.

87 Ortega Alba 1981, 106. 88 Lombard 1959, 238.

89 On pastoralism, see Melis 1974; Tapia 1965, 317. On timber: *Historia de Andalucía* (1982): 2:84. Tomaselli (1977, 34) says Spain was "still covered by forests" in the ninth century. Bauer Manderscheid (1980, 191–5) notes the Arabs' high standards of forestry, botany, and agronomy.

90 Davillier 1876, citing Marmol 1573.

the skills to resurrect the Moorish landscape, and indeed they were too few in number to do it, by a factor of about ten. Instead they did what their numbers permitted and what they knew best: sowed wheat and herded sheep. What they knew best could scarcely have been worse for the land. It required much more space than the Moors had used with their system and required felling and torching forests as far up as wheat would grow. The winter rains quickly carried off the bared soils, adding to the erosion deriving from the collapse of terraces below. In the late 1570s the small villages in Granada and Almería complained they could not pay their taxes: sand and silt had inundated their fields since the trees in the Sierra Nevada had been cut down, and agricultural yields were less than half of what they had been when the taxes had been assigned. The coastal deltas (see Chapter 7) date from this epoch. Indeed the erosion problem of southeastern Spain generally may be said to date from the sixteenth century.[91] The "human transfusion"[92] of politically reliable Christians into the Alpujarra helped to undercut the wealth of the valley, the river valleys below it, and the coast, all of which began to suffer from erosion and floods as never before. Just at the time that western Andalusia began to quicken economically with the wealth of the Indies, eastern Andalusia slipped into obscurity and poverty, helped along by the ecological barbarity of the reconquista and resettlement.

When the Alpujarra acquired maize and potatoes, it could feed itself with far less arable than when it had depended on wheat and barley. By the late eighteenth century maize had become the leading crop, at least in some places, permitting the abandonment of the high-altitude unterraced fields of barley and wheat.[93] As population remained well below levels of Moorish times until the mid-nineteenth century, the new American crops permitted the recovery of spontaneous vegetation in the Alpujarra and throughout the old Kingdom of Granada. As stock raising remained unimportant, where soils permitted even tall forest might have recovered.

Enough existed in southeastern Spain that in 1751 the navy sent inspectors on a tour from the Guadiana River to Almería to survey forest resources. They calculated that 14 percent of the trees they encountered were deciduous oak (a far higher proportion than one could find today). They also found tall timber in the Alpujarra, but noted that

[91] Iglesías Casado 1982, 70–1.

[92] Mignon 1981, 124. On the resettlement and its consequences, Mignon 1981, 123–60; Luna 1984, 27; Navarro Alcalá-Zamora 1979, 38–9; Oriol Catena 1935.

[93] In 1751–1841 in the lower Alpujarra irrigated area expanded 12% and maize, wheat, and barley were planted in proportions of 20:12:4 according to Tapia 1965, 324.

opening new roads through difficult terrain would be necessary to exploit it. In the end, the navy left the Alpujarra and all of southeastern Spain's (depleted) forests alone, turning instead toward their American empire and especially Cuba.[94] Once again inaccessibility served to preserve what remained, and what perhaps had regenerated, of Alpujarran forests. But the respite provided by naval disinterest and the revival of more appropriate irrigation agriculture in the Alpujarra would come to an abrupt end toward the end of the eighteenth century, as Chapter 6 will explain.

The Rif Landscape

Long controversy surrounds the impact of the Arab conquest upon the physical environment of the Maghreb. Most of the discourse is worthless, based on little or no evidence and motivated by political concerns. Classical authors report forest formations in the sixth century where none exists today, but so do Arab writers in the thirteenth and fourteenth centuries.[95] It is hard to establish a reliable chronology for the great forest clearances of North African history. But lately palynology and satellite imagery have provided some evidence.

The Maghrebi coast was certainly not shaded from Tangier to Tripoli before the Arabs as some have claimed,[96] but pollen evidence does imply considerable clearing in the seventh and eighth centuries, in both mountain and plain.[97] Presumably the expansion of stock raising accounts for this, as in the history of Anatolia and Castile. But forests remained plentiful enough that Fez could serve as a major lumber trade center in the sixth through twelfth centuries, deriving its timber from both the Rif and the Middle Atlas. In the days of the Umayyad Caliphate, the western Rif supplied not only Fez, but Almería, Málaga, and the shipyards of Gibraltar with naval timber, construction wood, and fuelwood for metallurgy, ceramic firing, and sugar boiling.[98] Small

[94] Bauer Manderscheid 1980, 131–2. The inspectors found that scrub oak accounted for 43% of the trees on their tour, pines for 30%, and deciduous oak for 14%. It is not easy to find a deciduous oak in southeastern Spain today. The Spanish navy had no shipyards between Málaga and Cartagena (Bauer Manderscheid 1980, 165).

[95] Shaw 1985, 392.

[96] E.g., Buffault 1941, 473; see also Roda y Jiménez and García Figueras 1950–5, 1:347–8.

[97] Pons 1984, 435–6.

[98] Lombard 1959, 238–41. Le Tourneau (1949, 390) says that Fez got its construction wood and charcoal from the Middle Atlas, not the Rif, but Lombard's work is more careful and complete on this subject.

shipyards existed on the Rif coast itself.[99] The flourishing economy of the western Mediterranean in the heyday of Moorish civilization no doubt depleted forests everywhere accessibility permitted, including the Rif. Presumably those slopes nearest to Fez and nearest the sea felt most keenly the impact of the woodsman's ax. Where grazing followed in the wake of timber cutting, maquis and garrigue would have resulted, and doubtless much of the widespread maquis formations of the western Rif date from these times. Pollen of *Quercus ilex,* a typical maquis species, appeared for the first time in the tenth century.[100]

In these centuries cultivators also arrived in the Rif in number. The chronology of settlement is obscure (more on this in Chapter 5), but literary and palynological evidence suggest settlement in the mountains began around A.D. 1000. The geographer al-Bekri in the eleventh century, al-Idrisi in the twelfth, and Leo Africanus in the sixteenth all mention very much the same settlements, so one may presume that by al-Bekri's time if not before the Rif had acquired an agricultural population. Archaeological evidence, of which little exists, also points to the eleventh century.[101] By the thirteenth century villages in the mountains of Gomara (western Rif) were so numerous "only God could count them."[102] The pollen evidence, of which there is a great deal, shows considerable forest clearance in the tenth through twelfth centuries. Olive culture arrived at this time, and perhaps other tree crops as well. This medieval deforestation ended in the thirteenth century. Possibly arboriculture had filled up all the appropriate locations by then. Or perhaps some disaster checked population growth. By the fourteenth century the Black Death probably reduced population, cultivation, and thus pressure on forests and soils.

Whatever the case, when Leo Africanus visited the Rif in the early sixteenth century neither the timber trade nor cultivation had reached large parts of it. He found the Rif "covered with numerous forests where one sees very tall and very straight trees." This would have been the cedar forest, now confined to a few thousand hectares around Ketama. At Chechouan, which photographs show as quite bare by 1900, he found "great forests and countless springs." At Beni Grir, "one finds lots of wood." At Tegassa, "the ground is all mountainous and in forest." At Beni Chelid (Khaled), the mountain "is very cold, covered with forests, with frozen springs." At Beni Mansur, "woods and springs exist

[99] Fernández de Castro 1945, 44–5. [100] Reille 1977.
[101] Gozalbes Cravioto 1982, 44–55; Reille 1977.
[102] Ibn Said al-Magribi, cited in Gozalbes Cravioto 1982, 14.

in very great quantity." At Terga he noted "forests that cover the harsh and cold mountains."[103] Similar comments apply to a few other settings. Such testimony leaves no doubt that the forests of the Rif were much more extensive in Leo's day than in the 1920s, when the next extensive commentary is available. Not until more recent times would the Rif acquire its stripped and degraded appearance.

The decisive era in the Rif's botanical history seems to have come between 1650 and 1900. Careful and extensive palynological study shows two bursts of deforestation in the western Rif.[104] The first began in the tenth and eleventh centuries and extended as high as perhaps 1,200 meters. The second surge came as early as 1650 on some mountains and was underway by 1680 generally. This left the species composition of Rif vegetation very much as it is today. Reille, the historical botanist who has worked on the Rif, attributes this burst to population growth. No evidence exists for or against this explanation (see Chapter 5).

My suspicion is that it is correct: in the seventeenth century the Rif Berbers acquired maize, refined their irrigation techniques to suit its needs, and found that they could produce much more food than ever before. Their numbers expanded, and so in turn did their fuelwood, timber, and perhaps pasture needs. Cedar and pine forest retreated on all fronts, while *Q. ilex*, always a sign of anthropic disturbance in the Rif, colonized the newly cleared landscapes. Rifians either pioneered or extended the modern pattern of long-fallow shifting agriculture (Chapter 2). This forest clearance process lasted for two hundred years. In favor of this hypothesis is that the timing of the forest clearances corresponds with the introduction of maize throughout the Mediterranean. Against it is the fact that while maize pollen is easy to detect, Reille never encountered it in his sediment cores in the Rif. Whatever the truth, by 1900 species composition in Rif vegetation seems to have frozen, although scrub versions of forest trees continued to replace the taller ones. A new pattern, which Reille refers to as "almost desert," has asserted itself in this century.[105]

[103] Leo Africanus 1956, 273, 274, 276, 278–9, 282. My translations from the French version.

[104] Reille 1977.

[105] Reille 1977; and Reille, personal communication, 7 January 1991. Large pollen grains such as maize do not circulate far in the winds. Perhaps no maize patches were close enough to Reille's sediment core sites to show up in his cores. Maize was certainly present in the Rif in the late nineteenth and twentieth centuries, and yet made no appearance in Reille's cores.

Conclusion

Until agriculture, landscape change moved at its natural pace, without acceleration from human agents. Indeed in the early millennia of agriculture, human numbers and powers did not permit strong changes except in limited zones, generally lowland ones. With the rise of ancient civilizations in the Mediterranean, the tempo of forest clearance and soil erosion quickened, but again these developments touched the coasts and low plains far more than the uplands. In the Taurus and the Pindus evidence suggests thicker population in ancient times than in other mountain areas, and consequently larger environmental shifts when those populations first settled and then were removed, by disease or war. In the Lucanian mountains the vibrant history of the surrounding lowlands and foothills had a modest effect. In the Alpujarra and the Rif, ancient civilization appears to have made no difference, permitting the slower rhythms of climatic, botanical, and geologic time to govern vegetation and soil history until the Arab expansion.

The end of classical civilization, like its beginning, had environmental repercussions all around the Mediterranean in the coasts and lowlands. In certain mountain zones too, including the Lucanian Apennines, it may have mattered. But in the Taurus the key appears to have been disease history rather than civilizational collapse. At any rate Byzantine authority followed smoothly upon Roman power in Anatolia; there was no real collapse of ancient civilization. In the Pindus, the Romans crushed all civilization. And in the mountains of the western Mediterranean, ancient civilization had never been established.

Eventually, by the tenth century, new forces in political and ecological history breathed life into the Rif and Alpujarra, and perhaps renewed life into the Lucanian mountains. Centuries later, a revival took place in the Pindus. All these movements of mountain colonization had common roots. They soon surpassed the levels of settlement reached in ancient times. New opportunity and renewed danger elsewhere brought the mountains of the Mediterranean substantially greater populations in medieval and early modern times than ever before. The agricultural involution combined with sweeping political disruptions and lowland malaria to improve the opportunity and strengthen the motive for mountain life. Four infiltrations into the Mediterranean, of pastoral tribes and kingdoms, of malarial mosquitoes, of irrigation technique, and of American food crops, pushed people toward the mountains. Highland populations increased from the eleventh through the thirteenth centuries, and then again more sharply in the sixteenth through eighteenth. These centuries saw considerable forest clearance in lowland Europe, and a general expansion of population and cultivation in both Christendom and western Islam. The beginnings of colonization in the mountains seem broadly consistent with the general trends of economic and demographic history of the age.

Mountain colonization was a slow development, connected to population growth at modest rates more than to migration, and it continued for many centuries. It inaugurated a new age in the mountains, but one that in most places did not reach its height until the period 1700 to 1900. People learned to make good use of the forest environment only through trial and error: cultural evolution as adaptation to a new niche takes time. Agriculture alone could never support large populations in the mountains, so people had to hit upon various auxiliary activities to make their livings. Mountain folk eventually created a viable combination of agriculture, as they fused Asian technique and American crops; of pastoralism, as they took part in the large-scale transhumance systems (and wool markets) favored by the Arabs, Turks, and Castilians; and of migrant work, as a symbiosis formed with the economies of lowland districts. In most Mediterranean mountain zones this took until about 1600 or 1700, although variation is so great that any date is perhaps more misleading than useful.

The new human infiltration of the mountains, though initially destabilizing, eventually gave way to a slightly stabler situation. This stability depended on the balanced exploitation of forest, field, and pasture. Depletion or degradation of forests or soils menaced this mountain world, especially when climate cooled after the sixteenth century, gradually shortening the growing season and heightening the need for fuelwood.[106] Thus its stability was a delicate one. This mountain system is the subject of Chapter 4.

[106] Interestingly the first surge of mountain settlement in the Alpujarra and the Rif came at a time of slowly warming climate (c. A.D. 800 to 1200). The second surge of mountain settlement, in the Pindus, the Rif, and to some extent Lucania and the Alpujarra, came as temperatures cooled significantly: the Little Ice Age of 1550–1850. For details of climate history, see the excellent work of Grove (1988).

4

MATERIAL LIFE IN THE MOUNTAIN ENVIRONMENT, 1700–1900

Cultivation when it progresses spontaneously and is not consciously controlled . . . leaves deserts behind it.

Marx to Engels, 25 March 1868.

Forest precedes civilizations and deserts follow.

Chateaubriand (1768–1848)[1]

By 1700, and in some cases well before, the Pindus, Apennines, Sierra Nevada, and Rif supported considerable peasant populations. So did other Mediterranean ranges, such as the Lebanon, the Dinaric Alps, and the Algerian Tell. The Taurus did not, and would not, until the late nineteenth century. These mountain populations lived precariously in the face of hunger and sometimes famine, sickness and sometimes epidemic, brigandage and sometimes war. In material terms, their existence, whether in the Pindus or the Rif, consisted of much the same elements: the quiet daily struggle for food, water, shelter, fuel. They fought these battles against the same enemies and in much the same terrain. So, although I will often point out differences from place to place, I will treat the material base of mountain life in the Mediterranean as a whole.

The period from 1700 to 1900, and even 1950 on occasion, I will also treat as a whole. This does not imply the view that history was somehow immobile in these centuries – merely that in material life continuities strongly outweighed changes. I would not make this claim for the centuries before 1700, when mountain settlement in the Pindus and resettlement in the Alpujarra brought radical changes. Nor could I make it for the twentieth century, certainly not after 1940, when new economic relations, war, and overshoot shook and emptied many mountain zones.[2]

[1] Marx to Engels, 25 March 1868, in Karl Marx and Friedrich Engels, *Selected Correspondence* (Moscow: Progress Publishers, 1965), 202. The Chateaubriand quotation is cited in O. Elorrieta y Artaza 1948, 23.

[2] Mountain settlement in the Taurus, and evidence concerning the Rif, scarcely existed prior to the nineteenth century. Hence most of what I say of these areas deals with the

Mindful of these difficulties of chronology, and the differences in local geography, I believe a mountain system existed in all upland areas of the Mediterranean. This system rested on exploitation of mountain soils, forests, and pastures (agro-sylvo-pastoralism). Auxiliary activities such as migrant labor or craft work complemented these pursuits, often indispensably. As long as soils, pastures, and forests held out, mountain villages could usually survive, if rather wretchedly. To survive rather better, and sometimes to survive at all, they needed markets for their labor and perhaps crafts as well. With these conditions met, the mountains yielded a fairly stable existence. Hard work in the fields and gardens, long hours in the forest, and lonely days in the high pastures won a measure of security. The diversity of activities involved enhanced the stability of the mountain system, at least when viewed over a period of years rather than centuries. Mountain folk invested their fate in a wide portfolio, doing different things at different altitudes, to protect themselves against disaster. Good and bad harvests made for easy and hard times, of course, and a string of bad harvests could imperil the survival of a community. But this was quite rare.

Less rare were several consecutive years of war. The auxiliary activities on which the mountain system often depended required rural peace. Without it, thrown back on their own economic resources, most mountain villages quickly came to the brink of starvation. Lowland villages were far more likely to get in the way of campaigns and armies, but if unscathed by battle itself they were far better positioned to survive conditions of insecurity than were mountain villages: they produced more than enough grain to live on in good years and often could survive indefinitely on their own resources. Mountain villages were generally remote from combat but suffered greatly from the interruption of trade and migration that war entailed. This systematic difference might help account for the widespread recruitment of mountaineers into armies: in wartime they needed outside work more desperately than anyone else. It certainly meant that war punctuated the history of mountain communities more memorably than it did the lowlands – except for those unhappy villages visited by warring armies.

In the short run the vagaries of war and the wrath of the climate defined the economic rhythms of mountain life. In the long run depletion of soils and forests corroded the system from within while global

early twentieth century. In the Rif some aspects of material life surely remained much the same into this century. The food crops that Leo Africanus mentioned are those still raised today, for example, with only a couple of additions. But patterns of seasonal migration, the importance of fishing to coastal tribes – these may well have changed over the centuries. There is very little evidence by which to judge.

economic change gnawed at it from without. The mountain system, as we shall see, was stable over generations, but not in either shorter or longer time frames. Frequently, as environmental degradation exacted its cost, the auxiliary activities became more and more important, only to be disrupted by the economic upheavals of the Mediterranean world since 1880, and the political violence of 1914–50. But for several generations in the seventeenth, eighteenth, and nineteenth centuries, the mountain system worked.

The Land and the Seasons

Land and Peasants

From ancient times forward major landlords have owned much of the agricultural land in the Mediterranean world. Latifundias and çifliks were the rule almost everywhere, from the Pillars of Hercules to the Cilician Gates. But not in the mountains. There smallholdings held sway, and the great majority of peasant families owned the land they worked. Indeed, the absence of landlords formed a large part of the attraction of mountain life, one that helped compensate for its many hardships. Minifundia in the Alpujarra date back to the time of earliest settlement, as do the free villages (*elefthochoria*) of the Pindus.[3] Only rarely did overlords manage to extort a share of the crop in Zagori, and never in the Rif. The Lucanian mountains constituted a partial exception.

In Lucanian mountain villages, although smallholding was the rule, local notables did enjoy some feudal privileges, as did some of the great magnates based in Naples.[4] A noble family resided in many of the larger Lucanian villages, although with time more and more became absentees in favor of Neapolitan *palazzi*. Generally they rented out their lands, and almost never engaged in direct management. Although most of the *comuni* (or *università* prior to 1806) had a latifundia with more than 200 hectares, peasants usually held long leases on parts of these estates.[5] The comuni without latifundia at all generally lay high in the mountains of western Lucania. A *relazione* of 1736 spells out the fiscal responsibilities of several mountain villages to various dukes who held title to much of the village arable. Even remote Terranova di Pollino

[3] AHN, Estado, signatura 771, "Taha de Luchar en la Alpujarra vaja." Théophilou (1983, 161–76) covers property regimes in Zagori and especially in Vitsa: 97% of holdings amounted to less than 5 hectares in the 1950s (p. 172). In the early nineteenth century Ali Pasha sometimes managed to extort a share of the harvest from Zagori villages.

[4] ACSR, Arch. Parl. (Jacini), busta 3, fasc.: Lagonegro, ff. 68–9, on land tenure in Lucanian mountains. A study that details these matters for Naples and Lucania is Astarita 1992.

[5] Nitti 1909–10, 16–23. Eighty of 125 comuni had latifundia.

owed 500 ducats a year to Prince Duke Pignatelli.[6] Calvello in the 1820s had more than a dozen gentlemen living principally off rents from their landholdings.[7] These dukes, and lesser *galantuomini* elsewhere, enjoyed the patronage of the villagers and, in return, ostensibly looked out for the interests of those who needed it, through charity, loans, and employment. Patriarchal relations began to break up in the late nineteenth century with emigration and with the development of more strictly capitalist relations between employer and worker, landlord and tenant. Emigration and remittances allowed more and more families to buy land. Moreover, wages rose in the years of emigration, encouraging wider ownership of land and discouraging large estates. Landlords took a very dim view of emigration.[8] For Lucanian peasants, it was the quickest route to what they all wanted: a house and land of their own.[9] Lucanian peasants, even those who owned no land, lived in a much different world than did those of Apulia, Terra di Lavoro, or any of the classic lands of latifundia.

In mountain areas outside of Lucania, most families had a couple of hectares to call theirs,[10] but this by no means made life easy. Often inheritance and marriage meant these smallholdings were scattered around and about in tiny parcels. In Jbala (western Rif) around 1960 the average family holding was 2.4 hectares divided into nineteen plots. In the central Rif an average family had as many as eleven small fields. Parcellation made for great inefficiencies. Cultivators lost much time walking between plots, and could not easily use draft animals or, later, agricultural machinery. Mountain agriculture required great amounts of labor of itself; the parcellation of minifundia merely aggravated the problem.[11]

[6] Biblioteca Nazionale, Napoli, ms. XIV–II–39. Relazione Gaudioso.

[7] Pedio 1961, 15. Terranova di Pollino had only one or two gentlemen in residence.

[8] On landholding, tenancy, wage contracts, and credit in Lucanian agriculture: Azimonti 1909–10, 5, pt. 1:42–71; Istituto Nazionale di Economia Agraria 1932, 8–48. Contracts were made *viva voce* in the piazza, as they were also in lowland Andalusia and are today in California's Imperial Valley.

[9] Nitti 1909–10, 17–18.

[10] In the Alpujarra landless families were not always rare however. In 1752 Mecina Bombarón had one hundred. Navarro Alcalá-Zamora 1979, 99. As in Italy these were the first to emigrate, so by the twentieth century such *peones* were few.

[11] In the Alpujarra an average family owned nine different plots. Ortega Alba 1981, 127. See Navarro Alcalá-Zamora 1981, 34–7, on minifundismo generally in the Alpujarra. King (1973, 119) shows that parcellation hampered Lucanian mountain peasants. Fay 1979, 88; Fay 1972, 2:349; G. Maurer 1968b, 38–45; and FAO Library (Rome) DP/MOR/73/016 (1976), 10 on the Rif. Théophilou (1983, 173) says that Zagori villagers (from five surveyed communities) held their lands in an average of six separate plots and often had to walk 5 to 10 kilometers to reach their fields.

Most every mountain village held land in common, usually pasture and forest. This posed another problem. Typically, proprietors of flocks rented this land on a seasonal basis, or else it was kept open to all villagers. Watchmen prevented strangers from exploiting common lands. But they could not easily prevent the land from ill use – overgrazing or deforestation – at the hands of peasants, sheep owners, and later, lumber companies whose civic conscience could not stand up to their immediate self-interest. Sometimes villages restricted access to common lands in order to preserve their patrimony, as Samarina did in the 1890s, but it often took a crisis of land degradation to achieve this restraint. Before the Spanish arrived, Rifian peasants sometimes observed conventions restricting use of lands belonging to the tribe as a whole – but often did not.[12] Italian investigators emphasized the "little love" that Lucanian peasants had for public goods.[13] As we shall see, selling off these public lands, at least as it was done in Spain and Italy in the nineteenth century, produced land speculation and even more acute ecological damage.

Seasons of Work

The succession of the seasons defined the sequence of tasks facing mountain peasants. Agriculture and pastoralism had the strongest seasonality of all peasant economic activities. In most cases agriculture was the most important, since between 80 and 98 percent of the population lived primarily from it.[14] So I will begin where the agricultural year begins: fall planting.[15] When the autumn rains come and soften

[12] G. Maurer 1968b, 38, 58.

[13] ACSR, Arch. Parl. (Jacini), busta 3, fasc.: Lagonegro, f. 69. On Samarina: Wace and Thompson [1913], 1972, 45.

[14] In the DERRO area of the Rif in the early 1960s, 79% of the people depended on agriculture; DERRO 1965, 17. In the Alpujarra before 1940 more than 90% of the population lived from agriculture, Navarro Alcalá-Zamora 1979, 87. In the 1927 Turkish census, 98% of the population of Eğridir district is described as peasant (Turkey, İstatistik Umum Müdürlügü 1929, 1:93). In Lagonegro in 1879, 90% of the population was in agriculture, ACSR, Arch. Parl. (Jacini), busta 3, fasc.: Lagonegro, f. 13. In 1950, for the twenty thousand families in the Arachthos watershed (in the Pindus), 63% of all income came from agriculture, 32% from stock raising, and 5% from forest-related activity: see Greece, Iperesia Syntonismou [C. A. Doxiadis] 1950, 82. Today all these figures concerning the importance of agriculture would be far lower, and in earlier centuries they might have been a little higher.

[15] This account is drawn chiefly from Navarro Alcalá-Zamora 1979, 93–9; Consejo Económico Sindical de la Zona de las Alpujarras 1970, II–5; Coon 1931, 49–50; Hart 1976a, 53–4; G. Maurer 1968b; Lévi Provençal 1918; a cursory account of the agricul-

the baked earth, usually in October, all available men prepare the ground for planting wheat and barley and, in the higher reaches, rye. This work can take several weeks because it is often done with hoes and not plows. In the Alpujarras, barley is planted in October and wheat in November. In the Rif cereal planting goes on until the end of December, and legumes (chickpeas, lentils, beans) are planted in the same season. Onions, an Alpujarran staple, are also planted in the late fall. Women and children begin to collect chestnuts in November and keep at it through January; in the Alpujarra and Lucania this harvest is very important. By December it is time to harvest turnips. Garlic must be planted in January. In the Alpujarras, in February chickpeas and tomatoes are planted, the wheat needs harrowing, and beans need weeding. In the Rif olives and oranges are picked in December and January. None of these tasks takes much time, and the winter season is slack. The days are cold, wet, and short, enlivened by intensified social activity. Most of the major celebrations and social rituals, weddings for instance, take place in the winter.[16]

Forest chores beckon in wintertime. Men, women, and children cut and collect fuelwood from the maquis and tall forest. In the Rif, gathering is women's work, and every winter day small bands of women and girls roam the maquis-clad slopes, looking for acorns, forage for their goats, and firewood. In Italy men tend coppices, pruning for the household stoves.[17] Charcoal making also takes place in the forests. Winter is also the season for craft work. Women knit and weave, men work in wood and clay. Clothes and tools are repaired, and all the countless household tasks that could be put off during busier seasons now get attention. Where sufficient fauna survived to make it worthwhile, men also hunted game in the winter. None of this brought in much food or cash, however, and so winter, the *stagione morte*, is also the season of migration to the lowlands.

Mountain folk move down to the olive harvest in November and December. In the nineteenth century Lagonegro women went down to

tural calendar appears in PRO FO 195/751, f. 19, "General Report on Epirus," 1863. The calendar of the Sarakatsani pastoralists in Zagori is in Campbell 1964, 19–24.

For the rhythms of the year among Greeks in the Isparta area of Anatolia, Vogiatzoglou 1986, 203–42.

16 Those that fall in busy times of year, such as Holy Week in the Alpujarra, are largely ignored. In the Rif the marriage season is between the August harvest and October planting. In Samarina, because there was no agriculture (after about 1870 at any rate), summer was the social season. Wace and Thompson [1913] 1972, 50.

17 Pavari 1955.

the Ionian Sea coast to pick olives while men dug roots and worked the vineyards.[18] Others from Potenza went to Apulia. In the Taurus, winter still sends people to the Aegean coast and the Antalya Plain for the citrus harvest, and jobs as root diggers and ditch cleaners. The cotton harvest in Antalya now brings people down as early as September, just after the maize harvest in the uplands.[19] Alpujarrans went to Jaen and Córdoba for the olive harvest in December and January, or (more recently) to lowlands everywhere to work the citrus orchards. In February and March the sugar harvest beckoned along the coast, around Salobreña and Motril, and entire families left their mountain homes for up to two or three months. A few Alpujarrans hired on as fishermen, at least in the early nineteenth century.[20] For about a century before Moroccan independence in 1956, Rifians used to go to French Algeria to work in the winter (and again in the late spring). Everywhere they went during the winter, mountain folk were welcomed (by employers) for their labor but despised for their backwoods manner and feared for their roughness and clannishness. In most cases the migrants were mostly young men, temporarily removed from some of the constraints of village life. Plainsmen regarded them as dangerous, if lamentably necessary.[21]

The slack season ends in March and April. Mountain folk trudge homeward from their winter work, carrying a little money and the few purchases earned by their wages: pots and pans, perhaps a gun – items not manufactured in the villages. In March and April the wheat and barley require weeding and hoeing. Then in April and May some of the garden crops, beans for instance, need harvesting, a hard job requiring many hands. In the Rif, as soon as beans and lentils are harvested, another crop of beans, or peppers, onions, or potatoes is planted. Potatoes and peas are planted as early as March. Rifian peasants wean their newly born lambs and kids in March, take them out of doors, and loose them on the slopes. They shear the animals in April. In April and May the first irrigation begins in the Alpujarra, another painstaking job. In Lucania, where there is little irrigation and few garden

[18] Azimonti 1909–10, 5, pt. 1:42, 85. ACSR, Arch. Parl. (Jacini), busta 3, fasc.: Lagonegro, f. 85. Foerster 1924, 532–3, and Italy, Ministero di Agricoltura, Industria e Commercio 1907, both summarize seasonal labor migrations in Italy as a whole.

[19] Planhol 1958, 168–72, and Planhol 1952, 590–2.

[20] González Montoya 1821, 21. Sermet 1953, 157, on the poverty of Almería's and Granada's fishermen.

[21] Not one in ten farms in Mediterranean Andalusia could survive without migrant labor. Mignon 1981, 220. The unpopularity of mountain migrants in Calabria – and no doubt elsewhere – derived in part from the fact that their desperation drove down latifundia wages and inhibited collective action by lowland peasants. Arlacchi 1983.

crops, May is a slack season: after hoeing the cereals there is little to do until the harvest. In the Rif May is devoted to the vegetable harvest and to tree crops, which need manure, water, and pruning. Spring in general is a much busier season than winter, with people out and about as the days lengthen, tending to a wide variety of agricultural tasks.

Summer is the season of the longest hours and hardest work. The cereal harvest starts in June in the Rif and Lucania, usually July in the Pindus and the Taurus. It can last through August where there is wheat, the slowest of the Mediterranean grains to mature. Threshing goes on through late September. Peas, more beans, and lentils are harvested in the Rif in June. Maize is planted in the same terraces in early July, and needs irrigation and manure. Rif soils yield two crops of maize between July and December, which helps explain the high population densities there. In the Alpujarra and Pindus, maize is planted in May and harvested late in the summer, together with potatoes, onions, and other garden crops. Figs are harvested in August in the Rif, together with almonds and walnuts. Everywhere but Lucania (where there is almost none) irrigation requires delicate attention from May on, as not a drop of water must be lost. At different moments the water is needed for different crops, and sluices and ditches must be kept in perfect working order. In August or so – it is not as urgent as a harvest – the sheep need shearing again; the wool must be washed and in some cases sold.

Peasants generally work from dawn to dusk in the summer, which can mean sixteen-hour days. Time is almost as precious as water, because there is so much to do and nature penalizes procrastinators. In the Alpujarra peasants whose plots are not close by the village spend much of the summer in their *cortijos*, small slate huts well up in the Sierra. One might say they practiced transhumant agriculture, exploiting the various elevations of the Sierra at various seasons. In Lucania peasants often sleep in the fields in the summer, or in makeshift summer dwellings. This can save four or five hours a day. It is not generally necessary elsewhere, because fields and homes are not so far apart.[22] In summer, formal gatherings and social life cease. They begin again with threshing, and expand once the grain harvest is securely in. In late September it is time again to think of preparing fields for winter barley and wheat. Everywhere but the Rif, the flocks head down to the plains, and the cycle begins anew. In the Rif, where animals stay put during the winter, the cycle begins anew when peasants burn the *matorral* (scrub

[22] At Laurenzana around the turn of the century, peasants had to walk four hours to reach their fields. At Viggiano they walked for three. Spera 1903, 43.

vegetation) to prepare the way for the planting of their long-fallow plots to cereals. They bring their goats indoors when the snows fall.

Naturally variations exist in the calendar from place to place and from year to year. The Rif has something of its own rhythm, having a growing season long enough to get two crops of maize and most garden vegetables. Everywhere a tardy autumn rainy reason would delay planting. In the Rif, or at least parts of it, some spring activities waited on the appearance of migratory birds, which would vary somewhat with the weather in northern Europe. But the main outlines of the year are much the same throughout the Mediterranean: time and water are scarce in the summer and plentiful in the winter. From the point of view of the intensity of labor, there is a winter season of comparative ease (except for the migrants) and a long season from March to October of hard work. From the point of view of the intensity of village social life, there is a summer season of dispersion and feverish work, and a long season from September through May when everyone (except the migrants) is concentrated in the village.

The pastoral calendar was much simpler.[23] In the Mediterranean pastoralists were of two sorts: sedentary and seminomadic or transhumant. No true nomads roamed the Mediterranean mountains, except perhaps in the eastern Taurus. Sedentary pastoralists, really peasants who kept sheep or goats, grazed their animals around their villages, usually charging children with the task. In winter they either entrusted their flocks to shepherds who went down to lowland pasture, or else they stabled their animals indoors (usually in their homes), bringing them fodder. A long mountain winter killed livestock, as could a summer drought. Building one's flock back up again might take several years.

Transhumant pastoralists shuttled seasonally between summer and winter pastures, seeking the best grazing for their flocks of sheep and goats. In Spain this involved long treks from north to south, but elsewhere in the Mediterranean the prevailing patterns exploited differences in altitude rather than latitude. In southern Italy, the lowlanders (or the state) rented pasture to mountain folk (including Lucanians) who owned the flocks. At its height, this transhumance involved several million sheep. But Lucania participated only in a marginal way; the busy sheepwalks lay between the Abruzzi and Apulia. In the Pindus too the seminomads practiced what is called "inverse transhumance," where

[23] On Mediterranean sheep and goats: Mason 1967; French 1970. On pastoralism: Carrier 1932; Wace and Thompson [1913] 1972; Kavadias 1965; Campbell 1964; Planhol 1958; Roux 1970; Marino 1988; Cribb 1991; Sprengel 1975.

the sheep and shepherds reside in the mountains and visit the plains. ("Normal transhumance" is that which was practiced in the Alps, where the valleys, not the mountains, were home.) Where malaria reigned, inverse transhumance made good sense.

The Taurus *yörüks* moved up to the mountains in April. In centuries past they dallied between low and high pasture, stopping for as much as a week at various points en route. More recently they sped from one home to another, as fast as their animals could go. (Indeed, today transhumant pastoralists in the Taurus and the Pindus and the few in the Alpujarra use buses and trucks for their migrations. Longer transhumant routes in Spain and Italy have used the railroad since 1900 or so, and rely heavily on trucks today.) Taurus yörüks planted vegetables in their summer yaylas as soon as they arrived, no later than May. In June they began the cheese-making season. In July, when the lowland plains became torrid and parched, entire villages might remove to the upland yaylas. In August and September they harvested and threshed what grain had ripened; in September they harvested their fruits and vegetables. They planted next year's wheat and barley in early October and then headed back down to the plains, where winter pastures beckoned and wage work could be had during the fall planting season. In the winter, new lambs were born and yearling males slaughtered before the spring migration up to summer pastures. This way of life characterized the Taurus until the 1890s, when transhumant pastoralism began to decline.[24]

A similar rhythm held sway in the Pindus,[25] where annual migrations took place punctually on chosen saints' days: the first shepherds went up from Thessaly to the Pindus on Saint George's day (6 May) and all had trudged back down again by Saint Demetrios's day (8 November). The migration from Larissa to Samarina took ten to fifteen days because shepherds and flocks walked for only about five hours a day. The rest of the time went to grazing. They always stopped at Grevená on the way up to lay in supplies for the summer. Shearing the sheep, a raucous social occasion for men, occupied a few days just before or after the spring migration. In some cases, it was done both in March and in May. Cheese making took place from June through the end of July, until pastures began to grow dry and thin. In August and September there was little to do around the sheepfold, and Samarinots, who did not

[24] On the transhumant life in the Taurus, Özbayri 1972, 57–61; Tunçdilek 1963–4, 65–6; Cribb 1991; Roux 1970, 68. Cribb (p. 165) suggests Taurus herders also practiced inverse transhumance.

[25] Psichoyios and Papapetrou 1987; Sivignon 1968; Wace and Thompson [1913] 1972.

practice agriculture, turned to forest jobs or idleness.[26] Women, as I will explain shortly, had plenty to do in any village that produced wool.

Transhumant pastoralism existed on a much smaller scale in the Lucanian mountains[27] and the Alpujarra, following approximately the same seasonal rhythms as the Taurus and the Pindus. Sedentary pastoralism predominated in these peasant massifs. In the Rif pastures are poor and transhumance has never amounted to anything. Mountain villagers of course did keep sheep and goats (sedentary pastoralism) as everywhere throughout the Mediterranean. In the western Rif and Jbala, richer villagers kept cattle (and sometimes grazed them in cedar groves). In the Rif's pastoral calendar, spring events came a little sooner and autumn ones a little later than on the northern side of the Mediterranean. Although Rifians have never practiced transhumance, and pastoralism is not a way of life, nowhere are goats regarded as more essential than in the Rif. Every peasant household cherishes them, because their manure greatly improves yields in maize and kif.[28]

Vulnerabilities of Mountain Life

The key feature to these systems of mountain peasant and shepherd life is its seasonal inequalities. Although there is little to do in winter, the myriad tasks of summer require huge quantities of labor. In the Alpujarra irrigated polyculture needs seven to nine hundred man-days a year per hectare, about twenty to thirty times what wheat requires.[29] Most of this work must be done between April and September, perhaps five hundred man-days. Hence it is very difficult to get by with fewer than two people per hectare in the summer. A land-tenure system of smallholding, characteristic of mountain agriculture, thus depended on lavish population as the source of labor. Every household needed children. A family with 2 or 3 hectares needed several children, unless it could draw on other relatives not otherwise employed in summer. Every three hundred sheep needed someone to tend them – four hundred if the shepherd had a good dog.[30] Hired hands were often indispensable, no matter

[26] Wace and Thompson [1913] 1972, 48–50. This refers to Samarina. Most people came up later in May and went down in September or October.

[27] Marino 1988, 40–2, 53–4, on the pastoral calendar in southern Italy.

[28] Lévi Provençal 1918; Fay 1984, 10–12; Deil 1987, 248; see Chapter 6 on the connection between goats and kif.

[29] Mignon 1981, 191–2. Olives require only 30 to 40 man-days per hectare. Vines need about 150 to 200, roughly the same as sugar cane on the Granada coast. Modern European wheat cultivation needs only about 30.

[30] On sheep to shepherd ratios: PRO FO 195/935, ff. 189–217, "Report on the Position of Artisan and Industrial Workers in Epirus," November 1870. Sivignon (1968, 16)

how desperately each family wanted to keep all the fruits of its harvests for its own consumption. In the Alpujarra and the Rif, peasants devised systems of labor exchange among themselves to get the big jobs done without having to pay anyone wages.[31]

Although mountain agriculture depended on heavy population, it could not easily support heavy population. Except for irrigated areas, yields were generally lower than in the plains (by about half in Lagonegro), and the growing season shorter. Often food crops produced only enough for six or eight months of the year, in the Pindus sometimes less. In Zagori, agriculture supplied only one-third the necessary food.[32] So forest gathering, stock raising, and migrant labor in the dead season were often essential to survival. Even so, winter often brought idleness (at least for men) and hunger.[33]

The mountain way of life, for all the ruggedness of its practitioners, was a delicate thing. It was vulnerable on many fronts, because survival depended on success in several enterprises, on access to several different sets of resources. Should mountain folk find themselves without access to spring and summer irrigation water, to massive summer labor, to forests in fall and winter, to lowland wage work or lowland pasture in the winter, life would become very difficult indeed. So mountain villages were vulnerable to drought, as any agricultural community is,

reports that in the Pindus a shepherd could watch 200 to 300 sheep. In the Taurus each nomad family kept about 30 to 80 sheep or goats (Gould 1973, 29) although Cribb (1991, 28) reports optimal herd size is 250 to 400 per shepherd. In the Pyrenees today, where wolves and sheep stealers are not a problem, an active shepherd can manage 1,400 sheep. This requires good dogs and great stamina, however, and most flocks are much smaller. What I know about sheepherding in the Pyrenees I owe to Bernard Obelliane, a (rare) young shepherd who showed me some of his business in 1986.

31 In the Alpujarras a labor migration to the higher villages took place in July and August for the threshing season. Floristan and Bosque 1957, 383–92. In Lucania, where agriculture was more extensive than anywhere else in the Mediterranean mountains, the ratio of people to cultivated hectares was about 11:10 in 1879. ACSR, Arch. Parl. (Jacini), busta 3, fasc.: Lagonegro, f. 13. In the Cosentino (Calabria) in 1939 the ratio was about one person per 2 hectares; Arlacchi 1983, 24. In the Pindus hired hands were very rare, PRO FO 195/935, ff. 150–68, "Report on Land Tenure. Epirus," 12 January 1870. Labor barter, called *tornapeón* in the Alpujarra, was a distant memory in the Rif by 1971 according to Fay, 1979, 88–9.

32 Lambrides, cited in Théophilou 1983, 184. Leake reported that the grain harvest in Metsovo sufficed for only one to two months of the year, and in Sirako and Kallarites for only four. Leake 1835, 1:278, 296. Typaldos Forestis (1912, 14) says a good harvest in Epirus provided food for six months of the year. In the Rif tribes of Mtioua, Mernissa, Beni Ahmed, and Senhaja, the grain harvest provided enough for eight months of the year (Segonzac 1903, 65). Bab Taza Rifians in the late 1970s produced enough cereals to last 40% of the year; Fay 1984, 11.

33 Coon (1931, 56) says Rifians were usually on the verge of starvation by January.

but perhaps more so in that a shortfall in the winter snows could lead to trouble in irrigating the summer's maize. They were vulnerable to labor shortage (say in the wake of an epidemic), like any agricultural community, but perhaps more so in that slope, parcellation, and polyculture made cutting corners potentially disastrous. They were vulnerable to deforestation, because fuel, chestnuts and other food, and animal forage came from the woods. They were vulnerable to political unrest, which might prevent seasonal migration of men or animals.[34] They were vulnerable to labor market conditions in the lowlands: a depression in the sugar or cotton industry, or an oversupply of eager hands, could spell trouble. They were vulnerable even to changes in settlement patterns in the lowlands: if population grew on the Ionian Sea coast or the Antalya Plain, then the mountain folk would not be needed in winter; and their flocks would have nowhere to graze in the midwinter lambing season, when good pasture is essential. If any one of these problems arose, mountain folk would have to shift as best they could. In fact, as we shall see, several of these problems arose in the Pindus, Lucanian Alps, Rif, and Alpujarra in the nineteenth and twentieth centuries, making the mountain way of life difficult and in some cases impossible. In the Taurus only a few of these problems have arisen, and only recently.

Mountain life also depended on certain social conditions. Children above the age of eight or nine could not spend much time in school, because their labor was essential, not just in summer but in fall and winter as well.[35] Women had to accept a hard lot. In addition to managing households and raising children, they worked in the fields and gardens, fetched water and wood, and gathered nuts and berries. A French observer in the Pindus in the 1830s found a line from the poet

[34] Wace and Thompson ([1913] 1972, 150–71) detail some of the inconveniences of war for Samarina: pastoral migrations became too dangerous, the wool market sank. It is perhaps no accident that Greek mountain villagers take care to be well informed about politics. Théophilou (1983) points out the irony that many Zagoriots know nothing of agronomy and soil conservation, but know a good deal about what is afoot between the United States and the USSR. It is not as strange as it seems: historically knowledge of distant conditions could make a big difference to Pindus villagers, and the habit of attention to politics may have been born of practical concerns.

[35] Dolores Ibarruri, the famous Spanish communist leader, spent her early childhood gathering nuts, berries, mushrooms, and fruits in the Basque hills (Ibarruri 1984, 49). A Basque student of mine, Elena Garmendia, tells me that people of her father's generation know scores of mushrooms and still gather them, but that she and her contemporaries know nothing of this old art.

Hesiod appropriate: "la femme et le boeuf, nés pour le labourage."[36] Wheeled transport did not exist in the mountains, and women to a large degree filled the gap. In the Pindus in the 1860s conventional reckoning had it that a mule could carry 140 kilograms, a donkey 70 kilograms, and a woman 35 to 45 kilograms. Eighty years later, a British classicist fighting with Greek guerrillas against the Germans declared a certain mission out of the question because the guerrillas had no mules to carry heavy loads of dynamite. The guerrillas told him they did not need mules because they had women, and indeed women carried dynamite equivalent to half their weight on mountain tracks over snow. In Lucania, the political exile Carlo Levi noted how local women routinely carried 7-gallon barrels of water, equal to about 28 kilograms, on their heads.[37] Without hard labor from women and children, mountain villages could not survive. To some extent, then, the social "backwardness" of the mountains was a necessary feature of life, mandated by economic and ecological conditions.

Auxiliary Activities

Cultivation and pastoralism took up most of mountain peasants' time, but did not generally provide enough to see them through the year. Hence villagers turned to other pursuits, principally, but not exclusively, in the dead season of winter. In some settings, these pursuits attracted full-time specialists, but more often they drew peasants and shepherds with time on their hands and mouths to feed.

Brigandage and Soldiering

One of the oldest auxiliary activities of mountain folk was organized violence. Around the world mountaineers have found that their mobility, hardiness, and skill with weapons can earn them useful sums. Typically they have done this in two ways: as laborers, that is, as soldiers in someone else's army, like the Gurkhas of the British and Indian armies; or as entrepreneurs, that is, self-employed specialists in violence, extortion,

[36] AMAE-P, Mémoires et Documents, Turquie 1834, 120:21bis. Pizanias (1988, 480) notes that in the nineteenth century Greek peasant women had only a quarter as much free time as men, 15 days per year as opposed to 60.

[37] The classicist was Hammond (1983). Also PRO FO 195/751, f. 15, "General Report on Epirus," 1863; Levi 1963, 106–7. Best (1842, 25) also noted the tremendous loads Greek women in Epirus carried. Roda y Jiménez and García Figueras (1950–5, 2:443–71) describe women's work in detail for the Rif.

and robbery, normally called brigands or bandits. Brigandage could be combined with herding,[38] but normally it was a man's sole occupation, though scarcely a full-time job. It entailed feast or famine for its followers. If pursued skillfully, it could lead to wealth, as in the case of el-Raisuli of Jbala, a brigand enterprising enough to take hostage an American businessman sufficiently important to move Theodore Roosevelt to action. El-Raisuli eventually acquired hundreds of followers and became rich, fat, and finally governor of Jbala. Rifians before and after him found it remunerative to capture and ransom Christians, helping to earn their tribesmen an unsavory reputation among Europeans.[39] For every successful el-Raisuli, however, there were a hundred poor and desperate brigands. It was a very competitive market, because the barriers to entry were low. All one needed was access to weapons, sufficient nerve, and a flexible conscience. Desperation could easily provide the latter two. Ex-soldiers often had the first. The less successful brigands also suffered harassment at the hands of the state, so brigandage as a niche could only support a few men except when state control was weak, or when rival states permitted cross-border brigandage in the guise of patriotism. These conditions, however, were met frequently in the Rif, Taurus, and Pindus, and often enough in the Sierra Nevada and Lucanian Alps. More successful brigands suffered little harassment from the state because they in effect became retainers of the local authorities and magnates. In return for occasional illegal chores, brigands were permitted to batten upon the local peasantry. The most successful *klephts* in the Pindus became *armatoles*, rural gendarmes permitted wide latitude by Ottoman authorities in exchange for pursuit of less successful brigands and protection of state interests. In the Taurus, too, the Ottomans set brigands to catch brigands. Consequently brigands competed among themselves to become so dangerous to the public and the state that the authorities would be moved to hire them to keep the peace. This arrangement, far more common than the "social

[38] Carrier 1932, 42–4. Ethnic frontiers often made shepherds seem like brigands to their enemies, as in 185 B.C. when the Romans captured seven thousand brigand/shepherds around Taranto. Brenan ([1957] 1988, 160–1) reports encounters with out-of-work shepherds who had turned to brigandage in the Sierra Nevada in the 1920s.

[39] Windus (1725, 78) mentions brigands in mountains south of Tetouan. Hostage taking, a venerable tradition between Christian and Muslim in the western Mediterranean, is something Rifians did occasionally according to Ayache (1981, 109) and as a way of life according to Harris (1921, 246–7). Lévi Provençal (1918, 84) says Jbala mountaineers enjoyed wide repute for their mastery at brigandage. El-Raisuli is the subject of an admiring book by Rosita Forbes (1924).

banditry" that robbed from the rich and gave to the poor, suited all parties except the peasants.[40]

Ordinary soldiering was no way to get rich and fat. It did, on the other hand, provide steadier work. In ancient times, Pisidia exported mercenaries to Egypt and elsewhere in Africa, where they acquired reputations as elephant hunters.[41] More recently the Ottoman army employed both Albanians and Rifians.[42] In the twentieth century, Rifians have signed up in droves for duty in the French and Spanish armies. In times past, armies formed and disbanded more informally than they do now, and soldiering could be seasonal work. As such it fitted nicely into the calendar of mountain life, especially if one could get home for the summer harvests. More recently armies have insisted upon, and been able to enforce, long terms of service. This has undercut the appeal of soldiering for mountain folk, who can ill afford to be absent when the crops mature. After 1908 the Young Turks imposed conscription on Christian and Muslim alike in the Ottoman Empire, so young men from Samarina went to America as soon as they came of military age: if necessarily absent, they might as well make some money.[43] Modern conscript armies have deprived mountain villages of men in their vigorous years, during slack season and harvest alike, and so have made mountain life more difficult.

Artisan and Craft Work

Mountain folk have also found markets for their skills that entail considerably less risk. Rifians specialized in the eminently peaceful trade of gardener to the rich in the villas of Tangier and Fez.[44] They also toiled as railway construction crews throughout French North Africa.[45] Epirots

[40] On brigandage and banditry: Hart 1987, 6–26; Koliopoulos 1987; Sakellarios 1888; Davis 1874, 305–6; Wilson 1988; Gaudioso 1987; Pedío 1987; Bourelly 1987; Monnier [1862] 1986. On social bandits, see Hobsbawm 1959 and Hobsbawm 1969 and the debunking by Blok 1972 and by Koliopoulos 1987, 26–35 and 277–81. On the Taurus: Gould 1973. The novel by Yashar Kemal, *Memed My Hawk* (New York: Pocket Books, 1978), illustrates how brigands routinely worked for big landlords.

[41] Planhol 1958, 70.

[42] Poujade 1859, 105–6; Brooke 1831, 389–90. Poujade, citing Gibbon citing an unknown source, says Albanian musketeers were so skilled they enjoyed a game in which a man shot an apple or egg off the head of his mother or wife. Rifians and Epirots were also famed as marksmen, although perhaps more attached to their female kin. Some of the so-called Albanians were, however, Epirot Greeks: Holland 1819, 2:256. Braudel (1976, 1:48) treats Albanians and other mountain military recruits.

[43] Wace and Thompson [1913] 1972, 171.

[44] Harris 1927, 45. [45] *Abd el-Krim et la république du Rif* 1976, 57–8.

acquired considerable fame as skilled migrants. Many natives of Zagori in the eighteenth and nineteenth centuries became physicians, and found work far and wide – but rarely at home. Far more became bakers and vendors, carrying trays of baked goods on their heads, in Istanbul and other cities. Indeed, Epirots are said (by other Greeks) to be flatheaded because their mothers smacked them on the head at birth, hoping to smooth their careers in Istanbul. Other Epirots became nomadic tradesmen. In 1870, most of the carpenters, joiners, and masons of Epirus were itinerants, ranging throughout Macedonia, Albania, and Greece. Natives of Pramanda specialized in carpentry. Agnanda enjoyed fame for its tailors, Kallarites for its silversmiths and ecclesiastical painters, known throughout Greece. Greek mountaineers from Pirsoguiani and Aghia Paraskevi worked as itinerant masons all over northern Greece, as did Metsovo Vlachs.[46] Lucanians also specialized in masonry, and many of them followed this trade as immigrants to the United States. Natives of Viggiano, in Lagonegro, specialized as roving musicians and won fame for their singing voices. Like the Vlach tailors and smiths, these people formed a sort of nomadic artisan class, represented in the Pindus, Lucanian Alps, and the Rif, and plying their trades in upland and lowland alike, wherever they could find custom. When exiled to Lucania in the 1930s, Carlo Levi encountered an itinerant pig doctor who roamed from village to village spaying sows, surely one of the more specialized niches available.[47] In the Alpujarra I have found no evidence of this tradition of nomadic artisans. Migrant work there concentrated on agriculture and mining.

Sedentary artisans existed in all of these settings. The larger Lucanian villages even supported professionals – physicians and lawyers – as well as a large complement of carpenters, cobblers, tailors, and smiths. Every *comune* produced wool, linen, tools, furniture, and leather goods until late in the nineteenth century. Calvello artisans in 1811 made barrel hoops, chairs, and nails. Except in Lucania, mountain villages were

[46] Mendras 1961, 14; Wace and Thompson [1913] 1972, 72, 184; Greece, Iperesia Syntonismou [C. A. Doxiadis] 1950, 13–14; PRO FO 195/935, f. 193, "Report on the . . . Artisan and Industrial Classes," November 1870. Leake mentions Vlach smiths, but in some Vlach areas tin and coppersmithing were considered a despicable gypsy trade. Blacksmiths' work was acceptable, however. Wace and Thompson [1913] 1972, 72. Stuart, the British consul in Ioannina, wrote in 1870 that blacksmithing was gypsy work, Jews monopolized the tin trade, and Vlachs worked as silversmiths (PRO FO 195/935, f. 199). It is hardly likely that things changed so much between 1870 and 1910, but I do not know who is right here.

[47] Levi 1963, 189–92. Another such niche was that filled by the Montefegatesi of the northern Apennines, who made clay figurines and peddled them all over Europe. Sarti 1985, 37–8.

generally too small to support much of an artisan class, and peasants did most of their own craft work. But some eventually grew skilled enough to help family finances, and in cases handed-down skills permitted full-time specializations even in remote villages. In Tarzhoute, in the central Rif, some *douars* lived largely from craft work in metal and leather. They had to import iron and other metals, but their skill created chains, saws, and guns that found markets throughout the Rif and northern Morocco (until the 1960s). Over two hundred specialists in leather were still in business in 1967.[48] In the Alpujarra in the later nineteenth century women made blankets and shawls of wool, linen, and cotton, which earned a reputation outside the valley; Portugos even had a small export trade in textiles. In Samarina and throughout the Pindus women made carpets, blankets, and much else out of wool for domestic use, and often created a surplus that their menfolk sold far and wide, notably to German and Austrian cities. Sirako's woolen capes also found distant buyers. The briskness of trade at Kónitsa and other wool markets meant everything to Samarina and other Pindus villages. Almost every Vlach village home had its loom, and women spent long summer evenings reeling, carding, spinning, knitting, weaving, and sewing.[49] Samarina men learned carpentry, tailoring, and cobbling, useful trades for winter months in the lowlands. The highly skilled silversmiths and other artisans of Ioannina often came from the Vlach villages. Most everywhere shepherds developed high skill as wood-carvers, fashioning decorative figurines that earned a little cash, as well as practical devices like crooks and slingshots. Wood-carvers made many of the household utensils – spoons, bottles, barrels, bowls.[50] Normally, however, peasant crafts were for family use, not for sale. Even in Lucania, where the craft tradition was best developed, outsiders considered local artisans poor and untutored.[51]

48 On Tarzhoute: G. Maurer 1968b, 59. On crafts in the Rif generally, especially woodworking: Ricard 1926.

49 Around 1800 the Epirot Pindus produced over 200,000 okes (about 250,000 kilograms) of coarse wool per year. Kyrgiannis 1984, 132. Kyrgiannis also says Epirots learned candlemaking from the Venetians and became specialists.

50 Dimitriades 1980, 211–16, on crafts in the villages of Kónitsa (Pindus). Anastasiadou 1955, 29–30; Leake 1835, 1:274–5, and Wace and Thompson [1913] 1972, 41–2, 72, 80–4, on artisan traditions among Pindus Vlachs. Also PRO FO 195/935, ff. 189–217, "Report on the Position of the Artisan and Industrial Workers in Epirus," November 1870 (ff. 194–5 on textile work). The Potenza museum has some examples of Lucanian wood carving. BPP, Sezione Lucana, X. 18 (1833 pamphlet); Calvello 1982, 9–15; and Pedio 1963–4, 5–48, 267–71, discuss Lucanian crafts. Navarro Alcalá-Zamora 1979, 52–4, on textiles, carpentry, smithing in the Alpujarra.

51 Pedio 1963–4, 51–3.

Even the poor and untutored could make something of a living as woodcutters where tall forests stood. In the Taurus woodcutting was the special province of a minority of unknown origin, the (formerly) nomadic *tahtaçı*.[52] For at least seven hundred years they have felled cedar and pines on the south slope of the mountains for export through Antalya or Alanya. They were true specialists, rarely planting crops or keeping animals, but living from the proceeds of timber sales. In Vlach villages of the northern Pindus, Perivoli, Vovousa, or Samarina, logging provided important revenue for mountain folk, and many men devoted their working lives to it. By the 1890s all these villages had water-powered sawmills and sold planks, carried down by mules, to towns and plains.[53] Elsewhere lumber trades existed as well, notably in the Rif, whose cedar went into many a stately building in Fez or Tangier, and into many a humble boat as well. In the Senhaja, the tribes of Beni Hamed, Beni Seddat, and Beni Hennuch derived much of their food by selling cedar to the lowlands. Carpenters and loggers worked as specialists and developed remarkable skill.[54]

The forests supplied other important auxiliary activities as well. In the Pindus and elsewhere vallonea oak yielded a marketable dye. In the Alpujarra, *Quercus coccifera*, when affected by the right sort of insect, produced a valuable red textile dye. Eventually a cactus-derived dye from the Canary Islands put an end to this trade of the Alpujarra, but the dried leaves of kermes oak were still used to add color to red wine. Charcoal making also brought in additional cash, especially where cities lay nearby. Near Tetouan, some Rifians made livings as full-time charcoal makers. In Antalya Plain at the end of the nineteenth century, Türkmen from the Taurus spent their winters making charcoal. Wherever men found enough wood and water, they constructed tall cupolas of brushwood, preferably holm and kermes oak, to make charcoal, a chore that takes about a week per job. They used charcoal in the villages, but also sold it to the lowlands and to towns, always poor in fuel. Townspeople would pay the higher price that charcoal commanded to get fuel that burned hotter but with less smoke than wood. Glassmaking, metalworking, and other industries could not do without it.[55] Charcoal

52 Gronhaug 1974, pt. 2; Roux 1970.

53 Wace and Thompson [1913] 1972, 75–6; Lambrides 1887, 73–4, n. 1.

54 SHM, Monografías de Kabilas, 2–36, Senhaya y Serair, pp. 41, 44–6, 88–92.

55 Andrisano and Quartulli (1986) treat charcoal making in southern Italy. I have seen it done with the traditional methods they describe in the forest around Lake Monticchio in Basilicata, and similarly in Epirus. Lazos (1935) describes Greek methods, as does Newby (1984, 150). Charcoal sales to towns were an important part of the income of Jbala Rif families: Harris 1927, 87; SHM, Monografías de Kabilas, 2–27, Beni-Ziat, p. 14. Antalya: Cuinet 1890–5, 1:854.

making, although it does require skill, normally occupied peasants during the slack season, and not specialists.

One delicate job that mountain villages came to specialize in during the eighteenth and nineteenth centuries was silk production. Silkworms are very particular about temperature, humidity, and a host of other conditions. The mulberry tree, whose leaves provide the silkworm's food, prefers well-drained soil and elevations between 500 and 1,000 meters (although it grows at 1,500 meters in southern Spain). Both the silkworm and the mulberry tree found circumstances in the Mediterranean uplands to their liking. In the Alpujarra, the silk had sustained the economy in Moorish days, but the art dwindled after the Castilian conquest. In the early nineteenth century, the neighborhood of Tetouan grew silk for looms in Fez, perhaps another vestige of Andalusian influence. Smaller Rif villages raised silkworms late in the century, and on a tiny scale the art survived to the mid-twentieth century.[56] In the Pindus silk production first appeared in the seventeenth century when war cut Ottoman centers off from suppliers in Iraq. It expanded to the nineteenth century, driven by European demand, and reached 25,000 okes per year in Zagori alone in the days of Ali Pasha.[57] Koukouli, a Zagori village, takes its name from the silkworm cocoon. In Lucania, especially Lagonegro, silk prospered until the late nineteenth century. Peasants might spend as much as two months of the year tending their mulberry trees.[58] Silk, except in the Alpujarra, was not strictly a traditional craft necessary to get by, but an adaptation to straitened circumstances in the nineteenth century. It was always for market, never for home consumption.

Peddlers, Muleteers, Smugglers, and Snow Dealers

The mobility of mountain people naturally led to specialization in trade and transport. Transhumant pastoralists like the Vlachs are the most dramatic examples, with their occasionally international commercial operations. After the Treaty of Karlowitz bestowed relative peace on the

[56] Brooke 1831, 1:401; G. Maurer 1968b, 59; Soria Marco 1948, 91.

[57] Kyrgiannis 1984, 133.

[58] Navarro Alcalá-Zamora 1979, 52; Issawi 1980, 254; Casparini and Bruni 1845, 124–6; Azimonti 1909–10, 5, pt. 1:67; and BPP, Sezione Lucana, X.18. ACSR, Arch. Parl. (Jacini), busta 3, fasc.: Lagonegro, f. 46, says the height of silk came about 1860. Nitti 1909–10, 140. The real center of Ottoman silk production was Bursa, not Epirus and Albania, after the sixteenth century. Stuart says that an Epirot long resident in Odessa introduced silk culture in 1847; perhaps he reintroduced it. PRO FO 195/751, f. 20, "General Report on Epirus," 1863. On the mulberry tree: Ruíz de la Torre 1979, 281–3.

Balkans (1699), Pindus merchants developed routes extending into Austria and beyond; many amassed small fortunes.[59] Humbler and far more common were the myriad petty traders who hawked metalwares and exotic goods like coffee along the mountain tracks. Others combined production and processing expertise with commercial acumen. Sütçüler natives, for example, specialized in the milk trade, controlling its distribution in several Anatolian cities.[60] Every itinerant trader needed a pack animal and thus was himself a muleteer or camel drover, or else employed experts. Mountain folk often specialized in these niches: Vlachs and Pindus natives generally made the best muleteers in their part of the world; in Epirus fifteen hundred men made their living this way in 1870. Isparta residents regularly worked winters as camel drovers throughout Anatolia. During the lead boom in southeastern Spain, thousands of local men made their livings packing and guiding mules. And every muleteer tried to make a little money trading on his own account.[61] Various smuggling trades also attracted mountain merchants, with their knowledge of remote routes and disdain for long distances. Epirots routinely flouted Ottoman trade restrictions. Rifians are as expert in contraband today as in the past century. In the days of Protectorate Morocco (1912–56), they enjoyed a high reputation as smugglers between the French and Spanish zones.[62]

One mountain specialty now long extinct was that of snow merchant. Where mountains are tall enough, snow deep enough, and cities close enough, mountain people could turn a profit carrying mule-loads of snow down from the heights to people who needed it for food preservation or wanted it for sherbet and other luxury concoctions. Granada had a snow trade from Moorish times. By the early nineteenth century an ingenious rope-slide system carried snow down to Naples. In the Pindus, locals even traded snow to the island of Corfu.[63]

[59] Lawless 1977, 520–1; Stoianovich 1960. Lambrides (1887, 75–6) notes the names of Zagori merchants active as far afield as Russia, Egypt, India, the United States, and even Japan!

[60] Planhol 1958, 174.

[61] Planhol 1958, 174–5; on Pindus muleteers: Wace and Thompson [1913] 1972, 74–5; PRO FO 195/935, ff. 192–3, "Report on the . . . Artisan and Industrial Workers," November 1870.

[62] PRO FO 195/801, Stuart to Bulwer, 13 June 1864, on Epirot smuggling. Ayache 1981, 108. Today they smuggle kif out and all manner of consumer durables into the country without paying duty. The proximity of the entrepôt of Gibraltar made illegal trade to Spain all the easier.

[63] On the snow trade: Davillier 1876, 194; Sermet 1942, 741; Packe 1868, 117; Craven 1821, 398–9; AMAE-P, Mémoires et Documents, Turquie, 116:292–3, "Rapport du Consul a Janina," 24 June 1866. As late as the 1920s Ioannina bought snow from neighboring mountains: BSA, Clarke, notebook c,104. Mules carried snow down from Olytzika and

Hunting and Fishing

Another auxiliary activity now effectively extinct was hunting. In previous centuries, when woods covered more of the landscape, and weapons were less accurate than now, mountain areas of the Mediterranean supported considerable animal populations. (A few examples appear in Chapter 7.) Edible birds attracted attention too, especially in southern Italy and Spain, choke points on the flyways of migratory birds en route to and from Africa and northern Europe. Today urban sportsmen take aim at these passing birds, while villagers still traipse through the maquis in search of partridge. In the winter months hunting provided a few servings of meat and grams of precious animal protein. In the Rif it may have been an important food source.[64] Habitat destruction and too successful hunting removed a small but at times significant source of food.

Where the mountains run down all the way to the seacoast, as in the Rif and the central Taurus, fishing was and is another possibility for mountain folk. But the fishing is bad in most parts of the Mediterranean, except at its western extreme. Alpujarrans, as mentioned earlier, sometimes moved from winter harvest work along the coast into jobs as fishermen. Rifians built their own boats with the tall cedars and oaks of their mountains and organized their own fishing business in the richest waters of the fish-poor Mediterranean. By the turn of the century they used dynamite to improve yields, quickly diminishing their resource.[65] But elsewhere the mountains and sea were (and are) different worlds, leaving villages dependent on upland forests and the lowland labor markets to complement cultivation and herding.[66]

Peristeri. In the course of a night 65% of the snow melted, but the rest brought a profit. The wealthy of Teheran used to buy melons from the oases of central Asia, packed in snow and carried on camels for thousands of kilometers. Before refrigeration snow and ice had real value.

64 Becker 1909, 9. Peters and Lovejoy (1990, 362–4) note that hunters fell hundreds of millions of migratory birds every year in the Mediterranean region.

65 Roda y Jiménez and García Figueras 1950–5, 2:253–79. Gresa de Camps 1903, 76. On dynamite: Fernández de Castro y Pedrera 1911, 20–1. Colder water rushes into the Mediterranean past Gibraltar, bringing with it an undersea environment more suitable to large edible fish than is found anywhere else in the basin. Typically the Mediterranean is too warm and too low in plankton to support large fish populations. The only major exceptions are the north Moroccan coast and the waters of the Nile fan, both of which are flush with microorganisms, vegetable life, and everything that fish like to eat.

66 The situation in the Gulf of Antalya was typical. Here, although the Taurus is nearby, neither mountain folk nor anyone else fished until 1922, when Turkish refugees from Greece were settled near Antalya. Planhol 1958, 63–4. In Lucania a careful survey found

A last reflection on the auxiliary activities of mountain peasants is in order. Under certain political conditions, villagers had no incentive to produce wealth that they could not hide from the tax collector. Agricultural wealth, especially harvested grain, was especially hard to conceal, whereas the fruits of trade, crafts, and migrant labor might better escape notice. So wherever outsiders collected the taxes (nothing much could be hidden from one's fellow villagers), a reason existed to put one's efforts into auxiliaries rather than agriculture. In the Rif, where outsiders could scarcely ever collect taxes, no such reason existed. In the Pindus it did, and Ottoman tax collection policies were such that villagers everywhere did their best to appear poor.[67] On balance, I think, appearing poor posed no problems for most mountain villagers in the Pindus or anywhere else, and their extensive participation in auxiliary activities derived far less from fiscal oppression than from ecological circumstances.

Basic Necessities

All this work, agricultural and otherwise, translated into the basics of life for mountain peasant families: food, fuel, clothing, shelter. All of these were often inadequate for comfort; but inadequacy of food made for problems more severe than discomfort.

Food

Bread is the staple food everywhere in the Mediterranean, perhaps more so in the lowlands but also in the mountains. Peasants often ate it for breakfast, lunch, and dinner. It came in many varieties: wheat, rye, barley, maize, and combinations thereof. Moisture and temperature conditions largely determined what grains would grow where and hence what bread peasants ate. In the Alpujarra, for example, well-irrigated higher villages could raise wheat, but the drier villages of the valley bottom and the Sierra Contraviesa more often ate barley bread. Above the highest villages in the subalpine meadows, the short growing season prohibited wheat but permitted rye.[68] Flour was ground by hand, or in mills powered by water or donkeys. Everywhere women baked at home, although there were also professional (male) bakers in Lucanian

fewer than a hundred fishermen in the total population in 1813–15. Martuscelli 1979, clxiii–clxv.

67 Greek villagers often avoid conspicuous displays of wealth so as not to incite envy, but this habit may derive in part from the arbitrary nature of taxation under Ottoman rule.

68 Navarro Alcalá-Zamora 1981, 45–51; Carandell 1934, 657.

and Alpujarra villages. In the Taurus peasants ate the old flatbread favored by the nomads: *yufka*, which stays fresh indefinitely and is good for scooping up the rest of a meal. They also ate bulgur, especially in winter. In Lucania, peasants ate wheat or maize bread, baked with beans, potatoes, acorns, or chickpeas in it. By the 1920s, pure wheat bread became the norm in Lucania. Maize cakes formed a large part of the winter diet. In Calvello in the 1870s, maize was the principal food. In the Pindus, peasants raised both maize and wheat, but maize proved more reliable, and they ate it more often. In Chouliarades, for example, maize bread was the staff of life, supplemented by garden vegetables. This too has changed in the twentieth century. In the Alpujarra, rich people might eat wheat bread, most everyone ate rye-maize or rye bread, and poor shepherds ate black rye bread. In Lucania in the 1870s peasants ate about 3 kilograms of food per day, most of it bread. Rifians raise more barley than wheat, especially in the drier east. In the 1950s they ate about 1.5 kilograms of barley or wheat per person per day. In the wetter western Rif they ate maize bread as well. Throughout the Rif, women gather *cistus* grains to mix into their flour. In the 1960s, some families ate 100 kilograms of cistus and an equal amount of acorn flour. Mountain peasants occasionally consumed their cereals in forms other than bread: porridges, cakes, and, in the case of Taurus yörüks, in a juice made from boiled juniper cones.[69]

Mountain peasants rarely ate meat but loved it when they did.[70] On

[69] Viggiani 1946, 88–9; Azimonti 1909–10 5, pt. 1:53–4; G. Maurer 1968b, 57; Morocco, Ministère de l'Agriculture 1960, 25; Fay (1972, 2:324) reports that in Jbala Rifians ate only 200 kilograms of cereals annually. AMAE-P, Mémoires et Documents, Turquie, 22:397; Brenan [1957] 1988, 131. Benekos 1974, 153–4. A wealth of information on the diet of Lucanian villages is in the questionnaires that formed the basis of the Jacini *inchiesta*: ASB-P, Prefettura – Atti di Gabinetto, 1861–1934, cartella 439. On juniper cone juice: Bent 1890, 156.

[70] ACSR, MAIC, DGA, VI, busta 255, f. 4, the correspondence accompanying the undated document (c. 1892), "Costo e consumo delle carni di bue e di vacca," indicates that in Potenza villagers ate meat only in times of epidemics. I believe this means epizootics, when the animals were dying anyway, since other sources say Lucanian peasants believed no harm came from eating sick and dying animals. ACSR, Arch. Parl. (Jacini), busta 3, fasc.: Lagonegro, ff. 75–6. On 1950 diet, see Carlyle [1962] 1985, 40. ASB-P, Intendenza di Basilicata, cartella 1338, "Statistica . . . delle popolazioni nel Circondario di Chiaramonte," 22 June 1812; Lenormant 1883, 239; Pedio 1963–4, 257 (citing the Murat statistics from 1811); Caldora 1960, 22; AMAE-P, Mémoires et Documents, Turquie, 22:396–7, Boislecomte à Rigny, 4 September 1834; Brenan [1957] 1988, 128; Hart 1976a, 45–6; Roda y Jiménez and García Figueras 1950–5, 2:124; Planhol 1958, 175–8; Roux (1970, 47–8) says the Taurus tahtacı rarely ate meat; Mendras 1961, 15; Wace and Thompson [1913] 1972, 51; Théophilou 1983, 312; Stoianovich (1967, 165) explains that meat consumption throughout the Balkans shrank in the eighteenth and nineteenth centuries.

feast days, according to Hart, a Rifian might eat an entire sheep. In Samarina, on All Saints Day, five thousand people, including children, routinely ate two thousand sheep.[71] Mountain folk had far more sheep and goats than did lowland peasants, and often plenty of hogs too (not in the Muslim Rif or Taurus). But their animals were one of their best stores of wealth, a sure source of ready cash in time of need; they also produced milk, yogurt, and cheese. In the Rif, where they are brought home at night and kept indoors in the dead of winter, goats produce much needed manure. Thin maquis and garrigue is good for little without sheep and goats, the only creatures (except for bees) that can transform these landscapes into something of economic value. So the animals were precious, and eaten only on special occasions: when there was something to celebrate, when there was an epizootic, or when there was nothing else left to eat. Lucanian peasants in the early nineteenth century ate meat only when their animals sickened and died, except for a pig at Carnival. Lucanian mountain folk at the turn of the century, and in 1950, got meat only on feast days. In 1811 north Calabrian peasants ate meat on Sundays and holidays, but by the 1880s Calabrians ate meat only on Christmas. Pindus mountaineers ate meat almost never in the 1830s, according to one French visitor, but about once a week by the 1950s, according to another. In Vitsa (Zagori), however, as late as 1973 villagers ate meat only on Easter and Christmas – and when an animal fell ill. Samarinots ate meat regularly at the turn of the century, indeed according to some accounts every day. Undoubtedly villagers in the Pindus have usually had more meat available to them than their counterparts in Lucania, the Alpujarra, or the Rif, where pastoralism was and is a smaller part of life. Alpujarreños generally sold their famous hams or, in the nineteenth century, gave them away to political patrons. Indeed Trevélez ham became political currency in the 1820s and 1830s. The poor sold ham to the rich, who gave it to local politicians for favors. These politicos used the hams in Madrid as a medium of exchange in patron–client relations. Don Natalio Rivas, an early nineteenth-century deputy from the Alpujarra, first exploited the political potential of Trevélez ham.[72] All this left little for the villagers. In the 1920s, Alpujarrans ate meat only on feast days. The same was true in the Rif in the 1950s. In the 1930s, Spanish Morocco averaged 11 grams of meat per person per day, or about 3 kilograms a year; rural

[71] Hart 1976a, 45; Sivignon 1968, 29. I once spent an Easter on Crete with a Greek family, at which occasion nine people ate six sheep.

[72] Navarro Alcalá-Zamora 1981, 60–1.

Rifians got less than the average. In the Taurus peasants ate meat on feast days, and when an animal was injured – altogether about three or four times a year.

Animal protein came from other sources but in small quantity. Fish was always rare, least so in the northern Rif, where suqs distributed sardines as far as a day or two's walk inland. Alpujarrans also ate fish fairly often, usually sardines, but sometimes salted cod, carried by mule up from the coast. In western Lagonegro peasants occasionally got fish from Maratea in the early nineteenth century. Trout streams existed in the Alpujarra and the Pindus, and Greek peasants built wooden trout traps. Cheese formed an important part of the diet of Taurus and Pindus peasants (the latter exported it) but less so elsewhere. Peasants who had sheep ate yogurt in quantity, again in the Pindus and Taurus more than elsewhere. In the 1880s in the Alpujarran villages of Trevélez and Berchules, all classes ate cheese, but in Capileira only the rich could. In Lucania in 1811 cheese was too expensive for peasants, and few of them had the animals to make it themselves. Only the villages high on Pollino, with good pasture, ate cheese regularly. In this matter, as in diet generally, the peasant condition in Lucania improved after the emigration began.[73]

Vegetables, fruits, and nuts completed the mountain peasant diet. Irrigated patches produced potatoes, tomatoes, beans, onions, chickpeas, lentils, peppers, and many other foods. A few more, such as asparagus and chicory in the Alpujarra, peasants gathered from the wild. In the Pindus village of Chouliarades, cabbage was a mainstay and onions appeared at every meal. Potatoes appeared for the first time in the nineteenth century; previously villagers grew and ate a tuber called the "frankish apple."[74] By the early nineteenth century, some areas of Lucania lived on potatoes as a staple, but elsewhere in the province its cultivation was unknown.[75] In the Rif, Ketama raised potatoes in the 1890s, but elsewhere they did not catch on. Few but Spanish colonists

[73] ASB-P, Intendenza di Basilicata, cartella 1338, "Statistica . . . delle popolazioni nel Circondario di Chiaramonte," 22 June 1812. ACSR, Arch. Parl. (Jacini), busta 3, fasc.: Lagonegro, ff. 75–6. Brenan ([1957] 1988) has remarks on fish in the Alpujarras diet. Soler y Pérez (1906, 15) says in Lacalohorra, a Marquesado village, the rich ate fish carried up by mule via Ujíjar. On trout: Ford [1845] 1966, 2:596, and Anastasiadou 1955, 30. On cheese: Pino Artacho 1978, 154.

[74] Benekos 1974, 153–4.

[75] Murat inquiry of 1811, cited in Pedio 1963–4, 258. By 1879 potatoes were everywhere in Lagonegro, and used even for feeding hogs. ACSR, Arch. Parl. (Jacini), busta 3, fasc.: Lagonegro, ff. 19–22.

planted potatoes in the 1930s.[76] Potatoes are still rare in the Rif. They are common everywhere else, especially the Alpujarra, where all classes ate them in the 1880s. Orchards yielded figs (below about 700 meters, higher in the Rif) and apples, important winter food. In the Taurus dried pears helped get people through the winter. The tahtacı supplemented their bread with bulgur, onions, chickpeas, beans, and, as winter specialties, raisins and dried prunes.[77] Olives and citrus fruit had to be bought and brought uphill, and so rarely figured in the diet of those living above 800 meters (higher in the Rif). In Chouliarades olives were regarded as a luxury and carefully rationed.[78] Those low enough to have olive trees used the oil in cooking; those too high used lard. The forests provided acorns and chestnuts, which in Lucania and the Alpujarra often accounted for a sizable portion of nutrition. In Lucania in 1896 chestnut, acorn, and mushroom consumption together came to about 45 kilograms per person. In 1747 in several villages of the Alpujarra – Portugos, Busquistar, Pitres, Mecina Fondales – the chestnut harvest outranked any other single crop (by volume). Many villages produced 3 to 5 *fanegas* (170 to 280 hectoliters) of chestnuts per family per year. In northern Calabria, chestnut bread was the peasant mainstay in the nineteenth century. These areas recall the medieval *civiltà del castagno* of Tuscany in their reliance on chestnuts.[79] When all else failed, Rifians combed the maquis for the acorns of *Quercus ilex* and used them as a base for flour.[80] Almonds (admittedly not a forest tree) also supple-

[76] Muñoz Torres 1939, 33. Some Rifians, at least the Beni-Jaled, also raised potatoes: SHM, Monografías de Kabilas, 1–16, Beni-Jaled, p. 16. Mouliéras (1895–9, 2:96) reports potato cultivation at Ketama.

[77] Roux 1970, 47–48. Roux offers the interesting observation that the tahtacı probably ate better in the nineteenth century than when he observed them in the 1960s (pp. 72–3).

[78] Benekos 1974, 155.

[79] Craven 1821, 341, and Ramage 1868, 84, on Calabria. On Lucania: ACSR, PC, Zanardelli e La Basilicata, busta 7, table C; and Ministero dell' Agricoltura 1976, 151. On the Alpujarra, AMAE-M, ms. 67 [1747]. In Basilicata as a whole between 1935 and 1952 the chestnut yield varied between 17,000 and 65,000 metric quintals, or 3–13 kilograms per person. In Calabria the harvest was about ten times as great. In Calabria today the chestnut harvest is a joyous occasion for groups of young people to roam through the woods, playing as much as gathering. No other nuts or berries matter much compared with chestnuts in Italy. Pavari 1951. A healthy mature tree can yield 200 kilograms of chestnuts per year. A famous one in Lanjarón, just outside the Alpujarra, gave 700 kilograms per year. O. Elorrieta y Artaza 1948, 24. On Tuscan *civiltà del castagno*, see Cherubini 1981, and Wickham 1988, 139–64.

[80] SHM, Monografías de Kabilas, 1–4, Beni Ammart, p. 17. Deil (1987, 248) reports that acorns are no longer eaten in the central Rif.

mented the diet in the Rif and Alpujarra.[81] Where the forests gave way to garrigue, beekeepers produced honey, also an important food in both the Alpujarra and the Rif.[82]

Very little food came from afar. Oil, needed for cooking because animal fat was scarce, came from lower elevations, the domain of the olive. Coffee and sugar came from overseas (except for Granadine sugar in the Alpujarra). Lagonegro villagers in the 1870s imported almost nothing else but sugar. Today Rifians will consign a notable proportion of their cash to get sugar for their tea, and life without coffee is hard to imagine for Greek villagers; but it is unlikely that sugar and coffee have a long history in the diet of mountain peasants: they cost too much.[83] The chief imported food was cereal flour, essential for those villages that did not produce enough calories to get through the year.

Wine often had to come up from lower elevations, and consequently mountain peasants often could not afford it. In the Alpujarras some sheltered barrancas grew vines, and local wine was available until phylloxera hit in the 1890s. In Trevélez everyone drank wine, although in Capileira only the rich could. Around 1910 in Samarina wine was commonplace (villagers carried it uphill with them on their springtime return to the mountains), but in the Zagori villages it was not. In Chouliarades villagers drank apple juice but no wine. In the Rif vines are everywhere except the highest slopes, but the locals eat raisins rather than make wine, although this may not always have been so, since Rifians apparently enjoyed a special dispensation from Moroccan sultans permitting them to drink wine. Lucanian mountain peasants could not afford wine in 1811 but drank it in vast quantities by 1879, according to the stern moralists of the Jacini *inchiesta*.[84] As a rule, springs

[81] AMAPA 262–1, "Datos de la cosecha de almendra . . . 1895." Granada province ranked third in Spain in almond production.

[82] In the 1930s the Rif produced about 75,000 kilograms of honey per year. Spain, Alta Comisaria de España en Marruecos, Delegación de Asuntos Indígenas 1935a. In the 1960s, most every household ate honey. G. Maurer 1968b, 52.

[83] Poor women allegedly preferred a cup of coffee to a meal in Yegen in the 1920s; Brenan [1957] 1988, 132. Benekos (1974, 156) reports that tea and coffee were unknown in Chouliarades (at an unspecified time in the past). In 1870 even the poorest Epirots drank coffee for breakfast, the only luxury of country folk; PRO FO 195/935, ff. 203–4, "Report on the . . . Artisan and Industrial Workers," November 1870. On Lagonegro: ACSR, Arch. Parl. (Jacini), busta 3, fasc.: Lagonegro, f. 62. Wace and Thompson ([1913] 1972, 41) give a list of items regularly imported to Samarina from below: olives, wine, flour, onions, beans, apples, pears, and petroleum (for lamps). Samarina had by this time abandoned agriculture.

[84] Murat inquiry cited in Pedio 1963–4, 259. By 1910 Nitti (1909–10, 178) described them as sober, and Azimonti (1909–10, 5, pt. 1:62–5) did too. ACSR, Arch. Parl. (Jacini), busta

and wells provided enough uncontaminated fresh water that peasants did not need wine. The porous limestone of the Taurus and Pindus provided excellent water, if one knew where to find it. It was often a long way off. Females, whose job it was to fetch the water, learned very young. The Zanardelli inchiesta of 1902 detailed the water situation in Lucanian mountain villages: some had several sources close by, but in others water came from springs and fountains several hundred meters outside of the village. In summer women in San Paolo Albanese had to walk 2,500 meters for water, a chore they normally performed twice daily.[85] Goat's milk for small children and the occasional hard liquor for grown men rounded out the roster of beverages among mountain people.[86]

If the dried figs, apples, chestnuts, potatoes, and maize bread lasted through the winter, all this amounted to a well-balanced diet. Polyculture avoided dependency on a single staple, and the vitamin and protein deficiencies that afflicted populations relying heavily on wheat or maize never mattered much in the mountains. Mountain folk, with their beans and occasional cheese and meat, got more protein than did lowland villagers, except fishing communities. In Lucania, the Alpujarra, and probably elsewhere, mountain people were taller than their lowland cousins.[87] In general, it is likely that in the nineteenth century Rifians ate better than Lucanians and Alpujarrans: they were certainly taller.[88]

3, fasc.: Lagonegro, ff. 32–3, credits the Lagonegrese with an average of 80 liters of wine consumed per person per year. Wace and Thompson [1913] 1972, 52–3; Benekos 1974, 156. In Epirus, Zitsa (elev. 720 meters) makes famous sweet wine, but Zagori and Malakassi, true Pindus cantons, scarcely any. Doxiadis (Greece, Iperesia Syntonismou [C.A. Doxiadis] 1950, 47) noted that in southern Epirus "the recreation of the people is limited to excessive drinking." Alpujarras: Pino Artacho 1978, 154. Rif: Liger 1963.

85 ACSR, PC, Zanardelli e La Basilicata, busta 7, cartella 4, has questionnaires dealing with water supply, *frane*, agricultural credit, and more from almost every village in Lucania.

86 Rum was available in Viggiano in 1878: ASB-P, Prefettura – Atti di Gabinetto, 1861–1934, cartella 439. In Vitsa in the early 1970s, the local café dispersed about 2,500 liters of wine and an equal quantity of spirits (*tsipouro*) per year to a population of fifty-seven families. Because women and children did not drink to notice, this means each man averaged almost 100 liters per year, most of it in the dead season of winter: Théophilou 1983; 293. Epirot males drank raki, as did the Taurus tahtacı: PRO FO 195/935. f. 210, "Report on the . . . Artisan and Industrial Workers," November 1870; Roux 1970, 48. All classes drank *aguardiente* in the Alpujarra, but only the rich drank milk: Pino Artacho 1978, 154.

87 ACSR, Arch. Parl. (Jacini) busta 3, fasc.: Lagonegro, ff. 75–6. Arlacchi (1983, 176–83) discusses upland and lowland malnutrition in Calabria.

88 On deficiency diseases in the Alpujarra: Pino Artacho 1978, 159–63. Anthropometric data appear in Coon 1931, 182–3; Ruíz Albeniz 1930, 58; Pino Artacho 1978, 111–63; SVIMEZ 1954; Roux 1970, 71, citing K. Güngör 1941, 11–15; von Luschan 1891, 51–3. Male Rifians in the 1920s averaged 169 centimeters in height; in the 1890s male

Unfortunately the food did not always last through the winter. Even with careful scouring of the oak and chestnut forests, even with many people earning money in the lowlands, even with the slaughtering of an animal or two, winter often meant gnawing hunger. Mountain peasants love spring.

Shelter

Winter cold made winter hunger worse. But in general, inadequacies in fuel, shelter, and clothing proved less dire than food shortages. Mountain peasants made their own clothes, or, more accurately, women made clothes for themselves and their male relatives. They generally wore the same clothes year round, adding or subtracting a cloak or coat to suit the season. In the Pindus, wool and goatskin prevailed. Local manufacture held sway there until the end of the nineteenth century, when European styles invaded, as in the Taurus until the 1940s. In Lucania, most peasants had no shoes in the early nineteenth century. Rough homespun clothing persisted here too through the 1940s. Today mountain peasants buy and wear clothing made in factories, even, to a large extent, in the Taurus and the Rif. In the Taurus women wear clothes much like those of their ancestors, but men wear Western styles. In the Rif the jellaba remains the basic article of men's clothing, but many men get these from Fez rather than from their womenfolk. They may well be less insulated from the cold than were their ancestors.[89]

Housing, however, has certainly improved, although in most cases only recently. A 1783 book described Lucanian dwellings as miserable huts, dark and filthy. Only the most prosperous had wooden dividers separating the human from animal quarters. Most had only one room, with the dirt floor generally below ground level in the mountains.[90] The huts were of limestone and clay, and scarcely kept the weather out. They were cold and wet in the winter, obliging peasants to burn a lot

Alpujarrans averaged about 166 centimeters tall; Spanish males in general averaged about 162 centimeters; Spanish outdoor laborers averaged about 161. Lucanian males in the 1890s averaged 163 centimeters. Taurus tahtaçı in the 1880s averaged about 167 centimeters, and the Turkish population as a whole about 166. Average heights are an imperfect guide to nutrition, however. See Komlos 1989, 26–8 on height, nutrition, and European history.

[89] On clothing: Planhol 1958, 179–80; ASB-P, Intendenza di Basilicata, cartella 1338, "Statistica . . . delle popolazioni nel Circondario di Chiaramonte," 22 June 1812. Azimonti 1909–10, 5, pt. 1:57–8; Viggiani 1946, 89; Lenormant 1883, 1:240; on the Pindus, PRO FO 195/935, ff. 202–3, 213, "Report on the . . . Artisan and Industrial Workers," November 1870, and Wace and Thompson [1913] 1972, 60–9.

[90] Galanti 1783, quoted in Nitti 1909–10, 173.

of wood to stay warm. This situation began to change only with the great emigration and the remittances from America. Returned emigrants often built much better houses. But other families continued to live together with animals in single-room stone huts until the mid-twentieth century. Indeed many families lived in caves, and not merely in the celebrated *sassi* of Matera, until the late 1950s. All that has changed in recent decades.[91]

In the Alpujarra houses resemble whitewashed cubes. Their flat roofs, often covered with firewood, required stout timber. Animals lived on the ground floor, with people above. Villages with a prosperous minority, such as Mecina Bombarón, had a few large houses of brick and tile. But most were simple constructions. An observer in the mid-nineteenth century said many houses were worse than caves.[92] On the northern side of the Sierra Nevada, in the tableland just south of Guadix, people still live in cave dwellings carved out of the soft rock. Some have three or four rooms. In summer, many Alpujarrans lived in drafty *cortijos*, slate huts with a dirt floor, no windows, and a low doorway facing the southern sun. These are pleasantly cool in the heat of the day (when their owners were out working) but very chilly even on a summer night.

Houses in the Rif resemble those of the Alpujarra. This is no surprise because Rifians helped settle the Alpujarra a millennium ago, and then Alpujarrans took refuge in the Rif after the expulsion of the Moors from Spain. The houses were built of stone (or adobe in the eastern Rif) with roofs of clay or thatch. When wood was more plentiful, Coon reports, shingled roofs were common. Every house required a sizable log as a ridgepole. A large house had a small courtyard in the middle. Almost all were of only one storey but of two or three rooms. Goats slept on the ground level, people on a raised platform above. Among the Beni Ouriarhel at least, every house before Abdel Krim's time had a "pillbox" where a man could defend all he surveyed with his rifle. Rifian houses show great variation, not merely between richer and poorer families, but between tribes and regions as well. Wood was more plentiful in the west, with consequences for construction design.[93]

In the Pindus, by the nineteenth century at least, peasants built their

[91] On Lucanian housing: ASB-P, Intendenza di Basilicata, cartella 1338, "Statistica . . . delle popolazioni di Circondario di Chiaramonte," 22 June 1812. Azimonti 1909–10 5, pt. 1:55–7; Nitti 1909–10, 173–4; Lopreato 1967, 58; and the thorough work of Franciosa (1942) on houses outside of villages (no more than 10% of the total in Lucania). A modern study on housing in mountain villages of southern Calabria is Foti et al. 1983.

[92] Madoz, cited in Carandell 1934, 679. On Alpujarran housing: Voigt 1937; Navarro Alcalá-Zamora 1979, 75–81; Carandell 1934, 675–9; Brenan [1957] 1988.

[93] Coon 1931, 68–73; Blanco Izaga 1930; Hart 1976a, 33–8; G. Maurer 1968b.

houses of stone. Prosperous ones might boast two or even three storeys, square in plan, with a sloped tile roof, and many windows on the top floor. Locals now refer to them as little palaces. Ordinary ones had one storey, with few windows, and a sloping roof of flat stones, or, where wood abounded, planks. In Zagori and other hotbeds of brigandage, houses were built with defense in mind. The most substantial came from the proceeds of successful brigandage or commerce or, later on, after the 1890s, from remittances from America. The floor was usually dirt, except in the prosperous homes, the air always smoky, but the whole kept scrupulously clean.[94]

In the Taurus peasants built with wood. Their structures often resembled log cabins, were easy to build, but were scarcely warm. The nineteenth-century settlers there came from nomad stock, accustomed to life in black felt tents, and had no skills in masonry and construction. Like many throughout the Mediterranean mountains, Taurus peasants lived far more comfortably in their summer huts and sheds than in their houses in wintertime. The tents of a prosperous nomad could seem luxurious in comparison. All this has changed in recent decades, and village homes are built with stone, cinder block, concrete, and other more solid and modern materials. Since the 1950s, stone construction has advanced in place of wood. In Zerk and the Yılanlı Plateau, houses now are usually two storeys, with animals below and humans above.[95]

Fuel

By and large houses in upland areas did little to keep their inhabitants warm and dry. So peasants had to burn a lot of charcoal and wood. On typical winter evenings in the mountains of the Mediterranean, each family gathered tightly around a fire or charcoal brazier, with capes or blankets on their backs, sewing or carving in their hands, or small children in their laps. In these settings, faces faintly illuminated by the fire or an oil lamp, parents told stories, sang songs, and transmitted to their children their beliefs and wisdom, ensuring the survival of their culture. If they did not have enough fuel, they would not survive the winter.

[94] Pouqueville 1826–7, 2:267; Boué, 1854, 2:57; BSA, Clarke notebook c, 107; Wace and Thompson [1913] 1972, 94–9, 173, 198. Théophilou (1983, 138–53) discusses the houses of Vitsa (Zagori).

[95] Planhol 1958, 131, 255, 425; Roux 1970, 37–82, 184. On nomad tents, Özbayri 1972, 56–7; and Roux 1970, 62–7. Yashar Kemal's novel *Memed My Hawk* describes nomad tents in the Cilician Taurus.

Before coal and steam the Mediterranean world had four sources of energy available to it: water power, animal power, wind power, and wood. The spate and drought cycle of the rivers made water power impractical in most places; the mountains constituted the chief exception. Winds were considerable and fairly reliable for parts of the year, but the Mediterranean was no Portugal or Holland with Atlantic winds. Water and wind power helped to drive a few flour mills and oil presses and, in the Pindus, some sawmills. But neither amounted to much as an energy source, so animal power and wood were fundamental.

Animal power, including human muscle, provided almost all the mechanical energy. In the mountains cattle were rare, but occasionally saw service pulling a plow. Horses, mules, and donkeys carried loads and pulled plows, threshed grain, and milled flour. Heat energy came from wood.

Mountain folk needed heat for many purposes. First and foremost was keeping warm in winter. This required at least 5 cubic meters of fuelwood per year per family, and more likely 10. Among the Sarakatsani, Pindus pastoralists who liked to keep their huts so warm in winter that one could not comfortably wear a coat, fuel use would have been greater still. This volume of fuelwood equaled the annual growth of roughly 2 to 4 hectares of mature forest.[96] Thus a modest village of a hundred families needed reliable access to 200 to 400 hectares of tall forest. Without it, the village would run down its fuel supply and eventually have to burn dung (which deprived fields and pastures), import fuel (usually too expensive), or freeze in winter. Oak and beech were the preferred fuelwoods, especially *Q. ilex*, heavy and dense, ideal for charcoal. But when in need, peasants burned just about everything combustible, including the scrub plants of maquis and garrigue. In the western Taurus, where wood is not so scarce, peasants burned nothing else as late as the 1950s.[97]

[96] This figure comes from Kontos 1929, 136, and pertains to the fuelwood consumption of Byzantine Constantinople. Grispos (1973, 151) cites the same figure. In the mountain cold it would have been higher. Filangieri (1980, 108) says in southern Italy each person averaged 100 kilograms of fuelwood per year, more in the mountains, less in the plains. Sarakatsani: Kavadias 1965, 119. The annual growth of black pine in the western Taurus averages 2–4 cubic meters per hectare. For other forest types in the Taurus the figure ranges between 2–5: Akman, Barbero, and Quézel 1978, 339–41. Italian cherry oak when coppiced can grow 8–10 cubic meters per hectare annually; chestnut about 1 in Spain, or 5–6 if coppiced. Coppiced chestnut growth averages about 18 cubic meters per hectare in Calabria, and 12 on Mount Athos in Greece. In the Rif, cedar's annual growth averages about 3 cubic meters per hectare. Seigue 1985, 110, 128, 204. P. Boudy (1948–54, 2:580) gives 1.5 as an average figure for cedar's annual growth. I have no data on the annual growth of holm oak, a preferred fuel.

[97] Tunçdilek 1955.

Cooking also required fuelwood, although for roasting meat (a rare event) all preferred charcoal, which gave a hotter flame. Charcoal, which gives off little smoke, also suited mountain houses better, as they often had poor ventilation. Charcoal also generates about twice as much heat per unit of weight as wood, which endeared it to those, often women, whose chore it was to gather and tote wood. But charcoal consumes forest at a voracious rate. It is burned twice, once to create it and once to use it. All in all, it took 5 to 10 kilograms of wood to make 1 kilogram of charcoal, so although twice as efficient in calorie terms, charcoal use destroyed forest at two to five times the rate of direct woodburning.[98] Charcoal cost much more, whether measured in money, in effort, or in forest, so peasant families normally made do with wood and brush regardless of the wishes of those who fetched it. Pottery and metallurgy required charcoal, however. Bakers needed huge quantities of fuelwood.

The forest and maquis had to meet all these fuel needs. Where transport proceeds on the backs of mules and women, fuelwood cannot travel far. So every village had to have woods nearby, in which men, women, and children gathered fuel for the winter, beginning their labors as early as August. They kept at it, as the severity of the season demanded, until February or March. By and large, men cut wood and made charcoal, whereas women and children gathered branches and bushes, and carried the loads from forest to village. Where settlement is concentrated, as in Lucania, the local pressure on forests for fuelwood alone, irrespective of the needs for arable and pasture, put villages at risk. A place like Calvello in 1800 needed the equivalent of roughly 5,000 hectares of tall forest to meet its fuel needs. Where villages were smaller and dispersed settlement more common, the pressure on forests was distributed more evenly over a broader area, and fuel crises thus loomed less often. Perhaps this helps to explain why only in southern Italy did the practice of managed coppices catch on. Coppices yielded three or

[98] Charcoal provides about 6,500 calories per kilogram, almost as much as coal, while wood gives about 3,000: Seigue 1985, 285–6. Filangieri (1980, 109) gives 7,000 for charcoal, and 2,000 to 3,000 for fuelwood. Bechmann (1990) puts the ratio at 30:18. Today charcoal making is a bit more efficient, and uses only three times as much wood as does burning wood directly. In medieval France, one charcoal pit could consume 100 hectares of forest in forty days of operation. Bechmann 1990, 153. A calculation based on official Spanish statistics from the Rif in the 1940s (González Vázquez 1945, 72) suggests that it took only 4 or 5 tons of wood to make a ton of charcoal, twice as efficient as the data cited by Bechmann. In Lucania around 1910 (La Marca 1916, 383) it took 100 kilograms of wood to make 17 to 25 kilograms of charcoal, an efficiency ratio very similar to that in the Rif in the 1940s. La Marca says more efficient ways to make charcoal existed from the 1820s, but were not used in southern Italy.

four times as much fuelwood per hectare as forest, which is why Calvello could get by without 5,000 hectares of forest. But even this rational practice, as we shall see, did not prevent energy shortages in the Lucanian mountains. Mountain life depended on wood fuel as surely as it depended on food and water.

Seasons of Life

In the Mediterranean, lambs and kids are normally born in the winter months. Peasants were normally born in the late winter and early spring. A diligent demographer collected data on births in the Alpujarra and discovered that before the 1970s births had a clear seasonality to them, falling chiefly between January and June, and above all in February and March. In November and December births ran as much as 30 percent below the average rate. The further back in time, the stronger the seasonality.[99] This probably had to do with female nutrition more than anything else: when women ate well, in the late spring and summer, they conceived more often. When they ate poorly, in the winter, they conceived less often. The connection between nutrition and the seasonality of birth probably obtained throughout the mountains of the Mediterranean – indeed everywhere that nutrition has a pronounced seasonality to it.

Once born, most babies soon died. Childbearing and child burying went hand in hand, as an Italian epigram had it.[100] This was no different from most parts of the world before the twentieth century. Infant and child mortality rates specific to mountain areas are hard to find for the distant past, but a few scraps of data suggest that more than half of those born did not see their sixth birthday. In Trevélez in the 1880s, 56 percent of all deaths came to those five or younger.[101] In Lucania as a whole, only half the males born reached age fifteen in the 1880s, by which time economic conditions had begun to improve there.[102] Presumably fewer than half of the females born reached this age, because

[99] Luna 1984, 70–5.

[100] Foerster 1924, 440. These general remarks on the life cycle are drawn from Hart 1976a, 47–8; Planhol 1958, 346; Azimonti 1909–10, 5, pt. 1:59–66; Arlacchi 1983, 34–5; Bronzini 1964. The last is a study of folktales from Basilicata divided into the various stages of life from birth to death. PRO FO 195/936:196–7, "Report on the . . . Artisan and Industrial Workers of Epirus," has remarks on schooling and school leaving in the country districts of Epirus. Vogiatzoglou (1986, 163–202) covers Greeks in the Isparta area, c. 1890–1920.

[101] Olóriz's data, cited in Pino Artacho 1978, 106. [102] Mortara 1910, 441.

in Lucania, as everywhere, male children are routinely valued more than female ones, and consequently fed and cared for better. A common salutation put it clearly: "salute, denari, e figli maschi," meaning health, wealth, and male children.[103] Greek peasants used to wish one another female lambs and male children. The birth of a daughter, who (in the Christian Mediterranean) would one day require a dowry, imposed the strictest economy on peasant families, and several daughters were widely regarded as a disaster. The first census in Spanish Morocco, in 1929, showed 7 percent more males under fourteen than females, suggesting that in childhood girls died more often than boys.[104] This was probably the case everywhere, mountain and plain alike, where food was sometimes short and disease culled the malnourished.

Children who survived were sometimes faced with school. In the Pindus, locals put great store in education and every village had a school by the eighteenth century. A surprising number of villagers managed to attend secondary school. Zagoriots even tried to create their own university based on central European models, but the destruction and impoverishment that came with the Greek war of independence scuttled their plans. In the Rif in 1928 Spanish authorities counted almost five hundred Koranic schools with close to five thousand pupils. Schooling usually occupied children only until they could be useful, at age six or eight. Parents often had to pay school fees for their children, so the temptation to put one's offspring to work was hard to resist. Boys might get an extra year or two, because they were less use at younger ages than girls, and because learning, most villagers believed, more befits a man than a woman.[105]

By six or eight one's working life had begun, never to cease until age drained away one's strength. The boys helped their fathers in the fields, with the flocks, repairing tools, dealing with people from outside the family. Girls helped their mothers in the gardens, with the chickens, with fetching water, with gathering fuel, acorns, and nuts, with spinning

[103] Quoted from Hooker 1927, 202. Jeréz Perchet (1875, 228) wrote that a Rifian felt wealthy if he had land, animals, and sons.

[104] BNM, Sección de Africa, "Censo aproximado de Gomara," 1929. The 1787 (Floridablanca) census in Spain showed 2% more males than females under age sixteen in the Alpujarra village of Válor. This census, in manuscript, is in Madrid's Biblioteca de la Real Academia de la Historia. I have lost the precise citation.

[105] On schools in the Kónitsa area, Dimitriadis 1980, 228; on Zagori, Papageorgiou 1987. Despite Epirus's traditions of education, Doxiadis reported in 1950 that in southern Epirus most children did not go to school and most priests were illiterate; Greece, Iperesia Syntonismou [C. A. Doxiadis] 1950, 47. On the Rif: BNM, Sección de Africa, Colección García Figueras; Intervenciones Militares del Rif, "Estadisticas 1929."

or sewing. Children performed these tasks for many years, learning thoroughly all the skills necessary to scrape a living from the mountain environment. Their apprenticeship lasted until they married or emigrated – or until a parent died.

Age at marriage varied considerably from place to place and over time. Marriage, as throughout the peasant world, represented an economic, social, and political transaction. Men ordinarily had to wait until they inherited land, or could convince their fathers to retire into old age. Women in the Christian lands had to accumulate a dowry. Thus the demographic and economic rhythms of communities and families strongly influenced age at marriage. As a rough rule, men married between age twenty-one and twenty-eight, women three to five years younger. In the Pindus marriage was delayed more often, at least in the twentieth century. In Samarina between 1900 and 1915, men did not marry until age thirty-three on average, and women not until twenty-four. By 1951–66 these averages had climbed by one year. In Vitsa (1931–70 data) men married at ages between twenty-eight and thirty-three on average, and women when aged twenty-four to twenty-eight. Travelers report women marrying as young as age twelve in the Rif at the turn of the century and in Zagori in the 1840s, but these are likely exaggerations.[106] In the Muslim Rif and Taurus, where dowries were not an issue, women certainly did marry younger. Everywhere, almost everyone got married sooner or later, and those who did not generally left their villages to seek their fortunes elsewhere. Women especially had few options outside of marriage.[107]

Marriages usually took place in a slack season in the agricultural year, either after the harvest and before the planting, or the winter months. In Samarina, where no harvest interfered, all marriages were celebrated either on 15 August or 8 September. Women gave birth to five, seven, sometimes ten or more children. The Rifians say that God gives children. Neither they nor anyone else in the Mediterranean world wanted to refuse the gift. Children had economic utility, they ensured one's future (none but the celibate tried to save money except as dowry), and

[106] Sivignon 1968, 35 for Samarina; Théophilou 1983, 230–2, for Vitsa (the figures for Vitsa would be slightly higher except for an influx of gypsies, who married much younger than anyone else); Segonzac 1903, 49; Blancard 1846, 64; data on Lucania and the Alpujarra are in Luna 1984, 107–14; Azimonti 1909–10, 5, pt. 1:65–6.

[107] Naturally great variations existed. In the Greek mountains, for instance, celibacy was not so rare, especially as sons often were forbidden to marry until all of their sisters had been betrothed.

enhanced one's status. Every man, and most women, hoped for several sons.[108]

From their twenties to their forties and fifties, men and women ran their farms and households, attending to the myriad chores of mountain life, and raising many children. People stayed married until death parted them; divorce was rare everywhere, even in the Muslim lands where the legal obstacles to it (at least for men) were minimal. Those who lived long enough to lose their strength turned over their property to their children. For Greek peasants, assuring their children a respectable inheritance was the focus of their life ambition. Partible inheritance was the rule everywhere, although in the Rif sons got shares twice as large as daughters. New property came with every marriage, so that plots of land, fruit trees, and animals were redistributed with every generation. Once an aging peasant turned property over to the next generation, he or she had little to do. Old people stayed around the house, looking after little children, helping with small tasks as their powers permitted. In the winter, when the family gathered indoors, old people had company and were especially useful occupying children. In the other seasons, when much if not all of the family was dispersed over the slopes, old people were more isolated, condemned to their own society. If their legs and backs permitted, grandmothers could still fetch water and participate in the social life of village women that way; and grandfathers could assemble in the cafés and under the plane trees for a smoke and conversation. In Lucania at least, the old were tolerated but not respected.[109] From the economic point of view this made sense.

By the eighteenth century death took the very young and the old, but no longer many people in between. In Trevélez in the 1880s, people between age eleven and forty-five accounted for only 10 percent of

[108] Data on numbers of births per woman in almost every Lucanian village in the 1870s is in ASB-P, Prefettura – Atti di Gabinetto, 1861–1934, cartella 439. In Calvello and Viggiano, men married at twenty, women at eighteen. In San Paolo Albanese men married after military service, women somewhere between eighteen and twenty-five, and women gave birth to ten or twelve children. In Lagonegro as a whole, men married on average at twenty-three, women at seventeen, ACSR, Arch. Parl. (Jacini), busta 3, fasc.: Lagonegro, f. 80. By 1909 there had been no change among men, but women married a bit later, around age eighteen to twenty-two, and families usually had at least four children, often as many as ten; Azimonti, 1909–10, 5, pt. 1:65–6. In the Rif in 1950 the average household size was 7.0, or five children. Spain, Dirección General de Estadística, *Anuario Estadística* 1941–55 (1955, 61). On the marriage season at Samarina: Sivignon 1968, 34. On saving and celibacy: ASB-P, Prefettura – Atti di Gabinetto, 1861–1934, cartella 439: San Paolo Albanese and Viggiano.

[109] Azimonti, 1909–10, 5, pt. 1:66.

deaths.[110] But running farms and households wore people out quickly. In Lucania and the Alpujarra people died in droves between ages fifty-five and seventy. It was likely much the same in the Taurus, Pindus, and Rif. All in all this was a long and healthy existence, if a grueling one, by the standards of previous centuries and lower altitudes. Better nutrition and comparative freedom from disease made the difference.[111]

The Village and the Wider World

Many mountain villagers had experience or knowledge of the wider world, although this varied tremendously from village to village and person to person. Some exceptional people, like the Pindus long-distance traders or the Rifians serving in the French *tirailleurs*, managed to see a good deal of the world, far more than the average peasant of mountain or plain. Quite ordinary folk who took part in labor migrations or transhumance broadened their horizons somewhat, if only to the lowlands a day or two away. Mediterranean mountain people did not live in isolation equal to that of, say, Andean or Himalayan peasants. The sea was never too far away.

But by the standards of the cosmopolitan Mediterranean world, mountain villages were remote indeed. The Italian state used those of Lucania for political exile: the province of Potenza was best known as an "internal New Caledonia."[112] Although mountain folk might leave their native villages and thus acquire some sense of life beyond, no strangers lived among them to further their acquaintance with other ways. Of Kónitsa residents over 95 percent of males lived in the village of their birth; among females the proportion dropped to 80 to 90 percent, because women when married went to live with their husbands. In Lucania early in this century, 85 to 90 percent of villagers lived in the *comune* where they had been born.[113] In San Paolo Albanese, 87

110 Olóriz's data cited in Pino Artacho 1978, 106.

111 In the Arachthos basin (Pindus) in 1938, pneumonia accounted for 35% of all deaths and malaria for 10%. These were by far the leading causes of death, although, as Doxiadis notes, few died in the presence of medical attention and statistics concerning cause of death must be regarded skeptically: Greece, Iperesia Syntonismou [C. A. Doxiadis] 1950, 48. Around the world mountain people live a little longer than the rest of us. Baker 1978.

112 Fittipaldi 1880, 1–2. The fascist government (1922–43) also exiled opponents to Lucanian hill villages, including Carlo Levi, whose *Christ Stopped at Eboli* provides such a vivid picture of village life. Greek governments have preferred tiny Aegean islands for political exiles. In this, as in mobility, protein supply, and much else, the mountains and islets of the Mediterranean had much in common.

113 Dimitriadis 1980, 456; SVIMEZ 1954, 36. The Kónitsa data come from the 1970s, the Lucanian from 1911–31. In Eğridir district the 1927 census shows four people who

percent of marriages between 1820 and 1980 were within the village population. The figure would be close to 100 percent if recent decades were excluded.[114] Only 5 percent of Samarina men found brides outside the village, and most who did married women from nearby Fourka.[115] In every setting the great majority of men and women resided and married within their native village.

Although the distances separating mountain from plain or mountain from sea were not great, the absence of passable roads limited interaction. In the 1860s, 91 of 123 villages in Lucania had no road whatsoever. The only roads suitable for anything more delicate than a cart connected Melfi to Potenza, and Potenza to Naples.[116] The villages of the upper Köprü Su lay fifteen hours' walk from the nearest wheeled transport until the 1950s. Only one bridge spanned the river (it dates from Roman times and remains the only bridge today), which is too deep and swift to ford even in summer. In this area, money scarcely circulated, and the only outside goods were a few tools, matches, tobacco, sugar, and coffee.[117] In the Pindus, carriage-worthy roads existed in the time of Ali Pasha, who used forced labor to build and maintain them. But these fell apart after 1821, and only stone paths suitable for mules (*kalderimi*) remained. The first road linking Zagori to Ioannina opened in 1919. True roads by 1958 reached about half of the Kónitsa villages.[118] No roads penetrated the Alpujarra until after 1860, and the two largest settlements were connected only by the (usually) dry riverbed of the Guadalfeo. These *ramblas* served as the

had been born outside of Turkey: three from Iran, one from Austria, all women (Turkey, İstatistik Umum Müdürlüğü, 1929, 1:123).

114 Nicolini and Pettener 1984, 26–9. Village endogamy was probably highest in the larger villages of Italy, rather than the smaller ones elsewhere. Mountain villages tended to produce an untoward number of mental defectives, probably because of endogamy. Potenza in 1909 had a rate of cretinism about 2.5 times the national average (Germano 1909, 98). Orioli (1934, 26) noticed a large number of imbeciles on his visit to Terranova di Pollino.

115 Sivignon 1968, 34. Data are from 1900–25. The figure drops to 85% by 1951–67. Théophilou (1983, 226–9) has scattered data from Zagori, showing the tendency to marry within one's village was strong (over 80%) between 1870 and 1900, but declined throughout the twentieth century to the point where, by the 1960s, only a tiny minority found mates within their native villages.

116 Pedio 1983, 125–6; Di Gianfrancesco 1979, 34–64. Filardi (n.d., 2:462) says that in 1872 in Campomaggiore it cost 6 lire, several days' wages, to transport 1 cubic meter 1 kilometer. This is hard to believe.

117 Planhol 1958, 181, 422–5.

118 PRO FO 195/751, Stuart to Bulwer, 4 September 1863. Mendras 1961, 13; Théophilou 1983, 45. They all had telephones by 1958, although Théophilou reported that in 1973 Vitsa had one telephone and it did not work.

highways of the Alpujarra.[119] Brenan's Yegen had no road in the 1920s. Everything that moved went by foot – or more accurately, on back, either human or mule. In the Rif no road at all existed until after the end of the war in 1926. Most women's experience of the wider world extended no further than the suq.

Isolation from wheeled traffic meant that economically the interaction between mountain villages and the wider world consisted only of exchanges in articles that could bear the cost of human or animal portage. Most villages received occasional visits from itinerant peddlers and artisans. But the only major connection with the outside world came in the form of experience of labor migrants (including transhumant shepherds). Until the 1880s, this experience too was limited, since most migrants went only two or three days' walk away.

Intellectually, villagers' contact with the wider world was limited because so few of them could read. In Lucania in 1871, 88 percent of the population was illiterate, in 1911 65 percent, and in 1921 53 percent. In Eğridir district in 1927, 89 percent of men and 98 percent of women could not read. In Ioannina eparch in 1920, 80 percent of women were illiterate.[120] Data from the Alpujarra and the Rif would show much the same situation.[121] Very few women could read, and their sense of the world beyond the village must have been slender indeed, largely derived from what men told them.

Women were more isolated than men physically as well as intellectually. Fewer of them migrated to seasonal work – none in the Rif or Taurus. A survey of village women in the Kónitsa area showed that women had notably different ideas of distances and geography than men; in particular, distances always seemed greater to them.[122]

Information about the wider world varied between men and women, between migrants and stay-at-homes, between literate and illiterate. Access to goods from outside the village varied between rich and poor,

[119] Roads and mule tracks in the province of Granada appear in the Mapa de Granada y Almería (c. 1860) in AHN, Estado, signatura 562. Subsequent roadbuilding in the province is detailed in AHN, Fondos Modernos, Obras Públicas, legajos 172–80. A cart road had existed between the Alpujarra and Almería in the late eighteenth century, but by the 1840s there were only mule tracks and *ramblas.* Ansted 1854, 145, 156.

[120] SVIMEZ 1954, 769; Foerster 1924, 515; Turkey, İstatistik Umum Müdürlüğü, 1929, 1:85; Greece, Yeniki Statistiki Iperesia, 1928. In Italy and Greece, these data refer to people over age six. In the Turkish census this is unclear.

[121] The Alpujarra, strangely enough, had three libraries in the late nineteenth century, at Orgiva, Soportújar, and Ujíjar: Villar Amador 1984, 211.

[122] Dimitriadis 1980, 476. Perceptions on just about everything, wealth, time, pleasure, differed markedly between male and female: on pp. 448–76, Dimitriadis tabulates answers by sex to a variety of questions.

and between migrants and stay-at-homes. Within a single village it was normal to find a wide range of knowledge and ignorance. Conversation could only go so far in communicating information and evening out the discrepancies in knowledge. Today, in a mountain village like Pramanda, one can find shepherds who know more, and care far more, about world affairs than does the average American city dweller. But one can also find old women whose knowledge is so local that they assume all foreigners speak a single language, that the world consists of those who speak Greek and those who speak "foreign."

Conclusion

The mountain way of life, because of its comparative isolation and the inapplicability of machines, changed but slowly from the time maize arrived until the twentieth century. Wars, famines, epidemics punctuated mountain history, as they did rural history everywhere. But villages recovered from these disasters, and the basic structures of life remained much the same. This apparent stability masked a certain vulnerability, not characteristic of rural life everywhere. The mountain system depended fundamentally on many activities and could not survive long should any of them become impractical.

The impression of stability or instability of the mountain system hinges on the length of time one considers.[123] Season to season it was highly unstable, as activities and food supply varied notably with the calendar, more so than in lowlands. Considered year to year it was somewhat less unstable, but harvests did fluctuate, and political violence did come and go. In this time frame lowland villages showed greater instability than did the mountains, having a less diverse portfolio of crops and food sources. Considered over the generations, mountain life was very stable, far more so than life in the plains. Although eras of brigandage or a mining boom might in local cases make a big difference from one generation to the next, isolation and the absence of improvements to agriculture and pastoralism told strongly. Children lived very much as their parents had. In contrast, lowland agricultural productivity improved notably in recent centuries, especially with the arrival of machines, so ways of life and levels of prosperity (if one can use the term) clearly changed from generation to generation. Considered over the centuries, however, mountain life was unstable, because virtually unsustainable.

[123] It hinges also on one's frame of reference. If compared with modern American agriculture, with its dependency on petroleum and pesticides, a system that lasted for two or more centuries before losing its resource base would appear remarkably stable.

It depended on vulnerable soils, on fragile forests, and on the numerical stability of dense population. Over the long haul, these elements were not easily compatible. Fluctuation and growth in population imperiled soils and forests and ecologically undermined the mountain way of life. Life in the mountains also depended on seasonal migrations of men and animals, on lowland labor markets and political conditions. These proved subject to profound changes after 1880 or so, which economically undermined the mountain way of life. The population history of the mountains is next.

5

POPULATION, SETTLEMENT, AND LANDSCAPES

Sunk are the bowers, in shapeless ruin all,
And the long grass o'ertops the mouldering wall,
And trembling, shrinking from the spoiler's hand,
Far, far away thy children leave the land.
Ill fares the land, to hastening ill a prey,
Where wealth accumulates, and men decay:
Princes and lords may flourish, or may fade;
A breath can make them, as a breath has made;
But a bold peasantry, their country's pride,
When once destroyed, can never be supplied.
Oliver Goldsmith, "The Deserted Village" (1770)

The government is very keen on amassing statistics. They collect them, add them, raise them to the nth power, take the cube root and prepare wonderful diagrams. But, you must never forget that every one of these figures comes in the first instance from the village watchman, who just puts down what he damn pleases.
Sir Josiah Stamp[1]

The relationship between population history and landscape evolution is not a simple one. Each affects the other in an endless pas de deux. In the Mediterranean mountains, population growth in rural settings usually meant land clearing, vegetation change, often accelerated erosion. Population decline generally meant landscape neglect, often accelerated erosion, but eventual revegetation by aggressive, opportunistic species – except where erosion had gone too far. Population stability, a rare condition in times past, has generally meant stability in vegetation and more modest rates of erosion, although all this depended on the systems of husbandry involved, the slopes, soil characteristics, climate, and more. Hard and fast rules do not exist here. The best generalization is that both population growth and population loss put strains on the physical environment and raise the risk of rapid soil erosion,

[1] Cited in Janet Abu-Lughod, *Before European Hegemony* (New York: Oxford University Press, 1989), 27–8. Stamp was a tax collector.

whereas greater population stability minimizes that risk. Further, the more rapid the growth or loss of population, the more pronounced the strains and risk. Population growth has the additional complication that it can lead to fuel shortage as well as soil shortage, a problem normally confronted first by women who have to walk further and further as years go by and the forest perimeter retreats from a growing population. Whatever the precise relationship in a given setting between population change and landscape change, everywhere population history is a basic determinant of landscape history. Sometimes little else matters. Population history, of course, has its determinants.

In the Mediterranean mountains the principal long-term influence upon population levels has been the prevailing system of human ecology: people's way of life. Agriculture permits human beings access to the sun's energy through the mediation of plants. Pastoralism adds a second mediator, animals, and is less efficient by approximately an order of magnitude. Thus pastoral societies operate at much lower population densities than do agricultural ones. The balance struck between agriculture and pastoralism decided how many people could live in mountain environments, which were, in terms of resource potential, very similar.

The Taurus is the best example of a mountain chain dominated by pastoralism. It was a *shepherd massif*, and stayed that way until the twentieth century. The Sierra Nevada was a *peasant massif*, where stock raising, although perfectly feasible, never counted for much. The Pindus was a shepherd massif until the sixteenth century, when it began a partial transition to agriculture. The Lucanian Apennines and the Rif have always been peasant mountains, although to different degrees.

What determined the ways of life practiced in the Mediterranean mountains was not primarily geography and ecology. The Taurus could have been terraced, irrigated, and farmed. The Sierra Nevada could have been used for grazing. Had Iberia rather than Anatolia been exposed to the influence of Asian nomads, or had Arabs settled the Taurus rather than the Sierra Nevada, things no doubt would have turned out differently. The Taurus might look like the Lebanon, a peasant massif with terraces and no remaining forests. Population history in the Mediterranean mountains, in its main lines, boils down to this: peasant massifs carried several times the population of shepherd massifs. But of course the balance between agriculture and pastoralism could shift.

To understand the influence of population on landscapes, and vice versa, one needs details of both. This requires many, many numbers. All of those that follow are wrong. All the numbers will seem a bloodless abstraction of the quietly desperate lives of mountain villagers. The numbers' sole merit lies in that they provide the greatest precision possible under the circumstances. With a little imagination, perhaps, the reader can agree with the statistician who, when challenged about the dryness of his data, said that social statistics are frozen tears.

The Quality of the Data

(This section treats the many weaknesses and limitations of the statistical information available. Those unconcerned about the trustworthiness of the data from which conclusions are drawn can skip it without missing any of the argument.)

With respect to population, national or regional data generally will not do, except as the roughest approximations. Borders have changed too frequently; nations and regions include lowland as well as upland. Hence village-by-village data are the most revealing, because with few exceptions villages have remained as units for census takers, and they are small enough not to span both mountain and plain.

Specific difficulties arise with each set of data. In the Taurus, for instance, before the censuses of the Turkish republic (the first of which came in 1927 and in its published version lacks local data) village-by-village data are unavailable. Ottoman census or cadastral data exist for the nineteenth century, and for the fifteenth and sixteenth.[2] These go beyond the province level (*vilayet* or *sancak* or *liva*) to civil and military district units variously known as *kaza, nahiye, zi'amet,* or *divan.* Terminology was unstable, especially in the fifteenth century, but changed also in 1831 and 1864.[3] Hence it is not easy to know just what area was meant by the place name Eğridir on different occasions. Other difficulties compound matters. First is the matter of trustworthiness. Although some experts have lately emphasized that Ottoman statistics are the best sources for Anatolian population history, they may not be very good. The author of the most detailed work on the western Taurus, Xavier De Planhol, felt that population figures prior to the 1950 census were of little use.[4]

Second, sometimes the Ottoman population data come in the form of counts of the number of households or the number of males in an area. To get from this to population, some sort of multiplier is

[2] Published in Karpat 1985 and Cook 1972.

[3] Cook 1972, 56; Karpat 1985, 7–8. Pitcher (1972, 124–9) explains terminology for the sixteenth century.

[4] McCarthy 1983, on the utility of Ottoman population statistics; Planhol 1958, on the dubious utility of counts prior to 1950.

required, so a further element of inexactitude creeps in. This problem affects Ottoman data on the Pindus as well as the Taurus, and indeed besets early data in Lucania and the Alpujarra as well. When population counts include only males, I have multiplied by two, but at times in mountain villages patterns of male labor migration would have made this procedure inaccurate. Pindus villages of the nineteenth century, for instance, were often 60 percent female. There is, however, no more reliable procedure. Where a multiplier for households is called for I have followed the experts, whose judgment (in the Taurus) ranges from about 4.5 to 5 persons per household. The distortion involved in this multiplier is likely to be modest, unless family size changed significantly over time. This matter is unclear in the Taurus.

The Pindus data suffer from some of the same limitations but are on the whole better. Village-by-village data exist for frequent intervals from 1805 forward. Early in the nineteenth century British and French agents, William Leake and F. C. H. L. Pouqueville respectively, traveled all over Epirus and Albania collecting information and doing their best to assure that local politics favored their employers during the Napoleonic wars. Their books, and Pouqueville's dispatches, some of which are in the archives of the Ministère des Affaires Etrangères, include frequent estimates of village populations. Countless other western Europeans visited the Pindus in the nineteenth century and some recorded population data. Greek savants, of whom Epirus produced more than its share in the nineteenth century, produced notable books including population counts for scores of villages.[5] Some of these travelers and savants based their figures on official Ottoman statistics, whereas others simply guessed. The published Ottoman data do not descend to the village level. But the Greek censuses, which cover the Epirot Pindus from 1913 forward, list every village.

Naturally there are difficulties. As with several of the counts for the Taurus, many of the early ones for the Pindus include only males. More troublesome is the confusion arising from seasonal migration. Mountain people have not always been home when the census taker passed through. Vlach villages like Samarina, to take the most extreme case, might show a population of ten if the census taker arrived in March, or five thousand if in August. Since 1940 the Greek censuses report two numbers for each community: the actual population and the legal population. The legal population in mountain villages is almost always higher than the actual population, because it includes natives who have

[5] Aravantinos 1856; Lambridis 1870; and Lambridis 1887–90.

left to work in the cities or plains, and sometimes even their children. The actual population is a more useful figure – except for the villages of the seminomads where the season of the census makes all the difference. Most Greek censuses in the twentieth century have been done in the spring, but there is a world of difference between March and May. The figures I report are the actual population, unless otherwise specified.

Population data for the southern Apennines and southern Italy in general are very good. Regional estimates exist from the fourteenth century forward and comprehensive village-by-village data exist at frequent intervals after 1793. Fairly regular censuses date from 1861. Most of these data have been collected and published.[6] For Basilicata, manuscripts from 1736 and 1853 provide additional data points for most villages.[7] Some of these data come in the form of households and a multiplier is needed to extrapolate to population. As everywhere, I have followed the experts in choosing this coefficient.

In the Sierra Nevada and Alpujarra population history is also fairly clear. A survey of Alpujarra villages from the 1570s is preserved in the archives of the Real Chancillería in Granada.[8] A few estimates for the Alpujarra as a whole exist for the later sixteenth century and one for the seventeenth, but village-by-village data resume only in 1730.[9] Nineteenth-century Spanish encyclopedists provide data for 1826 and 1845, and regular censuses begin in 1857. For the sixteenth and seventeenth centuries, data generally provide only the numbers of *vecinos*, heads of household, and a multiplier once again is required. But otherwise the Spanish data are subject only to the ordinary limitations inherent in the inaccuracy of census takers and other population counters.

In the Rif, population history is obscure. A special problem exists in this case, because Rifians, on the rare occasions when their number was counted or estimated, were grouped by tribe, not by area. But worse still is the absence of counts. Of the Maghreb, Ibn Khaldun wrote that

6 Filangieri 1980; Pedio 1987, 1:245–354, is the most recent and complete compendium.

7 Biblioteca Provinciale di Potenza, Sezione Lucana, X. 29, "Quadro della popolazione . . . 1853"; B.N. Napoli, Ms. XIV–II–39, "Relazione Gaudioso, 1736."

8 Called the Libros de apeos. These are in effect cadastral surveys taken shortly after the expulsion of the Muslims and at the beginning of Christian colonization of the Alpujarra.

9 Village-by-village counts for 1730, 1755, and 1760 exist in the BNM and BRAH (see bibliography for details). A superb document, a 1747 cadaster, is preserved in, of all places, the Foreign Ministry archive in Madrid. AMAE-M, ms. 67, "Estado y consistencia de la Ciudad de Granada y todos los pueblos de la Jurisdicción de su Corrigimiento . . . communicada por Sr. D. Joseph de Carbajal y Lancaster . . . 1747." This is a statistical description of every village in Granada. It includes livestock, harvests, and much more.

God alone could count its inhabitants.[10] Apparently no mortals tried. The medieval Arab geographers did not include population numbers; Leo Africanus (sixteenth century) has only scattered figures. Servants of the Moroccan crown did not record demographic data for the Rif until after the colonial era. Foreign travelers have offered estimates for Morocco as a whole, which are wildly inconsistent, and for cities near the Rif, Tetouan or Melilla for example. But for the mountains no record remains (so far as I know) of population prior to a count made by Abdel Krim's Republic of the Rif in the 1920s.[11] The first detailed data on the Rif date from the establishment of Spanish control in 1928: the Monografías de Kabilas, compiled by the Spanish army and housed in the Servicio Histórico Militar in Madrid. Official estimates were made in 1930, 1935, 1940, and 1945; the first real census came in 1950–1. The details of Rifian population history may never be known; but the trend in the twentieth century is very clear from fragmentary and circumstantial evidence.

The Taurus

Although the available population figures for the Taurus do not permit one to pursue the history of individual villages back before 1935, some figures exist for districts. This is a decidedly second-best basis on which to proceed, because the boundaries of districts often change over the centuries. Nonetheless a rough approximation ought to be better than none.

Eğridir district (*kaza*) lies on the western flank of the Dedegöl Dağ, includes the high plain of Yılanlı, and the steppe lands around Lake Eğridir. It is not all mountainous, although most of it lies above 1,000 meters. In the fifteenth and sixteenth centuries it included eighteen to twenty-one villages. Since the 1935 census it has consisted of between forty-one and forty-five villages.[12] Its borders may or may not have expanded: certainly more villages have been settled. The names of surrounding districts seem to have remained much the same since the

10 Noin 1970, 1:23.

11 Bibliothèque General de Rabat, Section Arabe, manuscript by Mohammed Azerkane, Ad Dill al warīf fī Muhārabat ar-Rif, edited by Ahmed Skirej. I have tried unsuccessfully to find this manuscript. It is cited in Ayache 1981, 95. Noin (1970, 1:21) thinks that perhaps the Moroccan archives will one day yield early population statistics, but none existed when he wrote in the late 1960s, and I believe none have been found since.

12 Cuinet (1890–5, 1:846) credits Eğridir with sixty-three villages. It is unlikely that a third of Eğridir's villages was abandoned between 1889 (Cuinet's data) and 1935. Research into unpublished Ottoman censuses and tax records would probably resolve this matter.

Table 5.1. *Population of Eğridir Kaza (District), 1475–1985*

Year	Total population	Rural population	Town population	Source
1475–81	2,655			a
1523	2,475			a
1566–9	4,797			a
1831	8,518			b
1840			2,500–3,000	c
1877–8	20,766			b
1881	21,022			b
1914	30,002			b
1927	28,490	24,078	4,412	d
1935	22,435	16,662	5,733	e
1945	25,661	19,922	5,739	f
1955	30,032	23,486	6,546	g
1965	37,308	28,396	8,912	h
1975	40,921	31,122	9,799	i
1985	47,679	33,414	14,265	j

[a] Cook 1972.
[b] Karpat 1985.
[c] Hamilton 1842.
[d] Turkey, İstatistik Umum Müdürlüğü 1929.
[e] Turkey, Başbakanlık Devlet İstatistik Enstitüsü 1937.
[f] Turkey, Başbakanlık İstatistik Genel Müdürlüğü 1948.
[g] Ibid. 1957.
[h] Turkey, Devlet İstatistik Enstitüsü 1968.
[i] Ibid. 1977.
[j] Ibid. 1986.

sixteenth century, encouraging the view that the districts' boundaries have stayed constant. But any comparisons between the sixteenth century and the twentieth are risky. Table 5.1 summarizes the population history of Eğridir district.

The record of modern population growth in Eğridir is clear enough. Almost exactly the same trajectory since 1935 describes the population history of Sütçüler, the neighboring district to the south (Table 5.2). And a random collection of mountain villages from Eğridir and Sütçüler shows very much the same pattern of steady growth since 1935 (Table 5.3). In fact, in all these cases, the late 1930s showed little growth, occasionally some loss, but after 1940 villages everywhere in the western Taurus began to grow and did not stop until the late 1970s; and in some cases they have not stopped yet. Population density has doubled since

Table 5.2. *Population of Sütçüler District, 1831–1985*

Date	Population
1831	4,076
1935	11,103
1945	13,149
1955	17,037
1965	18,773
1975	21,701
1985	20,804

Note: Sütçüler consisted of twenty-six villages in the censuses from 1935 to 1955. From 1965 on it consisted of twenty-eight villages.
Source: See Table 5.1; for 1831, also see Karal 1943, 205–6.

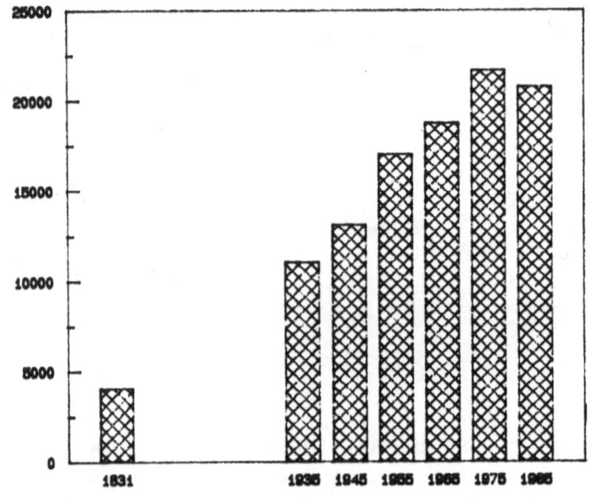

Graph for Table 5.2

1935, but improvements in agricultural productivity, and increased opportunity for wage labor in towns, or in the fields of Konya, Antalya, or even Aegean Turkey,[13] have prevented serious problems in these villages. They have yet to run dangerously short of wood or soil. Their good fortune derives in no small part from the lateness of settlement (or resettlement).

[13] Villagers from Koçular, Yakaavşar, Kasımlar, and others near Dedegöl and Anamas Dağları in the 1950s traveled to the Aegean to work as *kökçü* (root diggers) or *hendekçi* (ditch diggers). Planhol 1958, 173.

Table 5.3. *Population of selected western Taurus villages, 1935–1985*

Village	1935	1945	1955	1965	1975	1985
Zerk	455	488	704	660	646	624
Caltepe	821	527	773	729	511	649
Ayvalıpınar	1,355	964	1,464	1,675	2,382	1,901
Belence	435	632	792	895	1,144	1,062
Aksu = Yenice	408	468	1,087	1,782	1,782 (sic)	2,341
Koçular	745	857	1,037	1,133	807	644
Pazarköy	448	539	671	813	952	977
Yakaavşar	879	825	867	1,069	1,314	1,213
Kesme	1,076	1,085	1,355	1,712	2,212	2,131
Kasımlar	653	746	911	1,221	1,454	1,642
Sütçüler (town)	2,003	2,416	3,320	2,401	2,721	2,617
Total	9,278	9,547	12,981	14,090	15,925	15,801

Note: In the 1935 census Belence appears as Kartosasaği. On maps it often appears as Bolence. Kesme in the 1945 census appears as Kesince.
Source: See Table 5.1.

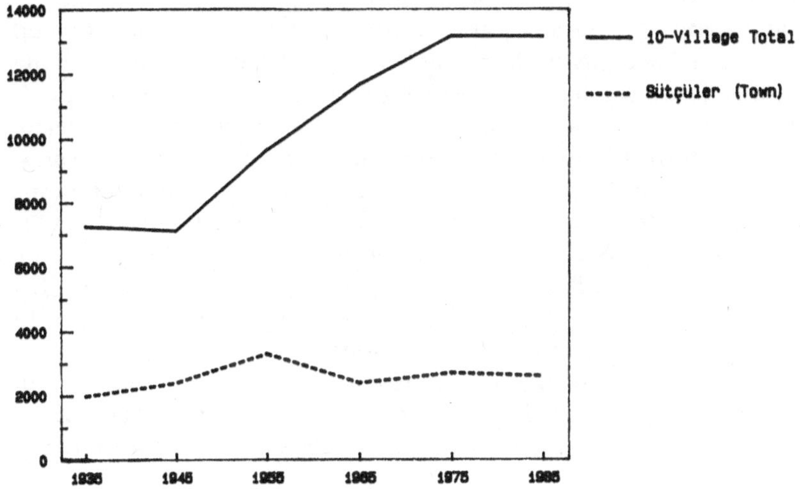

Graph for Table 5.3

A Turkish politician and wit once said that the Turkish republic had inherited southern Anatolia not from the Ottomans but from the Seljuks.[14] Exaggeration no doubt, but the statement has some truth to it. Southern Anatolia in general and the Antalya–western Taurus region in particular was depopulated and backward in Ottoman times. The principal reasons for this decline were a rise in nomad power, the shifting of the caravan trade and a decline in commerce in general, and the policy of the Ottoman state concerned to weaken the southern Anatolian power base of their newly conquered rivals, the Karamanids. Mehmet the Conqueror forcibly depopulated southern Anatolia in his attempts to settle European Turkey and the Balkans. So the ports of Antalya and Alanya became backwaters; the caravan crossroads of Eğridir dwindled. And rural population melted away, as peasant life came under frequent pressure from nomads.

Population figures from Ottoman registers show considerable growth in the province of Hamid between 1475 and 1569. In Eğridir, population doubled between 1523 and 1569, a growth rate higher than that of 1881–1985 (see Table 5.1). Toponymic evidence (place-names) indicates that the mountain villages of the western Taurus, unlike the adjacent Burdur district, were settled only after the arrival of the Ottomans.[15] In the sixteenth century, imperial authority reached its height, and security in the countryside encouraged peasant settlement. A subsistence economy flourished. But over the next two or three centuries Ottoman power could not control the nomads in southern and southwestern Anatolia.[16] They regularly sacked villages and caravans. State efforts at settling or deporting nomads came to naught, despite a concerted attempt in 1710 to remove all recalcitrant nomads to Cyprus. Peasant life survived only in a few pockets in these centuries, and in the mountains settled folk were certainly outnumbered by nomads. Under these conditions, the population of the western Taurus remained very thin, probably less than a tenth of today's.[17]

Local data are scarce for the seventeenth to nineteenth centuries, hence the gap in Table 5.1. But provincial (*liva*) totals of taxed households in Hamideli show remarkable stability between 1650 and

[14] Planhol 1958, 132. [15] Planhol 1958, 108. [16] Planhol 1958, 111–23.

[17] Cuinet (1890–5, 1:854) reports fifteen thousand nomads for the Antalya *sancak* and none for Isparta. This, admittedly, refers to a time when the nomads had begun to settle down, or to leave western Anatolia for the east, and a century earlier nomad numbers would certainly have been higher. And Cuinet, though he had access to Ottoman data, may have been wrong. Nomads are hard to count.

1834.[18] Hamideli included the villages of the western Taurus, and perhaps this evidence indicates the reality there. Local fluctuations can often disappear in aggregated data, so it is impossible to know for sure, but the apparent trend in the western Taurus is one of very thin and very stable population in these centuries. The surge of population growth in the sixteenth century seems a temporary aberration permitted by unusual political conditions. If true, this pattern is sharply different from those in the Pindus, the Lucanian Apennines, the Alpujarra, and the Rif.

Nomads lived in the high country for a few months every summer, but upland permanent population fell off from sixteenth-century levels and did not recover until, at the earliest, the late nineteenth century. Even the plains carried very little in the way of permanent settlement: Antalya Plain, especially its western half, had no villages as late as the mid-nineteenth century. Pine forest and sheep pasture prevailed. This pastoralism of the plains meant that no seasonal work existed that could help support mountain people – except perhaps military work.

In these centuries, most mountain landscapes in the western Taurus felt little more than the pressure of summer grazing. The lower altitudes on the south flank, as Chapter 6 will show, suffered the attentions of a logging industry. But elsewhere, without any winter population to speak of, the fuelwood demands on the forest came to almost nothing. No one carved permanent fields out of the forest. Summer nomads raised a few small crops, and they presumably burned forest when they needed more pasture. But their animals could subsist in forest, especially the open forest characteristic of the western Taurus. Thus when, in the late nineteenth century, after three hundred years of comparatively benign neglect, mountain resettlement took place, the forests and soils

[18] McGowan 1981, 118–19. The *liva* of Hamideli consisted of twenty-one *kaza*, presumably the same twenty-one, from 1650 to 1834. The number of taxed houses (*avanzhane*) is as follows:

1650	1,749	1718	1,733
1662	1,591	1755	1,753
1677	1,758	1786	1,753
1688	1,908	1834	1,746
1698	1,773		

It may well be that officials did not bother to count again at times, such as in 1786, but if these data are anywhere close to accurate, this implies a remarkable stability in population. Plague seems to have been inconspicuous in the history of the Taurus. Only one epidemic is recorded in Hamideli (1838–40) and one at Mut (1828). Panzac 1985, 122–3, 198–9. Indeed southern Anatolia in general suffered little from plague by the standards of the Ottoman Empire, 1700–1850.

of the western Taurus remained abundant. Since the fourth and fifth centuries A.D., the mountains had never carried much population, probably never more than they did in the sixteenth century. Population density at its sixteenth-century recorded maximum (1566–9) probably came to about one-fourth of its current magnitude. The Taurus remained a shepherd massif through the nineteenth century.[19]

In the nineteenth century, economy and settlement in the entire region began slowly to change (Table 5.4). First, trade networks began to reemerge, sparked by the demand for wheat and meat from British garrisons installed in the eastern Mediterranean during the Napoleonic wars; by the Egyptian development campaign carried out by the Albanian Pasha Mohammed Ali from 1807 to 1841; and by the monumental construction project of the Suez Canal from 1859 to 1869. Networks of Greek merchants in Egypt, Anatolia, and the islands facilitated this trade. Lowland agriculture expanded, depriving nomads of winter pasture. Nomad power waned in western Anatolia, and nomads began to settle, under extreme pressure from the Ottoman state. Of those who did not, many retreated eastward. The decline of nomads permitted the further expansion of cultivation, in both lowland and highland zones, and a consequent expansion of population. It is also true that an expansion of population helped bring about the decline of nomads: cause and effect were intertwined. Rural security improved to the point where brigandage was almost confined to mountain defiles. Traffic, settlement, and cultivation all expanded. This quickening centered on Isparta and Burdur, just west of the Taurus. In Eğridir district these changes took a little longer, perhaps because the high mountains of Anamas and Dedegöl eminently suited seminomadism. But generally throughout the nineteenth century the trend was clear. Here is Planhol, from whom I have drawn most of this account:

> On voit ainsi, tout au long du siècle, se développer considérablement l'occupation du sol dans ces montagnes jusque-là presque désertés. Nomades et semi-nomades s'y fixent, en général dans les quartiers d'hiver, par petits groupes dispersés ou par maisons isolées qui découpent des clairiers dans la grande sylve taurique.[20]

[19] Observers noted the large numbers of black felt tents, belonging to nomads, near Eğridir in the 1830s, and Yılanlı in the 1880s. Arundell 1834, 1:330; Sterrett 1888, 280; Ramsay 1906, 792: "Every year I notice new villages, where formerly were only nomad camps."

[20] Planhol 1958, 128. Roughly translated: One sees thus, throughout the century, settlement developing considerably in these mountains, until then nearly deserted. Nomads and seminomads settled down, generally in their winter quarters, in small dispersed groups or in isolated houses, carving out clearings in the great Taurus forests.

Table 5.4. *Antalya city population, 1530–1980*

Date	Population	Source
c.1530	3,450	a
1811	8,000	b
c.1870	10,000–12,000	c
c.1889	25,000	d
1945	25,037	e
1950	27,515	e
1960	50,908	e
1970	95,616	e
1980	173,501	e
1985	261,114	e

[a] Faroqhi 1990, 126.
[b] Beaufort 1818, 130.
[c] E. Davis 1874, 210.
[d] Cuinet 1890–5, 1:854.
[e] Turkey, Devlet İstatistik Enstitüsü 1988, 38.

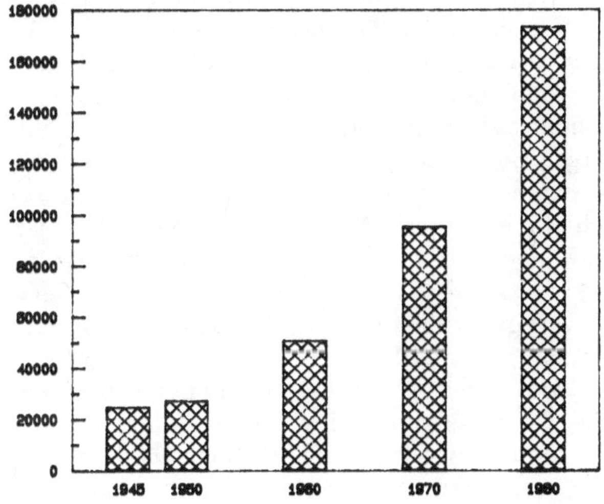

Graph for Table 5.4 (1945–80)

The nomadic tribes normally settled in their winter quarters, as Planhol says, although this often meant settling not in the Taurus forests but in the plains of Antalya or Adana. According to one estimate, about a quarter of the nomads settled in the Çukurova died from malaria in the middle of the nineteenth century.[21]

As nomad numbers and power ebbed, the Taurus underwent a change as fundamental as that which accompanied the arrival of the Turks centuries before. It slowly became a peasant massif. To some extent this represents a triumph of long-standing Ottoman policy. But it took some sweeping changes to make the Ottomans successful in breaking nomad power. For one thing, as military hardware grew more efficient and more expensive, armies and even gendarmes became better armed than their adversaries, nomads and brigands alike. And as Chapter 6 will show, as lowlands slowly filled up, transhumant pastoralists were squeezed out. Their flocks diminished, and their wealth and power declined accordingly.

Settlement in the Taurus continued into the twentieth century. With the exception of a population decline between 1914 and 1935 or 1940, demographic growth in Eğridir district has proceeded at roughly the same pace from 1831 to the present. The exceptional years were those of war, especially 1912 to 1923, during which Anatolia as a whole lost a larger share of its population than any other country involved in World War I and its many aftermaths. Anatolia's lost generation came to 30 percent of its 1914 population (20 percent died and 10 percent emigrated). The loss centered in those areas where Greeks and Turks fought in 1922–3, chiefly in Aegean Anatolia, and where Armenians came to grief, chiefly in eastern Anatolia. But in Antalya, far from major theaters of war, in 1927 more than 30 percent of women over age twenty were widows.[22] In Eğridir district in 1927, women accounted for 55 percent of the population and 16 percent of women were widows.[23] By 1935 the district still had only 74 percent of its 1914 population. These drastic losses affected the lands and cities near the mountains, but probably touched the mountain villages only through the loss of sons conscripted to fight (and, presumably, a decline in the birth rate), because no Greek settlement existed in the mountains. Before 1923 several hundred Greeks did live in Eğridir and Isparta and thousands in Antalya, but the brutal fighting that ravaged Izmir and Aydın, for instance, did not touch the western Taurus. So the check to population

[21] Gould 1973, 10–22. On nomad settlement: Bates 1973; Bates 1980; and Ramsay 1906, 791–2.

[22] McCarthy 1983, 118–21; Panzac 1988, 60–3.

[23] Turkey, İstatistik Umum Müdürlüğü, 1929, 1:85.

in Eğridir district, while real enough, probably reflects urban population more than rural.

Overall, twentieth-century population growth in the mountains has proceeded rather steadily, slowing only in the late 1970s and 1980s. The result is an unprecedented population density that may be unsustainable. Pressure on soils and forests is now greater than at any time since the fifth century A.D. At the moment no impending crisis is apparent. And perhaps the general trend toward urbanization in Turkey will prevent further growth of mountain communities, despite high levels of natural increase: emigration may prevent overshoot. Antalya (city) population has swollen with rural migrants since 1950 (Table 5.4). If the European Community after 1992 permits free movement of labor within its borders, mountain Turks will no doubt join the stampede to Germany. But for precisely this reason the European Community will probably not permit free movement of labor. The Germans will decline to play the same role for the Taurus that the United States (and other destinations) played for Pindus, Apennine, and other mountain folk a century ago. The capacity of the Persian Gulf states to absorb Turkish laborers has probably peaked – and the Gulf so far has attracted mostly those from southeastern Turkey. The ecological stress implied by the high rate of natural increase in villages such as Zerk and Koçular will thus more likely be felt in the villages themselves and in Turkish cities, where the *gecekondu* (shantytowns) are multiplying fast.[24]

The Taurus villages, however, may be able to withstand still higher population density. It is by no means clear that overshoot is at hand. More efficient irrigation is still possible. Some villages, like Zerk, have none. Increased agricultural yields could permit further growth in human numbers, to the point where fuel grows too scarce. And should it become feasible for peasants to use oil and gas instead of wood and charcoal, even this might not prove a strict limit. On the other hand, existing irrigation is the lifeblood of agriculture in most of these villages, and it depends on rural peace. Should that fail, as it did in the seventeenth and eighteenth centuries, and again between 1914 and 1923, then population thresholds in the Taurus will decline radically. All this, the unchoreographed pas de trois of population, landscape, and politics, remains to be seen.

[24] In Isparta province the population increase (1975–80) was 1.49% in the villages and 35.91% in cities and towns. Karpat (1976, 74–5) explains that rural migrants to Istanbul do not go for bright city lights, for education, but because village life is tough and getting tougher. (He is referring here to the mountains of Pontic Anatolia, not the Taurus.)

The Pindus

In the Pindus the dance is over. The trajectory of population history for the future is clear enough, at least as much so as that of the past two hundred years. The data on which the following remarks are based appear in Tables 5.5 through 5.8.

Pindus villages grew from insignificance into a major reservoir of Epirot population in the fifteenth, sixteenth, and seventeenth centuries. Available statistics are inconclusive. Ottoman sources credit the *liva* of Ioannina with 32,710 hearths in the early sixteenth century, or perhaps 150,000 people, less than a third the size of the Ioannina *vilayet* at the end of the nineteenth century. Even if the administrative boundaries of Ioannina did not remain constant through the centuries, mountain settlement must account for most of the population growth.[25] Lowland Epirus and western Macedonia were thinly settled.

Greek oral tradition and often Greek historiography attribute this increase in mountain settlement to Ottoman oppression in the lowlands.[26] Although there may well be something to this, the two centuries previous to Ottoman conquest (1430) had featured much civil war and rapine. And pre-Ottoman (Christian) landlords in the Despotate of Epirus had done their best to exploit the peasantry as well.[27] Ottoman rule, in its early centuries at least, may have created gentler rather than harsher conditions for Greek villagers. It is more likely that the growth of mountain population derived from the arrival of maize (see Chapter 3). Conceivably recurrent ravages of plague after 1348 encouraged people to move to isolated villages as well. The mountains provided a healthier environment than the plains and, because there were no landlords (Turkish or Greek), often more food and freedom as well. Around 1960 a UNESCO team found Epirus mountain villagers enjoyed a better diet than their brethren in the plains.[28] Fuel was far more abundant, and seasonal labor or entrepreneurship provided additional income or at least temporarily reduced the number of mouths to feed.

Architectural evidence suggests that Pindus villages grew large in the seventeenth and eighteenth centuries, which coincides better with the arrival of maize than of the Turks. The cavernous churches of Zagori villages generally date from the very end of the seventeenth century or

[25] The borders of the Ioannina *liva* fluctuated considerably from 1659 to 1834, including anywhere from four to eight *kazas*. McGowan 1981, 118.

[26] Vacalopoulos 1963, Vacalopoulos 1976. For a recent statement of this view, Théophilou 1983, 72–6. Panayotopoulos (1988) provides a more balanced assessment.

[27] Nicol 1984; Lawless 1977, 508. [28] Mendras 1961.

the eighteenth. The church in Negadhes dates from 1792, although it might be rebuilt on the site of an earlier one. It is about 50 meters long and 30 meters wide, and could easily hold over eight hundred people (Greeks normally stand in church). Unless the village saw fit to build a church much larger than they needed (not inconceivable) the population of Negadhes must have approached one thousand in the late eighteenth century. By 1850–80 Negadhes counted between six and seven hundred people. I think it likely the population had been as high or higher a century before.

In settling Pindus villages like Negadhes (or those of the Taurus a century or more later), human beings in effect moved into a new niche and prospered like a newly introduced weed species. They enjoyed all the advantages of opening a new frontier: wood aplenty, virgin soils, and no predators except microorganisms (which most likely affected them minimally until they established regular ties to the wider world). All that they needed was herculean labor: cutting timber, building terraces, eradicating other species that, like themselves, flourished in a newly disturbed environment. In all likelihood this was a time of early marriage, large families, and rapid population growth.[29] Numbers had not yet begun to approach carrying capacity. This is the first stage in the natural history of a newly settled mountain village (or one newly transformed from shepherds' summer quarters to genuinely settled village). In the Pindus this began in most cases in the sixteenth or seventeenth centuries, and lasted perhaps two or three or four generations. In some cases the terms of life began to change fundamentally as early as the end of the eighteenth century.

Painstaking research on village abandonment in Greek lands shows a rash of abandonments between 1780 and 1850 in Epirus generally and the Pindus in particular.[30] No doubt this had something to do with the political violence surrounding the demise of Ali Pasha and the Greek revolt of 1821. The Turks burned Sirako and Kallarites after destroying Ali Pasha's authority, and meted out similar punishment elsewhere.[31] And no doubt epidemic disease led to abandonment on occasion:

[29] Blancard (1846, 64) says that in Zagori men often married at age fifteen and women at age twelve. He spent only three days in Zagori before writing this and could well have been mistaken. Child marriage would have been radically inconsistent with later Greek peasant customs.

[30] Antoniades-Bibicou 1965, especially the maps.

[31] Stuart 1868, 320. Lambridis (1887–90, 5:57–70) has details on the sufferings of Sirako, Metsovo, etc.

Table 5.5. *Population of selected Pindus villages, 1805–1981*

	1805	1812	1820	1853	1854	1882	1890	1902	1913	1920	1928	1940	1951	1961	1971	1981
Dovra/Asprangeli	750			435	1,300	1,024			628	556	562	536	341	387	308	237
Vitsa		1,150		695	940	646			420	317	372	438	181	220	163	140
Vovousa				130	210	260	2,000		505	366	387	680	106	217	135	190
Vrysochori/Lesnitsa				410	1,270	1,600			488	314	330	339	205	215	83	152
Greveniti				515	1,500	1,524			786	714	676	803	439	429	343	295
Kepesovo				210	400	408			202	152	124	136	92	53	62	89
Kipi/Baghia		750		390	840	800			369	292	227	277	220	271	229	180
Koukouli		500		290	590	408			153	88	90	224	52	46	53	83
Makrino				215	450	500			133	97	99	171	87	88	67	114
Monodendri				270	700	506		500	110	186	223	247	104	133	126	136
Negadhes				280	690	608			334	169	205	187	129	132	61	79
Nea Amarousi/Doliani				210	450	408			192	157	157	160	85	89	56	75
Skamneli				465	1,000	1,224			418	264	245	396	144	140	178	202
Tristeno/Dresteniko				225	700	700			420	390	408	388		240	146	145
Tsepelovo	1,000			395	1,000	700		750	614	254	254	521	125	343	450	329
Flambourari				220	680	750			294	205	246	362	127	176	160	166
Frangadhes				475	900	700			367	273	291	269	192	203	139	158
Laista	1,000			765	1,800	2,210			892[a]	539	436	607	127	199	120	147
Sirako	2,790[b]	3,500[c]	2,500[d]	740		3,000	3,500[e]	2,700	2,620[f]	215	244[g]	1,388	47	18	11	77
Kallarites		2,500	2,650[h]	640		1,316[i]	842[j]		—[k]	709	424[l]	1,049	277	230	162	144
Pramanda						2,000[m]	1,642[n]			2,597[o]	2,344	3,021	1,700	1,760	1,687	1,585
Samarina		4,000[p]				5,400[q]	5,000		4,198	76	603	1,175	4	5	10	605

Sources:

1805: Leake 1835, 4:167.

1812: AMAE-P, Mémoires et Documents, Turquie, vol. 35, fasc. 313, Pouqueville to Bassano, 15 November 1812; and Pouqueville 1826–7, 1:153 and 2:182.

1820: Stuart 1868, 320.

1853: Aravantinos 1856, 2:328–41. Aravantinos's data are notably inconsistent with surrounding data. He provides numbers of families, which I have multiplied by five. Hammond (1967) considers Aravantinos's figures unreliable.

1854: Lambridis 1870.

1882: Blancard 1882, 131–4. Blancard's data, published in 1882, may be copied from the Ottoman census of 1866.

1890: Krystallis n.d., 434.

1902: Pagounis 1902, 21–2.

1913–81 Greek census data in Chouliarakis et al. 1972 and Chouliarakis 1988.

[a] Laista 1913 figure: Wace and Thompson ([1913] 1972, 199) give 400 houses for 1911.

[b] Sirako 1805 figure is for 1809. Leake 1835, 4:207.

[c] Sirako 1812 figure is for 1810. Krystallis n.d., 434.

[d] Sirako 1820 figure: Papakostas (1967, 132) gives 650 families for Sirako in 1821.

[e] Sirako 1890 figure: Philippson (1897, 243) gives 3,600 for 1893.

[f] Sirako 1913 figure: Papakostas (1967, 515) gives 700 families for 1912–13.

[g] Sirako 1928 figure: Hammond (1967, 178) gives 200 houses; he visited in the 1930s.

[h] Kallarites 1820 figure: Papakostas (1967, 132) gives 850 families for Kallarites. On p. 515 he states Kallarites's population as 4,476 in Ali Pasha's day.

[i] Kallarites 1882 figure: Greece, Ipourgeion Esoterikon 1884, has 1,459 for 1881.

[j] Kallarites 1890 figure: Philippson (1897, 243) gives 842 for 1893.

[k] Kallarites 1913 figure: Papakostas (1967, 515) gives 300 families for 1912–13.

[l] Kallarites 1928 figure: Hammond (1967, 178) gives 200 houses.

[m] Pramanda 1882 figure: Greece, Ipourgeion Esoterikon 1884 has 2,328 for 1881.

[n] Pramanda 1890 figure: Philippson (1897, 325) gives 1,642 for 1893.

[o] Pramanda 1920 figure: BSA, Clarke, Notebook c, 107, gives 5,000 for 1922.

[p] Samarina in 1770 had a population of 15,000 according to Wace and Thompson [1913] 1972, 154.

[q] Samarina 1882 figure: Wace and Thompson ([1913] 1972, 159) give 1,200 houses for 1877.

Table 5.6. *Population of Pindus towns, 1735–1981*

Date	Kónitsa	Metsovo	Zitsa	Source
1735		1,895		a
1754		2,185		b
1805		3,500	550	c, d
1812	8,000			e
1814	5,000	5,500	750	f, g
1816		7,000		h
1838		5,000		i
1946			500	j
1853	2,895	4,270	650	k
1878	8,500	7,000–8,000		l, m
1882			1,506	n
1887			975	j
1902	3,000	4,000	1,500	o
1913	2,620	3,429	1,148	p
1920	2,749	1,944	1,093	p
1928	1,959	2,156	1,220	p
1940	2,313	2,907	1,289	p
1951	3,716	2,798	1,332	p
1961	3,485	2,976	965	p
1971	3,226	2,823	951	p
1981	2,859	2,705	966	p

[a] Lambridis 1887–90 has 379 houses.
[b] Skafidas (1952, 660) has 437 houses for 1754–74.
[c] Stuart 1868, 321 (citing Leake 1835 for 1805, Pouqueville 1827 for 1814). Holland (1819, 306) has 1,500 houses for Metsovo in 1814.
[d] Leake 1835, 4:192.
[e] AMAE-P, Mémoires et Documents, Turquie, 35:313, Pouqueville to Bassano, 15 November 1812.
[f] T. Hughes 1820, 2:279. Pouqueville 1826–7, 1:170 has 600 houses for Kónitsa, and (2:230) 1,100 families for Metsovo.
[g] T. Hughes 1820, 2:231.
[h] Guillaume de Vaudoncourt 1816, 291.
[i] Boué 1840, 2:57.
[j] Lambridis 1887–90, 3:40.
[k] Aravantinos 1856, 2:328–41.
[l] Isambert 1878b, 260.
[m] Isambert 1878a, 172.
[n] Blancard 1882, 131–4.
[o] Pagounis 1903, 22.
[p] Greek censuses compiled in Chouliarakis 1988, 309–24, 419, 456–65.

Graph for Table 5.6 (1805–1981)

bubonic plague visited Greece as late as 1837.[32] But Antoniades-Bibicou maintains that war and disease had little to do with village abandonments, arguing for landlord rapacity instead. It is impossible to know for sure, but a case can be made (especially where landlords scarcely existed) that many villages were abandoned for the same reason that farms and settlements are normally abandoned in frontier agriculture: because yields had begun to decline and timber to grow scarce. Prospects came to look better a ridge or two away, where tall timber and virgin soil beckoned the young and energetic once again. If enough people chose to leave, then those wanting to remain behind probably found it impossible to carry on, as they were too few to attend to all the chores of labor-intensive mountain life. So villages fell to ruin. Many villages approached or even exceeded their carrying capacity at the turn of the eighteenth century, while others still had arable land, pasture, and forest to spare. Where human numbers began to press against available resources, villagers began to roam further afield to make ends meet (Chapter 4). Zagori and all the Pindus began to send out its legions of

[32] Stéphanos 1884, 511. Epirus and Albania suffered from plague almost as often as anywhere in the Ottoman Empire between 1700 and 1850. In more than a third of these years plague was reported somewhere in the vilayet of Ioannina. Most outbreaks occurred in Arta and the city of Ioannina. In 1816 half the population of Arta died from plague. The last visitation in this part of the world came in 1835, not 1837, according to Panzac 1985, 506; see also pp. 61–7, 109, 155, 164, 198–9, 448–9.

Table 5.7. *Population of Ioannina Kaza/Eparch, 1800–1981*

Date	Population	Source
c. 1800	109,000	a
1805	50,000	b
1812	70,000	c
1853	80,330	d
1866	72,334	e
1872	69,820	f
1880	90,719	g
1902	87,250	h
1905	95,000	i
1913	89,994	j
1920	85,992	j
1928	98,363	j
1940	122,651	j
1951	124,653	j
1961	155,326	j
1971	137,588	j
1981	125,250	j

[a] Lambridis 1887–90, 3:40.
[b] Leake 1835, 4:168.
[c] AMAE-P, Mémoires et Documents, Turquie, 35:313, Pouqueville to Bassano, 15 November 1812.
[d] Chasiotis 1887, 137–40.
[e] AMAE-P, Mémoires et Documents, Turquie, 120:89. PRO FO 195/751, "General Report on Epirus," 1 December 1863 has 70,000.
[f] Helle von Samo 1877, 47.
[g] Greece, Ipourgeion Stratiotikon 1880, 18.
[h] Pagounis 1902, 22.
[i] Christovasilis 1905, 29–33.
[j] Greece, Yeniki Statistiki Iperesia, 1928, 1929, 1946; Greece, Ethniki Statistiki Iperesia, 1955, 1962, 1972, 1982.

masons, muleteers, beggars, and doctors. These specialist niches removed some mouths from the villages and also brought back money and goods, helping many families survive and some to prosper.

Through these and other adaptations the Pindus villages hosted populations beyond what they could otherwise have supported, at the cost of depending on widespread peace and, to some extent, urban demand. Should war or brigandage, or economic depression, imperil the caravan trade, every village in the Pindus would find itself thrown back on its own resources, likely inadequate for its population. The prosperity of Kallarites, for instance, which peaked in the late eigh-

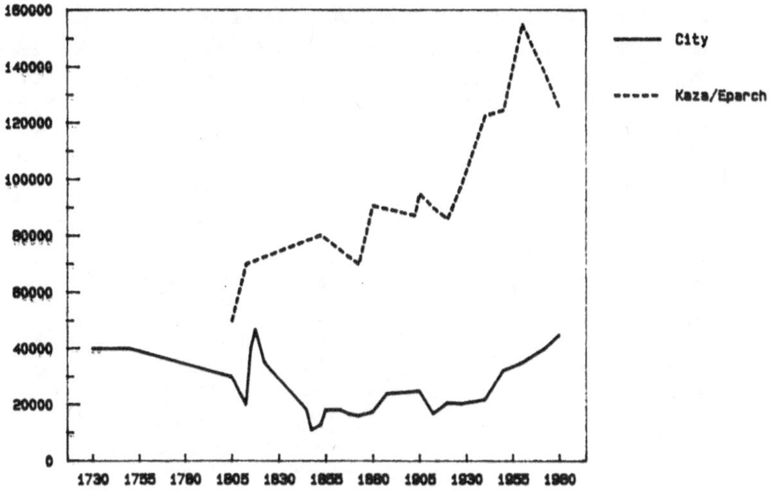

Graph for Tables 5.7 and 5.8 (1731–1981)

teenth century,[33] rested on fragile foundations. Kallarites, like many Vlach villages and some Greek ones, profited from the Napoleonic wars (Chapter 6). Seaborne traffic in the Mediterranean suffered as the French and British fleets struggled for control, but overland traffic increased. And the business of supplying meat and grain to foreign garrisons in Corfu (from 1797 on) fell to Epirots. In effect, distant war meant expansion for the mountain economy (and that of the plain as well), as long as peace prevailed locally. Kallarites had 4,467 people according to a census taken for Ali Pasha, a figure it never approached again.[34]

Local peace failed around 1820–1 in the mountains.[35] Before that the plains had suffered from the depredations of Ali Pasha, but most mountain communities made bargains with him rather than resist. The villages of Zagori did so as a unit. So for some mountain villages (in sharp contrast to lowland villages) the reign of the Lion of Ioannina proved good times: he enforced peace, stamped out brigandage (except for his own men's), built roads, and presided over regular markets in Ioannina, which grew rapidly under his control. But the death of Ali Pasha (he was shot by Ottoman soldiers in 1822) and the end of his

[33] Stuart 1868, 320.
[34] Papakostas 1967, 515. Other sources (see Table 5.5) give it 2,800–3,500 in the same era. Pouqueville (1822, 6) found only 900 there – perhaps he visited in winter.
[35] Papakostas (1967, 131–58) has details on local upheavals in the Tsoumerka district.

Table 5.8. *Estimated Ioannina city population, 1731–1981*

Date	Population	Source	Date	Population	Source
1731	40,000	a	1872	16,034	k
1751	40,000	b	1875	25,000	l
1805	30,000	c	1879	21,200	a
1812	20,000	d	1880	17,500	m
1814	40,000	e	1887	24,000	n
1816	45,000–50,000	f	1905	25,000	o
1822	35,000	a	1913	16,804	p
1842	20,000	g	1920	20,765	p
1844	18,000	a	1928	20,485	p
1846	10,000–12,000	h	1940	21,877	p
1853	13,000	i	1951	32,315	p
1856	18,265	j	1961	34,997	p
1862	18,250	a	1971	40,130	p
1866	16,690	a	1981	44,829	p

[a] AMAE-P, Mémoires et Documents, Turquie, 120:27. AMAE-P, Mémoires et Documents, Turquie, 119:382, has 50,000 for 1731.
[b] Marin 1913, 2.
[c] Leake 1835, 4:168.
[d] AMAE-P, Mémoires et Documents, Turquie 35:313 and 120:27.
[e] T. Hughes 1820, 1:455.
[f] Guillaume de Vaudoncourt 1816, 131.
[g] Best 1842, 20–1.
[h] Malherbe 1846, 1:337.
[i] M. Ubicini 1853, 1:46.
[j] Aravantinos 1856, 2:248.
[k] Helle von Samo 1877, 47, citing official Ottoman count; PRO FO 195/1002, 286–300, "Report on the State of Epirus," 26 April 1872, gives 16,100.
[l] Isambert 1878b, 203.
[m] Greece, Ipourgeion Stratiotikon. 1880, 22. Chirol (1881, 178) offers 20,000.
[n] Typaldos Forestis 1912, 5.
[o] Christovasilis 1905, 29.

pashalik burst the bubble on which mountain prosperity had depended. War and uncontrolled brigandage followed. The villages had, in many cases, overshot their sustainable populations and relied for their well-being on the maintenance of fragile political and economic conditions well beyond their control.

With the demise of Ali Pasha the population of Ioannina shrank dramatically, from perhaps forty thousand to less than fifteen thousand by 1850. Public order became less reliable, trade fell off, and villages had to produce a larger share of what they consumed. For most villages this was not impossible. Soil and wood still existed in quantity enough that drastic measures could be avoided. True, Kallarites never recovered, and Vitsa never regained the population it had in Ali Pasha's day, but for the other villages for which data exist (Kipi, Koukouli, Tsepelovo, Laista, Sirako) population in 1850 roughly equaled that of 1805 or 1812. A count for 1830 or so might reveal a decline in the 1820s in most cases followed by slow recovery, but it is impossible to know.[36]

Epidemic disease may have brutally reduced population pressure in these years. Scourges affected Filiates in 1813–14 and Arta, Parga, and Paramythia in 1815–16. Arta lost two-thirds of its population. Disease swept through all of Epirus in 1822, when eight thousand died in Ioannina alone.[37] These epidemics afflicted the plains more than the mountains, and the coasts more than anywhere else. Indeed they might have bypassed the Pindus villages altogether; it is uncertain. But because the enforced peace of Ali Pasha had promoted commercial traffic and probably labor migration too, it is quite plausible that these ravages made themselves felt in many villages. The lethality of these infections suggests that, whatever they were, they had been absent for some time so that most of the population carried no resistance. The roster of epidemics (no doubt incomplete) compiled by Stephanos lists none between 1736 and 1813. This respite, if real, would have permitted rapid population growth (which was real) and created an epidemiologically vulnerable population.

Famine, the frequent partner of epidemic, probably touched the mountains more than the plains. Although there is no clear record of it, the massive crop failures that distressed most of Europe after the record cold of 1816 probably blighted the mountains with special severity. Without summer heat, maize and garden vegetables might fail entirely, leaving Pindus villagers with only their animals and acorns to eat.

It is likely then, though not certain, that population growth throughout the Pindus received a check, perhaps beginning as early as 1812, and in places lasting until 1830. But only in Kallarites and Vitsa did this signal a permanent decline. If this was the first overshoot in the history of the Pindus, it was a small one, and the ensuing crash a mild one, felt strongly in only one or two places. It may not have been an overshoot at all: perhaps, even if population did approach carrying capacity around 1800 or 1810 in many villages, disease and violence reduced the pressure by 1830 so that no overshoot occurred, and no subsequent crash.

[36] An Ottoman census of 1831 exists, but Yanya kaza (Ioannina) appears neither in the summary published by Karpat 1985, 109–15, nor in the (otherwise) more complete Karal 1943. Quite possibly the Ottomans could not extend this census to Ioannina on account of warfare.

[37] Stéphanos 1884, 507–11. Pouqueville (1826–7, 4:441) says that after 1814 plague carried off one-sixth of all Epirots.

Whatever the case, the brutal reprieve or the mild crash did not check population for long. By the middle of the nineteenth century Pindus villages were filled to bursting once more. Virtually every village reached the maximum population in its recorded history between 1850 and 1880 (see Table 5.5).

Although the population in the town of Ioannina dipped considerably in midcentury, that of the villages mounted. As the economic system of Ali Pasha's regime disappeared, Ioannina's importance dwindled, and thousands of people left the city. Presumably some (if not most) had come from the mountains, and went home. Many travelers testified to the ruined state of Ali Pasha's capital, and that of the roads, bridges, and khans necessary for commerce.[38] Quite likely, more young people stayed in the villages under these conditions, and some of the itinerant traders and artisans of the Pindus villages stayed home.

Perhaps more important, no serious epidemics afflicted Epirus after 1822, except for a bout with cholera in 1881 (which left the uplands unscathed). In the past, epidemics had occasionally swept away a tenth or a quarter of the population, sometimes even more, providing what Malthus called a positive check. But this mechanism ceased early in the nineteenth century, allowing more rapid population growth, and deepening the danger of overshoot.

By the 1860s and 1870s the symptoms of overshoot emerged. Villagers had to summon their courage and their ingenuity in order to survive straitened circumstances. Brigandage increased, facilitated by the new borders (1881) that made Thessaly Greek and Epirus Turkish. Political restiveness multiplied, as dissatisfaction with life's prospects easily translated into nationalism. Overseas emigration, never before important in the history of Epirus, began to matter by the 1880s. Before undertaking these rather drastic measures, villagers presumably tried to cultivate every unused patch of land and tried to sell every imaginable manufacture. The village of Koukouli, for instance, came to specialize in the production of raw silk at the middle of the century. But economic conditions in Epirus remained unpromising as public order left much to be desired. In effect those youths who sought a living in brigandage made it more difficult for others who sought theirs in commerce and crafts. For most Pindus villagers, the decades after midcentury were hard times.

Population began to decline sometime after 1882. The explanation

[38] Spencer 1851, 2:181–4; Chirol 1881, 170, 186; Best 1842, 20–1, 127; Estourmel 1844, 1:40–1. Estourmel put the population of Ioannina at four thousand, one-third the size of the nearest estimate.

probably lies in local emigration and the violence surrounding the Greco-Turkish War of 1897 and the First Balkan War, through which Epirus became politically Greek.[39] No epidemics swept through. No great current of overseas emigration depopulated the mountains, although Epirots did begin to go to the United States after 1890. No sustained drop in the birthrate is at all likely, although there are no data one way or another.[40] Quite likely different villages lost population for different reasons; they certainly did so at different times. Sirako appears to have lost most between 1890 and 1902, whereas Monodendri and Tsepelovo declined between 1902 and 1913, perhaps suffering in the war. But in any event, an unusually large number of mountain folk left their villages, unable to continue at home, finding either new lives or early graves.

Population continued to shrink after 1913, suggesting that perhaps war itself may not have played such a large role in the depopulation of 1882–1913. Every village except Monodendri lost population between 1913 and 1920, although in most cases nothing like the numbers lost before 1913. Conceivably, war and overseas emigration took away enough young people to depress the birth rate – again no evidence exists for or against this idea – but just as likely is the idea that the current of migration to Greek cities and to lowlands continued to operate on a small scale even after the overpopulation of the mountains had disappeared. Men who found easier niches in towns might summon their brothers or nephews. And in some families, once the young and strong had left, labor was too short to carry on, so it made more sense for one and all to pack up and leave. All these motives might explain the continued depopulation after 1913.

The effect brought many mountain villages to the brink of undershoot, to the point where sufficient labor to raise crops, manage water,

[39] PRO FO 294/23 details troop movements in 1897 and reports on various Epirot villages. sacked. The British vice-consul in Ioannina (an Italian) reported anarchy in Epirus in 1904 and 1906. PRO FO 294/32, Millelire to Graves, 2 June 1904; 20 August 1906.

[40] Mansolas (1872 and 1878) published demographic data drawn from the Kingdom of Greece, 1838–73. These show a very steady birthrate for the period 1864–73 at 27 or 28 per 1000. The province of Acarnania, the closest geographically to Epirus, had the highest average family size in continental Greece (4.76 in 1870). Average household size in rural Greece remained at about the same level until the 1950s. Modern data appear in Daoutopoulos 1982, 14. Death rates in Greece for the period 1864–73 fluctuated slightly more, between 21 and 25 per 1000. Population increase for 1838–70 was 1.97% per year, a doubling rate of forty-five years. These rates are probably roughly comparable to those in Epirus. (Mansolas expresses the birth rate and death rate in persons per birth [or death], so his numbers are different from those given here.) See 1878, 13–27 especially. Sereléa (1978) presents similar data.

herd animals, and cut wood could not be found. Indeed some villages, such as Kepesovo or Doliani/Nea Amarousi, never recovered after 1913 and kept declining, becoming shell villages well ahead of their neighbors. But most Pindus villages, although certainly suffering environmental damage with the sudden depopulation, bottomed out around 1915 or 1925 and then began to recover.[41]

Their recovery could only go so far. They could never return to their former glories in the early nineteenth century, let alone that of the eighteenth century, because their landscapes had changed. Much wood and soil had disappeared with the expansions and contractions of population, and of the economy as well. So it took far fewer people now to fill the villages to the lowered bursting point.

Between 1928 and 1940 population grew in fourteen of the seventeen Zagori villages represented in Table 5.5. In some cases it grew with a speed that indicates immigration, or at least the return of those long absent. Natural increase alone could scarcely account for the growth in Vovousa, Koukouli, Makrino, or Tsepelovo. Perhaps the general depression of the 1930s helped send itinerant laborers and traders back to their ancestral homes, in the Pindus as elsewhere throughout the world. Whether numbers grew to the point where the local landscapes could no longer accommodate them – another overshoot – is unclear. War came in 1940 and shook the Pindus until 1949. Many villagers risked their lives, property, and that of those around them by joining the anti-German resistance, and then by fighting with the Greek communist insurgents. No doubt genuine convictions played a large role in these personal decisions: patriotism and revolutionary idealism. But it is also likely that many found such risks easier to take because they felt they had so little to lose: their prospects in life as things stood did not look good. But the German occupation and the whole decade of war made the mountains so frequently hungry that desperate decisions might well have flowed not from rational calculations about prospects but from the distress of the moment. So it is hard to say whether this population recovery before 1940 had brought the Pindus villages back to their (lowered) upper limit.

After 1949 every village population fell beneath the threshold required for efficient mountain agriculture. They all suffered from undershoot, and none has ever recovered, nor is it likely that any soon will. Nine years of war killed many, and the government relocated thousands in an effort to isolate the guerrillas. Ever since, population has

[41] After the Greek Anatolian adventure of 1922 the League of Nations arranged a swap of minority populations between Greece and Turkey. In this the eparch of Ioannina acquired 3,217 newcomers, equivalent to slightly more than 3% of its 1928 population.

slowly withered, as the elderly die off and their children stay in Thessaloniki or Ioannina, both of which grew rapidly in the 1950s and afterward. Many Epirots, like northern Greeks generally, went to work in the cold and dark of northern Europe after about 1953, finding the villages unworkable anymore. Upon their return to Greece they generally settled in towns and cities, although most villages in the Pindus today include a veteran or two of German factories or Belgian mines. They invariably hated the climate and usually felt ill-treated, but liked the money and had no way to make a living at home. In the twentieth century leaving the village world behind has become a common refrain in Epirot life – and music.[42]

Since 1949 the anthropic landscapes of the Pindus have lacked sufficient population for their upkeep. Fields and plots suffer from neglect, and now are colonized by aggressive maquis species. At Frangadhes, for instance, the cultivated area early in the 1980s amounted to about 3 percent of what villagers had tilled a century before.[43] The abandoned fields would require enormous applications of labor in order to become productive once more. Even should the allure of the cities and distant continents somehow abate, Pindus villages probably could not return to vigor. Their landscapes would need many people to work them but can support only a few. Wood is often scarce and soil always thin. A scissors crisis – to borrow a phrase from the USSR's early economic history – has destroyed the viability of the villages for a long time to come. The curve of maximum sustainable population is now too close to that of minimum sustainable population, and indeed is perhaps below it. Only remittances keep the few villagers going, as they now produce very little themselves. Centuries from now, where soils and vegetation have recovered (which in places they may if grazing continues to decline), the maxima and minima of sustainable population in these villages will again separate, and a human niche in the mountains will open once more. But in the meantime – and a long meantime it is, measured by the span of human lives – the mountain environments of the Pindus will remain almost deserted. The elder generation will die, and fewer and fewer in subsequent generations will

[42] Systematic evidence regarding emigrants' motives and conditions is in Moustaka 1964, who studied 423 Zagori migrants. I have spoken with dozens of such fellows as well, and conclude that their reasons for going while primarily economic were also a matter of morale: they felt their villages were dying and they did not want to remain behind with the corpse. Nonetheless no small number of them have returned in their old age to their birthplaces, usually because (they say) they intend to die where they were born. On Epirot emigration generally: Zotos 1935; Hadjigeorgiou 1958; Dimitras 1971. Hadjigeorgiou (pp. 65–90) and Mammopoulos (1966) explain the large place of emigration in Epirot popular culture, notably in songs.

[43] Sakellaridis 1984, 9.

have sufficient emotional ties to want to come to mountain villages in their latter years. The mountains will be left to herders and their animals, much as they were before the sixteenth century. Human settlement and cultivation will appear a passing phase in the natural history of the Pindus.

The Lucanian Apennines

The rhythm and motor of population history in the Apennines is, not surprisingly, very different from that of the Taurus and somewhat different from that of the Pindus. War and long-distance trade have played a lesser role in determining possibilities. Deforestation and erosion have played, if anything, a larger role. But the oscillation of human numbers in Lucanian mountain villages has been somewhat less dramatic than in the Pindus, chiefly for geographical reasons. The tyranny of slopes is a modest one in Lucania, so modern improvements to agriculture, such as machinery, are often applicable. This means that reduced population is less consequential: villages can produce adequate food with only a few laborers. Put another way, the minimum sustainable population has proved a more flexible boundary here than in the Pindus.

South Italian villages are normally bigger than those of the Taurus or Pindus, or indeed anywhere else in the Mediterranean mountains. Indeed, many would be called towns in Greece or Morocco. Outside of Italy most Mediterranean mountain villages have, and always have had, fewer than a thousand people. But in the Apennines, villages of one to three thousand people are more common. This gives them an entirely different character from villages elsewhere: more artisans, more formal government, more social life outside the family. In the early nineteenth century several Lucanian mountain villages had doctors and lawyers.[44] The usual reason adduced to explain this anomaly of size is the need for security. But the countryside in the Balkans and the Rif was, over the centuries, at least as dangerous as southern Italy. Patterns of medieval landownership and the extent of feudal domain may have played a role. And surely the firm hold that malaria has had in the Mezzogiorno obliged people to huddle together in those upland environments where malaria had a weaker grip. The larger villages are the hilltop ones

[44] Data on occupations were collected by the French in 1813–15. See Martuscelli 1979, 448–50 for Lagonegro *circondario.* See also Pedio 1961 (*Popolazione e prezzi*), 15 for data on Calvello, 1816–33.

Table 5.9. *Mountain village households in Lucania, 1277–1669*

	1277	1320	1521	1545	1561	1595	1648	1669
Abriola	78	71	194	292	376	228	500	421
Anzi	98	85	92	197	337	517	577	80
Calvello	165	148	149	248	476	498	498	256
Campomaggiore	15	20						
San Paolo Albanese						4	62	38
Castelmezzano	148	39	110	164	91	126	164	138
Castelsaraceno	48	103	227	387	437	486	400	167
Laurenzana	89	80	118	236	462	520	500	434
Pietrapertosa	280	80	434	521	543	332	332	172
San Costantino Albanese						59	109	55
San Severino Lucano								
Terranova di Pollino						25	88	117
Viggiano	205	185	218	410	482	578	695	381

Note: Part of the inconstancy in the apparent number of households derives from the fact that these numbers represent the number of taxed hearths (*fuochi tassati*). Local politics might strongly affect how many hearths avoided taxation. The discrepancy between actual and taxed hearths is generally supposed to be small in the sixteenth century, but greater later on.
Source: Pedio 1987–9, 1:132–68.

surrounded by lower lands where malaria would have been more lethal. However, it is also possible that Lucanian hill villages have been larger than those elsewhere in the Mediterranean mountain world because the land is gentler, and more arable could be found within a radius of two or three hours' walk from a given point.

The earliest data on population in thirteen randomly selected Lucanian mountain villages appear in Table 5.9. The modern population history of the same villages appears in Table 5.10. The population history of Lucania as a whole is summarized in Table 5.11.

Although no data exist to support the idea, quite likely Lucanian population rose in late medieval times, as it did in the Mediterranean and Europe generally. Norman, Angevin, and other monarchs struggled for control of what was surely a growing economy. But the Basilicata region felt the ravages of the Black Death after 1348. Naples served as a major crossroads of contagion and the modest connections between Naples and its hinterlands spread communicable diseases wherever ecological conditions permitted. Plague returned several times, devastatingly in 1394–5, leading to widespread abandonment of villages. By 1447, according to one estimate, Basilicata counted only 65,000

Table 5.10. *Population of selected Lucanian mountain villages, 1735–1981*

	1735–6	1793	1803	1806–8	1822	1828	1843	1853
Abriola	2,057	3,015	3,000	3,135	2,625	2,530	3,103	3,411
Anzi	2,000	3,092	3,092	3,430	3,419	3,297	3,968	4,351
Calvello	3,700	5,640	5,640	6,030	6,385	6,550	6,030	5,941
Campomaggiore		387	410	598	955	930	1,340	1,515
San Paolo Albanese	550	864	864	952	846	980	1,355	1,246
Castelmezzano	900	1,807	1,807	1,875	1,979	1,918	2,300	2,364
Castelsaraceno	1,500	3,288	3,288	2,947	2,863	3,245	2,837	3,075
Laurenzana	4,824	7,143	7,143	6,602	6,685	6,530	7,452	7,664
Pietrapertosa	1,300	2,566	2,566	2,719	2,933	2,836	3,174	3,135
San Costantino Albanese	475	1,040	1,000	1,013	1,034	1,181	1,522	1,591
San Severino Lucano	600	2,218	2,200	2,683	2,773	3,141	3,982	4,533
Terranova di Pollino	250	1,171	1,100	1,155	1,291	1,454	2,333	2,499
Viggiano	4,000	5,715	5,700	5,669	5,475	5,250	6,693	6,913
Total	22,156	37,946	37,810	38,808	39,263	39,842	46,089	48,238
Potenza (province)		303,377		262,947		333,002	383,909	418,847
Basilicata (region)		391,239		378,635		436,290	492,636	518,333

Source: Filangieri 1980; Italy, Istituto Centrale di Statistica 1967 and 1983–6.

inhabitants.[45] Disasters abated considerably and population grew vigorously until the mid-seventeenth century, when new rounds of plague, especially that of 1656, carried off as much as a quarter of the population.[46] Population probably reached its modern nadir in the 1690s, a bad decade throughout Europe, and one that brought another bout of plague to southern Italy. From the thirteenth century to the eighteenth, Lucania as a whole experienced sharp fluctuations in population (Table 5.11). The aggregate figures, however, hide even greater fluctuations at the village level (Table 5.9). Villages felt not only the effects of plague and famine, but the ebb and flow of local migration as well. The smaller the scale, the more volatile the oscillations. Unless the figures are grossly misleading, villages such as Castelmezzano and Pietrapertosa (two of the highest in southern Italy), swelled and shrank several times in these centuries. The consequences for landscapes and especially for soil erosion must have been grievous.

By the mid-eighteenth century a new secular trend took hold. The modern pattern of rapid population growth had emerged, in Lucania much as in the rest of Europe. The last serious epidemic of plague

[45] Da Molin 1979, 19–27; Pedio 1987–9, 1:49–53.

[46] Filangieri (1980, 196) says the Kingdom of Naples lost 21% of its population in the 1656 plague year.

1860	1871	1881	1901	1911	1921	1931	1951	1961	1971	1981
3,450	3,253	3,337	2,740	2,535	2,592	2,816	3,147	3,244	2,257	2,360
4,255	3,794	3,649	2,863	2,356	2,299	2,476	3,371	3,237	2,110	2,230
5,801	6,108	5,248	3,445	3,258	3,140	3,260	3,995	3,844	2,633	3,077
1,399	1,423	1,485	1,185	1,120	1,115	1,123	1,260	1,238	1,024	1,131
1,204	1,293	1,088	836	968	843	880	914	911	548	624
2,418	1,937	1,919	1,724	1,804	1,809	1,903	1,885	1,821	1,120	1,200
3,030	2,154	2,380	2,470	2,323	2,354	2,777	2,824	2,734	2,019	2,087
7,802	7,592	7,013	4,304	3,566	3,588	3,892	4,769	4,300	2,858	3,060
3,197	2,991	3,018	2,273	2,563	2,115	2,171	2,349	2,229	1,718	1,610
1,640	1,563	1,549	1,446	1,500	1,484	1,455	1,758	1,773	1,279	1,270
4,821	3,932	4,433	2,741	3,260	3,337	3,419	3,454	3,027	2,540	2,405
2,582	2,150	2,357	2,451	2,333	2,399	2,633	2,689	2,724	1,982	2,016
6,231	6,188	6,030	4,351	4,030	3,777	3,870	4,152	4,039	3,179	3,088
47,830	44,378	43,506	32,829	31,616	30,852	32,675	36,567	35,121	25,267	26,158
		411,266	361,980	355,859		373,672	445,188	444,166	376,103	406,616
		539,258	491,658	485,911	492,132	513,712	627,584	644,297	560,057	610,186

came (together with famine) in 1764, and the last of any sort in 1816, affecting only Apulia.[47] The rate of growth in Basilicata between 1736 and 1791, 256 percent in fifty-five years, is comparable to that of African populations after 1945, and notably higher than eighteenth-century growth rates in Britain or Germany (then the highest in Europe). Probably some of this exuberance was recovery from a specially bad seventeenth century: epidemics had opened so many niches in landscapes that large families could survive simply by bringing more land under the plow. Early marriage and large families made sense in these conditions. But whatever the reason, this was a time of extraordinary population growth by the standards of the age. In some villages, such as Terranova di Pollino in 1753, almost every family owned some property, a house, some land and vines, and several sheep, goats, or cattle.[48] This bespeaks a level of prosperity that even a relatively new and well-situated village (Terranova was founded in the sixteenth century in the middle of beech woods) could scarcely aspire to a century later. As the eighteenth century wore on, farm fragmentation and economic difficulty afflicted many in the mountains, and poverty deepened in the lowlands where landlords held sway and the price of labor dropped.

[47] Filangieri 1980, 197–8; Nicotra 1983 is a fine analysis of demographic crises and their abatement in eighteenth-century Sicily.

[48] Schilizzi and Viola 1981. Terranova had 58 households and 367 people in 1753, an average of 6.3 persons per household.

Table 5.11. *Basilicata population estimates, 1400–1791*

Date	Ranieri	Racioppi	Beloch	[fuochi][a]
1400	300,000[b]			
1505	123,000–200,000[c]	123,000	123,000	
1532			130,000	
1545			178,000	[32,318]
1561	194,000	213,000	213,000	[38,770]
1595			252,000	[45,881]
1648	196,000	216,000	216,000	[39,226]
1657	140,000			
1669		153,000	153,000	[27,795]
1732				[27,853]
1736		143,000		
1791	366,000	366,000		

[a] "Hearths," from Pedio 1987–9, 1:128.
[b] Boundaries slightly different at this date.
[c] Ranieri began with the same figure for households as the others, but offered a range of multipliers from 5.5 to 9.0.
Note: Converting *fuochi* into population is impossible to do reliably. Pedio 1987–9, 1:128–31, shows (with a comparison of 1732–6 data) that the ratio varies between 3 and 10. For Racioppi and Beloch I have converted household numbers into population numbers with a multiplier of 5.5 (recommended by Filangieri 1980, 142).
Sources: Ranieri 1961, 139–42; Racioppi [1889] 1902, 355–9; Beloch 1937, 215 (cited in Filangieri 1980, 140).

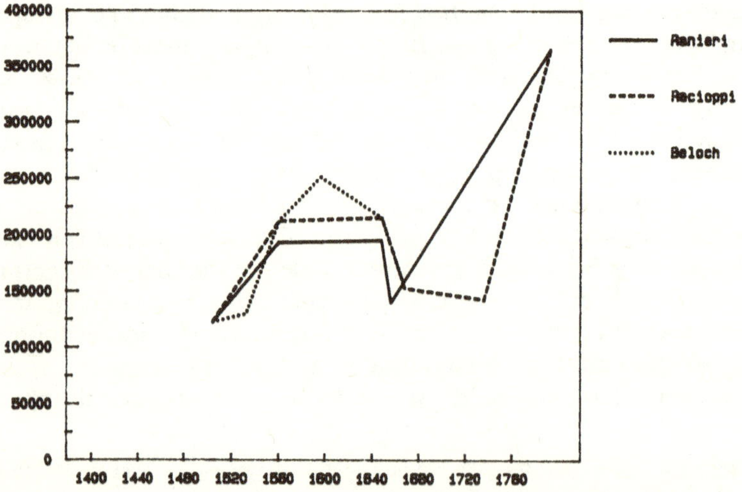

Graph for Table 5.11 (1500–1800)

Indeed, these were excellent times for the landlords of southern Italy and for the mercantile classes of Naples and other centers: architecture, and high culture generally, record some of their good fortune.

From the less exalted position of the mountain peasant, things looked less cheerful. By the early nineteenth century, population growth had slowed, and in a few villages (Abriola, Castelsaraceno, Viggiano, Laurenzana) population declined for decades after 1808 (see Table 5.10). In the thirteen mountain villages in Table 5.10, population stagnated between 1803 and 1828, perhaps attributable in small part to war but more fundamentally to the absence of niches: landscapes had filled up. Late marriage and smaller families made sense now, and hunger perhaps rendered more people vulnerable to infections. At any rate, the wide gap between birth and death rates of the late eighteenth century now closed.

The fate of the plains was no different until the 1840s. Lucanian population as a whole grew only slowly from 1791 to 1814.[49] The index numbers in Table 5.12 show the differential population history in Lucanian mountain villages, the province of Potenza as a whole, the region of Basilicata (which includes Potenza plus Matera), the adjacent low-lying region of Apulia, and all of southern Italy. From 1793 to 1843 the rates of increase are similar throughout, although the mountains lagged behind a little. But after the 1840s the lowlands (and urban areas) grew rapidly while the mountains stagnated. By the late nineteenth century clearly the mountains and the plains had very different demographic regimes, as they have had ever since. By 1981 Lucanian mountain villages had only 69 percent of their population of 180 years before, whereas Apulia, flat, agrarian, with only modest urban centers, had 399 percent of its 1793 population. The lowlands took part in the great demographic expansion of modern times; the mountains did not.

No doubt there are many reasons, but the most important is that the mountains had no more room after the 1840s. Mountain peasants lived as poorly as one could at midcentury. Their misery shocked travelers who had seen rural life in the Balkans and Anatolia.[50] Young men found little to look forward to as tillers of the soil, and large numbers of them turned to brigandage, and to the organized violence through which the Kingdom of the Two Sicilies (as the Kingdom of Naples was known from 1815 to 1860) became part of Italy. Southern Italy, while no stranger to revolt in the days of the Kingdom of Naples, saw risings in 1820 and

[49] The French made annual counts during their domination. Martuscelli (1979, clxii) gives the results for Basilicata: 392,000 in 1813; 390,000 in 1814; 396,000 in 1815.

[50] Urquhart 1838, 2:222: "... unparalleled misery in the south of Italy."

Table 5.12. *Differential population history in southern Italy, 1793–1981*

Area	1793	1843	1861	1881	1911	1961	1981
Mountain villages	100	122	126	114	84	93	69
Potenza (province)	100	126	130	136	117	147	135
Basilicata	100	126	130	138	124	165	156
Apulia	100	130	156	188	256	399	451
Southern Italy	100	128	138	154	183	255	399

Source: Elaborated from Filangieri 1980, 284–5; and Italy, Istituto Centrale di Statistica 1987, 49.

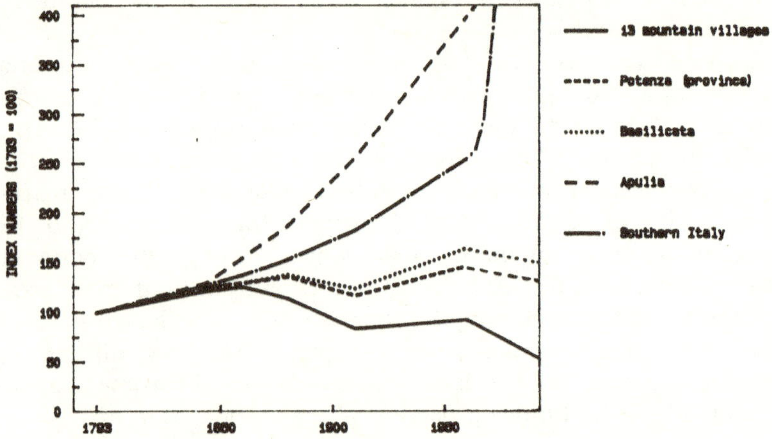

Graph for Table 5.12

1848, and then a brutal and prolonged war from 1860 to 1865. Though manipulated by Bourbon loyalists, the struggle derived its ferocity from its character as a war of the peasants against the classes that benefited from the Risorgimento. The Bourbon faction could find peasant lads willing to risk life and limb for God and King because so many had such dim prospects in life. A less crowded landscape would not, I think, have yielded so many recruits. The army of newly united Italy (not without its recruits from the Lucanian mountains, we may suppose) suppressed desperate guerrilla actions by 1865. The peasants of the southern mountains, while they may never have held central authority in any affection, have been permanently disaffected ever since.

The staggering poverty of the Mezzogiorno continued throughout the nineteenth century. The taxation and tariff policies of the central government did not help (Chapter 6), but the deeper problems lay in the landlessness of lowland peasants and the overcrowding among mountain peasants. Malaria, a scourge until the 1930s, sharpened the misery of millions and killed thousands. Earthquakes and landslides (*frane*) compounded problems. The earthquake of 1857 killed about 2 percent of the population of Basilicata, and permanently injured about 7 percent.[51] A landslide in the wet winter of 1885 demolished the fields and village of Campomaggiore, as 10 square kilometers of soggy landscape slid downhill.[52] Cholera hit in 1836–7, 1854, 1867, and made a final appearance in 1885.[53] The condition of the south attracted the attention of reformers and provoked several large-scale inquests, often headed by political figures who managed to become prime minister after conducting their inchieste.[54] These researches did not solve any of the problems of the south, but they have left monumental volumes of useful information. The solution for hundreds of thousands of peasants was "the hard bread of the emigrant."

The key problem was that lowland peasants had space but no title to the land they worked, whereas mountain peasants had title but no space. The lowland problem was chiefly political; the highland problem chiefly ecological. The mountain population had overshot by the 1840s. Without technological changes, mountain agriculture could not support mountain population at the densities of the 1840s for long, and could support it for a short while only by drawing down resources. The peasants performed the agronomic equivalent of eating their seed corn by cutting forests and plowing slopes. Thus they forestalled starvation in the 1840s, 1850s, and 1860s, but helped to diminish the carrying capacity of their mountains in the process. This diminution, however, was not their work alone. As Chapter 6 will explain, the Italian parliament, in effect, deepened the overshoot of the Lucanian Apennines.

[51] Spera 1903, 181–3. De Grazia 1921, 9 has statistics on the earthquakes of 1851 and 1857.

[52] Kayser 1963, 95. In this case the landslide reached to a depth of 50 meters.

[53] ASB-P, Prefettura – Atti di Gabinetta, 1861–1934, cartella 11, f. 4, "Rapporto sul colera" (1867).

[54] Zanardelli and Nitti for example. Zanardelli's inchiesta is in the ACSR, PC, Zanardelli e La Basilicata, buste 1–7; Nitti's is published (1909–10). In addition, the so-called Jacini inchiesta: Italy, Giunta per l'inchiesta agraria e sulle condizioni della classe agricola, *Atti della Giunta per l'inchiesta agraria* (Rome: Forzani, 1881–6), 15 vols. Volume 9 deals with Potenza. The raw data for the Faina inquiry, of which Nitti's volume formed a part, are in ACSR, Arch. Parl., Giunta Parlamentare d'inchiesta sulle condizioni dei contadini nelle Province meridionali e nella Sicilia (1904–10), buste 1–5.

Overshoot helped to fuel the desperate acts of 1860–5 in Basilicata, as well as the human hemorrhage of 1880–1914, when legions of mountain Lucanians headed for America. Certainly lowland peasants emigrated in large numbers too; they had every reason to exchange their *mezzadria* shares for distant shores. But mountain peasants, who in proportion outnumbered emigrants of the lowlands, fled not from social oppression but from ecological overshoot. Peasants explained to Nitti's investigators that the land no longer yielded sufficient crops.[55] Observers of the time found an obvious connection between the deforestation and depopulation of the Lucanian mountains.[56] While population in southern Italy as a whole, and in flat Apulia as well, continued to grow after 1880, that of the mountain villages declined sharply to 1914. Within Basilicata, population decline between 1881 and 1911 showed a clear correlation to altitude. Comuni between 300 and 500 meters actually gained population during the era of mass emigration, 1881–1911. Those between 500 and 700 meters lost slightly, while those over 700 meters lost heavily. Of 168 districts within Italy, emigration drained Lucania's mountain villages more severely than all but three – one of which was the Calabrian side of Mount Pollino.[57] Calvello and San Severino Lucano lost nearly 40 percent of their population to emigration between 1881 and 1901.[58]

As always, the young and the vigorous led the exodus. Men left agriculture to overworked women. The population of Lucania was 66 percent women about 1905; in the mountain villages the figure would have been higher still. In Lagonegro in 1881, male fieldhands outnumbered females by three to one. By 1901, in the fields of Lagonegro, women outnumbered men by eight to seven, and most of the men were too old or infirm to emigrate. In such circumstances, many less immediate tasks, such as soil conservation, were neglected.[59] Lands that decades of

55 Lagonegro peasant Nicolo Caiafa, aged fifty-four in about 1906: "The people go to America because the land produces nothing. I haven't gone because I don't have the money [for passage] and my mother is ill." In Latronico, Vincenzo Tucci, aged thirty-six: "It makes no sense to rent land because the land does not yield. If God gives me health, I will emigrate to America. Here we can't live." Quoted in Nitti 1909–10, 89–90. Endless such testimony is in Nitti's volume, especially pp. 86–119.

56 Franzoni 1903, 118; Franciosa 1930, 14; Ahlmann 1926, 122; Nitti 1909–10; Tichy (1962) also develops this theme. On motivations of southern Italian emigrants, see MacDonald 1956 and 1963; W. Douglass 1984 (on Molise); Cinel 1982.

57 On altitude and emigration in Basilicata, see the table in De Grazia 1921, 20; D'Aragona 1964, 72. On Italian emigration by district, Giusti 1943, 167.

58 Pedio 1987–9, 1:274–5.

59 Nitti 1909–10, 112–13; Azimonti 1909–10, 5, pt. 1:32–3 and 5, pt. 3:606–16. Italian authorities were greatly concerned about the moral consequences of the departure of young men, believing it led to heightened promiscuity and illegitimacy.

intensive labor had sculpted into delicate (if not highly productive) forms now were abandoned. Erosion carried off millions of tons of topsoil that carefully contoured furrows had hitherto helped to hold in place. In short, landscape degradation was both cause and consequence of emigration, deriving momentum from one another, as in the Pindus in the same decades, but here on a much larger scale.

Population did not grow or shrink significantly in the mountain villages between 1911 and 1931. World War I, the arrival of fascism, the great upheavals of Italian society – all these seem to have had no effect. Or perhaps the losses of the war, clearly evident on the plaques in every village churchyard, compensated for the reduced current of emigration abroad, dictated by the war and then the U.S. immigration restrictions of the 1920s. At any rate, numbers stagnated until sometime between 1931 and 1951 – no census could be taken in the conditions of 1941, so it is hard to know how to date the modest increase evident in Table 5.10. Perhaps, as suggested for the Pindus, in the depression of the 1930s those thrown out of work returned to their native villages. World War II brought much destruction and suffering to Sicily and the Campania; prudent migrants situated in Naples, say, might have gone home in 1944 and 1945. But the depression and the war were safely over by the year of the next census (1951), so the effect of far-flung returnees ought to have disappeared. Perhaps Mussolini's pronatalist policies had had some effect, although it is hard to imagine that mountain peasants took the urgings of the state into account in these matters.[60] The control of malaria in the 1930s and 1940s made a big difference to the coasts and lowlands but likely had little effect on the death rate in the mountains. The land reform pushed through after 1948 also affected the latifundia of the lowlands and touched the mountains only tangentially. Malaria and land reform may have helped the notable increase in Basilicata population between 1931 and 1951 (19 percent in twenty years, after sixty years of no growth), stanching the emigration and death rates. Perhaps mountain folk stayed put because those of lowland Basilicata stayed put, opening no new opportunities. Or perhaps modest improvements in agriculture permitted population to rise. No doubt several factors combined to produce the 12 percent population increase in the mountain villages from 1931 to 1951. Whatever combination it was, it did not keep people there long.

Mountain population declined slowly to 1961 and then cascaded to its lowest levels since before 1750. In part this came from the demographic

[60] For a delightful account of Lucanian peasants unmoved by Fascist exhortations, see Levi 1963.

transition: high birth and death rates, and a wide gap between them, gave way to lower rates and a smaller gap between them. This slow change had been underway since the end of the nineteenth century, but it accelerated in the postwar years.[61] But far more important was the second great current of emigration in modern Lucanian history. The Italian economic miracle of the 1960s required millions of workers. Those most willing to uproot, and work for modest wages, were the peasants of the south, and the mountain folk especially. Basilicata's total population held steady in the 1960s: the surplus of births over deaths balanced the outmigration. But the mountains lost 26 percent of their population in this decade, despite a surplus of births over deaths.[62]

This exodus of the 1960s was more torrential than that of the period 1880–1914.[63] As a proportion of those left behind, the emigrants of the 1960s were about twice as numerous as those of any previous decade. Even so it amounted to a modest loss compared with the 1960s exodus from the Pindus or (as we shall see) from the Alpujarra. And in the 1970s, population in the Lucanian mountains held steady: emigration did not leave behind unworkable communities. Undershoot has not arrived in the Lucanian mountains. The reasons for this are visible in the contrast between the Apennines and the Pindus (or the Alpujarra). First, the Italian mountain villages were larger. As such, they were more complex and diversified than others in the Mediterranean and contained artisans and other skilled people who produced for market. Every Lucanian village now has a paved road connecting it to the rest of the world, and with these roads village craftsmen can sell their wares here and there. They have not become rich; but they have stayed in business despite the reduction in local population in their villages. Calvello, where population grew by more than four hundred in the 1970s, is an example of a village where craft traditions have contributed greatly to the viability of the local economy.[64] Second, and more important, agriculture has survived the departure of many, if not most, of

[61] Data on the decline in vital rates for Basilicata region, and for Potenza province, appear in Basilicata (Regione), n.d. (1983?), 17–18. The death rate decline came more quickly after 1945, and the death rate dropped as low as 11.8/1,000 by 1951 in Potenza province. By 1982 it had declined to 8.8/1,000. The birthrate decline came more gradually: 26.9/1,000 in 1951, 23.4 in 1961, 18.0 in 1971, 13.8 in 1982 (Potenza province). See also Salvia 1956 and Locuratolo 1937.

[62] Natural increase in the 1960s in the Lucanian mountains ranged between 16/1,000 in 1960 and 12/1,000 in 1969. Basilicata (Regione), n.d. (1983?), 11, 17. Migration data for the period 1951–82 appear in this volume, pp. 10–15. See also Quaranta 1974.

[63] Cafiero and Rossi 1966; Formica 1975 and 1979; Italy, Ministero dell' Agricoltura e delle Foreste 1976, 216–34.

[64] Calvello 1982, 9–16.

the young. The agriculture of the Lucanian mountains never rested on labor-intensive horticulture, but instead on wheat complemented by garden crops. And, for the most part, the slopes are gentle, so the wheat farms can make good use of machinery, and thus survive the transition to scarce labor conditions. Hence, the lower threshold of population in the Apennines is much lower than in the Pindus or the Alpujarra, and the villages remain more viable. Those villages that continued to lose population in the 1970s, Viggiano, San Severino Lucano, Pietrapertosa, and (although it lost only nine souls) San Costantino Albanese, are among the higher villages in Basilicata, perched on slopes where agricultural machinery is least useful. They may be in undershoot, Viggiano especially.

High villages with steep slopes will either continue to dwindle, like Viggiano, or they may hang on if able to turn from agriculture to other pursuits. Terranova di Pollino actually gained a score or so of residents in the 1970s. Perhaps these were in part returnees from northern factories, home to spend their declining years in the land of their birth. But no great number left the village. Agriculture has dwindled at Terranova in recent decades, to judge by the abandoned terraces and fields round about, but the oak and beech forest, the sheep and goats, combined with remittances from those working elsewhere, with subsidies for small-scale agriculture from the European Community, and with government pensions awarded through the patronage networks of political parties, keep the village economy alive. Remittances and pensions pump money into many villages, as countless auxiliary activities did in the nineteenth century, allowing a continued lease on life. Most villages of the Lucanian Apennines, for the moment at any rate, resemble Calvello or Terranova di Pollino more than Viggiano in their population trends.

Whether this precarious viability will last remains to be seen. It could prove merely a brief stage in a demise slower than, but not fundamentally different from, that of Pindus villages. Only time will tell. Much will depend on the success of government in providing health, education, and other services to these communities. Much will depend on the ability and willingness of political parties to arrange state pensions for their followers. Much will depend on the job creation rate in the Italian economy as a whole, and perhaps after 1992 the European economy. Much will depend on the European Community's agricultural policy, which bolsters comparatively inefficient farming in southern Europe (although not enough to make unmechanized agriculture viable). In many cases the key will be the efficiency of agriculture itself. Lucanian mountain farms have managed to mechanize to some degree and so remain in business. If they can manage to make the adaptations required in the future, the villages will stay alive. If not, they will shortly empty out as thoroughly as those of the Pindus, and become true shell villages.

Since the burst of growth in the late eighteenth century that ushered in a new demographic regime, the population history of the Lucanian mountains has shown much more stability than other regions of the Mediterranean. Only two periods show any great departure from an identifiable normal level of population density: 1840 to 1890 or 1900, and the years since 1960. The first period fits the pattern of overpopulation and overshoot, producing desperation, violence, and emigration. The second does not; nor does it (yet) appear to be a case of undershoot. It is more likely a depopulation produced by rising expectations than by the prospect of deepening misery; few young Italians want to work the land anymore, and the Italian economic miracle permits them a choice. But where it will lead, and how the post-1960 population history of the Lucanian Apennines ought to be understood, will remain unclear for some time to come.

To an ecologist the comparative tranquillity of population history in the Lucanian mountains would suggest at least temporary equilibrium, a multipart harmony among birthrates, death rates, migration, agricultural efficiency, and the wider world's economy. People came and went from this area, and from this life, at rates that kept the ratio between population and resources comparatively stable, except for 1840–1900 and post-1960. One is tempted to conclude that this means that Lucanian mountain society was better adapted to the range of conditions it encountered than were the Pindus mountaineers, or the Alpujarrans: war, malaria, brigandage, *risorgimento,* earthquake all came and went, and Lucanian mountain folk survived, recovered, and quickly returned their numbers to a balance point. Ecologists generally think that relatively stable population curves show the fruits of a long process of adaptation to environment, whereas relatively unstable ones indicate either an early stage in the process, or a failure to adapt successfully.

Perhaps one ought to seek a cultural explanation. The Lucanian mountain villages do have a longer history than those of the Taurus or the Pindus,[65] and perhaps the extra centuries somehow have yielded customs and behavior that tend to keep population stable, as infanticide

[65] They have, generally, shorter histories than those of the Alpujarra, but as we shall see the cultural and agricultural continuity of the Alpujarra was shattered in the 1570s. Lucanian mountain villages have a longer history than do Christian villages of the Alpujarra.

did in eighteenth-century Japan, and as restrictions against marriage sometimes did among the European nobility. But one would be hard-pressed to identify such customs and behavior.[66]

Perhaps the relative stability of population in the Lucanian mountains derives not from better adaptation to conditions, but from less volatile conditions. With some justice, I think, one can claim that the perturbations in Lucania (earthquake, epidemic, and war) were less serious than those of the Pindus or the Alpujarra. But such things elude measurement.

What seems surest is that the nature of the mountains themselves helped Lucanian highlanders to withstand perturbations. The Lucanian mountains are lower and less steep than the Pindus or the Alpujarra; more land existed on which agriculture might be marginal but not impossible. Crop failure was less common, yields were more reliable, the future was slightly more predictable.

And the mountains were close to Naples, still one of the largest cities in Europe in the nineteenth century, a destination easily accessible to peasants from hill and plain alike. Naples did not promise an easy life to the peasant migrant, but it had many niches because it was so big, and because, as in most cities until quite recently, people died faster than others were born. So Naples could absorb Lucanian surplus population more effectively than could the far smaller urban centers of Eğridir, Isparta, and Antalya, or Ioannina, Grevená, and Thessaloniki. No other mountain zone had such a giant metropolis close at hand. With or without opportunities for migrants in the Americas or in Germany – which came and went – there was always Naples to absorb young people with poor prospects. These geographic realities must account for some of the comparative stability of population in the Lucanian mountains.

The Alpujarra

The Alpujarra had, it is safe to presume, only a tiny population until the Arab conquest in the eighth century. Syrian tribes first settled it.[67]

[66] Italian villages, for instance, developed institutions designed to help people through hard times, in effect an insurance scheme based on grain reserves. These were called the *monti frumentari*. In Lucania ten communities founded *monti frumentari* in the seventeenth century, ten more in the eighteenth century, and twenty-seven between 1800 and 1844, mostly in the 1830s, when times were getting tough. ASB-P, Prefettura – Atti di Gabinetto, 1861–1934, cartella 439: "Statistica dei monti frumentari nella Provincia di Basilicata," 4 April 1879.

[67] Taha 1988, 136, 166–82.

They built delicate irrigation webs, tapping the mountain's waters, and countless terraces to hold its soils in place. Cultivation and population grew but only slowly. Other areas of Moorish Spain remained more attractive, at least until the foundation of Almería in the tenth century. Almería quickly grew to be an important export center of silks, and thereafter the Alpujarra lived from a combination of horticulture and silkworm culture. The locals, now augmented by immigration from the Berber mountains of North Africa, covered the valley bottom and the lower barrancas with mulberry trees. It is likely that when Seville fell to the Castilians in 1248, and a rump Kingdom of Granada collected those Muslims who preferred not to convert to Christianity or transplant to Africa, another wave of immigrants came to the Alpujarra. The Alpujarra became sufficiently important and prosperous that at least two villages could claim descendants of the Prophet.[68] The bubonic plague probably carried off about one-third of the population, 1348–50.[69]

The Castilians meanwhile continued their advance, slowly and intermittently, and by 1492 had captured Granada from the Nasrite kings and thus broken Muslim power in Spain. The settlement reached between Ferdinand and Isabella and their foe, known in Spanish history as Boabdil, provided that he could remove to the Alpujarra with his followers, which he did. Another, perhaps rather small, wave of immigration reached the Alpujarra. By the sixteenth century the population of the Alpujarra had grown to about 64,000, about as much as it has ever had since.

Spanish Muslims after 1492 had to choose between conversion and expulsion, and although most chose the former they soon found the monarchy, on the advice of Isabella's confessor, determined to make their lives difficult. Rebellions of *conversos* flared throughout the Kingdom of Granada, culminating in a great rising begun on Christmas Eve 1568. This struggle began and ended in the Alpujarra and was a disaster for both the population and the landscape. The war lasted three years, of mostly guerrilla operations. The Alpujarrans got some help from Ottoman Turkey. The Castilians eventually devoted considerable resources, including the leadership of the half-brother of Philip II, to the cause. The campaigns, by all accounts, featured savagery and destruction of an intensity that perhaps only a conflict that is both a civil war and a religious war can achieve. When it ended, in 1571, the surviving population of the Alpujarra was deported, with the exception of two families per village, permitted to stay to keep alive the arts of

[68] Brenan [1957] 1988, 186. [69] Laborde (1809, 2:132) gives this figure for Almería.

silk-raising and irrigation. New settlers, recruited from more securely Christian parts of Spain, came to take the places of the dead and deported, but they numbered only a few thousand, and three-fourths of the Moorish villages of the Alpujarra remained empty.[70] This sudden depopulation of a landscape heavily dependent on skilled labor for its maintenance led to ruin (Chapter 3). Whatever their ecological crimes, the Castilians at least left good records of population. Statistics date from shortly after the suppression of Alpujarra in 1571 and show a widowed land (Tables 5.13, 5.14).

If the statistics can be trusted, population fell further after the initial resettlement by Old Christians from elsewhere in Spain. Perhaps many of the new arrivals found that the crops they knew would not grow well in the subtropical mountains, and they moved on or moved home after a disappointing season or two. In 1587 no village had more than about 380 people. Considerably more appear by 1594, far more than natural increase could account for, so either the counts have major errors in them, or new people poured into the Alpujarra in these seven years. Data are scant for the next 150 years, but the likeliest inference is that numbers grew rather slowly through the seventeenth century. The only estimate, which is for 1600–50, suggests a decline from 1594.

The Alpujarra clearly shared in the general spurt of population growth of the eighteenth century, and by its end numbers approached those of Muslim times. No doubt the reduced impact of epidemics played the key role in this transition,[71] but new economic opportunities brought by the expansion of the wine trade helped as well (see Chapter 6). The spread of maize to the Alpujarra may also have contributed to its growth. It came sometime in the late seventeenth or eighteenth century, and expanded greatly around 1790–1820. From 1751 to 1841 maize covered

[70] According to Navarro Alcalá-Zamora 1979, 37–9, of 155 settlements in 1568 only 38 remained in 1572. A huge literature exists on the rebellion. A recent assessment by a leading scholar is Vincent 1982a, 4:191–223. The depopulation was most severe in the Alpujarras, only slightly less so in the Marquesado. Elsewhere in the Kingdom of Granada, depopulation between 1561 and 1591 came to only 30%. See also the new edition of Hurtado de Mendoza 1970, which has population data in appendix 3.

[71] Carrascosa (1960, 105) says bubonic plague "covered the Alpujarras with mourning" between 1572 and 1752, excepting Orgiva for no apparent reason. After 1752 it abated (actually it had probably vanished at least forty years before that – it made its last appearance in western Europe (Marseilles) in 1718–21). Vincent (1982a and 1982b) lists plagues, locust invasions, and other calamities for the Kingdom of Granada, 1500–1800. The years 1676–1681 appear to have been the worst. Vincent (1969) treats plague in the kingdom in the sixteenth and seventeenth centuries, without mentioning the Alpujarra.

Table 5.13. *Population of selected Alpujarra villages, 1586–1986*

	1586	1587	1594	1730	1745	1747	1755	1760	1826	1845	1857
Pampaneira		34	162				729			992	866
Capileira	216	54	279	1,800	2,592	1,985	796	2,524	2,276	936	1,310
Bubion		126	279				328			558	708
Juviles	58	58	207	390	405	437	495	477	698	718	[700]
Busquistar	63			225	450	378	535	531	614	540	1,063
Berchules	81	131	378	2,250	1,516	1,517	1,660	1,588	2,853	2,285	2,562
Trevélez	67	67	288	390	711	702	828	823	1,976	1,185	1,399
Soportújar	49			225	315	342	441	432	610	499	669
Portugos	148	148	553	390	630	639	688	679	729	581	713
Total (9)	682			5,670	6,218	6,000	6,500	7,054	9,756	8,294	9,990
Orgiva	229			1,350	3,445	1,125	3,910	1,489	2,820	3,296	4,806
Pitres	135	135	598	675	855	945	936	913	976	918	995
Laroles	225	225	517	675	643	644	774	760	1,237	1,249	1,500
M. Bombarón	238	63	882	2,250	2,043	2,025	2,119	2,079	3,914	2,630	2,542
Nechite	58	58	256	225	450	437	432	414		359	471
Ujíjar	283	283	828	1,350	1,575	1,800	1,831	1,777	2,787	3,002	3,308
Valor	378	378	918	900	1,260	1,274	1,408	1,363	1,800	1,644	1,914
Yator	72	72	261	360	499	315	463	454	499	464	736
Yegen	72	72	238	450	585	765	999	963	1,148	1,013	1,261
Total (18)	2,372			13,905	17,573	15,330	19,372	17,266	24,937	22,869	27,523

Notes: The nine villages or towns listed last in this table have undergone adjustments since 1930; generally the census takers have chosen to fuse smaller villages to larger ones. The changes have been as follows:

1. Orgiva incorporated Bayacas as of 1930, and Alcazar y Frexenite in 1981.
2. Pitres was fused with Mecina Fondales, Ferreirol in 1981 and renamed Taha.
3. Laroles was fused with Mairena and Picena to form Nevada in 1981.
4. Mecina Bombarón was fused with Yegen to form Alpujarra de la Sierra in 1981.
5. Nechite was fused to Valor in 1950.
6. Ujíjar incorporated Cherin and Jarairator in 1981 (I have included them with Ujíjar from 1960 forward).
7. Valor incorporated Nechite and Mecina Alfahar in 1950.
8. Yator was fused into Cadiar in 1981.
9. For Yegen, see note 4.

Source:

1586: Hurtado 1970, appendix.

1587: Pino Artacho 1978:74–9.

1594: Hurtado 1970, appendix.

1730: BRAH, Ms. 9–6233 "Relación de las Ciudades . . . Granada."

1745: BNM, Ms. 2,785, "Compendio de las Ciudades, Lugares, Aldeas. . . ."

1747: AMAE-M, Ms. 67, "Estado y Consistencia de la . . . Granada y todos los Pueblos de la Jurisdicción."

1755: BNM, Ms. 2,785.

1760: BRAH, Ms. 9–6258.8, "Vecindario de la Provincia de Granada."

1826: Miñano y Bedoya 1826–9.

1845: Madoz 1845–50.

1857–1981: Spanish census data as presented in: Spain, Instituto Geográfico y Estadístico 1858; Spain, Instituto Geográfico y Estadístico 1877; Luna 1984; Navarro Alcalá-Zamora 1979; and García Fernández 1985, 71–3.

1986: "Mapa de la Alpujarra, escala 1/200,000" [printed by Colectivo "Trevelez" de Maestros de la Alpujarra].

1860	1877	1887	1900	1910	1920	1930	1940	1950	1960	1970	1981	1986
883	814	691	785	756	779	816	892	1,063	1,021	693	645	358
1,317	1,203	1,167	1,161	1,217	1,261	1,315	1,403	1,737	1,299	902	713	598
642	585	518	595	631	683	723	730	820	703	566	377	327
[700]	687	548	567	529	525	548	517	595	576	501	234	203
1,080	1,070	1,145	1,247	1,201	1,235	1,104	1,214	1,281	1,290	659	596	434
2,512	2,491	2,490	2,107	1,900	2,460	2,326	2,742	3,145	2,868	1,971	1,525	1,014
1,341	1,418	1,385	1,268	1,296	1,331	1,464	1,419	1,722	1,486	1,506	894	877
688	595	667	603	571	612	624	741	837	775	475	246	274
648	649	733	694	637	621	675	807	873	781	527	522	438
9,811	9,512	9,344	9,032	8,738	9,507	9,595	10,465	12,073	10,799	7,800	5,752	4,523
4,897	4,554	4,450	4,371	4,595	5,051	5,619	6,203	7,219	7,572	6,005	4,859	5,155
918	904	880	915	814	931	955	1,048	1,207				
1,524	1,420	1,484	1,305	803	1,019	1,313	1,452	1,467				
2,412	2,205	2,191	1,535	791	805	1,662	1,962	2,195	1,798	1,326	1,460	1,484
415	415	585	216	216	215	328	350					
3,432	2,792	2,944	2,903	3,557	3,080	3,026	2,660	2,863	4,066	3,544	2,945	3,149
1,952	1,718	1,644	935	775	853	1,266	1,452	2,169	2,232	1,982	1,229	1,255
717	683	738	476	432	561	616	666	695				
1,210	1,162	1,024	915	601	746	780	880	995				
27,288	25,365	25,284	22,603	21,322	22,768	25,160	27,138	30,883	26,467	20,567	16,245	15,566

20 percent more land in the Alpujarra than did wheat and barley combined.[72] It has been a staple ever since, thriving in the irrigated *huertas* of the higher villages.

No record exists concerning population in the Alpujarra in the turbulent years of the French invasion and occupation. The reputation that Alpujarrans had acquired in the sixteenth century as freedom-loving warriors (or brigands, depending on one's sympathies), they fully vindicated in guerrilla actions against the French, which the invaders never could stamp out.[73] Presumably many Alpujarrans lost their lives in these years, and others chose to get out of the way and head for the security of Granada or elsewhere. Whatever the effects of the war, by 1828 the Alpujarra was almost as full as it had been in 1568, really fuller, if one considers that the soils and forests of 1828 were not equal to those of 1568. Natural increase swelled the villages as year followed year with no major epidemic until 1834. Net immigration, an exception

[72] Bernal and Drain 1985, 420; Tapia 1965, 324.

[73] Moral Villalobos 1908; Carrascosa 1960, 117–22, on resistance to the French.

Table 5.14. *Population of the Alpujarra, 1568–1981*

1568	64,250[a]	1887	60,458
1572	6,810	1900	50,414
1587	4,650[b]	1910	46,838
1594	17,164[c]	1920	47,034
1600–50	12,750	1930	53,288
1752	38,475	1940	55,481
1828	58,952	1950	57,223
1845	54,234	1960	52,248
1857	68,674	1970	39,841
1860	67,940[d]	1981	39,791
1877	64,170		

[a] The 1568 figure comes from Luna 1984, 27. Bosque Maurel (1969, 169, and 1971, 79–81) also has it, and cites Simonet. Navarro Alcalá-Zamora (1979, 37) has 64,520, no doubt a transposition of two of the digits.
[b] Pino Artacho (1978, 74–9) has 9,774.
[c] Pino Artacho (ibid.) has 26,316. Vincent (1982b, 5:378) gives 2,765 *vecinos* for 1591, which corresponds to perhaps 11,000–12,000 people. Vincent (389–90) uses 4.0 as the multiplier, perhaps low.
[d] For 1860, Bosque Maurel (1971, 82) has 76,912. The reason for the difference I suspect lies in a wider definition of the Alpujarra, although since neither he nor Navarro Alcalá-Zamora explains his definition of the term, it is hard to say.
Source: Navarro Alcalá-Zamora 1979, 37–9; Spain, Ministerio de Agricultura, Pesca y Alimentación 1986, 63 (for 1981).

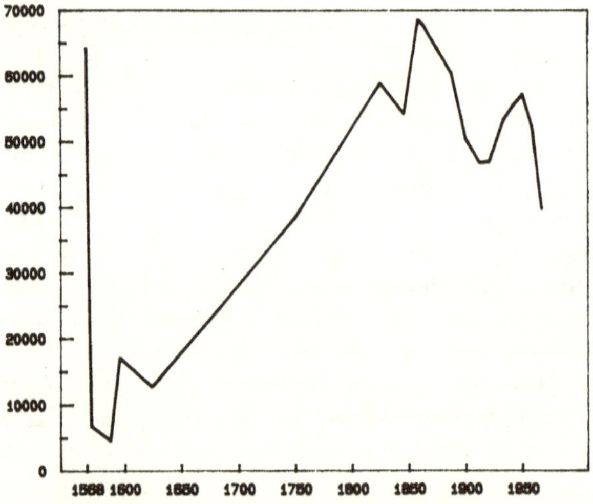

Graph for Table 5.14

to prevailing trends, coincided with a mining boom in the 1820s (see Chapter 6).[74]

Over the next two generations, the population of the Alpujarra first overshot and then crashed. It grew until 1860 or so, following much the same pattern as the villages in the Lucanian mountains, but with a slightly different chronology. Cholera attacks in 1834 and 1843 may have slowed the growth, but by the 1850s population in most villages approached the limits of the possible – given existing agricultural efficiency.[75] From at least 1857 to 1877 the Alpujarra was overcrowded: more people lived there than the land could support without diminishing soil fertility. Agriculture crept into every nook in every barranca; it climbed up the mountainsides to the snowlines, where very low yields of barley came at the price of very high rill erosion. Peasants cut and burned forest in order to extend arable. And with the laws privatizing communal lands (*desamortización*), cagey investors bought tracts of timberland and quickly logged it over (Chapter 6). The whole Sierra Nevada range, indeed much of upland Spain, lost large areas of forest in these decades. The carrying capacity of the Alpujarran landscapes shrank.

Population had begun to decline, perhaps as a result of environmental deterioration, and certainly as a result of the cholera of 1885,[76] when in the late 1880s phylloxera reached the Alpujarra. This infection killed vines throughout the western Mediterranean world in the 1880s and 1890s and caused great hardship wherever people depended upon vineyards. The consequences in the Alpujarra varied greatly from village to village, because some villages scarcely had vines at all while

[74] In the late eighteenth century the Alpujarra had provided migrants to the city of Granada, no less than 18% of the city's nonnative population in 1787. Sanz Sampelayo 1981, 2:502. Granada city had more deaths than births as late as 1900–20 and was sustained only by immigration from the countryside. The Alpujarra supplied more migrants than any region other than the Vega, both in the 1780s and 1900–20. Floristan and Bosque 1957, 371–6.

[75] Bosque Maurel (1971, 82) writes that the Alpujarra by 1860 had "saturated all possibilities of life." Cultivated area in the high Alpujarra doubled between 1750 and 1900; Mignon 1981, 201. The first cholera epidemic came in the spring of 1834 to Orgiva, and within a few months raged all over Andalusia. It killed those over age thirty disproportionately (65%), and women (63%). It was worse in Granada province than anywhere else in Spain, and probably anywhere else in Europe. Annual mortality for 1834 in Granada was 100/1,000, three times the normal. The total loss to cholera for 1834–40 may have been as much as 20% to 25% of the population. Rodríguez Ocaña 1983, 36–40, 81–95.

[76] This was the worst ever in the Alpujarra. Orgiva alone lost 1,476 dead. Sanz Sampelayo 1981, 2:509.

others had scarcely anything else.[77] Mecina Bombarón lost two-thirds of its population between 1887 and 1910; Laroles, Válor, and Yator lost about half in the same years. But the villages high up in the Poqueira barranca, Bubión, Pampaneira and Capileira, while having lost population after 1860, actually grew slightly in the phylloxera years. As a whole the Alpujarra lost about a fifth of its people between 1887 and 1910, most of the decline occurring in the higher villages of the eastern part of the Granadine Alpujarra. The loss took the form of emigration: natural increase remained strong, close to 2 percent per year.[78] Large numbers of ecological refugees went to the Americas and especially to French Algeria.

After 1910 population slowly grew again. Soon many of the villages were as crowded as in the 1860s. Villagers made every effort to expand food production, including planting rye and barley in subalpine meadows at elevations up to 2,850 meters and potatoes at 3,000 meters.[79] The rapid

[77] AMAPA, legajo 258–1, has an untitled *estado* from about 1890 on vines in the province of Granada, showing the area and yield of vineyards in every village.

[78] Data for 1880–9 appear in Pino Artacho 1978, 103. The excess of births over deaths in these years ranged from a low of 5.3/1,000 in Pampaneira to 18.1/1,000 in Soportújar. Data for a major part of the Alpujarra, 1900–78, appear in Luna 1984, 65, 159.

Years	Birthrate	Death Rate	Natural Increase
1900–4	54.40	44.27	10.13
1905–9	55.21	33.22	21.99
1910–14	49.98	31.67	18.31
1915–19	45.17	36.96	8.21
1920–4	40.04	27.87	12.17
1925–9	34.59	19.16	15.43
1930–4	35.88	17.47	18.41
1935–9	30.86	18.42	12.44
1940–4	30.23	15.88	14.35
1945–9	29.05	11.32	17.73
1950–4	26.15	9.10	17.05
1955–9	24.00	8.70	15.30
1960–4	23.68	8.30	15.38
1965–9	20.52	8.49	12.03
1970–4	15.42	7.62	7.80
1975–8	11.69	7.70	3.99

The birthrate figures concern only four zones of the eastern Alpujarras; the death rate figures probably do as well, although Luna identifies them as Alpujarran in general. No systematic difference is likely between the four zones and the whole.

[79] Carandell 1934, 652–7. These elevations are very high; I wonder if Carandell's altimeter was accurate. In 1910 Carandell found cereals at 2,700 meters in the Alpujarra, and 2,100 meters on the north face of the Sierra Nevada. A colleague found barley at 2,500 meters in Trevélez shortly before 1908, and maize at 1,600 meters, well above its ideal range.

expansion of industry in Catalonia and Vizcaya during World War I did not attract Alpujarrans. Some continued to head to South America. In Mecina Bombarón (which has a street named for it in Mendoza in northern Argentina), enough went that some arable was abandoned before 1920.[80] The 1918 influenza pandemic (the "Spanish flu") cut population back somewhat in all the villages; it caused the only annual excess of deaths over births in the Alpujarra, and in the province of Granada since 1900, and probably since well before that.[81] Nonetheless every community except two expanded between 1910 and 1930; most of the growth came in the 1920s. All but three grew again in the 1930s, despite the Civil War that killed or exiled over a million Spaniards. While most of Granada province either sided with or fell to Franco's Nationalists, in the Alpujarra resistance to Franco simmered until 1950. Once again the mountain folk – or at least a handful of them – displayed their preference to go their own way. But in contrast to the war of 1568–71, the guerrilla actions of the Civil War in the Alpujarra were minor. The real fronts in the war lay elsewhere, and its demographic consequences came to little in the Alpujarra, a reduction in the rate of natural increase of perhaps 2 per 1,000.

Every village grew in the 1940s, a difficult decade for most of Spain. By 1950 the Alpujarra had more people than at any time since the 1890s. The economy still rested on the same crops and same methods as in the 1750s, whereas lowland Spain, and especially coastal Spain, had begun a process of technical improvements by the 1950s. The Alpujarra villages remained largely self-sufficient, while the rest of Spain grew increasingly specialized and interdependent. Thus the Alpujarra became poorer and poorer in comparison with other parts of the country, while in absolute terms its standard of living merely stagnated, or perhaps declined modestly. As the economic trajectory of the Alpujarra and the rest of the country increasingly diverged, the stage was set for massive emigration, abandonment, and eventually undershoot.

In the 1950s Alpujarreño population declined slightly, by about 9 percent in all, despite a rate of natural increase of 1.7 percent per year. By 1957, over 30 percent of those born in Pitres lived in the city of Granada. In Trevélez, over 40 percent of the population left during the winter months, seeking agricultural work in the lowlands or factory work in the North.[82] Some, especially young men finishing their army

[80] Floristan and Bosque 1957, 398–9.

[81] Natural increase data for the province 1900–50 appear in Spain, Instituto Nacional de Estadística 1956, 122–3. In the Alpujarra the 1918 death rate reached 56.39/1,000 (Luna 1984, 159).

[82] Floristan and Bosque 1957, 376–99.

service, never returned. In most villages, emigration in the 1950s eased land shortage and improved the situation of those who stayed behind. But in the 1960s the Alpujarra lost a quarter of its population, and this at a time when births still substantially exceeded deaths. Thousands left, for Granada, for Catalonia, for the coast. Numbers continued to decline sharply in the 1970s and 1980s, as the same ecologic and emotional factors that spelled doom for Pindus villages threatened the Alpujarra. Irrigation and terraces require a great deal of hard labor, in the Alpujarra as in the Pindus, and once the young and vigorous have left it becomes very difficult for communities to manage their water supply, and for families to keep up their fields and gardens. Emigration, as always, took those whose strength was most needed, leaving the old, the very young, the infirm. Agriculture retreated in every village, first at the margins, the unirrigated fields of barley and wheat, which had yielded poorly anyway, and then in the huertas, where the maize, beans, and melons grew. As in the Pindus, shortage of labor made village agriculture increasingly unviable. As in the Pindus, the sense quickly took hold that the old villages were dying, and no one with any spunk wanted to commit his future to them. Of course, the birth rate declined sharply as fewer and fewer people of reproductive age remained behind in the villages. In this way ecological and emotional factors combined to drain the villages of their vitality, leaving only the shell behind.[83]

The western barrancas of the Alpujarra emptied out a little later than those to the east. Iron mining at Busquistar kept population up until the late 1960s. In the Poqueira barranca the largest hydroelectric facility in the province provided employment for a few, slowing population decline there. But the mines of Busquistar, once the second largest iron mining operation in the whole Sierra Nevada (the largest was at Aldeire in the Marquesado), peaked in 1966 and rapidly declined. Today they are idle. And the Poqueira *central* has since been outstripped by others in eastern Andalusia that are more efficient and closer to centers of demand. These nonagricultural pursuits gave the villages of the western Alpujarra a final fling before undershoot corroded their viability.[84]

The differences between the population history of the Alpujarra and the rest of the province of Granada, and the rest of Spain, help to elucidate the importance of the constraints on mountain life (Table 5.15). A comparison also shows the greater oscillations in the popula-

[83] On emigration: Mignon 1970; Reparaz 1962; García Barbancho 1967; Jutglar et al. 1968; Floristan and Bosque 1957; C. Carvajal Gutierrez 1986, 180–3.

[84] Bosque Maurel 1971, 87. Busquistar produced 200,000 metric tons of iron yearly at its peak, and exported to northern Spain and to Italy. An aerial cable carried the iron down to Orgiva, whence it went by truck to Motril on the coast.

Table 5.15. *Comparative population growth, 1845–1986: Alpujarra, Province of Granada, Spain*

Year	Alpujarra	Group I	Group II	Granada	Spain
1845	100	100	100	100	100
1857	127	120	120	119	107
1877	118	115	111	127	115
1887	111	113	111	131	121
1900	93	109	99	133	129
1920	87	115	100	155	148
1940	102	126	119	199	180
1960	96	130		207	211
1981	46	69		205	255
1986		54			

Note: Alpujarra here refers to Navarro Alcalá-Zamora's Alpujarras. It includes more valley bottom villages, and some from the Almería Alpujarra. Group I is nine Granadine Alpujarra villages for which administrative boundaries have not changed between 1845 and 1986. Group II consists of nine more villages for which adjustments have been made, mostly in the 1981 census, making comparisons with earlier years problematic.
Source: Elaborated from Table 5.13.

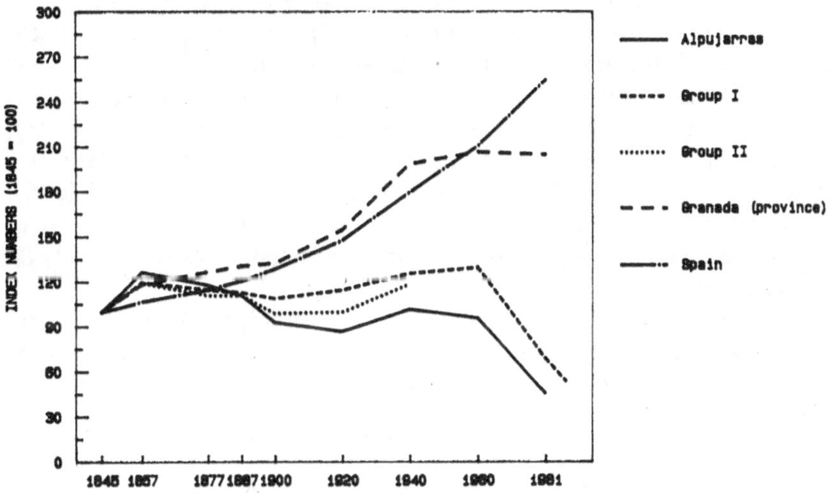

Graph for Table 5.15

tion history of the mountains (although this is also a reflection of the size of the units involved). While Alpujarran population peaked twice, in the mid-nineteenth century and again about a century later, that of the province, and the country, grew slowly but surely. In the mid-nineteenth century the Alpujarra villages grew much faster than Spain as a whole, and slightly faster than Granada. But the decline from the 1880s on in the mountains only modestly affected the rest of the province, and did not register at all nationally. The province continued to grow steadily to 1960, whereafter population stabilized, a result of declining natural increase and rising emigration to Catalonia, Vizcaya, and, to some extent in the 1970s, France and Germany. But the sharp decline in the Alpujarra after 1960 was not general in the province. And no decline or stagnation affected the nation as a whole.[85]

The explanation for these diverging fates lies in the pattern of migration from the Alpujarra, and from mountain areas generally in southern Spain. While many people left the mountains, they often did not leave the province. In both the emigration of the late nineteenth century and that of the late twentieth, three areas within the province showed rapid population growth, based on immigration of mountain people. These areas are the coast and adjacent lowlands; the *vega* de Granada, a rich plain north and west of the city; and the city of Granada itself. The coastal populations had long raised sugar cane and done a little fishing, but North African raiders and malaria kept their numbers down until the eighteenth century. Peace came to the coast after the death of the Moroccan sultan Sidi Mohammed in 1790, permitting a more secure existence by the sea. But population did not truly surge until the wars of independence in Cuba and its final detachment from Spain in 1898: that deprived Spain of its principal sugar source and sparked a boom on the subtropical coasts (see Chapter 6).

The career of the vega shows how open-ended population history can be in the plains. Twice in its modern history, in the 1780s and the 1880s, population pressure has threatened the peasants with ruin, or at least

[85] Rural exodus of course affected most of the Spanish countryside. In general it most strongly affected those areas closest to strong poles of attraction (the hinterland of Barcelona, e.g.), and poor, dry regions (Andalusia). With respect to interprovincial migration, Granada ranked sixth out of fifty in its rate of emigration for 1931–60. Neighboring Almería ranked first. Within the thirteen districts of the province of Granada, the Alpujarran ones of Ujíjar and Orgiva ranked second and fifth, respectively, in their rates of extraprovincial migration, 1931–60. In the period 1901–30 they ranked third and fifth. Albuñol, on the south slope of the Sierra Contraviesa, ranked first in both periods. García Barbancho 1967, appendix table A.2.

the necessity of mass emigration. Both times the combination of new technologies and new markets rescued the vega. In the first instance, the vega shifted from primarily cereal production into a combination of cereals and *cañamo,* the hemp used in ropemaking. With the expansion of the Spanish navy in the late eighteenth century, a strong and reliable market for Spanish-grown hemp developed. In the second instance, in the 1880s, after iron ships had begun to replace wooden ones in the Spanish navy and the market for hemp suffered, and the population along with it, the lowly sugar beet came to the rescue. As with cane sugar on the Granada coast, the decline of Cuban production meant opportunity for the vega. Once again, as in the 1780s, a stagnant population began to grow again, as a new niche opened up just when needed.[86]

From the 1960s on, the coast enjoyed another boom. New technology improved the yields in citrus and many vegetables, now often grown under plastic and multicropped. Tourism flourished as well, although generally more along the Málaga coast than that of Granada province or Almería. The population of the Campo de Dalías, a low-lying plain just behind the coast, doubled between 1960 and 1975.[87] Other coastal areas grew nearly as rapidly. At first Alpujarreños went down to these fields only in the winter, but eventually many of them moved house and family for good. In the same years the city of Granada blossomed, doubling between 1900 and 1940, and almost doubling again since; and this after 150 years of stagnation.[88]

All this is to be expected. On the coast and in the vega population densities could rise and rise again because improvements to agriculture took place, and other economic opportunities presented themselves too. In the mountains this flexibility in the maximum sustainable population did not obtain to equal degree. The terrain almost ruled out improvements to agriculture. Other economic opportunities did arise, as we shall see in the next chapter, but they rapidly destroyed vegetation and soils. So, in the 1890s, people had to leave to avoid starvation. And in the 1960s and 1970s they left in order to avoid further impoverishment (by this time Spanish peasants could expect the state to prevent them from starving). As in the Pindus, the long history of environmental degradation, again chiefly forest and soil depletion, meant that the maximum and minimum population

[86] Martín Rodríguez 1982. See also Ocaña Ocaña 1974.

[87] Luna 1984, 52.

[88] From 76,000 in 1900 to 264,000 in the early 1980s. García Fernández 1985, 72. From 1752 to 1900 the city's population grew by less than a third. Martín Rodríguez 1982, 63.

thresholds pinched together as time went on. A population as large as that of Islamic times could not sustain itself in the Alpujarra of the 1850s, and had to draw down existing stocks of resources to survive in the short run. And the sustainable population of the 1850s Alpujarra (perhaps a fifth less than the real population) was no longer sustainable in the further degraded landscapes of the 1950s. By the 1960s the overshoot threshold had sunk and the undershoot threshold had climbed, so that the Alpujarra moved from one to the other almost instantly.

The future of the Alpujarra will probably parallel that of the Pindus. Population will continue to wither, as the elderly die and no young replace them. The bottomland villages will probably survive, since their soils are richer and their fields flatter. A couple of the higher villages can endure through tourism. Perhaps craft traditions can keep a few families solvent as well. But it is hard to imagine many niches in the Andalusian economy that Alpujarreños can fill without leaving home. The Sierra Nevada will not likely remain a peasant massif much longer.

The Rif

Demographic history in the Rif is murkier than almost anywhere else in the Mediterranean world. Systematic numbers date only from the late 1920s and reliable ones from the 1950s. Prior to that the ups and downs of population are matters for speculation. A few things, however, are clear.

People have lived in the Rif for a long time. Like the Lucanian mountains and the Sierra Nevada, the Rif has always been a peasant massif, the domain of sedentary cultivators. While the prehistoric and ancient imprint seems concentrated along the coast, from Arab times forward the interior mountains have supported an agrarian population. Settlement in the Rif is probably about as old as it is in the Alpujarra. A considerable kingdom existed in the fertile valley of the Nekor beginning in the eighth century, known to us through al-Bakri.[89] Vikings partially destroyed the kingdom in 859–60, and a Fatimid governor of Tahart in Algeria did so again in 916.[90] The kingdom fell about 1015 to a local tribal confederacy, and the city was throughly destroyed, never to recover, by the Almoravids in 1080–1. Somewhat later, another

[89] *Description de l'Afrique septentrionale* (Algiers, 1913), 180–96. The city of Nekor, about 25 kilometers south of al-Hoceima, has never been excavated and nothing remains to be seen.

[90] In Leo Africanus 1956, 277–8, the attack is attributed to a Fatimid caliph of Kairouan (Tunisia).

population center emerged in the Rif, Badis, a coastal city now in Spanish possession as Peñon de Vélez. It served as a conduit between the Kingdom of Fez and the Kingdom of Granada, and may have handled some of the traffic between the Mediterranean world and West Africa. It flourished especially between the fall of Ceuta to the Portuguese (1415) and the fall of Granada (1492), but had existed from at least the thirteenth century as a commercial and shipbuilding center. When Leo Africanus passed through in the early sixteenth century it had six hundred households, perhaps three thousand inhabitants, all of whom, thought Leo Africanus, considered wine a divine liquor, and spent their evenings drinking and singing.[91] Spain gained permanent control in 1564. Other than Nekor and Badis, Rif population consisted of rural tribes, occasionally governed by the Moroccan dynasties based in Fez, but often not. The tribes, with one or two exceptions, have remained the same since al-Idrisi.[92]

All evidence (which admittedly is not much) suggests, however, that Rifian population did not.[93] Palynological evidence (recall Chapter 3) shows forest clearance and presumably population growth in the tenth and eleventh century. Perhaps the Arabs pushed the Berbers up the slopes. Or perhaps the growing commercial networks, linking Muslim Spain, North Africa, and the trans-Saharan routes stimulated demographic growth in the Rif.[94] Whatever the case, demographic catastrophes surely punctuated this growth. The vulnerability of the Rif to drought made for many famines, and its occasional connections to the great seats of civilization such as Egypt brought repeated epidemics. Leo Africanus, in his survey of Rif communities, says of several that they are well peopled. But he gives specifics in only a very few cases. Of the Beni Ouriarhel he says they could field twelve thousand warriors from sixty-three villages, which implies a population about half of today's.[95]

[91] Leo Africanus 1956, 274–6.

[92] Hart 1976a, 347. This summary of medieval Rifian history is from Hart 1976a, 343–53.

[93] For the sixteenth century information is better than for any period before 1900. Leo Africanus, born in Granada, raised in Fez, son of a Moroccan imperial functionary who owned property in the Rif, was captured by Sicilian corsairs and wrote a book published in Venice in 1550, based on his knowledge of Morocco. His Rif description dates from about 1508 to 1516. Much less reliable is the information provided by Luis de Marmol (1573). See Noin 1970, 1:23–8, on sources for early population history in Morocco.

[94] Miège (1961, 2:137–9) explains that Tetouan and Chechouan at a later date participated in long-distance caravan trade.

[95] Leo Africanus 1956, 287. The numbers of Beni Ouriarhel have recently gone over 100,000. Extrapolating from the number of warriors to total population by the method recommended by E. Carette (add a quarter and then multiply by three) gives 45,000

But one cannot know just what area the Beni Ouriarhel occupied in 1516; its population density might be greater or lesser today than it was then. Of Beni Oulid (Gualid) he says it had sixty rich villages.[96] The general sense that Leo Africanus provides is that the northern Rif had few residents, involved mostly in fishing (sardines) and trade. Many of the northern communities raised no grain at all. But the southern flank, the villages connected closely with the imperial capital at Fez, these he paints as teeming with activity, and full of figs, vines, almonds, and barley.

In the late seventeenth century Rifian armies took Tangier and settled many small coastal villages, perhaps evidence for population expansion in the northern Rif. It coincides with pollen evidence of increased forest clearance at higher altitudes. In the eighteenth and early nineteenth century the Moroccan Sultans often attacked and sacked the Rif, which did not voluntarily pay its taxes: in 1813 sultan Mulay Sulaiman left in his wake "a series of burned villages and grain fields."[97] The Rifian tribes in the nineteenth century distanced themselves from the sultans, and enjoyed a fairly full independence, chiefly because the Rif's poverty did not reward punitive expeditions. This perhaps saved the tribes from further ravages such as those of 1813, but internal violence, with disease and famine, probably helped keep population in check. An undated French manuscript, probably from the 1830s, gives 80,000 to 100,000 as the number of males over age ten in the Rif, implying a total population of perhaps 300,000. But the same author says Rifians lived in caves and tents, raising strong suspicions that he had never set foot there.[98] As a guess, population in the Rif at the beginning of the nineteenth century may have roughly equaled that of Leo Africanus's day.

During the nineteenth century the losses to epidemics probably began to shrink. The evidence is unsatisfactory. Plague disappeared after 1818. Cholera arrived, reaching Tetouan in 1834–5, but its ravages did not match those of earlier epidemics.[99] Rifian population began its long-term growth, although slowly at first.

for the early sixteenth century, about the population in the early twentieth century. See Noin 1970, 1:237.

[96] Leo Africanus 1956, 283. The English edition of 1896 translates this as more than six hundred rich families.

[97] Hart 1976a, 351. After 1813 the Sultan installed a representative in the old city of Badis, but the Rifians paid little attention. The sultans left them alone until 1898.

[98] AMAE-P, Mémoires et Documents, Maroc, 9:98–100.

[99] PRO FO 636/1, Diary of the British Vice-Consul for Tetuan, 1830–43. Bois (1949) lists population disasters for the Rif but gives no clear indication of their duration or severity.

Until the 1920s, Rifians, like many other mountain peoples, practiced blood feud.[100] They put great store on honor and on a man's prowess at acts of violence. In consequence, crimes and slights, real and imagined, often led to fights, sometimes to deaths, and these in turn required vengeance, which could be taken out not only on the killers but on anyone in the killers' patrilineage. Under these circumstances, all adult males had deadly enemies. Alliances and enmities between tribes, and between fractions within tribes, dominated political life.[101]

In all likelihood the blood feud is an old tradition among the Rifians. But it probably became demographically significant only when they acquired rifles, in the nineteenth century. While the Rifians themselves explain blood feud in terms of honor, it had another (unintended) function: to keep population in check. It was, in other words, an adaptation to a local environment that did not easily permit agricultural or economic expansion. It conceivably kept the Rif sufficiently poor and unattractive that the sultans left it alone for most of the nineteenth century, although no doubt much else entered into the *politique* of Fez. The death tolls in the nineteenth and early twentieth centuries probably mounted high, although no numbers exist: "The line between

[100] On blood feud generally: Black-Michaud 1975.

[101] Hart 1976a, 313–38. Ayache (1983, 199–227) disputes Hart on the question of the prevalence of blood feud. Pascon and van der Wusten (1983, 75–86) emphasize its importance for the Beni Bou Frah. Westermarck (1926, 2:12) wrote that a Rifian who had not killed another man was not considered a man himself. Montagne (1973, 49–50) emphasizes blood feud. Mouliéras (1895–9, 83) took the most extreme position, claiming that normally at a Saturday suq four or five battles would break out and fifty or sixty men die. Mouliéras never got closer to the Rif than Oran and got his information from an itinerant Algerian holy man who had passed many years in the Rif. Tales of wild men in the Rif certainly suited the purposes of outsiders between 1890 and 1930; a grain of salt is required, especially when reading casual tourists like Mark Twain – "swarthy Riffians from the mountains – born cutthroats" – in *Innocents Abroad* (New York: Harper & Row, 1966), 58. But the extremely subtle distinctions made in Rifian customary law concerning killing and kin connections (Blanco Izaga 1939) suggest that blood feud was an ordinary and accepted part of life, at least by the early twentieth century. Leo Africanus (1956) does not mention blood feud, but does say the inhabitants of Merniza mountain often fought their neighbors (over runaway wives), and that the Beni Ouriarhel regularly fought their neighbors (pp. 284–7). Perhaps the last word in this debate should go to Abdel Krim. In an interview shortly after the war he said: "even my most faithful supporters, and those of the greatest knowledge and intelligence believed that after the victory had been won I would allow each tribe to return to complete freedom despite their realization that this would return the country to the worst conditions of anarchy and barbarism." Pennell 1987, 117.

bloodfeud and massacre is at times rather thin."[102] Some revered survivors had allegedly killed fifty men.

In another remote mountain district of the Mediterranean, now in southwestern Yugoslavia just north of Albania, Montenegrin tribesmen engaged in blood feud until 1851. Montenegrins lived effectively beyond the control of the Ottoman Turks, as the Rifians lived beyond the control of the Moroccan sultans. Both mountain zones were, in the language of Christopher Boehm, tribal refuge areas, in which blood feud helped to maintain a rough balance between population and resources:

> The population of the tribal refuge area would remain close to the maximum that natural subsistence activities plus raiding could sustain, but it would constantly go through cycles. In traditional times, when the population would rise above this carrying capacity, a few Montenegrins would emigrate. But the vast majority would stay at home to compete for scarce resources and would increase their raiding. Because of severe crowding, feuding and warfare would then become intensified, and excess male population, in particular, would be reduced. This would provide short-term relief. Once the population had fallen below this carrying capacity, then warfare would become less frequent and feuding would decrease as well, insofar as the causes were economic. While natural increases in population eventually would have led to serious overpopulation again, immigration would make this happen much more rapidly. Thus a low point in fighting strength would soon be compensated for; but it also helped the next overpopulation crises to come along more quickly.[103]

Blood feud killed adult males. It may have reduced fertility as well as increased mortality, although polygamy and widow remarriage ought to have kept fertility from declining far. More important, blood feud interfered with food production. While women perform most of the agricultural tasks in the Rif anyway (a division of labor that blood feud may have reinforced), at times the labor of men, women, and children is needed. But men may often have been unavailable, in hiding, in pursuit, or just unwilling to put their rifles down to pick up sickles.

[102] Hart 1976a, 324.

[103] Boehm 1984, 179. The Albanian mountain tribes also practiced blood feud (PRO FO 195/1065/234–6, Blunt to Bulwer, 11 June 1860); as did the Sarakatsani of the Pindus (Campbell 1964). The Sarakatsani limited vengeance to the perpetrator of blood crimes and refrained from attacking his kin. Boehm's term for Montenegro, "mountain refuge area," which he applies also to the Rif without discussing it, is similar to one of the three types of mountain areas treated in Planhol (1968): "mountain refuge area," mountains of ancient settlement, and "bedouinized" mountains, taken over by nomads.

In this way blood feud, when it raged most ferociously, may have in effect temporarily reduced the carrying capacity of the Rif. If so, it would have temporarily increased the rates of malnutrition, infant mortality, and disease. Quite possibly, blood feud softened the fluctuations of population among those people who practiced it. By lowering population, however brutally, *before* genuine overshoot occurred, blood feud could have prevented more extreme swings of overpopulation and underpopulation.

By lowering population, it may have prevented the Rif tribes from becoming anything more than a negligible nuisance to the sultan in Fez. In Montenegro, Boehm claims, blood feud, with occasional exceptions, kept the tribesmen from posing a serious threat to the sultan in Istanbul, and hence preserved their safety. When on rare occasion they did grow so numerous that their raiding seriously annoyed the Ottomans, the Turks mounted "genocidal" expeditions that sought to kill as many Montenegrin males as possible, and to carry off all women and children. These disasters could cost the Montenegrins well over half their population; avoiding the wrath of the Turks while maintaining their freedom was a difficult business. Blood feud, according to Boehm, allowed them to do it: a vaccination of internal violence prevented epidemics of far greater violence. Thus the apparent barbarity of blood feud may have purchased the security of a steadier relationship between population and resources, characterized by less pronounced cycles, and greater freedom from demographic disaster at the hands of powerful empires.

This is a lot to lay at the door of blood feud. The real extent of its demographic impact is impossible to measure and hard to guess. But it seems plausible, indeed likely, that the practice limited population growth in the Rif, as in Montenegro, and perhaps regulated human numbers in ways that may be construed as beneficial.

Factors other than blood feud also shaped population history in the Rif. Drought, famine, and epidemics, as elsewhere, no doubt provided powerful (if perhaps diminishing) checks through the nineteenth century. Emigration also must have played a role. The cities of northern Morocco required immigrants for their survival, and some of these came from the Rif. In nineteenth-century Tangier, for example, Rifians worked as gardeners and water carriers.[104] Migration abroad amounted to little before the twentieth century.

In the twentieth century the population history of the Rif becomes a little clearer. Certain epochal events are obvious, as is the general trend

[104] AMAE-P, Mémoires et Documents, Maroc, 9:99 (undated, untitled manuscript).

of accelerated growth. The great events are three. First is the success of Abdel Krim in suppressing blood feud. It has never resumed, and although enmities are still remembered in the Rif and can influence elections, no one kills anyone anymore. Even the flourishing kif trade, which in other settings has led to much violence, Rifians manage fairly peacefully. Second is the Rif War, 1921–6, which killed untold thousands and, as wars do, led to widespread hunger and malnutrition, suppressed fertility, and inflated infant mortality. Third is the terrible drought of 1944–8, well remembered by all those then alive, and alive itself in folk memory.[105] Of these three events, two served to check population growth, whereas one served to promote it. The checks have had only a temporary effect, for the clear trend, since the 1920s at least, is of rapid growth. People are dying slower than ever before, and others are being born at roughly the same rates as always.[106]

Systematic population data begin with the effective Spanish occupation, in the late 1920s. Early counts were probably very approximate, so fairly reliable numbers date from about 1935. They indicate a community in overshoot. Population doubled between 1935 and 1970, and since then has almost doubled again. The Rif, beyond any doubt, now contains far more people than it can sustain with the existing economy. It has the highest population densities of any mountain area in Morocco, and the eastern part, the dryer part, has the highest densities of any rural area in Morocco, including the fertile plains of the Rharb and Marrakesh. The stress on the landscape is visible at every turn, with cultivation crawling up the mountains, with fuelwood shortages, and with uncontrolled erosion at rates among the highest in the world. The Rif badly needs an economic productivity revolution, a completed demographic transition, or mass emigration. If none of these happens, the prospects for political violence are discouragingly good. Rifians traditionally do not have the passive patience of Javanese or Egyptian peasants, who seem to accept increasing poverty as long as it affects one and all. The Rif rebelled most recently in the late 1950s, immediately after independence from Spain.

The evolution of population in the Rif in the twentieth century consists of accelerating growth marked by the two setbacks of war and drought. Combat, massacres, and deprivation associated with the Rif

105 Plague, famine, and drought are all remembered in song and prayer by the Beni Bou Frah. Pascon and van der Wusten 1983, 58.

106 In Morocco as a whole, the crude fertility rate in the 1960s was about 47/1,000, and in the 1980s 45/1,000. The crude mortality rate was about 27/1,000 in 1945, 19/1,000 in 1962, and 15/1,000 in 1982. Bisson et al. 1985, 120–2. Death rates are slightly higher in the mountains, however.

War of 1921–6 may have lowered population by as much as 25 percent, with great variations among tribes. Among the Beni Ouriarhel, Abdel Krim's people, losses may have approached 35 percent.[107] Even with this disaster, Rifian population grew between 1900 and 1936 at about 0.5 percent per year, and population density climbed from 29 to 35 per square kilometer. The drought and famine of 1944–8, an atavistic return to the old demographic regime of subsistence crises, cost the Rif about 10 percent of its population. The disaster fell fairly evenly throughout the Rif (see Tables 5.16, 5.17 and 5.18), since the drought affected all of the Maghreb. Population grew at about 1.3 percent annually between 1936 and 1960, despite the famine, and density reached 48 per square kilometer. Growth may have been fastest in the central Rif, the most mountainous sector.[108] Since 1960, in the absence of serious drought, epidemic, or war, population has grown at almost 3 percent per annum, and even higher around Ketama in the cedar zone.[109] Only in the artisanal area of Tarzhoute has population been anywhere near stable since 1960. Population density is now well over 70 per square kilometer in the western Rif, up by half since 1952. It is over 65 per square kilometer in the central Rif, where it was 30 in 1950. A quarter of the natural increase in the central Rif is lost to emigration – two-thirds in the eastern Rif. Emigration accelerates, but population still climbs.[110]

The Rif has a notably different population history from the other mountain chains of Morocco. Not only is it two or three times more densely settled, but its growth has kept pace with that of the country as a whole into the 1990s. The Middle and High Atlas chains, although

[107] This is a maximum estimate derived from the figure for the Spanish Zone for 1910 (760,000 or 766,000) given in Woolman 1969, 18, and Harris 1927, 22, and the figure for 1928 of 551,247 given in Roda Jiménez c. 1940s, 3. The Spanish army figured that Beni-Hozmar (near Tetouan) had over 8,000 people before the war, and 5,579 after it. SHM, Monografías de Kabilas, 3:51, Beni-Hozmar, p. 11. See the note in Table 5.18 on the Beni Ouriarhel.

[108] Noin 1970, 1:41–2 and 2:119, for growth rates and densities for a Rif that he does not define, but which clearly includes almost all of the Spanish Zone. In the central Rif, 1930–50, population growth came to 1.8% per annum. G. Maurer 1968b, 21.

[109] Pascon and van der Wusten 1983, 111–14. G. Maurer (1968b, 21) gives 2.9% to 3.0% for the central Rif in the 1960s, 2.5% for the 1950s. Fadlouliah (1987, 465) gives slightly lower rates for the 1960s and 2.5% for 1971–82. Ketama in the same years grew by 3.9% (p. 467).

[110] Fadlouliah 1987, 466–7 and 481–2; Bisson et al. 1985, 55. In 1960 Rif population density was 62 per square kilometer and in 1952 it was 54. FAO Library (Rome) SF: 18/MOR (1965) 1:16; Morocco, Ministère de l'Agriculture 1960, 24.

Table 5.16. *Population of selected Rif tribes, 1933–1982*

	1933	1935	1945	1950	1960	1971	1982
Tribes							
Beni Bouzra	3,231	3,165	5,106	4,172	5,010	6,544	8,013
Beni Rhzen	3,503	3,819	3,467	3,901	5,150	6,734	9,066
Beni Grir	2,951	3,143	2,885	2,725	3,839	4,793	5,886
Mtioua	6,355	6,679	6,324	5,878	9,532	12,510	15,270
Beni Ammart	7,933	8,012	11,876	11,008	13,246	16,393	18,578
B. Bou Frah	5,731	6,171	6,542	5,497	6,689	8,632	9,019
B. Ouriarhel	46,488	47,529	66,179	58,291	75,988	98,363	133,212
Total	76,192	78,518	102,379	91,472	119,454	153,969	199,044
Towns							
Ketama	8,333	8,391	11,293	11,268	7,115	9,464	14,476
Targuist	1,448	3,413	2,795	2,662	2,297	1,932	3,926

Note: The 1960 census has slightly different figures for 1960 than those reported in the 1971 census. I have used the latter source, assuming that it involves some corrections. The differences are always less than 5%.

Sources:

1933: *Boletín de la Sociedad Geográfica Nacional* 74 (1934): 339–40.

1935–50: Spain, Dirección General de Estadística, *Zona del Protectorado en el Norte de Africa. Anuario estadístico*, 1941, 1946, and 1955.

1960–71: Morocco, Direction de Statistiques 1971.

1982: Morocco, Direction de la Statistique 1982.

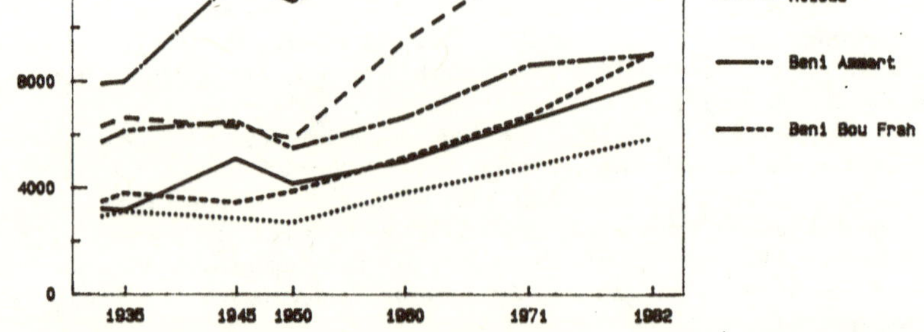

Graph for Table 5.16

Table 5.17. *Beni Bou Frah population, 1926–1979 (according to Pascon)*

1926	1932	1935	1940	1945	1958	1970	1979
3,479	5,348	5,791	5,382	6,586	6,051	8,036	9,047

Source: Pascon and van der Wusten 1983, 111–14.

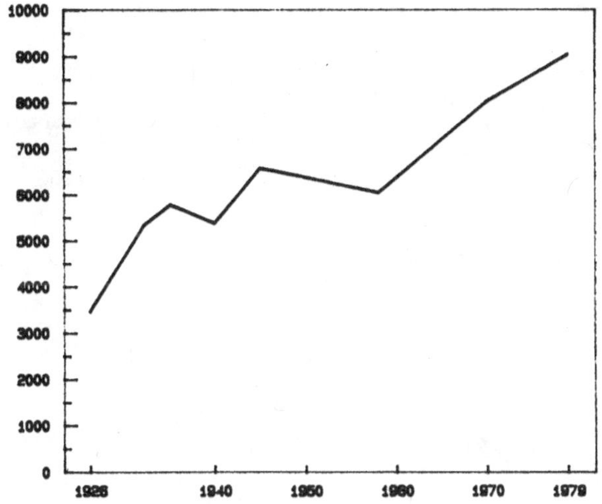

Graph for Table 5.17

their population has grown in this century, have not kept pace.[111] Emigration has absorbed less of the Rif's natural increase than in the sparsely populated higher Atlas chains. It is possible that by uniting the tribes and ending blood feud, Abdel Krim inadvertently prepared the way for overshoot in the Rif. Whether this is true depends on the degree to which blood feud regulated population: there is no doubt that after Abdel Krim internal violence in the Rif was demographically insignificant. No doubt diminished impact of epidemics also played a role, in the Rif as elsewhere.

Emigration from the Rif has allowed the region to survive to the present without fundamental changes. (Kif sales help too.) If the Rif is

[111] In 1900 the Rif accounted for about 11.75% of Moroccan population, and 10% in 1982. Other mountain areas totaled 21.4% of the Moroccan population in 1900, but only 11.75% in 1982. Bisson et al. 1985, 147.

Table 5.18. *Beni Ouriarhel population, 1929–1960 (according to Hart)*

1929	1938	1945	1952	1960
39,537	47,629	65,982	58,291	75,988

Note: La Martinière and Lacroix (1894, 1:338) give 35,000 for Beni Ouriarhel in about 1880. The Spanish army estimated 24,000 or 25,000 in 1927 (SHM, Monografías de Kabilas 2–24, 2).
Source: Hart 1976a, 17–18.

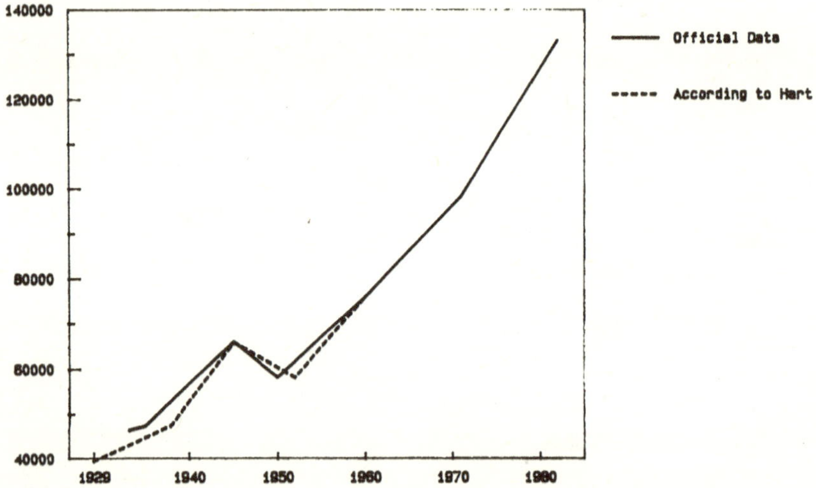

Graph for Tables 5.16 and 5.18 (Beni Ouriarhel, 1929–1982)

living on borrowed time, as I think is likely, the lenders are the lands, Moroccan and foreign, that accept Rifian immigrants. The Rif has long exported surplus people, to Fez, Tetouan, Tangier, and other Moroccan cities. Ordinarily the numbers involved before the late nineteenth century must have been small, although perhaps at times overpopulation drove battalions of hungry Rifians to the cities. At any rate, this early migration history is completely unknown.

In the late nineteenth century long-distance (and probably larger-scale) patterns of migration developed. In the 1850s Rifians began to apply for papers at Melilla to allow them to work on the farms of

Oranie, in Algeria.[112] By the 1880s a regular pattern crystallized that would last to the 1950s, interrupted only by the Rif War.[113] Rifians from the Nekor Valley east went to Algeria for several months a year, usually December through February, when the grapes were picked in Algeria, and April through June, when the cereal harvest came in. The rest of the year, and certainly for autumn plowing and summer harvesting, the Rifians were back home. By the end of the nineteenth century, this seasonal migration involved 40,000 to 50,000 men.[114] During the Rif War the *colons* of Oran found their labor force cut by two-thirds, as Rifians stayed home to fight. But with peace, things returned to normal, so that by 1931–2, one-tenth of the total population of the central and eastern Rif were working in Algeria.[115] By the early 1950s, about 35 percent of the adult male population of the Beni Ouriarhel migrated to work in Algeria in normal years. More went when the Rif harvest was poor, fewer went when it was good. Most of these men were young, all left their families behind, and all sought simply to make as much money in as short a time as possible. They eagerly worked overtime; the colons thought them much better workers than the Algerians, and the Rifians agreed. In the 1950s an ordinary laborer could send home about a quarter of his wages. All this came to an end quickly between 1955 and 1962, as the Algerian War made crossing the border difficult. A few Rifians worked for *pieds noirs* in Corsica after Algerian independence, but with this brief and small exception, this gambit ended forever in 1962.

Rifian migration to Europe has taken its place. This involves far fewer people, but the pay is so much better that the effect on incomes in the Rif is comparable, perhaps greater. The first significant labor migration to Europe came with the Spanish Civil War, in which (in 1938 alone) 50,000 to 60,000 men from the central and eastern Rif fought for Franco.[116] The total population of the Spanish Zone at this time came to less than 800,000 (Table 5.19). Perhaps one in three adult males from these areas fought in Spain, and their enthusiasm made a lasting

[112] AHN, Estado, 8357, "Sobre la Resolución Adoptada por las Autoridades Francesas Acerca de los Marroquies que llegan a las Puertas de Argelia."

[113] Milliot (1934), Benedicto Pérez (1949), and Hart (1976a, 88–95) detail this migration.

[114] *Abd el-Krim et la république du Rif* (1976), 57–8, commentary by J.-L. Miège.

[115] Milliot 1934.

[116] Pascon and van der Wusten 1983, 105, citing Seddon 1981, 130–1. Rifians fought in the Spanish army before the war, crushing the Asturias miners' rising of October 1934 for instance.

Table 5.19. *Population of Morocco, Spanish Zone, and Fez, 1550–1988*

Date	Morocco	Spanish Zone	Fez	Source
c. 1550	3,450,000		70,000	a
1820			88,000	b
1858			80,000–90,000	c
1900	4,800,000		95,000	d, e
1928		551,247		f
1933		720,273		g
1935	6,775,000	795,336		b, f
1952	8,612,000	1,003,817		b, h
1960	11,450,000		216,733	i
1971	15,400,000		325,327	i
1982	20,419,000		448,823	j
1991	26,200,000			k

Notes: (1) For Morocco 1935 figure is actually for 1936. (2) For Spanish Zone, 1952 figure is for 1950; Noin estimates 954,000 for the Spanish Zone in 1952, feeling the official tally is high. Also Woolman (1969, 18) has 760,000 for the Spanish Zone in 1912, high, but not impossible, given the cost of the 1921–6 war. (3) For Fez, Le Tourneau (1949, 155) gives 115,000 for c. 1900.

[a] Noin 1970, 1:235–40.
[b] Noin 1970, 1:26–34.
[c] Aguirre 1858, 22.
[d] Noin 1970, 1:249.
[e] Noin 1970, 2:301.
[f] Roda Jiménez, c. 1940s, 3.
[g] *Boletín de la Sociedad Geográfica Nacional* 74(1934): 340.
[h] Roda y Jiménez and García Figueras 1950–5, 3:299.
[i] Morocco, Direction de Statistiques 1971.
[j] Ibid. 1982.
[k] World Population council estimate.

impression on Spaniards. This employment came to an end in 1939, although many Rifians continued to serve with the Spanish Army in Morocco until independence in 1956. Many others served with the French armies in Africa.

Another avenue to Europe, far safer and far more remunerative, opened up in the 1960s and 1970s. Rifian migrants went first to German factories and mines, along with Turks, Greeks, Yugoslavs, Italians, Spaniards, and others. There they could earn wages ten times what work in Morocco brought, and soon they began to return with radios, cars, and cash. Eventually they worked in France and Holland as well, so that by 1979 250,000 Moroccans were at work in Europe, a large

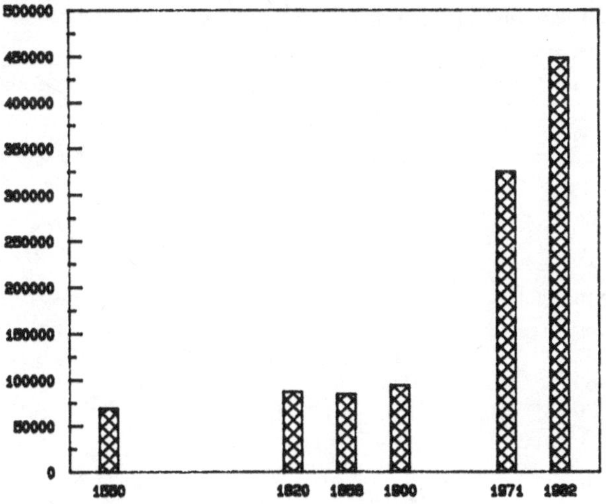

Graph for Table 5.19 (Fez, 1550–1982)

portion of them Rifians.[117] Dutch factories became especially popular destinations, in part because workers could claim social security benefits for family members even if they were still in the Rif. In 1968, 95 percent of Moroccans working in Holland (about 15,000 in all) were Rifians.[118] Some tribes had 10 percent of their population working in Europe by 1966.[119] Among the Beni Ouriarhel at least, those from the higher altitudes predominated, true to the pattern in other mountain chains around the Mediterranean. Rifian workers in Europe tend to stay on the job for at least six months at a time, longer than in the old Algerian migration. They often stay a year or two, and sometimes linger on when their working papers expire, living on the fringes of European society. No Spanish city is without its Moroccan street traders, selling wallets and belts when they can, scurrying from the police when they must. But probably the majority of migrant workers move back and forth from the Rif to their northern factories within the law, with one notable exception: they manage the kif export business.

Moroccan cities have also received plenty of Rifians fleeing poverty in the twentieth century. No migration numbers exist (to my knowl-

[117] Gazzo 1979. In the mid-1960s the Rif *cercle*, in Nador province, had more of its natives outside the country than did any other *cercle* in Morocco. Noin 1970, 1:304–5.

[118] Hart 1976a, 94, n. 29.

[119] G. Maurer 1968b, 25.

Table 5.20. *Population of northern Morocco cities, c. 1500–1982*

Date	Tetouan	Chechouan	al-Hoceima	Source
c. 1500	5,000			a
c. 1540		5,000		b
c. 1690	15,000			a
1858	12,500			c
1885		7,500		d
1900	22,000			e
1913	18,519			f
1927	36,085			f
1933	48,347	6,618	3,965	g
1936	55,269	7,749		f
1940	76,420			f
1945	70,078	12,302		f
1950	80,732	14,476		f
1960	101,000	13,712	11,262	h
1971	139,105	15,362	18,686	h
1982	199,615	23,563	41,662	i

[a] Latham 1965, 405–6.
[b] Marmol, cited in Noin 1970, 2:300.
[c] Aguirre 1858, 23 (he gives 12,000–13,000).
[d] Foucauld 1888, 1:9 (he has 7,000–8,000).
[e] Noin 1970, 2:301–2.
[f] Roda y Jiménez and García Figueras 1950–5, 3:300.
[g] *Boletín de la Sociedad Geográfica Nacional* 74 (1936): 341.
[h] Morocco, Direction de Statistiques 1971.
[i] Morocco, Direction de la Statistique 1982.

edge), but urbanization since the 1950s has proceeded rapidly in Morocco (although less so than in Algeria and Tunisia), drawing on rural exodus (Table 5.20). Those in the northern Rif have headed for the northern cities, Tetouan and Tangier, whereas southern Rifians have gone to the larger centers of Fez and Meknes. Many of these migrations are permanent, whereas those to Algeria or western Europe were seasonal or annual. And they are generally far less lucrative movements, since the wage scale in Moroccan cities, for those lucky enough to find work, compares poorly with that in Europe.

All these migrations, to Europe, to Europeanized Algeria, and to Moroccan cities, have three important features in common. First, the movements arose for economic reasons: all surveys show that migrants took the plunge because economic conditions back home looked so un-

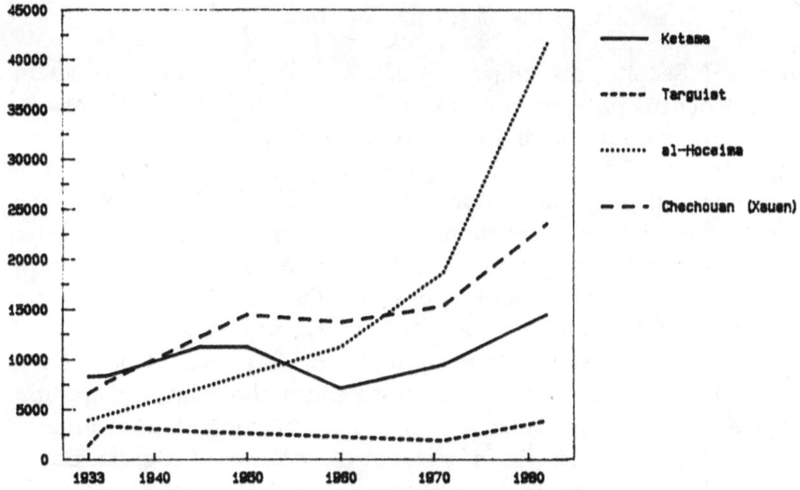

Graph for Tables 5.16 and 5.20

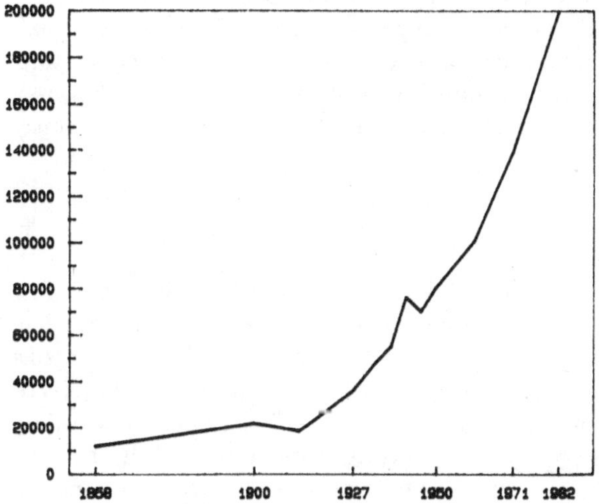

Graph for Table 5.20 (Tetouan, 1858–1982)

promising.[120] Second, the migrants were, and are, very young (median age of Moroccans moving to cities in the early 1980s was twenty-two), and, with the exception of the movement to Moroccan cities, overwhelmingly male. Third, again with the exception of the movement to Moroccan cities, these migrations allowed men to send home large amounts of money to their families. A regular, if informal, courier system existed from Algeria to the Rif to repatriate wages, and it now exists from Europe. This money permits Rifian families to survive on less and less land. In effect it has perpetuated an old, ineffective, and ecologically destructive system of agriculture, which would have been abandoned by now without remittances. Remunerative labor migration to Europe has probably prevented, or at least postponed, a far greater exodus to Moroccan plains and cities. In the Kabylie of Algeria, three-fourths of income derived from emigrants in the early 1980s.[121] It is surely less in the Rif, but it is true nevertheless that the Rifian economy is now fragile in a new way. It is dependent not merely on the rains, but on the hospitality of distant lands. A slump in the Dutch economy would be disastrous for the Rif: it would oblige far more Rifians to seek their fortunes in Fez and Casablanca. The occasional shiny Mercedes and spruced-up dwelling earned by long hours in foreign lands mask a new form of insecurity for the Rif. It remains a peasant massif, dangerously in overshoot.[122]

Conclusion

These five mountain areas have distinct population histories, but a few common threads exist. In the past two hundred years population has grown in all these highlands, and everywhere except the Taurus it has grown so much so fast as to produce acute hardship and desperate emigration. The rhythms of emigration show some similarity in many cases: in the Pindus, the Apennines, and the Alpujarra, and the eastern Rif, the late nineteenth century saw the beginnings of large-scale mountain emigration. It generally stagnated in the early twentieth century

[120] Bisson et al. 1985, 171.

[121] Bisson et al. 1985, 320–22.

[122] G. Maurer (1968b, 66) wrote of the central Rif: "People [are] far too numerous for natural resources." Montalembert (1972, 52–3) maintained twenty years ago that population growth and extension of cultivation in the Rif had gone as far as they could. They were right in the fundamental sense but failed to see the degree to which remittances could keep the Rif afloat – at a cost. Plit (1983, 79) argued that population pressure was the fundamental cause of environmental degradation in all the Maghreb. I think it likely that the entire Maghreb is destroying its resource base for the future in order to survive in the present.

(the Rif is an exception here), and accelerated in the postwar period. This has effectively emptied out the Pindus and the Alpujarra, but not quite emptied the Apennines, where agriculture is hanging on and other activities better complement cultivation. But nothing of the sort has happened in the Rif or the Taurus. Part of the explanation is that birthrates remain high in these Muslim mountains, so that even substantial emigration has not stopped population growth. But beyond this there is the matter of the prevalent system of human ecology in the mountains, changes in it, and the timing of the changes.

The Taurus is something of a special case in the Mediterranean context, although it may be typical of mountains of southwest Asia. It carried little population from the tenth to the nineteenth century because it had been conquered and controlled by nomads. The Pindus too was a shepherd's massif from ancient times until the fifteenth or sixteenth century. But the other chains, the Apennines, the Alpujarra, and the Rif, were all peasant mountains, unaffected by nomads. Under the control of nomads or seminomads, the Taurus and the Pindus suffered limited ecological damage. Nomads went up and down with the seasons, burning forest to expand pasture and trampling vegetation with their herds and flocks. But they stuck to the same routes and were few in number (see Chapter 7). As long as they regarded their animals primarily as subsistence rather than commodities, they had no reason to expand their operations beyond what they and those depending on them required. The peasant mountains, on the other hand, suffered from the constant degradation that fluctuation in population brought. Agriculture and timber cutting, combined with stock raising, put far greater stress on mountain ecosystems than seasonal grazing, except perhaps in conditions of population stability and careful husbandry. But no matter how conscientious and knowledgeable the peasantry, epidemics, droughts, war, and brigandage ensured that these conditions were rarely met. The ecological costs varied with the character of vegetation and soils, of course, but everywhere in the peasant mountains wood and soil vanished at higher rates than in the shepherd mountains.

The Pindus changed from a shepherd massif to a peasant one in the sixteenth and seventeenth centuries, and thereafter its landscape history was governed by different factors. The Taurus made this change only in the nineteenth and twentieth centuries, and the stress inherent in peasant exploitation has yet to become serious, and perhaps never will, if other factors prevent overcrowding and overshoot.

Political violence entered the equation as well, both as a response to overcrowding and as a source of landscape degradation. In general, it has had only a limited effect on population. Blood feud among the Rifians probably checked demographic growth more effectively than did the sporadic outbursts of warfare elsewhere, excepting the disaster in the Alpujarras of 1568–71. Elsewhere war generally produced a moment of exuberant fertility in its wake – a baby boom – and had no more durable effect than a famine or epidemic, and usually less. But war and brigandage could lead to landscape neglect and outright destruction, from the Hannibalic campaigns in ancient Lucania to the Greek Civil War (see Chapter 7). In the fragile environment of the Mediterranean mountains the consequences endure in a way they might not elsewhere.[123] Crowded landscapes implying diminished opportunity fueled war, brigandage, and perhaps blood feud, although the proximate causes of violence must be sought elsewhere.

Overcrowding and overshoot created ecological refugees by the thousands, peasants who fled their mountains for plains and cities, for northern Europe and America. At first they left because they correctly discerned that life in the village held diminishing prospects for them. Soon the success of some led to the departure of many more, and in the Pindus and Alpujarra, where agriculture needs many hands, undershoot ensued. In the Christian Mediterranean, America (North and South) absorbed the first wave of immigrants, and industrial Europe the second. For Rifians, Algeria was their America, conveniently close but sufficing only for labor migration and not as a permanent destination. Hence the Rif was more overcrowded in 1920 than any of the other mountain zones. This condition helped persuade Rifians to shelve their differences and internal struggles and unite behind Abdel Krim in the Rif War. The Taurus mountaineers did not need an America, as their numbers remained low until quite recently. They may or may not find one in Germany or the Gulf states.

Population is by no means the only factor in landscape change and the human condition in the Mediterranean mountains. Local and global shifts in economic organization have had strong effects as well. These are the subject of Chapter 6.

[123] Toynbee (1965, 2:35) says that the scars of the Hannibalic war still remain 2,200 years later in Lucania. It takes careful observation in northern France to detect the battle fronts of 1914–18, where bombardments destroyed all life except for men, lice, and rats. Today forests, fields, and meadows disguise the damage, indeed have repaired it.

6

POLITICAL ECONOMY AND MOUNTAIN LANDSCAPES

> Now in earlier times the world's history had consisted, so to speak, of a series of unrelated episodes, the origins and results of each being as widely separated as their localities, but from this point onward history becomes an organic whole: the affairs of Italy and Africa are connected with those of Asia and Greece, and all events bear a relationship and contribute to a single end.
>
> Polybius (died 118 B.C.)

> In place of old wants, satisfied by productions of the country, we find new wants, requiring for their satisfaction the products of distant lands and climes. In place of the old local and national seclusions and self-sufficiency, we have intercourse in every direction, universal interdependence of nations.
>
> Marx and Engels, *The Communist Manifesto*

The industrial revolution changed economic life in corners all over the world. So did the extension and tighter integration of market systems. World trade grew rapidly, and the mountains of the Mediterranean, like many other places, now felt the impact of economic events and trends originating on different continents. Long-distance trade of course had had a long history in the Mediterranean world. Caravans had connected the Mediterranean with China, India, and West Africa for centuries. Ships had made the connection with the Americas since 1492. But the scale of things changed with the worldwide population and economic growth that began in the eighteenth century and hit full stride in the nineteenth. By 1880, for instance, the falling price of wheat in the American Middle West affected mountain villages in southern Italy. Wool and refrigerated meat from Argentina changed the prospects for pastoralists in Spain. The world of the steamship and the railroad, of integrated markets and efficient information exchange, strongly affected communities beyond its frontiers, except for the very remote.

Villages in the Mediterranean mountains were generally not remote enough.[1]

Disequilibrium between population and natural resources threatened the survival of mountain populations, dependent as they were on their soils, pastures, and forests. They depended as well on their auxiliary activities – seasonal labor, petty trade, crafts, and the like. Economic shifts entirely beyond the control, and often beyond the understanding, of mountain peasants menaced villages in two ways. First, some contributed to the ecological degradation of the mountains. Mining and logging, for instance, often undermined the villages' resource base. Second, some shifts undercut the auxiliary activities of mountain folk. The decline of cottage and household industries, in northwestern Europe a quick and direct result of the greater efficiency of machines, eventually came to the mountain districts of southern Europe, North Africa, and Anatolia.

Both of these economic changes represent the forces of market integration so characteristic of the nineteenth century. Ecologically destructive activities focused concentrated demand of ever-larger economic units upon small landscapes. And efficient producers could market their wares farther and farther afield, displacing competitors throughout ever-widening economic spaces. The spectacular improvements in transport and communications, as well as in production techniques, drove these shifts. This is a large reason why hard times began to visit mountain villagers more often in the nineteenth century than before.

The nineteenth century was also crucial to mountain villages for political reasons. Political forces, once again well beyond their control and sometimes their understanding, altered the circumstances of material life for mountain villagers. The ideology of private property in land triumphed, most conspicuously in Italy and Spain, with distinct consequences for forests and pastures throughout these countries, and for mountain ecology in particular. In international politics, war became the affair of populations and not merely of professionals, and so took on a new importance for mountain peoples and landscapes.

Every locality had its own chronology, but on the whole, the environmental decay of mountain life proceeding from economic and political causes was an affair of the nineteenth century in Mediterranean Europe and of the twentieth century in Mediterranean Africa. In

[1] The exceptions were more frequent in the Rif and Taurus than in the Christian Mediterranean. Once again evidence is very thin for the Rif and the Taurus. So the appearance of comparative isolation may be deceptive.

Anatolia the processes began early but have only recently acquired major importance.

The Loss of Auxiliary Activities

Crafts

Throughout the Mediterranean, mountain people augmented their meager incomes by craft work. Women specialized in cloth, and men in wood, metals, and leather. In the course of the nineteenth century, northwestern European factories came to produce both cloth and metal goods of better quality and at lower price than could manual labor anywhere. Soon, railroads and steamships brought these goods to the Mediterranean, driving local people out of business. This change mattered most where metalworking and textiles were vital parts of village economies, as in the Pindus and the Lucanian Apennines. I will tell the story only once, for the Pindus.[2]

In the seventeenth and eighteenth centuries, the villagers of the Pindus had developed high skills in metalworking and wool processing. Eventually Ioannina came to dominate the metalworking trade, although the larger Vlach villages, such as Kallarites, retained reputations for fine handiwork in gold and silver. The spinning and weaving of wool remained the province of women in the high Pindus. The Balkan caravan system dispersed woolen goods throughout the Balkans, Austria, and beyond. Great quantities of spun wool traveled north to looms in Vienna, Leipzig, and Breslau. Woven blankets and cloaks found buyers throughout the Balkans. Istanbul bought exquisitely embroidered woolens from the Pindus. Local skill with metal and wool helped invigorate the trade fairs at Ioannina, Kónitsa, and elsewhere. Even in the rather slack years of the 1840s, Ioannina's bazaars reminded a French visitor of Les Halles's "pandemonium of commerce."[3]

The vigor of craft production in the Pindus, and the trade that rested on it, decayed from the 1830s. The Continental System of Napoleon's empire had preserved central European wool markets from British and

[2] For Lucania: Racioppi 1902, 363–71; Spera 1903, 36–7. On Jbala, see Fay 1979, 89; and for the central Rif, G. Maurer 1968b, 59. In the Rif the decline of local crafts is a story of the mid-twentieth century.

[3] Malherbe 1846, 1:311. Issawi 1980, 144, on the Ioannina trade fair. Mekios 1909, 118–19, on Kallarites metalworking. Hobhouse 1813, 1:67–75; AMAE-P, Mémoires et Documents, Turquie, 53:16–17, Boislecomte to ?, 8 July 1834; Hahn 1854, 54–7. For the Balkan caravan system: Stoianovich 1960; Anoyatis-Pele 1988.

overseas competition until 1815. It had kept out British manufactures of all sorts. But after 1815, British competition began to undercut Balkan cotton manufacture, in eastern Thessaly for example, where Ambelakia and other villages had sold cotton cloth throughout the eastern Mediterranean.[4] Some of these villages employed mountain folk during the winter months, like Tirnavos, a Vlach center especially favored by Samarina natives. There two thousand handlooms in 1812 dwindled to only two hundred by 1830 as German markets for muslin were lost.[5] In the Pindus, remoter than Ambelakia and the other textile villages of Thessaly, distant competitors seem to have made little inroad until the 1830s (perhaps because Epirus collapsed into war from 1821 until 1832). In 1834 the French consul in Ioannina reported that English and Austrian cloths, cottons especially, had begun to capture local markets, to the dismay of local producers.[6] By 1859 a large part of Ioannina's imports were metals and manufactures imported via Trieste from Britain, Austria, and Switzerland.[7] The British consul in 1863 reported that Ioannina's trade in general had declined sharply since the establishment of steam navigation on the Adriatic in the 1840s.[8]

The power of steam had begun to undermine both the caravan system (of which more later) and the textile and metalworking trades of the Pindus – and indeed many other parts of the world, from Scotland to India and beyond. Soon the wool trade to central Europe declined as well, because good wool in huge quantities from Argentina and even Australia permitted western European production to undercut central Europe's. The steam power that brought distant continents into regular commerce with western Europe, and at the same time revolutionized western European production techniques, converted the Pindus into a zone of almost exclusively raw material production. This was a familiar story around the world, from the Andes to Angola, wherever market integration put local textiles and metallurgy into competition with British factories.

Indeed the scale of the adjustment required in the Pindus was small compared with that in former textile centers, like Tlemcen in Algeria.

[4] By 1850 or so these villages were deserted. Stoianovich 1967, 88. AMAE-P, Mémoires et Documents, Turquie, 35:315, Pouqueville to Duc de Bassano, 15 November 1812.

[5] Stoianovich 1967, 88. Lawless 1977, 520–1.

[6] AMAE-P, Mémoires et Documents, Turquie, 22:395–402, Boislecomte to ?, 4 September 1834.

[7] Schläfli 1859, 16.

[8] PRO FO 195/751, "General Report on Epirus," 1 December 1863. Stuart made the same point in his "Commercial Report for the Half-Year ending June 30, 1864," PRO FO 195/801/155–63.

But in the Pindus, and in ecologically marginal zones generally, the adjustment was harder to make because the options for making a living were so few. Locals could ill afford to abandon these activities and did so with great reluctance. Eventually they turned their time and ingenuity to other pursuits – or emigrated. Pindus natives were fortunate in that many of them had wide experience of the world, or at least of the Balkans, and could find other work close to home more easily than could Lucanian or Rifian mountain folk.

Men often found other ways to compensate for their lost custom. But women and children could find few auxiliary activities to substitute for the textiles. Among the effects of the strong competition from machine-made cloth was to deprive families of some of the fruits of female and child labor.

In turning more fully to primary production – livestock, timber, raw wool – the Pindus found itself in competition with the great frontier lands of European settlement. Argentina, Canada, and other landscapes enjoyed all the advantages in this competition, as long as regular navigation and communication on the high seas proceeded smoothly. By 1864 Epirots imported hides from Buenos Aires.[9]

In silk, economies of scale mattered less. Raw silk weighed little but cost a lot, so it, far better than wool, could bear the expense of overland transport from mountain districts to Mediterranean cities. Tending mulberry trees and raising silkworms was painstaking and labor-intensive work, unsuitable to the frontiers of the world economy that did so well with cotton and wool, but quite appropriate wherever female and child labor existed in oversupply. For these reasons, silk production proved very suitable employment for mountain women and girls. With population growth, and the decline of wool manufacture and other crafts, the Pindus, Lucanian Apennines, and the Rif turned to silk in the early and mid-nineteenth century. Many other places in the Mediterranean world did so at the same time, including northern Italy and the hills of Palestine. Of course, the Alpujarra had specialized in silk from Moorish times. The story of the decline of silk in the Mediterranean mountains consists of two chapters, the first of which concerns the Alpujarra.

The Arabs, who imported silk culture from Persians who had gotten it from the Chinese, raised it to a high and prosperous art in Granada

[9] PRO FO 195/801/155–63, "Commercial Report for the Half-Year Ending June 30, 1864." Imports through Arta consisted of British cloth, sugar, coffee, and copper goods. Exports were predominantly livestock, horses, and mules to Greece, sheep and mules to the Ionian Islands, but also skins and wool to Trieste and silk to Italy. Babaretos 1929, 11–13, has additional Epirus import–export data, reported by Russian consuls. Typaldos Forestis 1912 has Epirus import–export figures for 1911.

and the Alpujarra. Every village had its looms, and Granada, Córdoba, Málaga, and Almería produced great amounts of silk, much of which, by the fourteenth and fifteenth centuries, found its way to Italy. Perhaps because the art was so closely identified with the Moors, Spanish crown policy after the reconquista made it a target. The mesta acquired privileges in the province of Granada in 1520, bringing sheep into mulberry zones, disrupting the fussy silkworms. Facing financial crises, Charles V and Philip II put heavy taxes on silk, helping to foment the great rebellion of 1568–71. In Mecina Bombarón, for example, silk paid three-fifths of the village's tithe in 1553.[10] After the expulsion of the Moriscos, royal order mandated the cutting of mulberry trees and forbade irrigation of tree crops.[11] In the Alpujarra, Orgiva, Ujíjar, Juviles, and the Poqueira barranca all were affected. After these setbacks, silk culture struggled along for centuries as a minor activity in the Alpujarra. A royal chartered company (Compañia de Granada) tried to revive manufacture in Granada beginning in 1747, and encouraged silk culture throughout southeastern Spain. This small example of Bourbon state capitalism gave the city of Granada an economic boost but developed rival sources of raw silk, notably in Valencia, to compete with the Alpujarra. The company went under in twenty years, ending official sponsorship of silk. The state abolished the mesta in 1836, to favor the development of agriculture. But nothing could resuscitate silk. In the 1840s about half of Alpujarra villages raised silkworms but in small quantities. The total production of the Granada factories came to about 6 percent of what it had been in the days of the Nasrids.[12]

At this point silk in the Alpujarra became very much like silk in the Pindus or Lucanian Apennines. Its long and glorious tradition meant nothing, although quality remained high (the German imperial family bought Alpujarra silk in the late nineteenth century). It was merely one of several auxiliary activities that brought in a little cash for a few families who had the trees, the silkworms, and the labor. Perhaps it had revived a little because of the strong growth in demand from the manufacturers in Lyon (the Granada of the nineteenth century). But in the next fifty years, by 1890, it withered in the face of competition

[10] Vincent 1982a, 186. On silk, Sanz Sempelayo 1980, 73–92; Garrad 1956; Ortega Alba 1981; Garzón Pareja 1972.

[11] Garrad 1956, 91–6; Ortega Alba 1981, 59.

[12] The company's woes are recounted in Sanz Sempelayo 1980, 77–84. The affairs of the company appear in Archivo General de Simancas, Hacienda, legajo 857. It never got the raw silk it needed: in 1752 only 40% of Granada's silk workshops were active: BRAH, ms. 9–5. 724, f. 796, "Noticia de los Telares de Seda . . . que ai en el Reino," 19 February 1752. See also Navarro Alcalá-Zamora 1981, 44–5; Bernal and Drain 1985, 423; Díaz Lobón 1982, 47.

from East Asia, the original home of silk culture. This is the second chapter in the decline of Mediterranean silk.

After 1863 Japanese silk came by ship and rail to the looms of Lyon and other centers, causing a crisis among Italian producers by 1881. The price of raw silk fell by 50 percent in the 1880s in Italy. Tuscany, the biggest producer, suffered most, but the southern mountains of Calabria and Basilicata lost a crucial source of income.[13] In the Alpujarra, a traveler of the 1880s reported the silk industry extinct, only a small exaggeration.[14] A French company tried to revive silk culture in Ujíjar, but got nowhere with it. By the 1920s it had all but disappeared.[15] Silk production in Japan and China was too large and too skilled for Mediterranean peasants. Those at any distance from the centers of manufacture, notably Lyon, quickly went out of business once the Americans opened Japan to European commerce (1853) and British navigation linked Japan to Europe. Thus ended a business that helped mountain people get by for generations.

Textile crafts of every sort suffered severely from the competition of machines and cheaper raw materials from distant continents. Even in rural areas tastes quickly adjusted and machine-made cloth and other goods became preferred items. Only when rural crafts came to the brink of extinction, in the twentieth century, did they acquire a market value, based on a nostalgia among urban people more than anything else. This has served to keep some of the skills alive even today, generally those that contribute to the creation of decorative objects. The Pindus shepherd's cloak (*flokati*), for instance, enjoyed a certain vogue that kept Samarina women in business after other local crafts had disappeared.[16] Objects for use come from factories.[17]

Seasonal Migrant Labor

Wage labor in the lowlands had long helped mountain folk to make ends meet. In the nineteenth century several new niches opened in the

[13] ACSR, MAIC, DGA, VI, busta 359, f. 29, "Tableau des récoltes soyeuses françaises. Tableau des récoltes soyeuses italiennes." On the decline of mulberry cultivation in Lucania, see Italy, Giunta per la Inchiesta Agraria [Jacini] 1883, vol. 9, fasc. 2, p. 11.

[14] Rubio 1881, 329.

[15] Brenan [1957] 1988, 95–6. Ujíjar's silk factories declined, employing by 1900 no more than two hundred workers (all women). Soler y Pérez 1906, 31–3.

[16] Sivignon 1968, 22.

[17] An exception is the plow used in the Rif. It is made of local wood, by hand. To some degree today's market for mountain crafts may reflect a return to roots on the part of descendants of mountain people, parallel to the reassertion of ethnicity among descendants of assimilated immigrants.

lowlands, as new crops boomed and new industries started up. But for mountain people more niches closed, as the low coastlands acquired their own permanent populations, and as distant competition provoked a general crisis of lowland agriculture.

One of the venerable forms of seasonal labor for mountain folk had been military service. However, when the Italian, Spanish, and Ottoman governments introduced universal military conscription, the opportunity for seasonal military work diminished sharply. Armies sought to maintain constant levels of manpower throughout the year, as the old tradition of a summer campaigning season disappeared. These modern armies paid their recruits next to nothing and eliminated almost all chances of booty from the battlefield. Military work no longer suited the needs and the calendar of mountain peasants and became something to avoid rather than to seek. Alpujarreños of the 1880s resented no demands of the state more than military service.[18] The chief exception was the enthusiasm with which Rifians served in the Spanish and French armies between 1927 and 1956. In this case the differential wage scale between European armies and other opportunities open to Moroccans (and the promise of a pension) made even long years of service attractive.

Brigandage also dwindled late in the nineteenth century, removing another auxiliary activity. After 1865 the newly unified Italian government hunted down brigands with considerable success in Basilicata, although they continued to flourish in Sicily and Sardinia into the twentieth century. After 1880 emigration gave potential brigands another field for their talents. In the Alpujarra, where brigandage had rarely amounted to much anyway, the state had equally good success. One might meet a lone bandit in the mountain passes, but brigand gangs had disappeared by 1910. In the Pindus brigandage flourished after 1881 because bands could easily slip across the frontier between Greece (Thessaly became Greek in 1881) and Ottoman Epirus. The ruggedness of the mountains may also have contributed to the longer career of brigandage here. But by the turn of the century this option had largely disappeared,[19] although to some extent the guerrilla bands of 1941–9 functioned as brigands.

In the Taurus and the Rif, evidence of brigandage is less complete. It appears that brigands held sway around the western Taurus in the early seventeenth and again in the late eighteenth century, but weak-

[18] Olóriz, cited in Pino Artacho 1965, 17.

[19] Koliopoulos 1987. The last "king of the mountains" (brigand chief) in the Pindus was killed in 1924; Théophilou 1983, 88.

ened notably throughout the nineteenth. No doubt the decline of nomadism contributed to this: from the point of view of those keeping records, nomads and brigands were hard to tell apart.[20] In the confusion surrounding the death throes of the Ottoman Empire and the establishment of the Turkish republic (1915–30), public order weakened in the eastern Taurus (and elsewhere), lawful opportunities diminished, and brigandage temporarily resurged. True brigandage seems to have been rare in the Rif, el-Raisuli notwithstanding. Possibly the tribal social structure discouraged it, as any act of theft or violence invited determined revenge against one's kin. The blood feud was not brigandage: it concerned honor, not property.

The eventual decline of brigandage throughout the Christian Mediterranean and Anatolia shows the increasing power of the state, and perhaps a change in the balance of military technology: as weapon technology changed more quickly, and state of the art weapons became more expensive, brigands found themselves at a growing disadvantage against the forces of law and order. Their decline does represent the closing of a niche once open to bold or desperate men, but its closing opened many other niches. With brigandage under control, ordinary folk could more freely move up and down to the harvests, or to construction work, or to wherever a wage could be had.[21]

Seasonal agricultural work had always helped mountain folk far more than brigandage, or military service for that matter. But in the nineteenth century the opportunities dwindled in some places, and the rewards shrank almost everywhere. A few examples will suffice.

One widespread improvement, undertaken at the end of the nineteenth century in Italy and a bit later elsewhere, was malaria control. Decades before insecticides, state drainage programs removed the anopheles's breeding grounds, sharply curbing the incidence of malaria. Quinine and its equivalents curtailed mortality. This had the effect of making coastal lowlands habitable throughout the year. The result, first in lowland Lucania and then later in lowland Epirus and Anatolia, was a sharp increase in permanent population, driven by immigration from many sources (including mountain villages) but also by rapid natural increase. Soon the Ionian Sea coast and the plains of Arta and Antalya had their own populations equal to the task of gathering the olives, grapes, fruits, rice, and cotton in which these lowlands came to specialize. The need for mountain folk abated, slowly and

[20] Planhol 1958, 112–25; Gould 1973.

[21] Wace and Thompson ([1913] 1972) give a good sense of how outbreaks of brigandage disrupted migration routines.

haltingly perhaps, as malaria control and lowland population growth sputtered here and there before 1945, but in the end decisively.

Lowland agriculture underwent other changes of great importance to mountain people. The first was a general crisis brought on by the competition of American and Russian wheat. With the railroads in America, and the shrinking of the Atlantic by improved shipping, Minnesota wheat could compete in Naples with that of Campania and Basilicata. The opening of Odessa helped bring the grain of the Russian and Ukrainian steppes to the Mediterranean at competitive prices. Although this was good news for urban populations, it undermined the wheat latifundia and çifliks of the Mediterranean. The prices they could command, and the wages they paid, declined, to the detriment not only of the big landowners, but also the migrant labor they hired in the harvest season. In Basilicata real agricultural wages declined from about 1860 until emigration provoked a labor shortage by the 1890s.[22]

Lowland crops other than wheat also suffered from overseas competition. The cotton of Thessaly had employed Pindus mountaineers before the mid-nineteenth century, but American and Egyptian producers, enjoying economies of scale and labor costs at least as low, undercut and destroyed Thessaly's cotton.[23] Sugar along the Granada coast, a source of wage work for Alpujarreños for centuries, suffered from the tremendous surge in Cuban sugar production from the 1790s to the 1840s. Granadine sugar stagnated until the 1840s, while Cuba emerged as the world's largest producer. After a brief revival (1845–80), the Vega de Granada, and much of Europe, began to yield important quantities of sugar beet, a further blow to Motril and Salobreña. The rebellions and eventual independence of Cuba (1898) brought another, smaller revival to Granadine sugar, but it has stagnated since 1914. The superior productivity of the world's agricultural frontiers (achieved often at tremendous ecological costs) reduced the viability of several lowland crops from which mountain people derived seasonal wages.

Landlords throughout the Mediterranean adapted to the general crisis of agriculture in several ways, among them lowering production costs through the adoption of agricultural machinery. In the mountains mechanization made little sense.[24] But in the broad latifundias of

[22] On the agricultural crisis in Spain: Garrabou and Sanz 1985; Spanish agricultural prices are recorded in Sánchez Albornoz 1975.

[23] Sivignon 1977, 395; Stoianovich 1967, 88.

[24] In the Alpujarra in 1968 there were over eight thousand plows, twelve tractors, four threshers, and zero harvesters. Banco de Granada 1970, 169. In Lucania in 1946 over 90% of the agricultural machinery belonged to farmers in the flatland of Melfi-Rionero.

Granada or the Campania, it did, and late in the nineteenth century progress-minded landlords invested in threshing machines, reapers, and the like. The first agricultural machine appeared in Granada in 1862, an English reaper. Steam-powered threshers spread quickly in Apulia and the Materano about 1870.[25] Most of this early machinery applied only to the grain harvests. Not until well into the twentieth century did machines appear that could take on tasks associated with other crops. Agriculture in southern Italy became fully mechanized, with thousands of American tractors, only after 1945. The Konya Plain in Anatolia still had only one tractor for every two villages in 1951. But a tractor and combine in the 1950s in Konya allowed one man to do the work of four hundred men and women. In a backwater like the Marquesado, machines triumphed only in the 1970s, a triumph delayed in part by Franco's commercial and agricultural policies. But by the 1880s mountain people started to find machines working the lowland grain harvests in their place. As unloved migrants, hired on only for reaping and threshing, their jobs were the first to go.[26]

Fortunately for the migrant workers of the mountains, the decline (and gradual mechanization) of Mediterranean wheat was partly redeemed by the rise of tropical and subtropical export crops. The same improvements in transport and markets that brought American and Russian wheat to the Mediterranean also permitted the export of Mediterranean fruits to northern Europe, and eventually even further afield. Most of these crops – oranges, limes, lemons, apricots, peaches, melons, figs, tomatoes, raisins, and more – required seasonal migrant labor, at least until coastal populations grew large enough to meet the demand for hands. By and large these export crops concentrated in zones with the mildest winters, such as coastal Granada and Antalya, where harvests have augmented throughout the twentieth century. Many of these crops had grown for centuries on the southern slopes of the Rif, supplying markets in Fez and Meknes. In the Rif the market integration of the Mediterranean with northern Europe made no difference; but then neither had the crisis in wheat.

While wages and opportunities shrank in wheat and grew in citrus,

Calvello had two threshers and nothing else; Castelsaraceno had one thresher; Laurenzana had four; Terranova di Pollino had no machines at all. ASB-P, Ispettorato Provinciale dell'Agricoltura, "Elenco delle macchine . . . 1946," fasc. 4.

25 AMAPA, legajo 226–2, Memoria de la Junta de Agricultura, Industria y Comercio . . . 1862; Morano 1981, 529–30.

26 Moral Ruíz 1979, 38–49; Bernal and Drain 1985, 435–7. On the Marquesado: Arias Abellán 1984, 281–7. On Italy: SVIMEZ 1954, 210. On Konya: Helling and Helling 1958, 24–30.

they probably improved slightly in a major lowland activity for mountain people: the olive harvest. The harvesting of olives, unlike wheat, comes during the slack season in the mountains, and hence was especially attractive to highlanders. No labor-saving devices affected the harvest until late in the twentieth century. And no other zones of the world have rivalled the Mediterranean for olive production. So this work remained open to mountain folk. Indeed, opportunities may have expanded somewhat, as in this case market integration worked in favor of a Mediterranean crop. With the overseas migration of Spaniards, Italians, and Greeks, new outlets for olives and olive oil opened up in Argentina and the United States, beginning late in the nineteenth century. Soon the tastes of the Mediterranean emigrants began to filter through to other populations in these lands, and Mediterranean hillsides and low plains could sell to millions of New World immigrants and natives.

Mountain folk had also spent winters tending vines, sometimes their own, as in sheltered barrancas of the Alpujarras, but more often those of someone else at lower altitude. This line of work remained open to them through the nineteenth century, again until lowland populations grew sufficiently to handle these chores. The market for Mediterranean wine strengthened, and the area under vines expanded, most notably in southern Spain. But then late in the century, phylloxera ravaged the vines of Spain, France, Italy, and Algeria. As it spread from vineyard to vineyard, a crucial source of cash disappeared for mountain folk. In the Alpujarra, phylloxera so damaged the mountain economy that many people emigrated in the 1890s, going to Algeria (where phylloxera would soon follow them) and to South America (recall Chapter 5). By 1910, when the crisis had passed, malaria control had made considerable progress, at least in the western Mediterranean, and lowland populations were growing fast, filling the jobs mountain villagers had depended upon.

Market integration and improved systems of transport chipped away indirectly at seasonal wage work of mountain people in several ways. A more direct blow came at the expense of muleteers. In both the Alpujarra and the Pindus, mountain villagers had worked as specialists in overland transport, both as seasonal work and as a full-time pursuit. In Anatolia, mountain folk had worked as muleteers and camel drovers, especially in the winter months. The power of steam captured many of the lucrative transport routes, and shunted those dependent on animal power to the marginal ones. Long-distance caravan trade in the Balkans had flourished before 1815, when decades of war had stymied maritime commerce. The north–south route in the Pindus and Dinaric Alps acquired formidable direct competition when an Austrian steamship

6.1. The caravan system of the western Balkans required bridges like this one near Kipi in Zagori. This bridge was built in 1853, a busy time in Pindus commerce, although earlier bridges may well have existed on the same site.

company inaugurated business on the Adriatic in the 1840s. By 1871 a steamer from Trieste called weekly at Préveza, year round in fair weather and foul.[27] It took the place both of the overland caravans and of Greek shipping on the Adriatic coast. Railroad construction began in the 1880s in Macedonia, further shrinking the opportunities for muleteers. Only local distribution systems remained to them. They controlled the capillaries of Balkan trade flows, but steam power appropriated the arteries. With roads and trucks in the mid-twentieth century, the muleteers lost the capillaries too.

Something similar happened in the late nineteenth century in southwestern Turkey. Wheeled carts and suitable roads emerged as the dominant means of transport when Turkish refugees from the Crimea and the Black Sea lands settled in Anatolia. Camel caravans gradually disappeared from the region, their departure hastened by their vulnerability

[27] PRO FO 195/935, 312–42, "Answers to Foreign Office Circular of 26 August 1871" (November 1871). The Trieste steamer carried 18,000 tons annually to and from Préveza, the equivalent of about 125,000 mule-loads.

to brigands.[28] The railroad linking Eğridir, Isparta, and Burdur to the Aegean coast further undermined the caravan trade, and the motor road to Antalya, opened early in the 1920s, completed the process.

Alpujarreño muleteers had connected the coast to the Marquesado. They brought fish, fruit, and wine over the passes in the Sierra Nevada, and returned with grain from the wide, level fields north of the mountains. The distances involved were short, but the routes difficult, and known best to natives of the high Alpujarra. This business flourished especially in the heyday of the lead mines of the Sierra de Gádor, when oats from the Marquesado fed the tens of thousands of mules used to carry lead from the Sierra to the coast. Indeed raising and selling mules during the lead boom helped sustain the Alpujarran population. All this withered when a rail line connected the Marquesado to Almería in 1899. Iron from Alquife justified the railroad, and production climbed considerably. Both the iron and the railroad were owned by a British company, which early in this century employed about a thousand people. This development may have served the interests of the Marquesado – it stopped outmigration there – but it damaged the Alpujarra. The line, which followed easy gradients well to the east of the Sierra Nevada, exercised a magnetic attraction on all the goods produced in the Marquesado, and all those imported to it as well. The Alpujarreño muleteers lost their most important route.[29]

Lowland Pasture

Almost everywhere in the Mediterranean mountains, sheep and goats moved down to lowland pastures for the winter. When malaria kept many lowland areas sparsely populated and often uncultivated, the sheep and goats found plenty of room. But when the transformation of the lowlands began, generally late in the nineteenth century, mountain people found it increasingly difficult to pasture their animals in winter. In many cases, such as that of the Vlachs in Thessaly, the rents they paid for winter pasture increased. Many Pindus Vlachs, in fact, found it convenient to change their official residence to Thessalian villages so as to secure land there as it became scarce. Owning one's own winter pasture provided insurance against rising rents, but it often was the first

[28] Planhol 1958, 133. Camel caravans made fine targets for brigands. Carts traveling alone were less attractive. Camels traveling alone would have been less attractive too, but they carried less, and forfeited the labor efficiency of the caravan, where one man could manage several camels.

[29] And they did not find work in the Alquife mines: at no time did more than twenty-two Alpujarreños work there. A. Cohen 1987, 229, 271.

step toward abandoning seminomadism and the life of the mountains.[30] Taurus pastoralists and nomads also found themselves squeezed out of lowland pastures, and increasing numbers of them took up farming (or agricultural wage labor) in the plains from the nineteenth century on. The Alpujarreños who had used the plain of Dalías for winter pasture lost access to it entirely, as new crops took over the landscape, providing the local landowners with far more income than renting out their land to shepherds could ever yield.

The tale of the Campo de Dalías is instructive. Its development is typical of its kind, if more successful than most.[31] This area, measuring 300 square kilometers, is a low-lying plain along the Almería coast. Once it served as winter pasture for flocks from the interior, notably but not exclusively from the Alpujarras. The Moors had cultivated the plain, through careful irrigation. But after 1570 it returned to spontaneous vegetation. In 1830 a visitor found the Campo "quite uncultivated."[32] Lack of precipitation prohibited rainfall agriculture except in the extreme northeast of the plain. The condition of the Campo began to change when, in 1898, irrigation returned, in the form of a conduit from the Sierra de Gádor. Then, in the 1940s, wells and pumps began to distribute the copious groundwater derived from the nearby mountains. The land now became valuable, capable of producing garden crops at all seasons of the year. Enterprising locals and townspeople bought up plots. They applied techniques that had come from the Canary Islands in the late nineteenth century for raising beans, tomatoes, and potatoes on sandy soils – really on sand. Eventually they covered over their plots with plastic to reduce evaporation. Now they could produce up to five crops a year, and sell their produce out of season to northern Europe. This whole development soon revolutionized the Almerian and Granadine coast (except for the sugar zone of Motril and Salobreña), making it far richer than ever before. But the coast's gain was the mountains' loss, as the Campo de Dalías and other coastal plains no longer supported flocks in winter.

Intensified agriculture, often commercial agriculture responding to opportunities opened up by more efficient markets and transport, also

30 Sivignon 1968, 21–2. When in 1922 tens of thousands of Turks left Thessaly, new land became available to Greek citizens. Pindus shepherds bought all they could. Since 1957, when the Greek Forest Service took over Samarina's communal forests, shepherds have found their summer pasture limited too, although this has been a less onerous constraint than the loss of winter pasture.

31 Mignon 1982, 317–23, 333–8; Rodríguez Martínez 1985, 69–97; Sáenz Lorite 1977, 254–5.

32 Widdrington 1834, 1:47.

took over former lowland pasture at Berja, either near or in the Alpujarra, depending on one's definition. Much the same thing happened at Metaponto and Maratea in Lucania, at Arta and in Thessaly on either side of the Pindus, and in the plains of Antalya and Adana. Rarely did it happen as suddenly as in the Campo de Dalías, where tapping subterranean water quickly opened new possibilities. Normally the end of piracy was the first prerequisite (accomplished fairly well in the western Mediterranean by 1805) and malaria control the second, followed by transport links to cities and markets. Everywhere it happened, transhumant mountain pastoralism suffered a devastating blow.[33]

The great changes in the character of the world economy of the nineteenth and twentieth centuries worked against the villagers of the Mediterranean mountains. They, more than most rural people, depended on specializations and sidelines to make ends meet. Market integration closed down many of these auxiliary activities, although it temporarily opened a few, such as kif. But that too will end, leaving Rifians where market integration left other villagers of the Mediterranean mountains: with little to rely on but their fields, flocks, and forests. Market integration also undermined agro-sylvo-pastoralism by corroding the mountain environment.

Economic Change and Environmental Decay

The great economic expansion of the nineteenth century sharpened the appetites of cities and Europe in general for goods of all sorts. This heightened demand had strong consequences for landscapes around the world, including the mountains of the Mediterranean.

Mining

For millennia Spain has provided rich ores to great powers. Carthage and Rome exploited abundant veins of metals both precious and merely useful, almost all of which lay under mountain soils. But the ancients,

[33] In the twentieth century, national forest services have appropriated mountain areas that once served as pasture. This has deprived pastoralists of some of their summer grazing land, especially in the Sierra Nevada and the Pindus. Shepherds often point to this as the main reason behind the diminution of their flocks and are inveterate enemies of reforestation schemes. Not normally taking the ecologist's view, they regard reforestation as a waste of land, which cannot be said of the cultivation that deprived them of low pasture. But the ambitions of the forest services are only the most recent, and most obvious, of several factors working against transhumant pastoralism.

and indeed the moderns until 1820, missed the richest pockets of lead in Europe. In the province of Almería, immediately south of the eastern Alpujarra, the slopes of the Sierra de Gádor concealed lead ore (galena) so rich it would quickly drive out of business other lead mines in Germany and Britain. The Sierra Lújar near Orgiva contained rich galena, 12 percent lead.[34] Two events provoked a spectacular lead rush in the 1820s. First, the new Spanish state, emerging from French occupation, and then the loss of its American empire rewrote the laws governing extractive industries in Spain. Mining had been a state monopoly until 1817. The old law favored mining in the American colonies rather than Spain. And the law prohibited foreigners from mining anywhere in the Spanish Empire. All this changed by 1820.[35] Second, the burgeoning urban growth of northern Europe expanded the market for lead, especially after cities began to use lead pipes for municipal water systems.

Lead mining began in the Sierra de Gádor in 1818, boomed in the 1830s and 1840s, dwindled in the 1850s, and died in the 1880s.[36] The mines reopened between 1940 and 1955, when impoverished and ostracized Spain could not easily get lead from abroad. Since then the old lead zone has been deserted. In its day, around 1840, the Sierra de Gádor mines employed as many as twenty thousand men and fifty thousand mules. The men dug four thousand open shafts, usually to a depth of only 60–100 meters, and the mules carried away 1.5 million metric tons of lead to the coast, a distance of about 20 kilometers. For many years Spain's second most valuable export (after wine) was lead. It was traded as far afield as the United States and China. For two decades the Sierra de Gádor was the world's most productive lead zone.

The lead rush had its effects on the villages of the Alpujarra, the closest agricultural region of any consequence. From the 1830s to the 1850s these villages could sell all the surplus food they could produce. In particular they expanded their oats, sowing fields high up the Sierra, because fifty thousand mules need a lot of feed. Alpujarreños also found wage work whenever they wanted it – poor men came from all over Spain to work in the Sierra de Gádor, a devourer of labor. As the masters of the routes over the top of the Sierra Nevada, Alpujarreños could trade between the mining camps and the Marquesado, which

[34] Consejo Económico Sindical de la Zona de las Alpujarras 1970. Sánchez Picón (1983, 40–4) says the ore in the Sierra de Gádor had a metallic content of 60% to 80%.

[35] Nadal 1982, 182; Rubio 1881, 326–8.

[36] Sánchez Picón 1983, 36–44, 231–2; Ruz Márquez 1981, 55–6; Sermet 1942, 741–2; Madoz 1845, 8, 484–5; Ansted 1854, 156–9; Consejo Económico Sindical de la Zona de las Alpujarras 1970; *Enciclopedia universal ilustrada*, vol. 45, p. 818.

produced much of the grain eaten by the miners. And they could sell as much wood as they could cut, because at the Sierra de Gádor mines lead was smelted on the spot. The lead rush bolstered the economy of the Alpujarra (and the Marquesado) for a generation, quite a long run as mining booms go. In strictly economic terms, feeding and fueling the mines was good business while it lasted, reflected in the population growth described in Chapter 5. Indeed the prosperity of the Alpujarra inspired one observer to suggest that a road connecting the Alpujarra and Granada would revive the provincial capital's flagging economy.[37] But in environmental terms, mining was bad while it lasted and little better when it ended. The lead boom was an ecological disaster not only for the Sierra de Gádor and the Sierra Lújar themselves, but for the Alpujarra as well, reducing its capacity to support life.

Mining always means a lot of construction projects. Buildings of all descriptions spring up. Mineshafts normally need braces and props. The Sierra de Gádor had 400 kilometers of mineshaft by the mid-nineteenth century, which must have required plenty of timber. But this was only a small part of what mining meant for local forests.

The fuel needs of the smelters denuded landscapes all around the mining zones. To smelt 1.5 million metric tons of lead takes a lot of fuel, especially with the primitive smelters in use before British companies reorganized operations in the late 1830s.[38] One smelter, the San Andrés, burned about 10,000 metric tons of coal in 1841. Those that burned wood, of which scores dotted the Sierra de Gádor and environs, were less fuel-efficient. And twenty thousand men need a lot of firewood, as much as twenty-five or thirty average-sized Alpujarran villages. In the late 1830s, the mines on the Sierra de Gádor employed two thousand men cutting and hauling fuelwood. Over the decades of the lead boom, several million tons of fuelwood must have been extracted from the vicinity. Everything combustible was at risk: wood, brush, coal. Soon the smelters imported wood from the Balearic Islands. By 1851 the smelters used esparto grass, a sure sign that wood had grown scarce. A British mining engineer who had experience of mining zones on four continents wrote that he had never seen ground so barren as the Sierra de Gádor.[39] Fuel shortage was one of the reasons the mines declined in the

[37] Giménez-Serrano 1846, 20.

[38] At Le Pilar foundry, a day's work yielded 4 metric tons of lead. Laborers earned 7 reales per day; fuelwood cost 90 reales per day. Didier 1845, 817. I do not know how much fuelwood it took to smelt a ton of lead.

[39] Ansted 1854, 170. He wrote there was not a tree, not a blade of grass. Widdrington (1834, 1:48) found the Sierra Gádor "perfectly bare." Didier (1845, 510–12) concurred.

1850s, when the lead deposits had not yet been played out. Indeed as early as the 1840s the Sierra exported lead as ore for want of fuel to smelt it.[40]

If fuel was short in the Sierra de Gádor and Sierra Lújar, where foreign companies were willing to pay for it, its price must have soared in the villages of the Alpujarra. The forest gathering economy must have suffered, as woodsmen cut trees to cash in on fuelwood demand. During the lead rush, Alpujarran families could do without the gathering economy, and they might be able to afford high fuelwood prices, or spend extra time searching for fuel, because they too profited from the boom. But when the boom collapsed in the 1850s, the forest did not spring back to life. Fuelwood was still scarce, food gathering more difficult, and now families had none of the former compensations. They also found their pastures, fields, and *acequías* damaged by accelerated erosion in the wake of wide deforestation.

The lead rush also provoked a rapid expansion of cultivation in the Alpujarra, followed two or three decades later by a swift contraction. Almost all of this took place on high slopes above the villages, where barley and oats could grow without irrigation.[41] Peasants either cut trees, burned garrigue, or plowed up grasses to make room for their oats. This opened these high fields to the winter rains, inviting accelerated erosion. Terraces were too much work up here. When the lead rush ended, peasants abandoned much of this new oats frontier, leaving what soils remained open to the rains.

The environmental consequences of the boom in the Sierra de Gádor resembled those of frontier booms, where a tide of people washed in, consumed local resources with no thought for tomorrow, then washed out. People from afar, with no knowledge of local conditions, tried to cash in by raising crops on marginal lands, or cutting timber wherever they could find it, regardless of whatever traditional functions the forest might have fulfilled. When they departed, they left behind their slag heaps, tree stumps, and washed-out soils. This rapine approach to land use has characterized many frontier booms, from Brazilian coffee to Thai logging. The Sierra de Gádor, while not a frontier in the

[40] Ford [1845] 1966, 2:597; Sánchez Picón 1983, 44; Pérez de Perceval Verde 1983, 166–8. A lead-mining town in eastern Anatolia quickly shrank from three thousand people to three hundred when the fuelwood ran out in the late nineteenth century. Cuinet 1890–5, 2:232–5.

[41] The oat fields above Capileira extended for more than an hour's hike above the village. This is far more oats than the people and animals of Capileira could need (especially since the people did not normally eat oats), and must represent production for the mules of the Sierra Lújar lead mines, active until the 1890s. Rubio 1881, 201.

conventional sense, was exploited like a colony. The fact that the mining companies involved were British and Catalan, not local, only encouraged such an approach.

The lead rush of the Sierra de Gádor went a long way toward deforesting the slopes above the villages in the eastern Alpujarra. The smaller lead-mining operations of the Sierra Lújar affected the western end of the valley. The Sierra Lújar mines, or the Orgiva mines as some of them were known, never disgorged the same massive quantities of lead as those of the Sierra de Gádor. Opening in 1845, they yielded a total of about 400,000 metric tons of galena by 1970. They had an environmental effect similar to but smaller than the Sierra de Gádor mines.[42]

Small mining operations took place all around the Sierra Nevada in the nineteenth century, in the Alpujarra proper and in the Marquesado. In 1862 iron mines operated at Capileira, Lanteira, Aldeire, and Jerez del Marquesado. Trevélez produced copper, and Capileira silver.[43] By 1900, iron mining in the Alpujarra and Marquesado employed between 2,000 and 2,500 men and yielded about 200,000 metric tons annually.[44] Most of this took place in the Marquesado, where the iron is (or was) more plentiful and the railroad lowered transport costs. Once the rail link connected the Marquesado to the wider world, mining in the Alpujarra proper dwindled to insignificance. The last iron mine closed at Busquistar in 1973. None of this needed the enormous loads of fuelwood that lead smelting required, so its impact on vegetation was limited. In their environmental and socioeconomic consequences, the lead mines dwarfed all others.

Directly across the narrow neck of the western Mediterranean from the Sierra de Gádor is the mining region of the Rif. It was, and is, centered on the hinterland of Melilla, and about 200 million years ago, before the Mediterranean Sea was born, the Melilla hinterland and the Sierra de Gádor were adjacent. The lead and iron of the Rif (there was a little copper, chromium, and graphite too) is probably part of the same general formation as that of Granada and Almería. It too was exploited in a small way by the Carthaginians and Romans, and in modern times by Spanish and British companies. The Spanish began at the turn of the century, and found the iron a higher grade than that

[42] The Sierra de Gádor mines opened again in the twentieth century, but their productive years were 1825–50. Data on Spanish mining production appeared frequently in the *Revista Minera*; export figures for 1795 to 1841 from Granada and Almería appear in Madoz 1845–50, 8:485.

[43] AMAPA, legajo 226–2, "Memoria de la Junta de Agricultura, Industria y Comercio . . . 1862."

[44] Ortega Alba 1981, 129–32.

still available in Spain. They operated especially between 1914 and 1930, producing as much as 800,000 metric tons of iron per year, most of it exported to Germany. A British company extracted the lead.[45]

Mining in the Rif was a good deal more efficient than in southern Spain, chiefly because it took place a century later. It employed far fewer people (most of them Spaniards, not Rifians)[46] and consumed far less energy per ton of metal. Most of it was iron, not lead, which needs less processing. And Melilla's mining zone is well to the east of the steep mountains, so that although the mines did strip vegetation, as mining does everywhere, the consequences in terms of erosion were less than in the Sierra de Gádor and Alpujarra. So the socioeconomic and environmental effects of Rif mining amounted to little next to those in the Baetic mountains of Spain. For lack of exploitable deposits, mining amounted to nothing at all in the western Taurus, the Pindus, or the Lucanian Apennines.

Logging, Charcoal, and Fuelwood

Like mining, timber cutting can be ecologically brutal when dispersed demand is concentrated on a small source of supply. Even fuelwood gathering and charcoal making, when undertaken by people eager to sell to urban populations and heedless of the consequences, can deplete forests severely, disturb watersheds, and promote erosion. In many settings around the world, market integration in the nineteenth and twentieth centuries, together with population and economic growth in timber-poor regions, has led to such ecological calamities. The major cases come from other parts of the globe, since by 1800 the Mediterranean world almost entirely lacked accessible timber of good quality. But accessibility changed with technology, and a few corners of the Mediterranean felt the intensified search for timber that is still in progress elsewhere today. And the growth of Mediterranean cities meant more people had more money to give to men who would sell them charcoal and fuelwood, regardless of the consequences to rural landscapes.

Commercial logging dates back to ancient times in the Mediterranean basin. Navies, from the Phoenicians to the Venetians, had always taken a share. By the eighteenth century, both France and Spain, timber-rich countries by Mediterranean standards, sought naval timber far

[45] Agard 1957; Morales Lezcano 1976, 69–87; Soria Marco 1948, 65–70; Spain, Dirección General de Marruecos 1931, xv.

[46] In the 1950s the Melilla mines employed about 650 Moroccans full time, and another 1,200 to 1,400 irregularly. Seddon 1981, 136, 241.

and wide, in their American colonies where it abounded, and in the western Balkans where their enemies could less easily prevent its delivery. Even the British, with Canadian forests at their disposal, had timber agents negotiating with Ali Pasha.[47] By 1800, naval and other construction demands had stripped the Mediterranean coasts, hinterlands, and river basins of all suitable timber that cultivators had not already felled. Only the inaccessible mountains retained their forests. Only continued isolation could protect them.

In the nineteenth century, railroads, cable lifts, funicular cog railways, hydraulic winches, and other devices made mountain forests accessible as never before. Eventually, in the mid-twentieth century, roads and trucks penetrated almost everywhere up to the timberline. But by this time the various forces foreclosing on mountain life had done their work (the Rif is an exception), and few people were left to suffer from the disappearance of forest.

The general expansion of trade in the Pindus that came with the Napoleonic wars extended to timber. By the late eighteenth century, coastal Epirus and Albania sold tall oak to European navies. With the British naval hegemony that followed 1815, Epirus began to sell twenty to thirty shiploads of wood per year to Malta, now a British naval base.[48] Several of the larger rivers draining the Pindus, such as the Kalamas, Acheloos, and the Arachthos, floated logs down to the coast for sale, often to Corfu or Patras. In the 1830s, Préveza was a good place to buy wood, according to the French consul.[49] Late in the nineteenth century, the Ottomans governed Thessaly and Epirus as a province they expected soon to lose, and encouraged Turkish timber merchants, such as the father of Mustafa Kemal Atatürk, to bring timber to Salonika. After the upper Aspropotamo (Acheloos) became Greek, Bulgarian woodcutters employed by Greek merchants felled all the fir they could. "Speculators" in Patras and Trikkala easily secured permits from authorities and fitted out timber-plunder expeditions to the high Pindus. The stern opposition of the local inhabitants, who depended on this forest, obliged these parties to travel and work fully armed. The woodsmen succeeded to the point where floating fir trunks in the 1890s made fording the rivers a dangerous business. Philippson, who reported this

[47] PRO FO 196/3/158, letter of unknown author to Foresti, 28 September 1811. AMAE-P, Mémoires et Documents, Turquie, 22:398, 405, Boislecomte to Rigny, 4–5 September 1834.

[48] Holland 1819, 1:122.

[49] AMAE-P, Mémoires et Documents, Turquie, 22:409, Boislecomte to Rigny, 5 September 1834. Floating timber made it dangerous in June 1853 to ford the Peneus. Tozer 1869, 163.

information, expected that the ancient fir forests of the Acheloos would shortly disappear, with grievous consequences for the population.[50]

Away from the rivers, poor transport checked timber exports throughout the nineteenth century. Mules still carried timber down from Malakassi to Thessaly in the 1930s, and from Vovousa to Ioannina as well.[51] As late as 1955, when roads had begun to penetrate the Pindus, total Epirot lumber production came to less than 33,000 cubic meters.[52] Probably the majority of timber cutting in the Pindus, in the nineteenth century as in the twentieth, served the needs of Ioannina (which needed rebuilding after a fire in 1821) and other small local centers, to the consternation of generations of Greek economists who lamented the lack of exports.[53]

Fuelwood and charcoal demand, functions of local population growth, probably mattered as much as or more than timber for distant markets in changing the Pindus landscape. Ali Pasha required all Epirot villages to provide fuelwood for his capital, which was regarded as "one of the severest oppressions to which the villages in the district are subject."[54] Figures for the 1930s and 1950s show that lumber production came to only about a quarter of that of fuelwood in Epirus. Forest extractions (both timber and fuelwood) affected central Zagori more than anywhere else in the Pindus in these years.[55] Today little tall timber remains there, and commercial loggers have moved on to eastern Zagori, nearer the Pindus spine. Between them timber demand, even if generally not from distant lands, and fuelwood needs, almost always local, have surely done far more to the Pindus landscape than has clearing for agriculture. Indeed fuelwood cutting and charcoal making probably rank as the principal agents of Pindus deforestation, although any such judgments are necessarily unreliable.[56]

[50] Philippson 1894, 326; Philippson 1897, 288–9. Atatürk's father operated in the Mount Olympos area around 1880, just before Thessaly became Greek (Volkan and Itzkowitz 1984, 22). On the character of Ottoman administration after the Congress of Berlin: Berl 1909, 26–7.

[51] Papakostas 1967, 203; Hobhouse 1813, 1:88. Hammond 1967, 261, 264–5; Andreades 1905, 8; Philippson 1894.

[52] EHM, Lamour (1957?), 25. This estimate was based on official data, always suspect in the case of logging.

[53] Lazos 1935, table B. The economists: Kassandreas 1966, 49–50; Andreades 1905, 13. Foreign foresters also have regarded remaining Pindus forests as an untapped resource: Food and Agricultural Organization 1968.

[54] Leake 1835, 4:164, cited in Turrill 1929, 198.

[55] Lazos 1935, table B; EHM, Lamour (1957?), 25.

[56] Recall Chapter 3, in which Grispos's view on this question for Ottoman times is presented.

In southern Italy large-scale logging dates back to Roman times. But little took place between the second century A.D and the nineteenth. The stands in the Lucanian mountains attracted scant attention, because better and (slightly) more accessible timber stood in the Sila Plateau of Calabria. The Romans and the Arabs had depleted the Sila forests in their day, but, because much of the plateau is fairly flat, soils remained and forests in large part recovered. Perhaps conifers replaced oak and beech to some extent. From Angevin times, laws had prevented forest clearance so as to preserve the best mast timber for the navy.[57] In the eighteenth and early nineteenth centuries, population growth and clearance for agriculture began to nibble away at the Sila forest. At the end of the century German, French, and local lumber companies took great bites. Between them they created a railway, a funicular, and road system to connect the Sila to Cosenza and the wider world. Local forest inspectors sold the wood they were (under)paid to protect. Chestnut stands were cut for tannin and construction wood, and sold to Bari, Naples, Foggia, and Salerno. Oak was converted for sale as charcoal. Beech went to Naples for use in barrels, furniture, and tools.[58] Goats attacked the underbrush and the new shoots, minimizing second growth. Norman Douglas, who bemoaned this destruction between 1906 and 1911, expected that the Sila would become a stony desert in a generation. Some of it had by 1910, when the *Inchiesta agraria* termed parts of it "squalid and unproductive," but fortunately for the forest and the local watersheds, World War I suspended logging operations.[59] A few stands of tall trees remain, helping one to imagine the forest of a century ago. Now part of the Sila is protected as a national park. The whole of it stands as a monument to the local effects of concentrated demand.[60]

The lead rush preempted commercial logging in the Sierra Nevada and Alpujarra. Inaccessibility had prevented timber operations before 1820. The Spanish navy's timber surveyors concluded the area was too remote (Chapter 3). Only the smelters of the Sierra de Gádor could provoke fuelwood gathering and timber cutting. After the mining boom, commercial logging touched the Sierra Nevada only very slightly. Very

[57] This meant that only mature conifers were (legally) felled. Filangieri 1980, 108.

[58] Azimonti 1909–10, V-II:193–4.

[59] Ibid., 204. N. Douglas 1915, 295–301. On the Sila deforestation see as well: Ahlmann 1926, 85; Ramage 1868, 79–80; Sciacca 1977, 80–1.

[60] Today the Sila is officially 64% forest, and the Sila Greca 39%. This includes much open woodland. The latest forest statistics (1986 data) appear in Italy, Istituto Centrale di Statistica 1989. Lucania's subregions have forest cover varying from 12% to 27% (p. 115).

little timber remained after 1860 or so, and the continued absence of roads or railroads made commercial operations impractical. Locals cut timber for their own use, but not with the fervor that linkage to strong, distant markets inspired elsewhere. Fuelwood and charcoal production existed in proportion to local population. After 1860 the Alpujarra never exported either item.

Commercial logging affected the Rif to some degree in the nineteenth century, but as in the Pindus charcoal and fuelwood demand have surely loomed larger. Nineteenth-century Moroccan and European records show no evidence of an export timber trade.[61] A small one probably existed, and certainly a market for cedar existed within the suqs of northern Morocco. During the rainy season woodsmen floated trunks down the swollen rivers to the Mediterranean coast and to the Ouerrha River. The demand from local markets occasioned a "veritable massacre of trees."[62] But without roads or forest sawmills, only the trees close by watercourses were at risk.

Systematic exploitation awaited the advent of the Spanish Protectorate. Tetouan and Tangier had always bought Rif cedar, but with the Protectorate came quick urban growth in Tetouan and Melilla and much stronger markets for timber.[63] In 1925, when the war with Abdel Krim was drawing to a close, the Spanish opened the cedar forest of Ketama to commercial use, and a brisk trade flourished with Tetouan, Targuist, al-Hoceima, Chechouan, and (on a modest scale) even Spain. More extensive logging began in 1937 (when war in Spain disrupted Spanish timber supplies) with the contruction of forest roads. The Centa–Tetouan railway used Buhasen oak from the western Rif for its ties, an extravagance. In the 1940s, sawmills opened in cedar stands near Targuist and Ketama, and Spanish railways used cedar for ties, an incredible extravagance.[64] The scale of logging operations remained fairly small; official exports reached only 37,000 cubic meters in 1955. Certainly larger quantities were cut for use within the Spanish Zone, and much may have escaped the notice of the record keepers.[65] But the charcoal market probably destroyed more timber in the Rif than did

[61] Miège 1961. Mouliéras (1895–9, 2:95) mentions cedar exports from Ketama to Europe. Mouliéras never visited the Rif and is often held to be unreliable.

[62] G. Maurer 1968b, 58.

[63] In this era Fez apparently got its commercial timber from the Middle Atlas, not the Rif. Le Tourneau 1949, 390.

[64] G. Maurer 1968b, 58. Using oak and cedar for railroad ties was extravagant, because Aleppo pine does perfectly well in that capacity. See Cerón Martínez 1879, 299.

[65] On logging in the Rif: Sánchez Cózar 1942, 11–14; Vázquez del Río 1935, 16–25; Spain. Dirección General de Marruecos 1931, lii; Spain, ACEM 1955, 52; Spain, ACEM, Servicio de Montes 1943, 3–6. This last is a typescript in the Sección de Africa, BNM.

commercial logging. A Spanish forester noted that under the excuse of cutting for local needs a large trade in charcoal had developed by the 1930s in the mixed oak and cedar forests of the western Rif. Catalonia imported Rif charcoal, but far more went to Tetouan, Tangier, and Melilla. Probably more still went to the Rif villages.[66] Official Spanish figures, which no doubt leave much to be desired, show the total annual fuelwood use of the Rif in the early 1940s at 60,000 metric tons.[67] Population growth, more than market integration, played the key role in the modern deforestation of the Rif.

It is in the Taurus that commercial logging had its strongest effect. The north flank felt little impact from logging, as no suitable means of transport existed. But the southern face looked down upon the sea and across to Egypt, a land without trees. In Seljuk times, as before, Antalya and Alanya exported Taurus pine and cedar and lowland cypress to Egypt.[68] After the sixteenth century this trade fell off, as did the Egyptian trade as a whole. But late in the eighteenth century the commercial situation in the eastern Mediterranean underwent a basic change. Foreigners and the Christian minorities of the empire, the Greeks especially, intensified their trade in Ottoman lands. The Greek state, formed in 1830, favored the merchant marine, and Greek shipping increased sixteenfold between 1821 and 1914.[69] Greek communities from Odessa to Alexandria served as brokers of cross-cultural trade, linking Russia, the Balkans, Anatolia, and Egypt more closely than ever before. Within this network a new trade revived between southern Anatolia and Egypt, breathing new life into Antalya and Alanya (and some other ports as well).

A significant part of this trade consisted of timber, generally headed toward Egypt. In Egypt, urban population grew rapidly in the nineteenth century (assisted slightly by immigration from mountain areas of southern Europe).[70] Construction boomed, but timber was unavailable locally, mandating extensive imports. The south face of the Taurus, from Finike to Adana, met the demand. Greek merchants based on the island of Castellorizo specialized in selling Turkish timber to Alexan-

[66] Vazquez del Río 1935, 24–5. He predicted that a free market in charcoal would doom the Rif cedar forests, and he might have been right. Because the forests have been protected (although sometimes ineffectively) since the 1930s, one cannot know.

[67] González Vázquez 1945, 72. Of this 60,000 metric tons, 40,000 were consumed directly as wood, and another 20,000 were converted into 4,000 to 5,000 metric tons of charcoal.

[68] Lombard 1959, 240–1. [69] Stoianovich 1967, 94.

[70] Emigrants from Kónitsa villages in the Pindus left for Egypt between 1840 and 1850 and also 1870 and 1920. Dimitriadis 1980, 441.

dria. In both Egypt and Syria, Antalya pine, prized for its durability, found a strong market.[71] Early in the century, the great reformer Muhammad 'Alī Pasha (not to be confused with Ali Pasha of Ioannina) built a navy at Alexandria. He acquired ship timber in Italy and southern Anatolia, but found the quality unsatisfactory. His invasion of Syria and Cilicia (Adana) in the 1830s was in part motivated by his desire for good cedar from the eastern Taurus and northern Lebanon ranges.[72]

The south face of the Taurus kept much of its timber through to the 1860s, because the Greek traders operated on a small scale and Muhammad 'Alī wanted only cedar. Southwest Turkey as a whole retained considerable forest, although less around Antalya and the coast than on inland slopes. Even the nearby island of Rhodes was well wooded in 1860, and the malarial plain of Adana in south-central Turkey too.[73] But the scale of timber operations changed when French enterprise realized an old dream and financed the digging of the Suez Canal (1859–69). In its time it was the second-largest construction project in the history of the world, after the Great Wall of China. The canal stretches over 168 kilometers. When first constructed it displaced 76 million cubic meters of earth (subsequent deepening and widening would double this figure). For countless other purposes, not least the construction of Port Said, the builders needed immense quantities of timber. They could get none in Egypt or the Sinai, little from Cyprus or the Levant. The nearest supply of tall timber lay in the Taurus mountains, and the Greek merchants of Alexandria knew how to get it. The tahtaçı of the Taurus had never been busier than in the 1860s.[74]

After the opening of the Suez Canal the growth of Egypt only accelerated, and timber requirements kept pace. By 1889, timber accounted for 19 percent of the exports of Antalya, the most diversified port in southern Turkey. Alanya also shipped wood to Egypt. Even Akseki kaza, elevated and isolated in the Taurus northwest of Alanya, exported

[71] E. Davis 1874, 215 (Davis lived in Alexandria); Leake 1824, 127. Beaufort (1818, 31) notes planks on the shores of Lycia awaiting export to Egypt.

[72] Personal communication from Professor Mike Reimer of the American University of Cairo. Egyptian pharaohs had launched expeditions, military, diplomatic, and commercial, in search of Lebanon cedar since at least 2600 B.C.: Meiggs 1982, 63–8. The Egyptian state sent a certain Wen-Amun on a timber-seeking mission to the Lebanon in the eleventh century B.C.: Toynbee 1969, 275.

[73] E. Davis 1874, 217. Meiggs (1982, 394) has assembled relevant evidence. AMAE-P, Correspondance Commerciale, Smyrne, vol. 43, "Etat du Commerce et de la Production de la Syrie," 1 September 1833, notes that good construction timber surrounded the Cilician (Adana) Plain.

[74] The Taurus felt "large-scale fellings" for the Suez Canal construction: *Unasylva* 6 (1952): 120 (anonymous untitled article). On construction: Marlowe 1964.

timber through the modest port of Manavgat. Through the 1880s, most of the construction timber used in Palestine – cedar, fir, and pine – apparently came from the Taurus. This trade to Palestine shrank in the face of competition from northern Europe after 1880, and disappeared entirely by 1918. It appears that in the twentieth century the eastern Mediterranean as a whole turned to more distant sources of supply for timber.[75]

On a modest scale southern Anatolia also provided fuelwood and charcoal for Egypt.[76] In Seljuk times, apparently, fuelwood exports were unimportant. But by the early nineteenth century, "Alexandria [was] supplied, in a great measure, with fuel from the woody mountains of Karamania."[77] The small ports west of Antalya sold firewood and charcoal to the Greeks of Castellorizo, who dealt it to Egypt. The populations of Antalya, Alanya, and the other ports required fuel. Further east, behind Adana, charcoal burners destroyed large tracts of forest for the city and for export.[78] But these were modest towns until recently, with no traditions of metallurgy, glassmaking, or other enterprise that required charcoal.[79] And in their climate one scarcely needed heating. Because the highland population (which did and does require considerable fuelwood for six months of the year) also remained small until the twentieth century, the impact on Taurus forests of fuelwood and charcoal use must have been minuscule compared to that made by logging.

Other timber operations thinned or clear-cut slopes in Lebanon, Algeria, Cyprus, and throughout the Mediterranean. In many respects 1850 to 1914 represented a golden age for timber merchants in the Mediterranean basin. New technology permitted exploitation of hitherto inaccessible stands. Strong demand from Europe and European colonial construction in North Africa assured a fine market. And forest protection rules and regulations scarcely existed or were easily avoided. In some mountain areas, growing population and proportional fuelwood demand exceeded logging in impact upon forests. Where cities lay nearby, charcoal manufacture also took a heavy toll. Given the slopes and the climate, the golden age for timber merchants, of course, corresponded

[75] Cuinet 1890–5, 1:858, 866–7, 870; Biger, Gideon, and Liphschitz, forthcoming.

[76] Lombard 1959, 241; Leake 1824, 127; Fellows 1838, 214, 219.

[77] Beaufort 1818, 10. By "Karamania," Beaufort, following Venetian usage, meant the whole south coast of Turkey, not just the Anatolian district of Karaman.

[78] Dunn 1905, 37.

[79] Antalya counted 8,000 people in 1811–12, and Alanya 1,500–2,000, according to Beaufort 1818, 130, 173.

to a dark age for forests, soils, hydrology, and the stability of Mediterranean mountain landscapes.

Regional Effects of Commercial Agriculture

Concentrated demand and intensified commerce changed mountain landscapes through agriculture as well as by mining and logging. Booms based on commercial agriculture, though sometimes temporarily beneficial, often left villagers with less soil and wood than before. Examples exist for every region. Here I will consider two, the Pindus and the Alpujarras.

The mountain way of life in the Pindus, as we have seen, rested in part on commerce, especially for the Vlachs. But, as everywhere, it rested fundamentally on soils, pastures, and forests. In the nineteenth century, expanded trade indirectly threatened the ecological balance of the Epirot Pindus. The key to it was the development of the Ionian Islands, and of Ioannina.

The Ionians belonged to the Venetian Empire until its demise in 1797. Under Venice they had functioned as one of the many outposts of maritime empire, useful as entrepôts above all. The largest, Corfu, had a population of 44,000 in 1802.[80] After 1797 the Ionian Islands became pawns in the struggles of the great powers in the eastern Mediterranean. They became French, Russian, French again, and finally British after 1815, and remained British until 1864 when Greece acquired them. Between 1797 and 1864 the islands housed army garrisons and frequently naval fleets. Civilian population grew to almost 71,000 by 1855.[81] All these people, and to an extent British fleets at sea in the Mediterranean as well, relied on Epirus for food, fuel, and other items. Corfu itself produced little but olives, wine, and fruit.

In Venetian times, Corfu and the Ionian Islands conducted only a modest trade with the Epirot coast. The islands got most of their supplies via Venetian ships coming from points all over the Adriatic, and exported their wine, olives, oil, wax, and honey to Venice and Trieste. But with the fall of the Venetian republic, political obstacles to trade with the mainland disappeared. Quickly the Ionians, and Corfu in particular, became a channel connecting Epirus to the wider world.[82] Indeed, the British forces there soon became dependent on Epirus's

[80] PRO CO 136/1261/335, Statistica dell'Isola di Corfu.

[81] PRO CO 136/1419/226–7. On the Ionians under British rule, Yannoulopoulos 1981.

[82] Guillaume de Vaudoncourt 1816, 73–5. Venice controlled five towns on the mainland, and through these supplied Albanians and Epirots who wished to rebel against the Turks or Ali Pasha (80–7).

Table 6.1. *Imports to Corfu from the mainland, 1811*

Item	Quantity	Value in French francs
Oxen	10,000	850,000
Sheep, goats	3,000	—
Hare skins	—	93,000
German cloth	—	81,000
Cotton and wool	—	46,000
Camel skins	—	26,000

Source: AMAE-P, Mémoires et Documents, Turquie, 35:320–321, Pouqueville to Bassano, 15 November 1812.

grain for their sustenance.[83] In exchange, Corfu channeled British manufactures to Epirus. By 1811, Epirus's trade to Corfu had grown dramatically. The French agent Pouqueville summarized the new trade figures (Table 6.1) for his superiors in Paris.[84]

Trade to Corfu continued to grow after 1815. It followed fairly regular patterns from the 1820s to 1864 (Table 6.2). Each year Corfu imported between forty thousand and seventy thousand sheep, goats, and pigs, and between eight thousand and sixteen thousand cattle. About two-thirds of these animals came from Ottoman Epirus, and the other third from Greece.[85] In terms of value, the most important imports were forest products: timber, fuelwood, and barrel stoops and staves. The figures in Table 6.2 represent Corfu's imports, not all of which came from Epirus, and not all from mountain zones. The cattle, for instance, must have come from the lowlands. The sheep and goats came overwhelmingly from the Epirus and the Pindus. Probably the timber and fuelwood did too, because no other forested zone lay quite so close by. The figures represent imports for the civil establishment of Corfu only. The needs of the garrison and navy would add considerably to these totals.

The consequences of this expansion of Epirot trade to Corfu were

[83] PRO FO 196/2 [Canning?] to Foresti, 24 December 1810: "Our numerous establishments in the Mediterranean, and especially the Ionian Islands, more or less derive their supplies of corn from this country."

[84] AMAE-P, Mémoires et Documents, Turquie, 35:320–1, Pouqueville to Bassano, 15 November 1812. Pouqueville estimated that the trade between Ali Pasha's dominions and the Ionians was about 13% greater than the foreign trade by caravan: 1.4 million French francs to 1.2 million. But he noted that estimating the value of the overland trade was sheer guesswork.

[85] AMAE-P, Mémoires et Documents, Turquie 23:225, Boislecomte to Rigny, 24 November, 1834. This proportion refers to the 1832 commerce, but probably holds throughout.

Table 6.2. *Imports to Corfu, 1827–1858 (in thousands)*

Item	1827	1830	1835	1840	1845	1850	1855	1858
Small stock	50	64	69	64	62	44	41	—
Cattle	17	14	9	11	11	11	11	—
Potatoes (lbs)	692	959	1,078	1,304	1,500	998	507	798
Wool (lbs)	27	5	20	30	19	21	9	38
Timber (£)	15	7	20	16	16	30	13	20
Firewood (loads)	12	8	15	14	11	12	4	10
Firewood (£)	2	3	7	4	5	6	1	6
Staves, etc.	615	128	763	1,360	888	1,700	772	397
Staves (£)	5	3	4	6	6	9	3	13

Source: PRO CO 136/1391/44–7; 136/1394/200–2; 136/1399/182–5; 136/1404/220– 2; 136/1409/222–5; 136/1414/222–3; 136/1419/258/61; 136/1422/160–279.

several. First, it helped to compensate the Pindus for the lost markets in woolens, preventing a more severe economic downturn. Second, it changed the routes by which the Pindus (and Epirus as a whole) communicated with the outside world. The overland routes through Macedonia and Albania lost traffic in favor of those leading to the seaports of Igoumenitsa, Parga, and Préveza. But most important perhaps, the trade to Corfu influenced production decisions and land use in the Pindus. It now became practical to raise as many sheep, goats, and pigs as possible, because not only the sheep's wool but the animals themselves brought a price. It also became rewarding to cut as much of the forest as possible, since oak and beech had a market as construction wood or fuel, and conifers as barrel staves and stoops. The most remunerative use of land imaginable would have been to cut timber wherever possible and then run goats through the stumps and shoots in the years to come. I believe a good deal of the maquis and garrigue of Epirus and the western flank of the Pindus was created this way.

Epirots also had a new incentive to expand crop production, because they could now sell flour and potatoes to the Ionians.[86] Potatoes require a great deal of labor and do best with irrigation, so it would have been unwise to clear much new land for them; it is more likely that the villagers replaced maize patches with potato patches. Wheat requires less work, and they may have cut or burned forest to seed new land with wheat. But certainly the major impacts of the Corfu trade upon Epirus

[86] Mountain folk on the eastern flank of the Pindus, in Kipurio, traded flour to the Ionians around 1810. Leake 1835, 1:303.

and the Pindus were the felling of timber, the collection and creation of fuelwood and charcoal, and an increase in flocks.

When Britain ceded the Ionians to Greece in 1864, the Greeks quickly severed the commercial connection between Ottoman Epirus and Corfu. Epirots, having adjusted production and land use so as to profit from the Corfu trade, had grown dependent on its continuation. The British consul in Ioannina, Robert Stuart, perhaps not an impartial observer, wrote in 1871: "The cessation of British protection in the Ionian states had a most damaging effect upon all the material interests in these parts."[87] Epirots had ravaged their forests and destroyed their soils, only to find themselves deprived of their former markets and thrown back upon their natural resources, never rich, now much depleted. Accounts of exports for 1871, sufficiently detailed to include tortoises, make no mention of wood.[88] According to the returns made to the Ottoman tax farmers, land went out of cultivation between 1860 and 1871 in the vilayet of Ioannina. The total numbers of sheep and goats reported declined from 2.6 million to 1.9 million in the same eleven years.[89] Local insurgencies multiplied in 1866–7, breaking out in eastern Zagori, in Agrafa, and elsewhere. Stuart believed them directly related to the loss of trade with the Ionians.[90] They may also have been connected to the forest depletion and land degradation that the commercial connection to Corfu (and to all Britain's Mediterranean posts) promoted.

In the Alpujarra livestock and commercial timber had less to do with changing the face of the land than did vines and sugar. Andalusia had long exported wine but never very much. However, in the late eighteenth century and throughout the nineteenth, exports of wine (mostly to Latin America) and raisins (to the United States) multiplied. Relaxed regulations on trade to Spain's American colonies (1778) and the growing English thirst for Spanish wine provoked a realignment of agriculture and trade along the coasts of Málaga and Granada, formerly forbidden to trade with the Indies. Catherine the Great of Russia so admired Málaga wine that she gave it a special customs exemption throughout all the Russias.[91] Málaga merchant houses, many of them English and French, quickly financed a transformation of the coastal

[87] PRO FO 195/1002/270–6, "Commercial Report for 1871. Janina." [88] Ibid.

[89] PRO FO 195/1002/286–300, "Report on the State of Epirus, 26 April, 1872." Peasants, of course, sought to hide livestock from the tax farmer, and the tax farmer had incentive to underreport the taxes paid to him. So these figures are sure to be too low.

[90] PRO FO 195/801/500–3. Various correspondence between Stuart and the Foreign Office.

[91] Sermet 1953, 158.

mountains, except for a few pockets with unsuitable soils. One of the areas most affected was the Sierra Contraviesa, the south flank of the Alpujarra. Here the Málaga merchants created an authentic monoculture of vines on slate and schist soils, good for grapes, but difficult to terrace. Vineyards hold soil poorly under the best of conditions. Under those prevailing, a tremendous surge of erosion followed the extension of vines in the Sierra Contraviesa, leading to soil impoverishment and swift abandonment. The deforestation even troubled the Consulado of Málaga, the organization of merchants whose enterprise caused it. Perhaps the spectacular series of floods in the city of Málaga attracted their attention to the matter. At any rate, in the mountains behind Málaga and in the Sierra Contraviesa both, the overseas markets for wine generated an internal frontier of rapine agriculture.[92]

The vineyard boom of 1780–1850 replaced marginal agro-pastoralism on the Sierra Contraviesa with a monoculture and made the local villages dependent on Málaga merchants, their financiers and marketers. It converted the Contraviesa from a shepherd massif to a peasant massif, but the peasants were of a distinctly modern kind, linked to cities and markets. It also provided seasonal wages for thousands of villagers from the north slopes of the Alpujarra (the high Alpujarra), right across the Guadalfeo River from the Sierra Contraviesa. In the course of the early nineteenth century these wages helped to allow the population growth of the high Alpujarra. But as numbers grew, the necessity of wage labor grew, because the food-producing capacity of the high Alpujarra had little elasticity. So when the phylloxera ravaged the vineyards in the 1890s, the Málaga merchants lost some of their speculative investments, the cultivators of the Sierra Contraviesa lost their livelihoods, and the peasants of the high Alpujarra lost a vital source of income. Thousands emigrated from the Sierra Contraviesa, and thousands more from the high Alpujarra, despite the fact that the vineyard expansion never reached its higher slopes. Wine production in the Sierra Contraviesa, indeed in all the province of Granada, never approached half the levels of the 1880s.[93]

Overseas markets also inspired a wide expansion in the area under

[92] García Manrique 1982, 48–9. Mignon 1982, 192–9. Laborde (1809, 2:132) and Martínez Ruíz (1977, 31) both have wine export statistics. Clemente Rubio (1807) treats the botany of Andalusian vines and several other subjects.

[93] AMAPA, legajo 19–1, especially "Estado demonstrativo de la superficie occupada por el viñedo de la inválida por la filóxera . . . Granada," 10 December 1891; Jiménez Blanco 1986, 2:607–34; Mignon 1982, 203–4, on the phylloxera in the Alpujarra. In the province of Granada, wine production collapsed: 1896–1900 levels were 6% of 1886–90. Of the area under vines in 1883, only 3% remained by 1900. The recovery from phylloxera

Ohanes grapes. Ohanes is a village in the Almerian Alpujarra known for its table grapes, accidental hybrids developed late in the eighteenth century. In Britain they go by the name of "Málaga grape," as Málaga was the usual point of export until about 1860. In Spain they are known as *uva de Ohanes, uva de Almería,* or *uva de embarco,* because they were grown first around Ohanes and after 1860 exported through Almería. The grape's distinction is its very tough skin, which can stand travel, even by mule over the rough tracks through the mountains to the coast. It could last through long voyages in ships' holds to Britain and, after 1904, to the United States. Locals say it even found markets in British India and czarist Russia. In 1834 a Málaga merchant saw the export possibilities of the *uva de embarco,* planted it in Lanjarón, and 300 hectares (1830) soon became 6,500 (1910). Aside from Lanjarón, the upper valley of the Andarax River, especially the villages of Ohanes, Canjayar, and Ragol, was the home of the *uva de embarco.* Any village between 500 and 1,000 meters could grow it and join the bonanza. Production climbed from 1,100 metric tons in 1865 to 7,700 in 1881. The phylloxera destroyed the Ohanes grape by 1892 (a tough skin proving inadequate defense), but by 1905 exports rebounded to reach 60,000 tons. Wherever enough laborers lived, land was put under the grape. Yields declined (35 percent between 1900 and 1930) as eager Alpujarrans tried to use less suitable soils, and erosion took its toll on the best. Through the 1920s exports stayed high, but the market disappeared with the Great Depression, Spanish Civil War, and World War II. Refrigeration, which allowed thin-skinned grapes to travel well, provided the final blow to the international position of the Ohanes grape. By the 1950s Spaniards had begun to acquire a taste for it, and production revived somewhat.[94]

The Ohanes grape traveled in barrels, made of pine or, after 1880, cork. After 1860, the villages producing the grapes also made the barrels, up to 2.5 million of them a year, and used sawdust in which to pack the grapes. Every such village, and there were several between Lanjarón in the west and Ohanes in the east, altered its local landscape to clear enough land for vines and to get enough wood for barrels.

was only 30% in Granada, 59% in Almería, and 69% across Spain (Jiménez Blanco 1986, 2:644–51). In the Alpujarra villagers turned to figs and almonds, but much land went out of cultivation.

The current speculative boom in tourism along the same coast and many other coasts throughout the Mediterranean may bring similar problems. Tourism will flag with economic or political hard times, and until it revives will leave locals with nothing in its place.

[94] Sáenz Lorite 1977, 194–210; Mignon 1982, 209–13; Jiménez Blanco 1986, 2:591–607; Abélanet 1969.

Both the production and the transportation of the Ohanes grape intensified the stress on forests and soils in and around the Alpujarra. When the market disappeared, as it did from time to time on account of remote economic and political events, local peasants found their landscape impoverished and their own survival menaced. The grape export boom had softened the consequences of the decline of the Sierra de Gádor mines for poor families who needed wages. Now those from higher villages found another source of wage work foreclosed. From the viewpoint of population ecology, the *uva de embarco* had provided a partial respite for a population in overshoot. The soils and forest could not easily be repaired, so only emigration – or a timely resurgence in the market – could solve the peasants' problem. Concentrated demand from Britain and America that was focused on a small production zone transformed local ecology in a way that could not be undone when the demand disappeared or, as in the case of America, was met by other sources (California).

A mighty plant in world history, sugar cane, played a role similar to that of the lowly table grape. The Arabs had brought cane to the Granada coast. It grew well in the hot and humid delta of the Guadalfeo, around Motril and Salobreña, and at neighboring Almuñecar. With the help of the Arabs' horticultural skills, it flourished in the sixteenth century when the European taste for sugar – a mild chemical addiction in fact – developed. The Spaniards continued its cultivation after the expulsion of the Moors, but production languished, especially after the rise of Caribbean sugar.

To crystallize sugar out of cane syrup required boiling. Through the middle of the nineteenth century in the Guadalfeo delta sugarmills used an open fire built with wood, an inefficient technology far in arrears of Cuba and other major sugar centers. In the 1830s the remaining nine *ingenios* on the coast consumed about sixty thousand loads of fuelwood every season. The sugar boiling consumed mountains of firewood larger and taller than the sugarmills themselves. As early as 1540 wood had run out around Motril. Late in the eighteenth century, proprietors spent money on larger sugarmills to economize on fuel. But by 1845 sugar had devoured all the wood within 10 kilometers of the ingenios.[95] Then, with the introduction of new refining technology in 1845, a small sugar revolution took place, abetted by warfare in Cuba after 1868. Protectionist legislation in the 1870s helped. The delta began to specialize in the industrial aspects of sugar production – the latter stages of refining – even importing raw sugar from Cuba. But cultivation

[95] Sagra 1845, 34–44; Almagro 1932, 10; Septsault 1987, 193.

increased too: sugar cane spread from 500 hectares to 5,000 by 1900, where it has stayed since, with some ups and downs.[96] All this new cane juice had to be boiled, requiring fuel. The Guadalfeo *rambla* served as the local highway, so its watershed suffered the greatest deforestation. Eventually sugar beet,[97] raised in the Vega de Granada, took over much of the market, but not before the Guadalfeo Valley had lost its combustible vegetation, opening its soils to the winter downpours. Soils washed down to the delta redefined the coastline, as we shall see in the next chapter. Another boom, driven by distant demand (although in this case chiefly domestic), degraded a local landscape, victimizing the villagers whose livelihood, once the boom ended, depended on forests gone up in smoke and soils now in the Mediterranean. As the soils washed out to sea as silt, the peasants flowed out in all directions as emigrants.

With the concentrated demand for wine, table grapes, and sugar, the mountains of coastal Granada underwent a fundamental change. They joined the world of international markets. The system of agro-sylvo-pastoralism that had slowly emerged after the expulsion of the Moors gave ground to several small, local monocultures. None of them was rooted in the high Alpujarra, but each bordered it. Each of them produced a reckless expansion of cultivation, accelerated erosion, and two of the three produced widespread deforestation. This did extend to the high Alpujarra villages and their communal oak and pine forests. The monoculture booms (like the mines) provided wage work for Alpujarreños at the same time as they degraded the landscapes of the Alpujarra and its neighboring mountains and valleys. This encouraged overshoot, already in progress for reasons more purely demographic, by lowering the carrying capacity of the land while temporarily sustaining, or increasing, its population. To some extent the arrival of one boom compensated for the collapse of another, but it was by no means easy for the villagers themselves, for whom one year without wage work often spelled acute suffering.

The high Alpujarra has the forbidding peaks of the Sierra Nevada to its north. After 1780 vineyards stripped the soils and vegetation of the mountains to its south. After 1825 lead mining scorched the earth of the Sierra de Gádor to its southeast. After 1845 sugar cane denuded the hills and valleys to its southwest. After 1860 the *uva de embarco* ravaged the landscape to its east. The Alpujarra became a chronically over-

[96] Mignon 1982, 208–9; Jiménez Blanco 1986, 2:690–726.

[97] Sugar beet processing required charcoal, which the Vega de Granada had to import in considerable quantities after the 1880s.

populated elevated island, itself short on fuel and land, surrounded by deserts-in-the-making. Mass emigration began in the 1890s.

Across the narrows of the western Mediterranean, and a full century later, a similar tragedy is at work today in the Rif. Like the Ohanes grape or silk in the nineteenth century, kif in the twentieth is raised by small-scale producers with an eye toward a market. It has lately become as important to the Rif as silk was to the Alpujarra in its heyday. In the central Rif, a third of the peasantry depended on it in the mid-1960s, and in some douars, it provided 85 percent of all income. And this was before it became big business. Although the kif trade is a highly visible fact of life in the Rif, accurate information is hard to come by.[98]

Kif (*Cannabis sativa*)[99] is a summer crop, planted in the spring and harvested in August or September. It requires much labor and careful attention. Growers seed it grain by grain, and must keep beds clear of all weeds. Weeding is generally women's work. It requires careful irrigation, especially during a two-week spell in May when seven or eight waterings must be administered. Kif also needs heavy applications of manure in order to yield well. When skillfully cultivated, plots can produce up to 2,000 kilograms of kif per hectare.

Rifians mix tobacco in with their marijuana before smoking it. They also consumed it, and perhaps still do, in beverages, or ground up with sugar and eaten as hashish. It may be that Rifians now consume less kif, in whatever form, than formerly, because the income it can provide is so great and so needed: the opportunity cost of smoking it is now very high. In less profitable times, Rifians used to let the plant mature and use its fiber for ropemaking. Now its only use is to yield kif, primarily for sale.

Kif may have grown wild in the Rif for centuries or longer, but as a crop it has a short history. It is an adaptation to new twentieth-century conditions, both market and ecological. Neither Leo Africanus nor Marmol mention kif in their extensive remarks on the sixteenth-century Rif. Mouliéras, whose informant visited the Rif in the 1880s, writes that everyone at Ketama smoked it. Other Rifians did too.[100] Oral interviews conducted in the 1950s reveal that kif amounted to very little in the 1890s, but by 1910 at least one douar, Azila, had come to specialize in it. It spread thereafter until the ascension of Abdel Krim, whose Muslim principles forbade the use of narcotics. He actively

[98] No studies exist since G. Maurer 1968b, 54–7, 66, from which most of what follows is drawn. Today inquiries are not welcome.

[99] Strictly speaking, kif refers to the blend of marijuana with tobacco. But Moroccans often use the word for the plant and the entire enterprise as well as the product.

[100] Segonzac 1903, 47; Mouliéras 1895–9, 2:96.

suppressed its use and cultivation during the short-lived Republic of the Rif (1921–6). After the war, its use and cultivation blossomed, because the Spanish permitted it in the Rif while the French, in their much larger Protectorate, did not. Hence, Rif producers supplied customers throughout Morocco, and smugglers found no end of opportunity for their talents. By 1960, twelve tribes in the Rif produced kif. In that year, the Moroccan state illegalized it in northern Morocco, the former Spanish zone, but enforced only the law against sales and consumption, which therefore went underground. Production, while illegal, remained open and tacitly permitted. In the early 1960s, no more than 2.5 to 3 percent of Rif households depended primarily on kif production.[101]

It began to thrive in the 1960s, when demand from Europe first stimulated production. Production figures do not exist, and those on cultivated area must be considered rough at best. The Spanish calculated that kif covered 21 hectares in 1934–5. It had spread over about 1,500 to 3,000 hectares in 1968, and over 25,000 hectares in 1988.[102] This is a sizable proportion of cultivated area in the Rif, and the lion's share of irrigated area. Kif has driven down the production of maize, because of its irresistible competition for irrigable plots. In the richer valleys of the western Rif, everyone produces kif, and the best land is turned over to it. Most of this land is in the dwindling cedar zone around Ketama. The best way to make money fast in the Rif is to cut and sell cedar, and then plant and sell kif.

All this is an eminently rational adaptation to circumstances by Rifian peasants. Nothing else can earn them anywhere near the same income. Should they attempt sustainable forestry with cedar as their cash crop, they could not feed themselves. Nor could they if they put all their irrigated land under maize.

Their land and their numbers now require them to concentrate on kif. It is a spectacularly successful adaptation of the same sort as silk in nineteenth-century villages. Together with remittances from those working in Europe, it allows the Rif to continue living – precariously – on borrowed time.

Dependence on kif income puts the Rif at risk in three ways, one economic, one political, and one ecological. First, tastes and habits change, and northern Europeans (the Dutch seem to be the best customers) may lose their appetite for Moroccan marijuana. Or they may

[101] FAO Library (Rome) SF: 18/MOR (1965) 1:xi. The FAO hoped to convert kif fields into something legal.

[102] On kif cultivated area: Roda Jiménez 1941, 15; G. Maurer 1968b, 56; *Economist*, 10–16 December 1988, 16.

turn to another source of supply, although it is hard to imagine anyone else matching Rifian producers in price, quantity, or, Rifians insist, quality.

The Rifian hold on the market, even should the market remain strong, requires the political cooperation of the Moroccan government. Should it choose to enforce its laws and eradicate kif production, thousands of Rifians would be impoverished. That, of course, would probably bring about another Rif rebellion, so for the moment the state turns a blind eye. As long as the Netherlands or the European Community does not put pressure on Morocco to dry up the source of supply, as the United States does with South American countries where coca grows, Rifians can expect continued official tolerance of their leading cash crop. But it is easy to imagine this changing: Morocco's quest to join the European Community, moribund for the moment but not abandoned, might oblige Morocco to conform to European expectations and laws. The government would not hesitate long before sacrificing the interests of Rifians if it eased entry to the EC. In short, kif production survives only at the pleasure of the Palace, where Rifians are not loved and scarcely represented.

In addition, kif production deepens the ecological crisis of the Rif. It indirectly undermines all other forms of agriculture. Kif needs fertilizer, and in the Rif that means manure, chiefly from goats. Indeed, the main reason Rifians keep goats is for their manure. The scrawny Rif breeds have very limited value as producers of meat or wool. Milk yields are "derisory," about 24 liters per year.[103] Rifians used to keep goats (and keep them stabled at night) so as to have manure for their maize and garden crops. Now they need more manure for their kif, which yields poorly without it. Maximizing kif production implies maximizing manure production, which implies raising as many goats as possible. Those with land suitable for kif must grow it. Those without can at least graze goats on common land and sell manure to the kif growers. Production costs are negligible, as children can herd goats and the land they roam is free. No family with hungry children can afford to do otherwise. The real costs are those inflicted by heavy overgrazing on the vegetation and soils – that is, the carrying capacity – of the Rif. Its great commercial success has reduced the ability of the Rif to sustain its population beyond the moment when market and political conditions are no longer propitious. Should the kif bubble burst, as did that of the Ohanes grape, Mediterranean silk, and so many others,

[103] G. Maurer 1968b, 52.

Rifians will have less to fall back on than they had a generation ago.[104] Distant demand, concentrated into a small economic space, is still contributing to the ecological degradation of Mediterranean mountain landscapes.

Politics and War

National politics as well as international economics affected the fate of mountain villagers. Again the nineteenth century represents a break with the past. It witnessed a sharp increase in the relevance of politics, as states grew in power (except in Morocco), and generally placated volatile urban populations at the expense of the countryside. In Spain and Italy, the liberal ideology of the day provoked land policies that led to ecological catastrophes. And almost everywhere, nationalist and other warfare took its toll on lands as well as peoples.

Land Alienations

In both Italy and Spain the church, state, and local communes held title to large proportions of land before 1800. Custom and law – and rental agreements – governed the degree to which anyone could cultivate these lands, use them for pasture, or fell their timber. In time of duress law and custom could be, and often were, ignored, but nonetheless this system of land tenure had the effect of keeping much land out of cultivation and in forest.

Such a system, often called feudal, struck eighteenth-century rationalists as inefficient, as indeed it was. A wealth of land and timber stood unused, while many people persevered hungry and cold. The rationalists – and, after the Napoleonic occupations, the governments – of Italy and Spain saw that full exploitation of this dormant wealth required an end to the feudal system of land ownership. In many stages, beginning in 1806 in southern Italy and some years later in Spain, the state sold off church, state, and village lands, hoping to put them in the hands of people who would put them to productive uses. More often than not, they directly or indirectly put them in the hands of speculators, pros-

[104] If chemical fertilizer were substituted for manure, this ecological problem might be avoided, or at least traded for another, perhaps lesser, one. But it is expensive, and as long as production remains in the hands of small growers who reap only a small share of the profits, manure will have to do. Those who make large profits from kif sales are not likely to lend, as an agricultural bank might if kif were legal, to small peasants in the interest of preserving what little remains of Rif soils.

perous urban elements who happened to wield considerable influence in the national parliaments and local governments.

Similar things happened elsewhere in Europe when governments privatized church and public lands. In Britain, for instance, such movements began in the 1760s and, after much suffering, succeeded in creating a more rational pattern of land use and a greater sum of national wealth. Capitalist agriculture spread at the expense of forest and pasture, hard on peasants in the short run, but certainly good for the cities and in the long run for the sum of national wealth. But in northern Europe, far flatter in relief than Mediterranean Europe, forest and pasture played a less important role in stabilizing soils. In southern Europe, especially its mountain regions, the economically rational abolition of feudal land tenure proved ecologically irrational.[105]

In Spain the sale to private individuals of state, church, village, or military orders' land was called *desamortización*. It began with the inauguration of the liberal era in Spanish politics in 1813. The officials involved showed keen interest in the cash revenue brought in by the sale of public forest lands, which helped allay the constant fiscal anxieties of the state far better than could preserving the forests.[106] The state disposed of about 4.5 million hectares in all, equivalent to half of Portugal. All over Spain it led to enormous deforestation, quick and ignorant plowing of all kinds of soils, and the destruction thereby of millennia of natural capital accumulation. Flash floods and soil erosion menaced lowland and upland alike. Every century in Spain has seen ecologically costly mismanagement of forests and soils, but the nineteenth century stands in a class of its own, because of the desamortizaciones.[107]

Land alienations meant a great opportunity for the alert and financially able. The economically rational bought only the lands that promised good returns on the investment. But for the socially ambitious, any land was a good investment. Hence, many mercantile men bought up land they could not afford, borrowing the purchase price. To pay their debts they had to squeeze revenue out of their lands quickly. This implied felling all the marketable timber, and renting out as much land as possible to tenant farmers, sharecroppers, or shepherds. The value of

[105] Bauer Manderscheid (1980, 82–4) compares the alienations in Spain with those of France and Prussia. The best and most recent work on Spanish forest policy is Groome 1990; pp. 46–55 deal with desamortizaciones nationally.

[106] The accountant's cast of mind is clearly visible in the documents concerning sales, 1882–87, in AMAPA, legajo 355–26.

[107] O. Elorrieta y Artaza 1948, 91–100. Bauer Manderscheid 1980, 66–89. Groome (1990, 54) offers estimates of 4 to 10 million hectares of forest land privatized.

the timber generally exceeded the annual income of forests by three to five times in Spain, so only the patient and unindebted refrained from logging their new lands.[108] Many succeeded in paying off their debts this way, and many others did the same to realize a good return on their investment. Those who could not pay off their debts soon sold their land to richer proprietors. Thus the desamortización intensified two trends in Spain: deforestation and concentration of landed property.[109]

The Alpujarra typified these trends, although there alienated lands belonged almost exclusively to the church and religious brotherhoods. Throughout the province of Granada the major alienations took place between 1858 and 1864, following the *desamortización civil de Madoz* of 1855, which put up for sale lands belonging to the state and the municipalities. It also confiscated and sold church lands without compensation, ignoring a Concordat of 1851, to the eternal dismay of the clergy. Of the eleven *comarcas* in the province, the Alpujarra ranked third in terms of the area of land privatized: about 7,000 hectares entered private hands, through 156 sales. A large proportion of the buyers were senators, deputies, or other government figures (one was a professor). Others were already large landowners. Only three resided in the Alpujarra.[110]

Although the desamortización in the Alpujarra led to a quick and ecologically unsound expansion of arable, much of the damage would have occurred anyway, with or without the new laws. By the 1850s, with population at a high tide and still rising, peasants had already begun to cultivate the public forests (*montes públicos*). The law of Madoz in effect ratified a process already in train. In the province of Málaga, the public lands available for sale in 1855 came to a fraction of those registered in 1753, suggesting that much had already passed to private owners. In all likelihood, possession was nine-tenths of the law in Granada and the Alpujarra as well.[111] Population growth doomed forest and pasture alike; the law served to expedite and legalize the process.

In Lucania land alienations began in 1806 with the French occupation of the Kingdom of Naples. Formerly the Bourbon kings of Naples

[108] Iglesías Casado 1982, 72.

[109] The evolution of latifundia in Andalusia is treated in Artola, Bernal, and Contreras 1978. See also Groome 1990, 55. Herr 1988 has much on early disentailments, especially in Jáen and Salamanca. See also Jordana 1870; Jiménez Blanco 1986, 1:408–16.

[110] Gómez Oliver 1985, 116, 158–64, 171–2, 215–218.

[111] Mignon 1982, 197–9. Sáinz y Gutierrez (1862, 31–2) emphasizes the ecological damage peasant clearing did to public lands in Granada province, including the Alpujarra, and lauds privatization as a solution.

had preserved many forests as part of an extensive royal domain, in which the rights of pastoralists and woodcutters were strictly regulated. Existing laws, however, did not suffice, and the Bourbons passed more restrictive legislation in 1759 and 1780.[112] The French restricted timber cutting so as to benefit their navy, and after the restoration (1815) the public domain was preserved by laws more comprehensive than anywhere else in Italy.[113] Although begun in 1806–15, the greatest alienations (often called *quotizzazioni*, parcellations or assessments) came after the demise of the Kingdom of the Two Sicilies in 1860, when Lucania came under the jurisdiction of the newly unified liberal Italian state. Laws of 1863, 1866, and 1867 opened additional public and church lands to sale. In Italy as a whole between 1877 and 1910, more than 2 million hectares of forest land passed into private hands, over half of the forest area in the country.[114] As in Spain, the local bourgeoisie took advantage of these opportunities, after having passed the laws that created them. At Ferruzzano, in Calabria, communal forest went to eighty-three people, including two brothers of the mayor, two of his brothers-in-law, two nephews, seventeen of his cousins, and several more of his friends and more distant relatives.[115] As in Spain, the process contributed to two trends already in progress: the concentration of landed property and the ecological degradation of the landscape.[116]

Privatization of forest land in Lucania led to wholesale clearances. Between the 1877 law designed to protect forests from further damage and 1900, 170,000 hectares were deforested. Only 120,000 hectares of forest remained. In the Camastra basin between 1861 and 1900 forest cover declined from 70 to 23 percent of land area. Speculators bought much of the land, stripped it for timber and charcoal, and then abandoned it. Landslides, a long-standing curse of Lucania, became even more common. Floods, malarial pools, and silt deposition menaced lowlands. The Italian state got 30 million lire for the public and church lands it sold in Basilicata, but soon ended up spending many times that sum for hydrological projects on the same land.[117] In addition, it soon

[112] Trifone 1957, 39–44.

[113] Napoleon's brother Joseph was king of Naples when land alienations began in southern Italy, although they went much further when Joachim Murat, Napoleon's brother-in-law and marshall, took over. Joseph then became king of Spain, preparing the way for the alienations in Spain.

[114] Nitti 1909–10, 213.

[115] Clark 1984, 16. For fraud in Sila privatizations, see Arlacchi 1983, 166–7; Franchetti 1875, 75.

[116] On latifundia and the alienations: Franchetti 1875, 124–7; Nitti 1909–10, 70.

[117] Spera 1903, 14–20; Filangieri 1980, 110–11; La Marca 1916, 390; Colangelo 1977, 15; Tichy 1962, 66–91.

spent more on malaria control, as this second curse of the south worsened throughout the nineteenth century. In Potenza and many other comuni, residents complained of the evil effects of the alienations, linking them to ecological problems and to emigration.[118] As landscape conditions worsened, the peasants' condition worsened, enhancing the appeal of emigration.

One Italian historian goes so far as to say that ecological deterioration "expelled" the emigrants, emphasizing that the poor lands of Basilicata were in effect man-made and the inevitable result of the political and social structures of the countryside.[119] This may carry things too far. Certainly, as in Spain, peasants had begun to deforest communal and state lands of their own accord, taking timber as they needed it and sowing wheat where they could. In the 1830s and 1840s some observers took note of "vandalous" deforestation in Lauria and Lagonegro districts, leading to heightened erosion and to hydrological problems in the mountains and the river valleys below.[120] To a large (but hopelessly indeterminable) degree, the environmental problems associated with the land alienations resulted fundamentally from the rise of population. As in Spain, the laws ratified changes already illegally underway. And, as in Spain, the laws ensured that local notables, rather than peasants, drew most of the profit from the ecological degradation of the newly alienated lands.

In any case, forests disappeared with especially deleterious consequences because of the character of Lucanian soils. To add to the distress, privatization restricted the customary rights (*usi civici*) of villagers to what forest land remained, limiting access to chestnuts, fuelwood, and the special food of hogs, acorns. Access to pasture may also have become more difficult.[121]

The land alienations of the nineteenth century caused incalculable damage to the sustainability of life in the Lucanian mountains, already under population pressure. The tax policies of the new state compounded the problems of the peasantry. The newly unified state, run largely by northerners, governed the south like a conquered province after 1865. Lucanian peasants had to come up with cash to meet new

118 ACSR, PC, Zanardelli e la Basilicata, busta 2. A petitition signed by 555 agricultural laborers of Potenza explained: "Una delle nostre ricchezze poi erano in passato i boschi che con una smania feroce si son venuti graduatamente distruggendo . . . e così va via un patrimonio che è esclusivamente popolare." This is an important document in that most of the complaint against deforestation, and the call for reforestation, in southern Italy came from landlords and the prosperous classes.

119 Colangelo 1977, 16. 120 Pedio 1961b.

121 Foerster 1924, 88; Arlacchi 1983, 20, 166–7.

obligations, much higher than those imposed on peasants in the north. Land taxes, livestock taxes, and many others undermined family finances in Lucania more severely than in any other province. Land expropriation because of debt was four times as common in Basilicata as in any other part of the kingdom. Those not expropriated often had to sell their lands, or the timber on it, to pay their taxes – and to avoid the land tax in the years to come.[122]

The miserable condition of the south provoked no end of studies and inquiries. Two of these, conducted between 1899 and 1908, recommended reconstitution of the public domain – politically out of the question – and massive reforestation among the basic steps necessary to address the poverty of the Mezzogiorno. Basilicata and Calabria alone required a minimum reforestation of 600,000 or 700,000 hectares. The Italian state never did this, because it would have cost 350 lire per hectare, or about 24 million lire for these two provinces. Rome had sold public lands too cheaply to afford such expenditure. Financially, as well as ecologically, the alienation of public land cost Italy dearly.[123] Short-sighted land and forest policy that benefits a few at the expense of the many is, however, characteristic of most countries.

In ecological theory it is often held that resources held in common are abused: the tragedy of the commons. In Chapter 7, I will argue that such situations did indeed exist. But in the case of the land alienations in Spain and Italy, public policy consciously aimed at ending communal ownership created a deeper ecological tragedy of privatization.[124]

The alienations of forest lands in Spain and Italy permitted quick-witted men to make money fast. For a modest fee the state presented them with the opportunity to privatize the profits and "commonize" the

[122] Germano 1909, 128. On the tax burden and its consequences: Nitti 1909–10, 76–85; Spera 1903, 20; Ranieri 1961, 96–7; Foerster 1924, 50.

[123] Nitti 1909–10, 211–36, especially 214. Between 1866 and 1909 the Italian government spent 610 million lire on watershed control and 280 million lire on repairing the damage of deforestation, according to the Touring Club Italiano 1913, 86. These are huge sums next to the 30 million brought in by the alienations, but of course hydrological damage and deforestation did not result solely from the alienations, so a precise balance sheet on these actions of the state is impossible. The second inchiesta to recommend reforestation was Zanardelli's.

[124] Here is the Spanish Minister of Development [Fomento], José Echegaray, in 1873: "No son estos los únicos intereses que el ministerio de Fomento debe proteger y debe en la medida de sus fuerzas desarollar: los usos comunales . . . los aprovechamientos de los pueblos, todos estas prácticas socialistas deben ir desapareciendo, y al disfrute, confuso, irregular, demoledor y primitivo del suelo, bueno es que se sustituya por la propriedad individual, germa de todo progreso, garantía de todo orden y correctivo eficacísimo contra esta especie de socialismo campesino" (quoted in Jiménez Blanco 1986, 1:409).

costs of logging and plowing forest land. The revenues from these actions accrued quickly, and exclusively to the new proprietors. The true costs accumulated gradually over decades and were paid by everyone who depended on forests and soils in the general vicinity and indeed, to some extent, by everyone who paid taxes.[125]

War and Brigandage

While land policy determined by considerations of national politics produced environmental calamities felt most sharply by mountain peasants, international politics also contributed to landscape degradation through war. Here the twentieth century proved more damaging than the nineteenth, but once again the nineteenth constitutes a break with the past.

For centuries before the French Revolution, conventions limited the destructiveness of war, at least in Europe. Armies were small by modern standards, rituals limited the violence, and civilians normally avoided harm. But warfare changed with the French Revolution. Armies became mass armies, war became (at times) total war, and nationalism meant that civilians frequently directly felt the destruction and suffering. In the Mediterranean, modern wars have more often than not been people's wars with guerrilla dimensions. Defeating guerrillas often involved mountain campaigns, destroying their sheltering forests. Caesar Augustus's generals had done this in their day, and others had done it before them.[126] But in the nineteenth and twentieth centuries such stratagems became more common than ever before, with unhappy effects for vegetation. Campaigns against brigands usually involved forest burning as well. Because the line between brigandage and irregular warfare is often hard to draw, I will treat them together.

In Spain the important wars since 1800 have all involved guerrilla campaigns. This is true of the uprising against the French of 1808–11, in which Alpujarreños played a part. It is true of the Carlist wars – civil wars that did not affect the mountains of the south but did those of the north. And it is true of the Civil War of 1936–9, which in the Alpujarra smoldered on to about 1950.

In Italy warfare has been a more regular business, more often conducted between armies and consisting of set-piece battles with only local consequences for landscape. Nonetheless the civil war in the south

125 On privatizing profits and commonizing costs, Hardin 1985, 106–14.

126 Carrier 1932, 378, on Augustus. In the suppression of the great rebellion of the Alpujarra (1568–71) every Spanish siege was followed by widespread destruction of vegetation (see Chapter 3).

of 1860–5 tempted the state to destroy the brigands' forest lairs, just as Baron Hausmann destroyed the urban labyrinths of Parisian rebels for Napoleon III. In 1863 Massari explicitly proposed the elimination of Lucanian forest and maquis so as to flush out brigandage. The next year a local prefect recommended that the minister of the interior destroy the forest around Lake Monticchio, which he described as a 7,000-hectare brigands' fortress. Ironically, the brigands beat the government to the punch, inflicting great damage on this forest shortly after the prefect's recommendation.[127]

In the Ottoman lands too, irregular warfare haunted the mountains throughout the nineteenth century. Taurus brigands (and nomads resisting state efforts to settle them) sought protection in mountain forests, and so state antibrigandage campaigns took on an antiforest character. How great an impact this had on Taurus vegetation is unclear. Other parts of the Ottoman Empire surely suffered more.

The Pindus forests felt the effects of brigandage, war, and antiguerrilla campaigns as deeply as anywhere in the Mediterranean. From the eighteenth century on in the Balkans, local populations engaged in nationalist uprisings against the Turks, with occasionally devastating results for landscapes. Ordinary brigandage and its suppression had similar effects. In the Pindus, Ali Pasha cleared woods around Ioannina because robbers used them as hideouts. The Lion of Ioannina campaigned against bandits and insurgents far and wide in the Pindus, burning woodland wherever he went.[128] The Ottoman suppression of Ali Pasha, and of the Greek rising in Epirus, brought violence to dozens of Pindus villages between 1820 and 1832. The Turkish soldiers (mostly Albanians as it happened) burned villages, crops, and forests to undermine the Greek bands. These years were catastrophic for Pindus forests,[129] especially as the years that followed encouraged pastoralism for wool exports. Other Greek risings occurred in Epirus in 1854, 1866, and 1878. The Balkan War of 1913 saw more than a hundred villages of Epirus destroyed, as the Greek army advanced on Ioannina and Ottoman forces retreated. The Turks, consistent with military traditions of modern times, sought to leave nothing of value as they lost Epirus, and fire was their best weapon.[130] Finally, the resistance to the Italians and Germans (1941–4) and the final stages of the Greek Civil War (1944–9) centered in the Pindus. These conflicts destroyed wide areas of Greek

[127] Prefetto al Ministro, 22 February 1864, ACSR, MAIC, DGA, I, busta 288, f. 4, Ispettore al Signor Amministratore delle Acque e Foreste, 21 April 1864 (in the same busta). See also Caldora 1962, 396; Pedio 1983, 191.

[128] Holland 1819, 1:189. Hobhouse 1813, 1:116–17.

[129] Zervas 1932, 18.

[130] Trapman 1913, 13.

forests, as much as 25 percent in 1941–4 alone, according to some estimates. Much of this damage, if not most, took place in the Pindus.[131]

In addition to forests burned down, the conflicts of 1940–9 in Epirus destroyed villages, fields, terraces, and livestock in huge quantity. In Greece as a whole, occupying forces between 1941 and 1945 burned 1,770 villages. A map published by the Ministry of Reconstruction in 1946 shows that the majority of these villages were in Epirus. Population fled this guerrilla war zone, abandoning all agriculture. Livestock populations declined by about half.[132] After this holocaust came the Civil War, with its atrocities, burning of villages and forests, and general destruction. This focused on the Pindus even more than the impact of the occupation, as both sides burned Pindus forests for tactical advantage.[133] In the final stages, the Nationalist government pioneered the use of napalm as a defoliant on Mount Grammos.

The whole period from 1821 to 1949 witnessed chronic violence in the Pindus. In addition to the outright destruction, this meant that people abandoned their fields and plots to get out of harm's way. To save their lives, they neglected their lands and terraces, opening them to unchecked erosion. Furthermore, the economic area of which Pindus villagers were a part fluctuated year to year with the political situation. Markets for various goods expanded and contracted intermittently. Production decisions had to take account of the fluctuating likelihood of war, brigandage, or insurgency, not merely in the immediate vicinity, but throughout the region as well. This meant that peasants brought land in and out of cultivation, permitting neither the stability of a carefully tended, humanized landscape, nor the stability of one left to nature. War and the threat of war have played a major role in the deforestation and soil erosion of the Pindus.

The environment in the Rif, and the Maghreb in general, has also suffered heavily from war in the modern era. In the nineteenth century the Moroccan sultan launched several expeditions to tame the Rif, including a brutal one of 1813 in which his troops burned several villages.[134] With the resurgence in Spanish imperialism, Spanish troops fought a number of engagements against Moroccan forces, culminating in the Rif War (1921–6). This was a true people's war, involving every Rifian. For most of the war Abdel Krim's forces pursued a guerrilla

[131] Letsas 1953, 33; Tsoumis 1964, 36; Doxiades 1946, map 26.

[132] Doxiades 1946, maps 25, 53, 57. In a rising in the Peloponnesus in the 1770s, three-fourths of the livestock was killed. Beaujour 1800, 94.

[133] Personal communication, Christos Raptis, of the Ioannina office of the Greek Forestry Service.

[134] Seddon 1981, 13.

strategy, using the mountains and forests to their advantage.[135] The Spanish and, from 1925 on, the French sought to destroy the refuges of the Rifians, which involved among other things dropping incendiary bombs on Rif forests. According to a Catalan botanist, immediately after the war French occupying troops felled "cedar in enormous quantity" at Tizzi Ifri. (The consequences proved durable: in the 1960s, no cedar, indeed no forest of any sort existed at Tizzi Ifri, only matorral.) In the *kabila* of Beni-Hozmar, near Tetouan, Spanish investigators found the war had greatly reduced the local pine and cork oak forests, to the point where only a few patches of woodland remained.[136] The Rifians lost this war and their independence in 1926, as the Spanish and French committed half a million men. After they got their independence back, in 1956, they revolted against the Moroccan government, in 1958–9. This time twenty thousand troops and full air cover sufficed to restore the authority of the new King Mohammed V. On this occasion too, a number of Rifians took to the mountains, and the government flushed them out regardless of the cost to the landscape.[137]

In other mountain zones of the Mediterranean as well, war took its toll on forests. The Ottoman campaigns against the Arabs and British in 1917–18 completed the deforestation of the eastern Taurus (or at least Anamas Dağ) and the Lebanon. With coal unavailable, locomotives ran on oak and cedar from the mountains.[138] The same problems that plagued the Pindus, nationalist violence and rural insecurity, afflicted most of the Balkans from the first Serbian risings (1804) through the guerrilla resistance and civil war of 1941–5, destroying untold forest wealth.[139] In Algeria, the French conquest and long antiguerrilla campaign against Abdel-Qadr (1830–47) involved considerable forest burning.[140] Further struggles in 1871–2 led to the burning of the forest cover in the mountains behind Algiers.[141] The war of decolonization (1954–62) cost Algeria much of what remained of its forest wealth in the Kabylia and the Tell, as French troops and jets sought to destroy

[135] On the war: Woolman 1969; Pennell 1986. Abdel Krim's forces succeeded so well that they served as inspiration for Ho Chi Minh, a cook in Paris during the early 1920s. He later employed the lessons he learned from following this war in Vietnam's struggles. When a Palestinian delegation visited him seeking advice on how to conduct a war of national liberation, he told them to study a fellow Arab, Abdel Krim. Ho forgot that Abdel Krim was a Berber, not an Arab.

[136] On Tizzi Ifri: Font y Quert 1927, 529; G. Maurer 1968b, 10. SHM, Monografías de Kabilas, 3–51, Beni-Hozmar, p. 11.

[137] Seddon 1981, 178–80; G. Maurer (1968b, 12) says in the central Rif the forest had historically been a refuge of brigands.

[138] Bates 1973, 10–11. [139] Turrill 1929, 204. [140] Boudy 1928, 1:17.

[141] Shaw 1985, 392.

the mountain hideouts of the Front de Libération Nationale. In the early 1980s, one could still see forest areas destroyed by napalm.[142]

Frequent political violence in the mountains also probably promoted certain economic activities over others. Chronic warfare and brigandage improved the logic of producing goods that were easily transported and easily hidden, and of crops that need not be harvested at any particular time. Thus the rise of popular wars helped spread reliance on livestock and potatoes, and discouraged wheat, tree crops, and most cash crops. Seminomadic transhumance, for instance, had two signal advantages in conditions of insecurity. It encouraged people to store their wealth in highly mobile forms, and gave them two places to call home. Vlakhs, Sarakatsani, Türkmens, and others always had a place to go when war (or brigands) threatened, and could usually take most of their property with them. It is impossible to know whether (and how much) the mountain people in their production decisions engaged in what is now called political risk analysis. But their behavior is consistent with the idea that they did.[143]

War is always destructive, but modern war, with its powerful weapons and its popular character, has brought far greater damage to forests and soils than did the limited warfare of earlier times. As nationalism infiltrated the Mediterranean world after the French Revolution, war became a direct factor in landscape history because it now involved masses of people and often assumed a guerrilla form. A curious irony exists here. As war furthered the degradation of landscapes, especially in the vulnerable mountains, the mountain populations became ever more restless. With their prospects for the future dimmed by the gradual corrosion of the mountain way of life, young people found the revolutionary doctrines of nationalism (and later fascism and communism) increasingly attractive, as these promised to make sense of their resentments and frustrations. This helped to provoke rebellions, insurgencies, wars of national liberation – the popular, guerrilla wars that inevitably brought further landscape degradation.

Conclusion

With the consolidation of world markets and transport systems in the nineteenth century many Mediterranean mountain villages acquired

[142] Sari 1979; Sari 1970, 351–64; Pons and Quézel 1985, 38–9; Plit 1983, 80–1. General surveys on the impact of wars on forests appear in Thirgood 1981, 58–62, and Kuhnholtz-Lordat 1939, 249–50.

[143] Sivignon (1968) argues that in the Pindus pastoralism flourished as long as it did because it was well suited to conditions of political violence.

new prosperity and new vulnerability. The kif-growing parts of the Rif are doing the same today. The prosperity has usually proved fleeting, based as it was on a natural resource easily depleted or a trade route easily deflected. Entry into the world of markets provided some new opportunities, chiefly for wage work but occasionally for entrepreneurship, helping to expand population. But with this expansion came a dependence on the maintenance of production and markets, something that the villagers themselves could never control. War, economic recessions, distant prairies, even tariff policies – all became direct menaces, able to bring hunger to the mountains. At the same time, integration into larger market systems generally transformed the mountains into zones of exclusively raw-material production, undermining what modest industry had once existed thanks to isolation. Many of these processes of raw-material production corroded mountain ecosystems, lowering carrying capacity while population grew. Once populations became unsustainable with local resources, village life became hostage to the maintenance of political and economic conditions completely beyond villagers' control. As a Burkina Faso (West Africa) proverb has it, he who sleeps on another's mat never sleeps easy.

The extension of market systems brought with it the triumph of the liberal state in Spain and Italy, and the demise of feudal traditions in land tenure. As public land became private, the new landholders cashed in by logging and plowing huge tracts, upsetting hydrological regimes throughout both countries. Naturally the mountains suffered the most. The new era of popular warfare, with its tendency toward forest destruction, completed the roster of forces contributing to the degradation of the mountain way of life in the past two hundred years. Chapter 7 will document the many changes to Mediterranean mountain ecosystems since 1800.

7

THE CHANGING LANDSCAPE SINCE 1800

> The upland landscapes of the Mediterranean are not in line with their geomorphologic situation. Bare rocks obtrude where they do not belong.
>
> Carl Sauer (1963, 146)

> Find me a new religion – a religion of which the chief rite is tree planting.
>
> Mustafa Kemal Atatürk[1]

Concentrated demand corroded many mountain landscapes from the eighteenth century on. The mountain way of life endangered many more by its cumulative effect over the centuries. It is well to summarize the range of potential effects of the local economy upon mountain ecology before evaluating the evidence of ecological change.

Economy and Landscapes

Agriculture

Of the economic activities in the mountains of the Mediterranean, agriculture, pastoralism, logging, and fuelwood gathering had thc strongest impacts on the land. In the peasant massifs – the Rif, the Sierra Nevada, the Apennines, and, after 1600 or so, the Pindus – agriculture lay at the heart of economic life. Cultivation destabilized mountainsides more completely than any other human activity. This is so in part because people practiced it far and wide, but chiefly because it required full substitution of existing plant cover. In the earliest stages of that substitution – land clearing – topsoils and humus long protected from wind and rain suddenly became exposed and vulnerable. Land

[1] Cited in Turkey, Orman Genel Müdürlüğü 1967, 13.

clearing always involved accelerated erosion. Its rapidity varied with slopes, soil mechanics, and much else. In the Mediterranean, cultivators normally cleared new land in the late summer, when parched vegetation burned most easily. In trying to clear 5 hectares, peasants often inadvertently burned hundreds or thousands.[2] On sown fields, between late summer and the emergence of spring cereals, the torrents of the winter rainy season carried off soils without impediment.

Once farmers established their plots and fields, they could limit erosion, if never stop it, through careful husbandry. Terracing, contour plowing, and eternal vigilance could go a long way toward maintaining soils. These practices required not only a certain basic agronomic knowledge but also a concern for the future and plentiful labor. Mountain villagers did not always have all three, and the absence of any one could result in high soil erosion.

Basic agronomic knowledge concerning soil conservation techniques in mountain settings must normally have guided peasant cultivation. Generally, only when newcomers entered a mountain zone, as in the Alpujarra after 1571 or in the Pindus 1450–1700, could ignorance account for unnecessary erosion.[3] Various conditions could make prudent cultivation techniques hard to follow. In Jbala, for example, the customary system of agriculture depended on a twenty-year fallow and the careful control of goats. Population pressure forced villagers to begin shortening the fallow toward the end of the nineteenth century, and a breakdown of tribal discipline after the Rif War weakened the controls on grazing. Swift land degradation has followed, and now only lentisk will grow on slopes that once supported people. Nature exacts an extreme penalty for unsound land use in the climate and on the soils of the Rif. Demographic pressure, not to mention war and rural insecurity, could make sound agronomy hard to practice anywhere, even when properly understood.[4]

Mountain villagers normally intended to bequeath their lands to their children, so their concern for the future must generally have been high. Only rarely did rental separate the cultivator from an interest in the future of the land, a common source of poor husbandry in lowlands. A peasant intending to emigrate might ruthlessly exploit his land

[2] Vázquez del Río (1935, 20–2) points this out for the Rif.

[3] Nonetheless I have seen sloping fields plowed vertically in Lucania and row crops planted not with the contour but with the descent of slopes in the Rif. Bruce Lansdale of the American Farm School in Thessaloniki explained to me that Greek fields are often plowed with the slope and not the contour because division upon inheritance creates long, narrow fields spanning elevations.

[4] Fay 1984, 9–10. Thornes 1976, 15, on unsound fallow practices in the Alpujarra.

in order to get up a stake to go – if he had a cash crop. But even in times of emigration, concern for the future of a family's lands usually survived, because rarely did migration occur *en famille*. And when it did, the emigrant might have refrained from abusing his land out of concern for its sale price.

But while concern for the future must normally have been strong, the immediate future always took full precedence over the more distant. Mountain peasants lived in a world of considerable insecurity, in which goods in the future (land included) meant much less than goods in the present. In effect, they discounted the future heavily – an entirely rational calculation in their circumstances – and behaved accordingly. Such behavior might easily include "unsound" farming techniques. Suppose, for example, that war in the lowlands temporarily raised the price of wheat. A farmer might well be tempted to cash in on that temporary condition before it changed, even if that meant sowing to wheat slopes that he well knew could not hold soil without terracing. Or suppose he had more children than he could reasonably have expected to survive infancy; again he might well mortgage the future to help his family survive in the present. Given the great uncertainty surrounding the future – drought, landslide, plague, or war might intervene before next year – mountain peasants very likely acted on the urgencies of the present more often than did farmers in more predictable environments.[5] Outside observers have often characterized mountain people (and peasants in general) as childish in their behavior, the principal aspect of this childishness often being an apparent unconcern for the future consequences of actions. This observation normally signals a difference in the mental discount rate of the observer and the observed.

Soil conservation required great investments of labor. Hoisting buckets of earth up hillsides to rebuild washed-out terraces is a task fit for Sisyphus. Indeed, irrigated polyculture on steep slopes is among the most labor-intensive agricultural systems in existence. A given family might fall short of the requisite labor from time to time, and prove unable to maintain terraces or to manage irrigation water carefully. In the Pindus, especially Zagori, while men scoured the Balkans for work and trade, agriculture often fell to overworked women. Everywhere in times of emigration a similar "feminization" of agriculture took place. Although women could call on the very old and the very young for help, it must often have been necessary to take shortcuts, even when that clearly imperiled the future of plots and fields. Furthermore, given the often strict segregation of labor along gender lines, it may frequently

[5] Jones (1981) discusses how environmental risk may have influenced peasant behavior on a world scale.

have happened that men left agriculture in the hands of women who did not fully know proper techniques for soil preservation.[6]

Even when the requisite knowledge, concern, and labor were all present, the problem of the commons could still lead to land degradation. Everywhere in the Mediterranean mountains (at least prior to 1860) most pasture and forest land belonged to the state or, more often, to local communities. In either event it was common land. Concern for the future of one's family might well entice peasants (and shepherds) to misuse land held in common. Acting in the rational interest of one's family indeed required maximizing the benefits one extracted from common land, even if that led to rapid depletion of forest or to overgrazing of pasture. Only where mountain villagers could trust one another not to take undue advantage of commons did it make sense for them to refrain from such action themselves.

Naturally villagers understood the problem and sought to regulate access to common lands. In cases, as in Samarina in the Pindus and Beni Bou Frah in the central Rif, such restraint checked landscape deterioration. Careful regulation also seems to have worked well in many instances in the Alpujarra, where villagers drew lots to determine who could have access to what from common lands. Elsewhere, an informal but sometimes effective coercion kept shepherds and peasants from abusing common land: a man who grazed too many sheep on village lands might find his sheepfold burned in the night. Only a family with many male relatives (as a deterrent) could risk informal coercion of this sort. So even where forest and pasture were open to one and all, custom and coercion could limit the temptation to overexploit. Naturally villagers ignored laws and agreements when they felt they could get away with it; regulation of common forests and pastures was far from perfect.[7] And where population had long been sparse, as in the Taurus, no such rules and restraints existed, so forest depletion went unchecked even in the late nineteenth century when, as we shall see, it began to denude and impoverish some areas.[8]

Where ethnic, religious, or class differences undermined trust and cooperation, villagers felt additional justification in exploiting landscapes before their neighbors did. In mountain villages such differences were rare but not entirely absent. By the nineteenth century the competing

[6] Physical strength may also have mattered here, since rebuilding a terrace is no easy business. Formica (1983, 93–4) discusses this in the context of contemporary southern Italy. But the stamina and strength of mountain women in the Mediterranean were often prodigious.

[7] As it is today: Cuadrado Iglesias 1980.

[8] E. Davis 1874, 132–3. Forest policy in the Ottoman Empire began only with the *tanzimat* of the mid-nineteenth century; and it never had much effect. Özdönmez 1973.

national identities of Greek and Vlach coexisted in Pindus villages. In the Taurus, Turkish peasants and recently settled Türkmen nomads saw real differences between them. In the Rif, neighboring tribes might feel themselves Arab or Berber, although for most of their history the tribal affiliation was the only one that mattered. In the larger Italian villages, a genuine class structure existed, and village notables sought to gain access to common lands for their own sheep, leaving the poorer to graze their animals where and when they could. Even in the smaller villages of the Pindus, considerable social distinctions existed among pastoralists (more so than among cultivators), and herding families formed temporary alliances among themselves to improve their ability to perform all the needed tasks and to assure access to common pasture.[9] In Trevélez, the village had three sections, each inhabited by peasants of different means. Village politics, and thus access to commons, came under the control of the richer villagers. With careful regulation (and harmonious politics) a village could prevent the "tragedy of the commons" from despoiling their land and forests. But careful regulation was hard to maintain. Inter- and intra-village rivalry, eventually combined with demographic pressure, led to many small and slow tragedies. In the Rif today civic restraint has broken down, and a pitiless competition for fields and pastures has led to a larger and faster tragedy. Wholesale privatization of common land (recall Chapter 6) could lead to giant ones.[10]

Absent the strong incentives to abuse land that market agriculture (and herding) supplied, subsistence economies alone could undermine landscapes. But when market opportunities encouraged limitless cultivation or grazing, landscape degradation accelerated. Eager peasants, often working on behalf of someone else, brought steeper lands into use, and often did so with no thought for tomorrow. The vines of Málaga and the kif of the Rif (Chapter 6) are but two examples.

Pastoralism

Pastoralism, whether on commons or private lands, also destabilized slopes throughout the Mediterranean. Shepherds burned large tracts

[9] Sivignon 1968.

[10] See Fay (1987, 405) on the Rif. The classic scholarly statement of this tendency to abuse common property is Hardin 1968. Aristotle (Politics 2.3) noted much the same thing: "What is common to the greatest number gets the least amount of care. Men pay most attention to what is their own; they care less of what is common." McCay and Acheson 1987 is a recent compendium of anthropological opinion on the commons question.

of forest every dry season in order to promote the growth of tender grasses for their flocks. Woody maquis usually follows a burn within two or three years, so shepherds often burned the same landscapes every year. This practice began shortly after mankind domesticated sheep and goats and has persisted until the present, although now it is much diminished.[11] After repeated fires, mountain slopes lose most of their deeply rooted vegetation (except for some fire-resistant species such as *Quercus coccifera* and *Erica arborea*) and eventually their soils as well. In temperate woodlands, the forest floor has plenty of underbrush suitable for goats, hogs, cattle, and sheep. But in Mediterranean forest, now often dominated by pines up to 1,000 or 1,200 meters, the forest floor is barren, so the forest is the enemy of the shepherd until one reaches the upper altitudes of oak and beech. Shepherds are the chief reason that Mediterranean vegetation, whether forest, maquis, or garrigue, is adapted to fire.[12]

The damage that herders do to landscapes would be no greater than that of the occasional forest fire if it were not for their goats. Goats can eat almost any plant, including those with spiny leaves, and thus can prevent second-growth forest. Where grazing is unrestricted, timber is a nonrenewable resource. If sufficient goats are present every summer, no substantial plant cover can develop to protect soils from the winter rains – hence the reputation of goats as despoilers of Mediterranean landscapes, although in reality it is men who destroy the forest and the goats that keep it from coming back.[13] The ecological consequence of prolonged pastoralism in the Mediterranean has been rigorous selection for noxious and fire-resistant plants, able to withstand both the attentions of goats and the fires set by their keepers.

Shepherds and their flocks, especially nomads, have consistently conflicted with one another and with peasant farmers. This too has had

[11] Boué (1840, 1:504) says throughout European Turkey locals burned land for their sheep. Hammond (1983) said that in the 1930s every autumn large clouds of smoke filled the air as shepherds set fires all over northern Greece. On ecological succession after burns: Papanastasis 1978. Thirgood (1981, 73) says in the Lebanon shepherds felled tall cedar and silver fir so that their goats could pounce on the needles!

[12] Barbero and Quézel (1983) refer to Greek landscapes as "pyropaysage" to emphasize the role of fire. Kuhnholtz-Lordat 1939 is a study of fire and landscapes in the Mediterranean and elsewhere.

[13] See Turrill 1929, 204–11. The goat has its defenders, however. See French 1970; Kolars 1966. Certainly there is a logic to grazing goats in garrigue and maquis that would be otherwise useless. But doing so every year, and with more than a certain critical number of goats, has proved very costly in every Mediterranean country. A pine tree must survive about fifty years to grow tall enough that goats cannot kill it. Samios 1900, 160.

7.1. Pindus goats in search of sustenance. Goats and fire are the chief agents through which humankind influences the spontaneous vegetation of Mediterranean mountain landscapes.

ecological consequences. Shepherds always tested the limits of what they could get away with in grazing their animals on common or rented land. Flocks, especially on migration, can trample crops; fires can spread to valued chestnut groves, to fields, or even to villages. In the Taurus and the Pindus, cultivators and pastoralists have struggled against one another for land, and at times for survival, throughout the centuries. In the western Taurus, cultivators triumphed only in the past hundred years, whereas in the Pindus they have dominated for three hundred. But distrust, rivalry, and animosity have often colored their relationship since. In Zagori, cultivators say the transhumant shepherds grazed their goats where they pleased, "rifle in hand."[14] In Spain and the Kingdom of Naples, conflicts (and royal financial interests) gave rise to official state policies delimiting the respective spheres of shepherd and pastoralist: the dogana and the mesta had their privileges and duties inscribed in law from the thirteenth and fourteenth centuries. In the

[14] Théophilou 1983, 179. Roux 1970, 69–70, for some details in the Taurus, where in the last six or seven decades violence between nomads and peasants has diminished sharply. See also Cribb 1991.

days of the Kingdom of Naples, the main sheepwalks in southern Italy were to be of a certain width (111.11 meters) and peasants who infringed upon it risked the death penalty.[15] But as with local law and custom, the strong could often abuse their privileges and avoid their duties. In the Rif, sheep and goats invariably have belonged to the cultivators themselves, obviating conflict of this sort.

Peasants and shepherds have a common interest easily realized in lowlands. Pasturing animals for the winter on the stubble left after the autumn harvest suits both parties nicely and can form the basis of a symbiosis. But if lowland peasants own their own sheep, they have no need of transhumant sheep or nomads. And if summer fires or sheep stealing gets out of hand, cooperation degenerates into conflict. Countless acts of valor recorded in the songs and folktales of the Pindus (and elsewhere) testify to the frequency with which this happened. Conflict between cultivators and shepherds, more common in the Taurus and Pindus than elsewhere, led to a level of instability in land use that furthered erosion and deterioration. It must also have sharply increased the number of fires.

Instability is an inherent condition of pastoralism. Flocks are vulnerable to drought, disease, predation. Independent of market factors, their size fluctuates greatly from year to year. Herders do their best, through breeding, culling, buying, and selling, to adjust their flocks to available grazing (and labor), but nonetheless stability in production, in land requirements, and in labor requirements is much lower among pastoralists than among cultivators.

This annual instability is compounded by a longer-term instability in the viability of pastoralism. By and large, conditions of peace and rural security favor cultivators and force pastoralists, especially nomads, to less favored lands. Political insecurity favors pastoralists, whose capital is mobile and whose way of life emphasizes valor, endurance, resistance. The rise and fall of states and empires (recall Chapter 3) affects the balance between pastoralist and cultivator in any given region.[16]

Among cultivators, both clearing and abandonment accelerated erosion; among pastoralists, only clearing did so. Pastoralism's greatest menace to soils and vegetation came when herders persisted in stable patterns of overgrazing.

[15] Marino 1988, 42.

[16] Cribb 1991, 23–34, 58. Today there is an intermediate-term instability as well, because, at least in the Taurus, nomads often settle down for a few years, even a generation, but then return to nomadism. They do so largely in response to perceived chances of making money. To what extent this pattern of punctuated phases of nomadism and settlement characterizes the past is unclear. Cribb 1991, 119–22.

Overgrazing has haunted the landscapes of the Mediterranean for centuries and probably millennia. Shepherds, whether fully nomadic, transhumant, or local, rarely own the land on which their flocks graze. Either they rent its use or they partake free of charge. In either event, it is normally in their interest to run as many animals as they can on any patch of ground. Only where local peasants graze their own animals for use by their families is pasture spared chronic overexploitation. Governments, even local ones, can rarely keep shepherds from following their own interest. Keeping goats from ranging where they please is more difficult still. In Yugoslavia some thirty years ago, the government spent twice as much per animal in repairing damage from overgrazing as shepherds realized in profit. No more illogical economy can be imagined, but shepherds, even in a socialist society, can easily "privatize profit" while "commonizing costs."[17] Goats, untroubled by illogic, do not resist serving as the agents of the destruction of their own habitat. For all these reasons overgrazing has long been commonplace.

In recent centuries it has become the norm. The temptation to overgraze is especially hard to resist when a strong market exists for wool or meat – or, as in the Rif, for goat manure. When transport nets concentrated distant demand on Mediterranean mountain pastures, locals responded to their economic interest by raising more sheep and goats – or by renting more land to eager transhumant shepherds. In medieval times the boom in woolens of England and Flanders had its effect on Spain; a later boom in northern Italy affected the Apennines; a still later one in central Europe encouraged overgrazing in the Balkans, including the Pindus. Woolen manufacturing in Morocco has promoted overgrazing in the Middle Atlas. No great booms in wool manufacturing have touched the Taurus. Indeed, by the late nineteenth century, foreign manufactures penetrated Ottoman markets so that the demand for Taurus wool might well have declined and the logic of overgrazing disappeared.[18] At the same time large quantities of cheap wool from Argentina and Australia helped diminish the logic of overgrazing in Spain and, to a lesser extent, in Italy.

At times, the demand for meat also encouraged overgrazing. As a rule, only urban populations and foreign garrisons could afford meat in sufficient quantity to affect production decisions. But in the nineteenth century, cities grew rapidly throughout the Mediterranean world, and in places such as Corfu foreign garrisons and squadrons exerted a

[17] Thirgood 1981, 73. This language is from Hardin 1985, 104–14.

[18] This hypothetical decline in demand came at the same time that cultivators were establishing their control over the lands surrounding the Taurus, at the expense of pastoral nomads. Perhaps the first helped bring about the second. On foreign penetration of Ottoman markets: Pamuk 1987.

strong influence on land use patterns. By late in the century, however, meat from far afield began to reduce the incentive to overgraze in the Mediterranean (see Chapter 6). Because meat has never traveled as well as wool, the power of distant markets for meat to encourage overgrazing has never equaled that of wool markets.

In short, the pastoral economy in the Mediterranean has always tended toward overgrazing, and the rise of wool markets (and occasionally meat markets) has often sharpened this tendency. Although these markets first became important with the rise of northern Europe in medieval times, their greatest effect came in the early and mid-nineteenth century, before the impact of North and South American wool and meat. Given the fragility of Mediterranean mountain ecosystems, the cost of the pastoral economy in terms of soil and vegetation has been far greater than the cost of similar practices in Scotland, Norway, Patagonia, or other lands of commercial pastoralism.

That said, however, a word is in order about the spatial limits of pastoralism's effects on landscapes. Although it is true that pastoralists burned forest to expand pasture, and that their animals prevented the return of forest once burned, they generally did so over clearly defined, and usually limited, swatches of earth. Transhumant and nomadic pastoralists were commuters, following familiar routes whenever possible between summer and winter pasture. They tended to make their camps in the same places, year after year. This is so because nomads had certain requirements. They needed grass for their flocks. They needed water for themselves and their animals. They needed fuel, meaning firewood, because nomads were too mobile to make charcoal or to build up stocks of animal dung. (They needed fuelwood even in the summer months, for preparing milk products.) And they needed level terrain for their camps, preferably sheltered from the elements by a cliff or protruding boulders. Often they chose sites near ancient ruins, because the ready cut stones provided material with which to build the bases of tent dwellings and huts; and some old buildings made fine animal pens. As a final requirement, or at least preference, they needed campsites far from cultivation, because conflict with farmers usually got pastoralists into trouble with the forces of law and order. Nomads and transhumant pastoralists could fulfill all these requirements only along a few routes, because (in the Taurus and the Pindus) level ground, water, and fuel are hard to find, and harder still to find all in one place. As commuters, flocks and their masters trampled, ate, and burned vegetation only along their chosen tracks.[19]

These tracks are still used in some cases, despite the rise of motorized

[19] See Cribb 1991, 134–8.

transhumance. Even where they are abandoned, it is easy enough to detect their traces, and archaeologists have done so. For both the Taurus and the Pindus fine maps exist showing the major sheepwalks.[20] Over long periods these routes did change of course, as pastoralists were squeezed out of winter pasture here, or managed to supplant cultivators there. But generally, for simple geographic reasons, these tracks got regular use for century after century. By and large, flocks affected landscapes severely within tightly limited boundaries, except in times of political disorder, when transhumant pastoralists, in Anatolia at least, could expand their operations at the expense of cultivators, and graze their animals where they pleased. Of course, in the shepherd massifs, political disorder was often the rule.

Logging

Agriculture and pastoralism have both had strong effects on Mediterranean mountain landscapes. In most areas logging pales in comparison. It has gone on for millennia, but in a narrow range of locations, and with comparatively short-lived effects – when not followed by agriculture or grazing. Until the age of roads and trucks, logging affected only coasts and river banks. In a few places mules carried timber down from the mountains, but only rarely was this practical. Selective logging, say for mast timber in the Pindus or fine cedar in the Rif, has scarcely any durable effect on landscapes. Clear-cutting is another matter, of course, opening soils to erosion with permanent consequences. The Sila Plateau in Calabria bears scars testifying to the impact of clear-cutting. In most locales, however, even the consequences of clear-cutting remain modest if grazing animals – goats in particular – do not follow in the wake of the axes. If left alone, a logged-over Mediterranean slope will lose considerable soil, but will normally stabilize in two or three years with the growth of maquis. And the maquis species will give way to forest trees (indeed some of them will become forest trees) if undisturbed by goats, fire, or hoes.[21] This condition has not often been met, given the incentives for overgrazing. But it is the continued

[20] On the Taurus: Cribb 1991, 128; Andrews 1989, appendix G (citing H. Saraçoglu, *Türkiye Coğrafyası Üzerine Etüdler. III. Akdeniz Bölgesi*, Istanbul, 1968). On the Pindus: Psichoyios and Papapetrou 1988. Braudel 1976 has a map of transhumant routes in the Mediterranean generally.

[21] Great differences exist between species. Rif cedar regenerates well after logging – if left alone. Cherry oak in the Taurus (*Quercus cerris*) does not, and is usually replaced by pines if forests recover after cherry oak is cleared. Most oaks regenerate poorly and often are succeeded by pines, which do better on degraded soils.

grazing, rather than the logging that fundamentally causes erosion and slope destabilization. Logging merely represents a more useful destruction of forest than does the shepherds' fire.

Fuelwood

Unlike logging, fuelwood gathering (and charcoal making) have always centered around villages. No peasant walked more than a few hours for fuelwood, although a big city like Istanbul imported it from afar. Generally, if fuelwood supplies ran out within a radius of two or three hours' walk, villagers found it preferable to leave for a new landscape. Since men normally would make such decisions, and women normally gathered firewood, it may well have been the case that villages were not abandoned for reasons of fuel shortage as quickly as they would have been had either the men or the women done both the gathering and the deciding. If so (I have no evidence), devegetation for fuelwood gathering was more concentrated on account of this division of labor within villages. In any event, fuel needs often helped to create circles of scant vegetation around villages. Where villages lay huddled together, the cumulative effect of fuelwood gathering could be considerable. Until the 1950s, when kerosene and other fuels began to reach the mountains (and population began to fall), Greek statistics show that fuelwood needs led to ten to twenty times as much woodcutting as did timber demand. In Ottoman times fuelwood extraction destroyed more Greek forest than did anything else.[22] But fuelwood gathering did not necessarily involve felling trees: it often required only cutting branches and picking up windfall. In Lucania, villagers carefully pollarded trees so as to create coppices for fuelwood. Cutting branches and picking up dead wood have negligible permanent effects on forests, unless done so thoroughly as to prevent nutrient cycling. Hence the impact of fuel needs upon vegetation cover and soils rarely came to anything like that of cultivation or pastoralism, even when population growth increased fuel requirements, or market demand for charcoal increased financial incentives, inspiring more abusive cutting. Both of these happened in the Christian Mediterranean after 1800 or so, but neither happened much before. Both have happened in the twentieth century in the Taurus and the Rif.[23] Everywhere these developments have come simultaneously

[22] Kentro Programmatismou kai Oikonomikon Erevnon 1976. For Ottoman times, see Grispos 1973, 150.

[23] Charcoal demand in Spain, for instance, destroyed vast areas of *thuya* in the eastern Rif in the 1920s and 1930s, leaving little vegetation behind. Vázquez del Río 1935, 16–17.

with expansions in cultivation and pastoralism, and the latter have usually produced more enduring effects on vegetation and soils. The chilly Pindus may be an exception.

In general, devegetation and soil erosion have varied with population pressure and the strength of connections to distant markets (Chapters 5 and 6). With respect to agriculture, often the single most important agent of mountain landscape deterioration, the situation is slightly more complex. Instability – the rapid succession of expansion and contraction, of land clearing and land abandonment – has damaged slopes far more grievously than has stable husbandry. In other aspects of the mountain economy with direct impacts on landscapes – pastoralism, logging, fuelwood gathering – instability had no great costs. High intensity rather than high instability produced degradation, and intensity varied closely with population and market opportunities. In the past two hundred years instability in agriculture has probably increased, and intensity in other activities certainly has. Thus it was that the nineteenth and twentieth centuries witnessed the most rapid landscape deterioration in the history of the Mediterranean mountains.

Evidence of Landscape Change

So far I have given little evidence of this supposed deterioration of Mediterranean mountain landscapes, except for a few instances in Chapter 6 associated with the extension of market nets and the impacts of the alienation of public lands. The evidence is great in quantity but poor in quality. After some cautionary remarks about the nature of the evidence, I will present some examples from specific locales. Readers uninterested in the nature of the evidence are invited to skip this section.

The evidence for historical landscape change in the Mediterranean mountains comes in several varieties, none of them satisfactory. Most direct (but perhaps least useful) is visual evidence. Even the untrained eye can detect many clues to recent landscape history. Eroded slopes that reveal a tree's roots can indicate the extent of erosion within a tree's lifetime. Abandoned terraces or threshing floors testify to former cultivation. Ecologists and geologists can of course see more than I have in the species composition of second-growth forests or the braided channels of silt-laden river beds. But even with proper training in landscape interpretation, the utility of visual evidence diminishes sharply with time: it is excellent for recent decades, but almost useless for two hundred years ago. And one must cover a lot of ground before the accumulated evidence amounts to much.

The oral or written testimony of others forms a more useful source of information. One is at the mercy of the honesty and accuracy of one's witnesses, a familiar problem to historians, but at least one's witnesses were present in past times. In the nineteenth century, legions of gentleman travelers scoured the Mediterranean for antiquities and wrote books when they got home. Some of these gentlemen were indefatigable and meticulous observers. Leake's descriptions of northern Greece 180 years ago are invaluable for landscape history.[24] Others of course had lesser standards. Oral accounts related by elderly locals of earlier landscapes have only one virtue: they can address a specific question. But memory is often weak, and most Mediterranean peasants and shepherds prefer not to trust strangers and often judge it safest to mislead them, however innocuous their questions may seem.[25]

More precise testimony from the past is available through photographs. For some mountainsides, like that of Mitsikeli, located picturesquely across the lake from Ioannina, several old photographs exist. But nowhere does the photographic record go back before 1900 in any systematic way. Sketches, drawings, paintings, lithographs, and the like do exist from the nineteenth century, and indeed well before. Edward Lear drew over a hundred sketches and watercolors of rural scenes in the Pindus about 1850. Tuscan and Umbrian painters from the fourteenth century provide a record of the landscapes in the hills of north-central Italy.[26] But artists have other goals than mere precision, so even careful ones like Lear, who took up to three days for a sketch and made notes on vegetation, must be treated with skepticism.[27] Even if such works were reliable, they are essentially anecdotal, treating only what the eye of the artist could take in.

Statistical evidence is not anecdotal but systematic. It can cover wide areas, and long spans of time. But governments have busied themselves

[24] Leake 1835.

[25] I have asked scores of them about features of local landscapes and have usually found old men very cooperative. But I have felt it imprudent to put much faith in their remarks, because they could not know whether to put any faith in the apparent innocence of my inquiry. For reasons of linguistic incompetence, I did little of this sort of work in the Rif and the Taurus.

[26] These scenes of the Pindus are preserved in the Gennadeion Library in Athens. On Lear as a meticulous "topographical" draughtsman: Hofer 1967, 52–8. The Italian paintings are visible in the Museo Nazionale in Perugia, or the Uffizi Gallery in Florence.

[27] I have found photographs helpful for the Rif and the Pindus more than elsewhere, and artwork most helpful for the Pindus, specifically because of Lear. I did not search widely for photographs and paintings of landscape in the Lucanian Apennines and the Alpujarra, for which other varieties of evidence were strong.

collecting data about forest area, crop area, pasture area, and so on only rather recently, from the 1870s in Lucania but from only the 1940s in the western Taurus. Ordinarily these data concern administrative units, often inconvenient for analysis of mountain areas. And, predictably, these statistics, available in official publications and archives, are often inaccurate.[28] Generally, the older the data the less reliable – but the more useful, since clues of any sort, reliable or otherwise, are fewer for earlier periods. With this array of incomplete and faulty evidence, and the help of generations of geographers, botanists, ecologists, palynologists, anthropologists, and others, I have tried to reconstruct landscape history.

Deforestation

The Taurus. In the western Taurus deforestation in modern times has not undermined the local population. Villages are few, and usually of recent origin, so that the cumulative effect of burning, logging, and grazing has not yet denuded most slopes above 1,000 meters. Considerable thinning of forests, but rather little thorough deforestation of the sort so visible elsewhere in the Mediterranean (and in Turkey), took place before 1860. The most rapid deforestation came only after 1915. I will review the evidence for the Taurus as a whole, proceeding from west to east, because that for the western Taurus alone is rather scant.[29]

In Seljuk and Ottoman times, most of the Taurus remained forested (see Chapter 3). Nomads passed through montane forest zones, occasionally setting fires, but their impact was incidental because they and their animals spent almost the entire year either above or well below mountain forests. The tahtaçı felled timber for Turkish cities and for export. Their efforts account for most of what deforestation took place in the western Taurus before the nineteenth century. Anecdotal evidence suggests it was not much.

In 1872 the Reverend Davis rode south from Isparta to Antalya,

[28] Special problems exist in the Mediterranean. Multiple use has led to land being counted twice, both as forest and as pasture for instance. And usually forest land is a residual category, including landscapes that might once have had forest or could one day have forest. Franciosa (1926, 71) points out that various agencies published estimates of forest area in Lucania that varied from 130,000 to 207,000 hectares. Crops grown in association also can produce double accounts of crop area. Jiménez Blanco (1986, 1:46–60) explains the weaknesses of Spanish agricultural statistics prior to the 1890s.

[29] By 1950 deforestation in Turkey came to about 100,000 hectares per year. Saatçioğlu 1952, 80.

marveling at the forests as he went. At one point, he gazed at Bozburun Dağ from the northwest, across the Aksu: "Nearly all the wide valley below was in thick forest; only a few scanty patches of green marked the cultivated or open spots." Looking east to the Taurus proper, he continued: "On the opposite side of the valley ranges of mountains rose one beyond another . . . and over all, except at rare intervals, spread the dark pine forest." Davis even found cedar in a valley near the headwaters of the Aksu. On one of his descents toward the Aksu he admired the thick oak and pine woods, and the "huge fallen trunks that lay across our path, as in some wild Canadian forest." While the mountain slopes towering above the Aksu carried thick forest in Davis's day, the environs of the small city of Isparta were already bare, wood was scarce, and city folk had to import charcoal from a "considerable distance."[30] Urban areas certainly had already had an impact on forest cover. But the cities near the Taurus remained small until quite recently. Only larger changes in the economies of Anatolia and the eastern Mediterranean could affect the remote woodlands of the western Taurus.

Pressure on western (and other) Taurus forests grew with the demands of the Suez Canal (Chapter 6), and the gradual settlement of areas previously left to nomads. Whereas nomad tents still dotted the Yılanlı Plateau in summertime in the 1880s, a few years later peasant cultivators – probably settled nomads – had moved in to stay. The tall cedars of the Davras Dağ that Arundell saw in 1830 and that Sarre remarked upon at the end of the nineteenth century had vanished by 1950. Arundell wrote of woods down to the edge of Lake Eğridir, and Hamilton around 1840 found oak on the slopes south of the lake, but it too had disappeared by 1950.[31] All the evidence concerning the mountains between Eğridir and Antalya is consistent with the view that forest cover thinned considerably between 1850 and 1950.

Davis, ever interested in forests, remarked in the 1870s:

> Rich as the country yet is in forests, it would appear to have been richer still at an earlier period, but the neglect of the Governments and people, the destructive wars of which it has been the scene, the incursions and final settlement there of the pastoral tribes, who every year destroy by fire a great extent of wood to obtain fresh pasturage – all have helped destroy the forests; nothing, indeed, but the scantiness

[30] Davis 1874, 150, 159, 179–80, 182, 185. Similar passages on the mountain forests south of Isparta, around the ancient city of Sagalassus, are on p. 178.

[31] Planhol 1958, 49; Arundell 1834, 1:332, 336; Hamilton 1842, 1:481–2; Sarre 1896, 131–70. Lucas (1714, 254) easily found enough fuelwood to stave off the January cold in 1707 between Eğridir and Beyşehir.

> of the population and the want of roads, &c., has in many districts saved those that still exist.[32]

The western Taurus was just such a district when Davis saw it. But meanwhile sedentarization of nomads and the creation of peasant villages began to change the area's character. As settlement thickened, forests began to thin. A road system came in the early twentieth century, and accelerated the thinning.

In the remoter parts of the western Taurus plenty of tall forest remains today, usually open forest or parkland. In some places it has shrunk since the 1950s, despite only modest population growth. Planhol, whose work gives the most complete idea of forest history in the western Taurus, described "épaisses forêts" on the mountains flanking the Aksu and Köprü Su.[33] Thin forests cover these slopes thirty-five years later. The difference is probably a result of logging, a vigorous business today made more profitable by new roads, rather than a result of the depredations of peasants or shepherds.[34]

On the south flank of the western Taurus, in Antalya (ancient Pamphylia), peasant settlement denuded low slopes that forests had covered as late as 1915.[35] In the middle of the nineteenth century, forest covered substantial parts of the plain of Antalya between the Aksu and the Köprü Su. One traveler, Charles Fellows, wrote that "trees grow, die, and fall unheeded." Another, Charles Texier, lost his way in the forest here. A third, Tschichateff, mentioned forest in several places along the coast. The plain has lost all its forest by now.[36]

Peasants found these low slopes good for maize and tree crops, some of which they could sell to the coast, where commercial agriculture (cotton and citrus) boomed, especially after 1945. Cultivators burned forests to plant their crops and cut others to get construction timber. Older houses are made of wood, while newer ones are not. Indeed this is the case in much of Anatolia, suggesting a wood shortage – reflected in higher prices – only in very recent times.[37]

[32] E. Davis 1874, 304. [33] Planhol 1958, 421 ("thick forests").

[34] See Turkey, Orman Genel Müdürlüğü 1969, for details of logging in Antalya in the 1960s. Akan (1950) covers logging in Eğridir district in the late 1940s.

[35] Blumenthal 1963.

[36] References in Rowton 1967, 264; and Planhol 1958, 42–50. (I cannot find the Fellows quotation in the edition of his book available to me, but Planhol used a later edition.) To the west of the Taurus, in the mountains south of the Menderes (Meander) River, extensive stands of pine existed when Philippson passed through about 1910, but had vanished when Planhol visited in the 1950s.

[37] Helling and Helling (1958, 18) write: "The Anatolian villager has never been called upon to adapt stone or mud into the domed village houses as have been the peoples

In the central Taurus the rhythm of forest history appears to resemble that of the western Taurus. Around 1820, William Leake traveled through one of the major Taurus passes, Mut, which links the coast to the Konya Plain. He found the pass and the mountains well wooded with tall oak, beech, and pine.[38] Tall oaks, good enough for "timber trees," as Leake put it, take at least a century to grow. If the pass at Mut had hundred-year-old oaks in 1820, the pressure on forests in the remoter parts of the Taurus must have been mild indeed. Some sixty years later J. T. Bent passed by (although not through) the same gap in the Taurus, and found the vicinity "densely wooded . . . [with] fine trees, oaks, planes and other."[39] The fine oaks were probably *Quercus cerris,* which grows up to 10 meters in height and ranges from 800 meters to the tree line in the Taurus. By 1960 its stands had been much degraded by fires and grazing; indeed most had disappeared to be replaced by maquis and *Pinus brutia.* Even rather modest human pressures sufficed to degrade and destroy *Q. cerris,* a sensitive species. The *Pinus nigra* forests in the central Taurus have fared slightly better: once closed formations, even on the Mediterranean flank, they now remain as open woodlands.[40] Many majestic stands survive today. Cedar and fir, however, are much reduced from a century ago.

Further east, in the Cilician Taurus, the once enormous stands of oak, cedar, and fir have felt more severe pressures. Near the Lamas River, both the great extent of forest and the merciless siege it was under impressed Bent in the 1880s. Locals were depleting country,

> very rich in tall, straight fir-trees, much prized for ship masts; then they lay bare whole tracts of country that they may have fodder for their flocks; and nothing is so destructive to timber as their habit of tapping the fir-trees near the root so that turpentine may flow to the wound. A deep notch is cut, and after a while the tree itself is cut

of Persia and Syria. The visitor to an older village house is likely to hear a remark like this: 'My grandfather built this house and he cut those trees from the hill you see over there. But the forest is gone now.' All of this goes to show that although Turkey is an ancient land its forest coverage apparently lasted up until recent times."

In Zerk a middle-aged resident told me his father had built his house, which was mostly of wood, but he hoped to build one of more modern materials. Rowton (1967, 276) writes of "catastrophic destruction of forest and woodland in the past century and a half" in Anatolia.

38 Leake 1824, 104–7.

39 Bent 1890, 460. Bent attributed the survival of forest here to an absence of nomads, but an absence of peasants is a more likely explanation.

40 Akman, Barbero, and Quézel 1978, 211–14; Zohary 1973, 2:360; Kürschner 1984, 100–1. *Q. cerris* regenerates poorly once degraded and cannot coexist for long with farmers and herders.

> down, and the wood in the vicinity of the notch is found to be saturated with turpentine, and used for torches, the only light they ever make use of. Again, they bark the cedars to make their bee-hives and roofing for their huts; in short they are the most inveterate enemy of the Asia Minor forests. Luckily, the vast extent of the timber and the sparseness of the inhabitants make the destruction less apparent, but at the same time it proceeds steadily, and must one day ruin the forests of the country.[41]

Only fifteen years after Bent's visit, the British consul at Adana lamented the deforestation: "The cedars of the Taurus to the north of Adana [the Cilician Taurus] supplied the markets of Rome in the days of Cleopatra, and up to recent times a remnant of their former splendour existed, but, as everywhere else, they are fast being exterminated."[42] Cuinet, whose information dates from the late 1880s and often derives from official Ottoman sources, wrote that the province of Adana had 490,000 hectares of forests, but fewer every day. "All this is exploited without forethought," he wrote, "and little by little . . . this province, once so rich in forests, will soon be without wood."[43] So far this prediction has proved extravagant. Deforestation continued in the Cilician Taurus through the twentieth century. Root structures show that trees existed as late as 1955–60 in upland yaylas, corroborating the claim of locals. Fuel needs of the villages (made up of recently settled nomads) caused the deforestation.[44] Here in the Cilician Taurus the combination of local needs and distant markets (Syria and Egypt) have bared the land much more thoroughly than in the central or western Taurus. The turning points came with the Suez Canal construction and the settlement of the plain of Adana (the Çukurova), which malaria had protected so well that it too was once covered with oak forest.[45] This brought distant timber demand to bear at roughly the same time that population in the region began to take off.

On the arid northern flank of the Taurus, north of Mut, plenty of

[41] Bent 1890, 455. Bent blamed this destruction on nomads, but nomads lived in black felt tents and did not build huts with cedar-bark roofs. Nor did they keep bees. Best probably encountered seminomads or recently settled *yoruks*. See Planhol's discussion of this passage (1965, 105–6).

[42] Dunn 1905, 36–7. Charcoal making is mentioned among the causes of deforestation here.

[43] Cuinet 1890–5, 2:16. The original: "Tout cela est exploité sans prévoyance, et peu à peu . . . jadis aussi riches en forêts, cette province sera bientôt déboisée."

[44] Bates 1973, 19–20. When Bates last visited them, these villagers grew poplars to provide themselves with construction wood. The villages of Yılanlı, at the edge of the western Taurus, do so as well.

[45] Zohary 1973, 1:292.

good forest remained in the 1870s when a French forestry mission, invited to rationalize forest policy in the Ottoman Empire, made a survey of the vilayet of Karaman. They found magnificant pines and cedars in the Taurus, in the provinces of Konya and Adana. The French too noted that the forests of Adana, recently rich, had become more sparse. Those in Karaman appeared destined for the same fate. The tahtaçı relentlessly pursued cedar, which active timber merchants sent to Egypt and Syria. Shepherds, peasants, and camel drivers, the mission reported, routinely burned trees to keep themselves warm, and often inadvertently set major fires. Rather few people caused extensive damage although by no means evenly distributed. Timber cutting flourished only within reach of rivers, and shepherds and peasants generally kept to defined sheepwalks and settlements.[46] Here peasant settlement, distant demand, and local railway building combined to scar the mountain landscape fairly quickly, and in places irreversibly.[47]

Far to the east, not part of the Taurus chain but subject to some of the same historical forces, is the upland region around Lake Van. Here a pollen diagram reveals a vegetation history that probably parallels that of the Taurus. Not until about six hundred years ago is there any clear sign of human influence on what was predominantly oak forest. In early Ottoman times, however, grasses quickly spread at the expense of oak, a sure indication of pastoralism. But a sharp decline in oak pollens, and in tree species generally, came only about two hundred years ago. A rise in the incidence of pines and the appearance of walnut pollen suggests peasant settlement, bringing widespread cutting and burning of oak and its replacement by cultigens and opportunistic pines.[48] Here in broad outline is the forest history of much of upland Anatolia: initial disturbance at the hands of pastoralists in the fourteenth and fifteenth centuries, and major changes with the extension of agriculture in modern times. In the Taurus these changes came perhaps a little later, but the general pattern is in all likelihood the same.

The forest history of the Taurus as a whole has been a rather uneventful one, at least until recently. The experience of the western

[46] Bricogne 1876, 365–81. The mission contrasted the cedar forests of the central Taurus with the much depleted cedar forests of the Lebanon (p. 383), a comparison Planhol made ninety years later (1966, 292). The key difference between what are physically very similar mountain chains was that the Lebanon had a long history of mountain agriculture – it was a peasant massif – while the Taurus did not.

[47] On peasant settlement in Karaman from 1870: Hütteroth 1968.

[48] Van Zeist and Woldring 1978. They say their data apply to eastern Anatolia as a whole, not merely the site of their cores (Lake Van area).

Taurus appears little different from that of the other sections. No doubt a fuller record would reveal some vicissitudes presently obscured, but not many, and no great ones: the Taurus was a shepherd massif from the fifth century until almost the twentieth, and the shepherds were chiefly nomads, whose imprint on mountain forests remained slight. Although often blamed for the destruction of Anatolian forests,[49] nomads spent very little time in forests. With their hardy Bactrian camels as transport, they passed their summers above the tree line, sometimes well above it, and their winters in lowland plains. Their grazing animals helped to limit lowland forest recovery (although both the Antalya and Adana plains supported forests a century ago). In summer the nomads exploited the highest forest from above, because they needed fuel in their high yaylas. Wherever the nomads clustered, this probably lowered the tree line artificially. But the broad belt of forest between, say, 800 and 1,500 meters did not interest nomads. They passed through it in a day or two, stopping at most once in a clearing.[50]

Intense pressure on Taurus forests came only when the nomads settled down and became peasants, in the late nineteenth and early twentieth centuries. Indeed the three great epochs of deforestation in Anatolian history correspond with eras of peasant settlement: Neolithic farmers began the clearing of the inner Anatolian plateau; ancient Greeks and Romans denuded the coastal lowlands; and sedentarized nomads and their descendants have partly deforested the mountain regions.

The Pindus. In the Pindus the evidence for rapid landscape deterioration in modern times is very strong. Population growth, the rise of a market economy within the domains of Ali Pasha, and the linkup of this system to wider market networks combined to produce rapid changes between 1800 and 1870. Intensified grazing and gathering, together with agriculture made unstable by political violence, led to accelerated deforestation and soil erosion. Pressure on forests continued, in fits and starts, until after 1950. Rapid soil erosion persisted until most Pindus settlements became shell villages in the 1970s.

In the 1730s or 1740s an English merchant based in Aleppo, one

[49] Planhol 1965, 101–2. See also Bent 1890, 445.

[50] Planhol 1965; Roux 1970, 60–1, 183–4. Nomads routinely covered great distances quickly on their migrations, sometimes as much as 300 kilometers in a week. The Bactrian camel, of central Asian origin, flourishes in cold weather and is as much the ship of the mountains as the ship of the desert. The dromedary camel of the Arabs, on the other hand, is a warm-weather beast. Perhaps this helps explain why the nomadic Turks made more use of the mountains for seasonal grazing than did the nomadic Arabs.

7.2. Mount Mitsikeli in October 1987. Two centuries ago this mountainside was well forested.

Arthur Rollinger, visited Epirus. He described it as "a barren soil and full of forests, but near the sea it is somewhat more favourable."[51] Unless Rollinger avoided the main routes and stuck to the most isolated districts, he saw an Epirus quite different from that of 1850.

The Ioannina valley was the first major area after the coast to lose its forest cover. The city's needs, and those of the villages surrounding it, put pressure on the local forests. Its prosperity in the era of Ali Pasha led to quick urban growth, and consequent increased demand for timber and fuel. Quite likely agriculture also extended up the sides of the valley, although Ali Pasha also drained marshes to facilitate the expansion of cultivation. Travelers reported the basin denuded by 1850, and the French and British consuls seconded the opinion in the 1860s, contrasting its treelessness to other parts of Epirus.[52]

The mountain across the lake from Ioannina, Mitsikeli (1,810 meters),

[51] BL AM, 61, 742, f. 15, "Geographical and Other Notes Relating to Greece and Asia Minor, compiled by Arthur Rollinger, Aleppo merchant," n.d. but between 1733 and 1749.

[52] Boué 1854, 2:43; Stuart 1869, 291; AMAE-P, Mémoires et Documents, Turquie, 116:285, "Rapport du Consul à Janina," 24 June 1866.

provided timber and fuelwood for the city. Both Leake and Pouqueville, who knew Ioannina well (as of 1805–15), reported that Mitsikeli had recently lost its forest cover. Oral tradition has it that Ali Pasha cleared its slopes to prevent rebels and brigands from hiding there. Just as likely, if more prosaically, the fuel needs of the growing city stripped the mountain. Hughes, who visited in the cold winter of 1813–14, found the fuelwood situation worthy of comment:

> The sides of Mitsikeli and other mountains in the vicinity, were formerly clothed with very fine timber. Within the last thirty years this has entirely disappeared, chiefly owing to the increasing size of the capital, whose houses are for the most part constructed of wood, and the large and numerous serais which Ali and the other members of his family have built. The wood-cutters are now obliged to go a considerable distance into the mountains of Zagori for fuel, which is carried on the backs of asses and mules to the borders of the lake, and thence transported to Ioannina. This length of carriage makes the article so dear that it is almost out of the power of the lowest classes to purchase.[53]

Its expense troubled the upper classes too. Hughes found that while traveling with his retinue in Epirus, "our expenditure in fuel was very near as much as that which all the other articles on board occasioned." Boué, who visited in 1840, reported that Mitsikeli was covered with stunted kermes oak and holly. A lithograph from about 1850, testimony from the 1870s and from 1895, and photographs from 1909, 1940, and 1958 all agree that Mitsikeli had no trees, only a few shrubs, chiefly junipers. In recent decades a small belt of planted pines has grown up on the lower reaches of the mountain, but most of it is bare rock, of little use even to goats.[54]

The main road to Ioannina from the south (Arta) also lost its vegetation early on. Ali Pasha had built the road to be suitable for carriages,

[53] T. Hughes 1820, 1:501.

[54] Hobhouse (1813, 1:74) thought that a few timbers suitable for shipbuilding remained on Mitsikeli. Leake 1835, 4:164; AMAE-P, Mémoires et Documents, Turquie, 33:194bis, "Mémoire sur la topographie du vallon de Janina, Pouqueville," December 1811; Hughes 1820, 1:501–2 (for the second quotation). Boué 1854, 2:44 (the original: "de quelques chêne verts rabougris et de houx"). The lithograph is in La Poer Beresford 1855; photographs are in Mekios 1909, facing 184; the 1940 photo is on the wall of the Paris Hotel in Ioannina; that from 1958 is in Ioannina's Dimotiko Museio (Popular Museum) housed in the Aslan Agha mosque. Testimony of the 1870s from Lambridis 1887, 1:14–15; that of 1895 is from Baldacci 1917, 180. Holland (1819, 1:136, 189) also notes the wood shortage around Ioannina. The lithographs in La Poer Beresford appear highly romanticized and may not be at all accurate. Today's pines represent reforestation efforts of the Greek forestry service.

and made it 10 meters wide. An anonymous visitor in the mid-1820s found it passed through a "stony desert" with "no vegetation beyond the dwarf holly." Spencer in 1850 found it much the same. The road made the gathering and transporting of fuel and timber comparatively easy, and so focused upon its margins the needs of the population for many miles around. Presumably sheep and goats used the road as well, as they do its modern counterpart, and kept vegetation down.[55]

Every traveler worth his salt crossed the Pindus at the Metsovo pass, usually known as the Zygos (which simply means "pass" or "crest"). Leake found it well wooded, with beech, fir, and oak, although he noted that the locals used the fir for fuelwood and consequently it was diminishing. Pouqueville in 1805 found delightful forests between Ioannina and Metsovo. Urquhart in the 1830s discovered the finest beech forest he had ever seen ("lofty beech, straight as arrows, dropped, like plummet lines, on the inclined sward"). Lear (c. 1850) noted beech forest on the eastern side of the pass, and "magnificent pine" at the summit. This forest wealth disappeared sometime in the twentieth century, perhaps after General Metaxas's government put a motor road through in the late 1930s. Today the pass and the route between Metsovo and Ioannina have a few tall pines, but most of the vegetation is scrub.[56]

Olytzika (1,816 meters) stands just west of the theater of Dodona, in ancient times a sanctuary dedicated to Zeus and now a very modest tourist attraction. It attracted its share of literate visitors, even before the theater was identified as the site of Dodona. Leake described the northeast face of the mountain (that facing the theater) as bare toward the top but with a belt of firs below that, and cultivation crawling up the lowest slopes. Hobhouse found "woody hills." Lear saw "immense clumps of trees" at the base of the mountain. Gilliéron (c. 1875) noted "yellow ravines, cut by waters," and some firs higher up. Chirol in 1881 wrote that Olytzika "still bears on its flanks black patches of oak forests." Baldacci in 1895 saw fir beginning at 1,200 meters. In the 1930s, grandfathers told Hammond that in their childhood Olytzika "was heavily wooded on its upper slopes, but many trees were felled in 1912 and afterward." Clarke in 1922 found the mountain mostly bare and stony. Photographs from the 1930s reveal a slope almost completely denuded of trees.[57] Today a narrow belt of pines crosses the mountain.

[55] Anonymous 1827, 26; Urquhart 1838, 1:154; Spencer 1851, 2:196.

[56] Leake 1835, 1: 414; Urquhart 1838, 1:282; Pouqueville 1822 (London edition), 72; Lear 1852, 382–3.

[57] Leake 1835, 1:263–4; Hobhouse 1813, 1:64; Lear 1852, 371; Gilliéron 1877, 86; Chirol 1881, 193; Baldacci 1917, 182; BSA, Clarke, notebook. Hammond 1967, 172. Photographs from the 1930s are in Hammond 1967, 169; and Chiliadakis 1940, 36.

7.3. Mount Olytzika in October 1987. This slope lost most of its forest cover in the nineteenth century, but vegetation has recovered somewhat in recent decades.

Olytzika slowly lost most of its forest cover between 1800 and 1940. About 15 kilometers over rough ground from Ioannina, it has remained little affected by the city. But villages (Dramsi, Dodoni, Madio) lie at its foot, and their slow incursions and depredations gradually achieved much the same effect as the demands of Ioannina upon Mitsikeli. On both slopes vegetation has made a small comeback since 1960 or so. Nonetheless Olytzika still bombards the villages below with mudslides and boulders. Indeed locals have had to build new houses, further away from the slope. The ecological sins of the fathers are visited upon the sons.

Kónitsa is a small town to the north of Ioannina, situated on the east side of a triangular plain watered by the Aoos River. The river flows out into the plain between two shoulders that in Leake's day were completely covered with fir. The mountain above Kónitsa (to the east) he described as covered with wood. Spencer some four decades later found it "a dense forest of noble oaks." Today the shoulders above the Aoos carry only low scrub, and the mountain behind Kónitsa has only stunted kermes oak. At some point between 1850 and today the forests around Kónitsa disappeared, probably at the hands of shepherds and fuelwood gatherers. These slopes are steep enough that agriculture has never been practical, and the plain below yields bountiful maize, and has since at least 1800 (and probably well before). So the denudation of the slopes around Kónitsa probably came about in slow degrees, and not through clearing for cultivation.[58]

Paramythia lies at the head of a long north–south valley drained by the Kokitos, a tributary of the Acheron. Holland visited the valley early in the nineteenth century and found many of its villages "enveloped in wood." Clarke passed through a century later, and was told by a priest that in "Venetian times . . . all these slopes were heavily wooded." But in Clarke's day (1922) the valley walls supported only maquis and garrigue, as they do today. Olives and fruit trees dot the lower slopes, as this valley is a low-lying one. Here cultivation has certainly expanded since Holland's visit and may well be the chief reason for deforestation. Much of the land is now barren, with neither crops nor trees.[59] As throughout Epirus, the eastern wall of the valley is more heavily eroded than the western – it gets more rainfall.

Kallarites is on the western flank of the Pindus spine, here called Peristeri (2,295 meters) to its north, and Athamanon or Tzoumerka (2,393 meters) to its south. Here is Pouqueville on Peristeri:

> On embrasse un horizon semé d'inégalités qui n'offre au loin que l'image de la stérilité. Cependant, autrefois, tout cette contrée était couverte d'arbres, que les pasteurs Valaques incindent chaque année aux approches de l'automne, afin de laisser mûrir les cendres sous la neige, pour ensemencer quelques champs en seigle au retour du printemps. Ainsi, pour un produit éphémère, car la terre s'épuise au bout de deux ans, ils détruisent ces bois majestueux, enfants des siècles, qui attirent les nuages autours de leurs dômes de verdures, et envoient aux plaines et aux vallées les pluies qui les fécondent.[60]

[58] Leake 1835 4:107 and 1:396–7; Spencer 1851, 2:175. The Aoos, not coincidentally, carries a huge load of silt and gravel onto the Kónitsa Plain. The river passes through braided channels in a bed up to 300 meters across. The channels cut into the bed to a depth of about 2 meters. The trained geologist no doubt could make more of these details than I can, but anyone can deduce that the Aoos has swept millions of tons of soil and gravel from the slopes of Zagori and Kónitsa. Most of the soil ends up in the Adriatic; the gravel gets no further than the Kónitsa Plain.

[59] Holland 1819, 2:251; BSA, Clarke, notebook c.

[60] Pouqueville 1826–7, 2:186. Roughly translated: "One takes in a jagged horizon which from a distance appears sterile. However, once this entire country was covered with trees, which the Vlach shepherds burned every year as autumn approached, letting the ashes ripen under the snow, so as to seed a few fields with barley when springtime returned. Thus, for an ephemeral product, they destroyed these majestic woods, the progeny of centuries, which attracted clouds to their verdant domes, and provided to the plains and the valleys the rains that make them fertile."

Pouqueville may have misunderstood the role of forest in the fertility of Epirot valleys, but he was right to point to the recklessness of the Vlach shepherds. Stuart visited some sixty years later and described Kallarites as

> situated on the southern slope of Mt Polyanos [Peristeri's ancient name] at the foot of a bare treeless declivity called "padourè maré" in Vlakh, "the great forest," whence it would appear that this declivity was originally thickly wooded like other parts of the mountains, but in course of time was stripped of its clothing by the ruinous process of burning and felling timber which is still in unchecked operation.[61]

Peristeri today is mostly bare rock, without even garrigue. Grazing is much more sparse than it was in the nineteenth century, when Kallarites had more than fifty thousand sheep. Water, which once trickled down the mountain, in small streams and aqueducts, to irrigate the garden agriculture of Kallarites, now tumbles uselessly to the bottom of the Kallaritikós gorge. Kallarites has abandoned all agriculture – for several reasons, not merely environmental ones – and its population has scattered.[62]

These accounts of mountains such as Mitsikeli, Olytzika, Peristeri, and even of lower hills at Paramythia and elsewhere suggest that the forest cover of the Epirot Pindus has shrunk considerably since 1800. Leake, Holland, Hughes, Pouqueville, and others from the early nineteenth century speak of forests elsewhere – in the Louros valley north of Filipiada, on the ridge of Mavrovouni between Metsovo and Grevená, along the mountainous banks of the Kalamas River, near Vasilikó (or Tjaraplaná) by the Albanian border. Apparently none of these forests existed when Hammond made his tours of Epirus in the 1930s; none exist today. Of Zagori, Lambrides wrote in the late nineteenth century that the oldest houses were made of wood, showing that pine, fir, and oak existed in abundance near the villages. Newer houses used stone.[63] Surveying Epirus as a whole in 1863, the British consul wrote:

> The forests which at one time covered the greater part of the country, have every where been much thinned, and in many places wholly cleared away. Nevertheless, although the process of felling still goes on, according to the demand for timber, firewood and charcoal, considerable tracts of forest still remain. Trees of great age and growth crown most of those heights which are difficult of access; the pine is

[61] Stuart 1868, 317.

[62] The Tzoumerka district fared no better: Papakostas 1967, 100–2.

[63] Cited in Vavaretos 1929, 19, without a precise reference. I have not found the original passage in Lambridis's works.

> chief of the soft-wooded kinds. Of the hardwooded, the oak, beech, ash, elm and more rarely the box. The walnut flourishes in every part of the Province and is in large demand for furniture, gun and pistol stocks, etc. The spreading plane, the chestnut, sycamore and cypress are common where there is sufficient depth of soil; the olive thrives in the coast district; the mulberry has been successfully introduced here and extends with every year; the orange and citron have found congenial places in the lowlands, together with several other fruit trees of warm latitudes. The apple, pear, pomengranate, fig, apricot are indigenous to the soil [Stuart was incorrect here], as well as vine, which is extensively cultivated, contributing largely to the wealth of the Province. The liquorice tree and several valuable dyewoods must also be specified in this class of the country's productions.[64]

In 1870, consul Stuart wrote of woodlands in Epirus: "These lands are very extensive in the mountain districts; . . . though great numbers of trees are every year cut down for timber, firewood, and charcoal, and the woods and forests are sensibly diminishing."[65]

The elm and ash have become exceedingly rare since Stuart wrote, and much of the forest wealth he saw disappearing vanished within a half century. In their survey of Epirus just before the end of Ottoman rule, the French pair Rolley and de Visme concluded that the areas around Ioannina, Kónitsa, and Metsovo had had fine forests as recently as the 1880s but that these had lately been destroyed.[66] They attributed deforestation and erosion to uncontrolled woodcutting. By 1908 the vilayet of Ioannina imported timber, chiefly construction wood, from Austria, Italy, Turkey, and even from Greece.[67]

In short, the period from 1800 to 1950 saw great deforestation in Epirus generally and in the high mountains especially. The expansion of agriculture and pastoralism played a major part, especially in an early (c. 1800–30) surge that centered around the Ioannina basin and the immediate hinterland of the Adriatic coast. Very likely the informal coercion that helped restrain overgrazing vanished when Ali Pasha assumed ownership of a large proportion of the sheep and goats of Epirus. No one could prevent his shepherds from overgrazing. Because he sold wool, cheese, and meat, it would have suited his interest to promote overgrazing, especially as he knew his position as local pasha

[64] PRO FO 195/751, "General Report on Epirus," 1 December 1863, by Robert Stuart.

[65] PRO FO 195/935/150–68, "Report on Land Tenure, Epirus," 12 January 1870.

[66] Rolley and de Visme 1911, 137.

[67] PRO FO 294/32, "Mouvement Commercial du Vilayet de Janina durant l'an 1908." See also PRO FO 294/23, ff. 288–9. "Tableau de l'Importation Commerciale du Vilayet de Janina pour le 1900."

was notably insecure. In his era (1788–1822) uncontrolled burning and cutting affected every part of the region, even the high mountains. Political violence and brigandage, especially late in the century, prevented systems of restraint from reestablishing themselves for any length of time. Since 1913 there has been more contraction than expansion in agriculture and pastoralism, but crowded conditions to 1940 ensured that fuelwood demand remained strong. Only after 1949 – and another round of fires attributable to the Civil War – have Epirot forests been granted a respite. But in forty years they have made but little progress toward recovery, despite efforts of the Greek forest service. Soil, as we shall see, has become too scarce, and goats continue to devour any and all saplings they can find.

This rough chronology of the deforestation of the Pindus and Epirus bears out the views of the Greek forester Lazos, who in 1935 published a series of four maps tracing the retreat of forest area in northern Greece.[68] These four maps depict forests at four moments in the past: an unspecified prehistoric moment, Alexander the Great's time, 1821, and 1935. These maps show lowland Thessaly and the Ioannina basin losing most of their forest by Alexander's time (late fourth century B.C.). Between Alexander and 1821 Thessaly continued to lose forest, while around Ioannina pines recovered. The biggest changes in northwestern Greece came between 1821 and 1935. Forest disappeared around Delvinaki, from the Kalamas River to the Alabanian border; along the Adriatic coast to about 10 kilometers inland; in the high Pindus, 10 kilometers north of Metsovo and 20 kilometers south of it; and finally in the corridors between Arta and Ioannina, the valleys of the Louros and the Arachthos rivers.

Just what Lazos based this reconstruction upon is unclear, but his analysis roughly corresponds with that of Bottema (see Chapter 3) for prehistoric and ancient times (although Lazos does not recognize the fluctuations Bottema discovered in prehistoric forests). It also roughly coincides with the impression I have: that the modern landscape of Epirus and the Pindus emerged in the period between 1800 and 1950.[69]

[68] Lazos 1935, between 68 and 69; 17–29 for the accompanying explanations. The maps, unfortunately, are impossible to reproduce here.

[69] Tsoumis (1964, 36) says of Greece as a whole: "An extensive forest cover was preserved until relatively recent years." Kandrelis (1980, 76) says Greece in 1880 was 48% forest, in 1980 only 19%. Kolokotronis, a great hero of the Greek revolution, said that in the Peloponnese mountains that carried forest during Turkish times became bare in the subsequent decades; see Kontos 1929, 176. Even the Aegean Islands carried tall forest until the 1820s according to E. Davis 1874, 6–7.

Table 7.1. *Forest area of Basilicata, 1800–1962*

	Hectares
1800	290,000
1860	240,000
1908	180,000
1930	126,000
1962	169,000

The Lucanian Apennines. Modern deforestation in southern Italy has been so obvious and costly as to have attracted considerable attention. Consequently estimates and statistics exist for the nineteenth century, as they do not for the Taurus or Pindus. The accuracy of these statistics is open to question, but the general picture of pervasive and catastrophic deforestation is clear.

Some of the more considered estimates follow. Pastina proposes that Italy as a whole had 15 million hectares of forest in 1500, but only 5.5 million in 1925. Franciosa wrote that woodland originally covered two-thirds of Basilicata, but by 1930 only 12 percent. Ahlmann says forest comprised 33 percent of Basilicata in 1800, but only 16 percent in 1925.[70] Rossi-Doria, in a careful reconstruction, offers the figures presented in Table 7.1.[71]

In 1800 forests covered almost a third of Basilicata, principally in the mountains. All sources agree that while deforestation proceeded throughout the nineteenth century, it surged after 1865, when public lands passed to private hands, reaching its apogee between 1887 and 1903.[72] The reason all agree is that they base their estimate on the careful calculations made by F. S. Nitti during the *Inchiesta agraria* published in 1910. He concluded that forest area in Basilicata shrank

70 Pastina 1986, 146; Franciosa 1930a, 20; Ahlmann 1926, 86–7.

71 Rossi-Doria 1963, 77. His figures are almost identical to those of Tichy 1957, 292. Because the total area of Basilicata is 9,992 square kilometers, one can convert to percentages very easily. For example, Rossi Doria's 290,000 hectares is 29% of Basilicata's total area.

72 Viggiani (1946, 91) says that between 1860 and 1900 at least 30% of Basilicata was deforested. Cagli (1910, 28) writes that 200,000 hectares of forest disappeared between 1806 and 1910. Vochting (1955, cited in Puglisi 1977, 93) put it at 170,000 hectares deforested after 1860. The city fathers of Potenza claimed that 174,000 hectares of forest had vanished between 1877 and 1901, and only 120,000 remained (ACSR, PC, Zanardelli e la Basilicata, busta 5, "Memorandum del Comune di Potenza, 1901").

by a quarter between 1860 and 1909.[73] Modern geographers and foresters endorse this general picture.[74] Most of the burning and cutting took place in the mountains of western Lucania, especially in oak stands. Wheat and sheep replaced oak and beech, but in time much if not most of the deforested land was abandoned.

Anecdotal evidence supplements these estimates and statistics. An account of the province in the eighteenth century described the mountains as "all covered with beech, oak, and ilex, and [abounding] in various sorts of game." Saint-Non, traveling between Lauria and Lagonegro in the 1770s, encountered "vast forests" and countless waterfalls. Just south of Castelluccio, near the border between Basilicata and Calabria, he found great forests. Neither great forests nor many waterfalls exist today. The botanist Tenore, whose acquaintance with the Kingdom of Naples dated from 1800 to 1815, wrote that between 300 and 800 meters of elevation the kingdom was "almost entirely garnished by trees." Craven, who toured the Kingdom of Naples just before 1820, found that timber "plentifully covers the Calabrian Apennines," some of it, he thought, good for shipbuilding. Lagonegro and western Lucania he considered "productive of fine timber, not much cultivated." Three botanists who visited Pollino in 1826 noted extensive forest of "colossal" fir and beech on the southern (Calabrian) flank of the mountain. They reported that "all is left to nature." Almost none of this Calabrian forest remains today. In the 1840s, Casparini and Bruni, perhaps indulging in a bit of local patriotism, insisted that "le regioni boscose del nostro Regno [the Kingdom of Naples] estendono dalle vaste lande presso il mare, fino alla sommità delle alte montagne." As late as the 1870s, observers found well-wooded valleys in the upper Bradano basin, north of Potenza, and "colossal" beech forest on the hills of the upper Basento. In the 1920s a disappointed visitor to Lagopesole, the thirteenth-century site of Frederick II's hunting lodge, noted not a scrap of forest, only "groups of trees here and there." Local gentry had hunted bears here as late as the eighteenth century, and in 1647 a vexed visitor found there "a most thick and dense wood of all kinds of trees. To enter it is easy, to come out again greatly difficult." No one has gotten lost in woods near Lagopesole for a long time now. Many place names in Lucania imply past forest now nowhere in evidence. Nemoli, for instance, was called Bosco under the Bourbons. But by the late nineteenth century, the interior of Lucania had become "terra fumante," as

[73] Nitti 1909–10, 5:39.

[74] Ranieri (1961, 43) speaks of the "vast scale" of deforestation between 1860 and 1880; Kayser (1961, 22), Puglisi (1967, 16), and Tichy (1957) all concur. Puglisi mentions demand for railroad ties, a point other observers ignore.

the principal landowner in Campomaggiore put it, because cutting and burning of woodland sent billowing smoke up over every hillside.[75] Today some of these hillsides are in wheat, some in pasture, and many in ruin.

Some Lucanian landscapes have fared somewhat better. In the far northwest of Lucania are two small lakes famed for their tranquil beauty. The Monticchio Lakes fill a crater of the extinct volcano of Mount Vulture. Their environs formed an "immense wood" when in 1862 they left the state domain and passed to A. Lanari and Company, the Banco di Roma, a Franco-Swiss bank, and others. Quick deforestation followed upon the first sale. The purchasers sold off the timber from 5,168 hectares, mostly cherry oaks, some of which measured 6 meters in circumference. They then tried to recruit peasants to farm the lands around the lakes. Eventually settlers arrived from Ancona, and then from Apulia, Avellino, and the Marches, all as sharecroppers. Plans to keep a certain portion of the "internal colony" in forest worked poorly, perhaps because forest guards and timber merchants collaborated to sell wood illegally to Foggia and Bari. By 1910, to judge from photographs, very little remained. But the internal colony – of which the Italian state encouraged many as an alternative to emigration – failed, the sharecroppers left, and the Monticchio Lakes recovered their former tranquillity. They also recovered their former forest. Apparently flocks have left it alone, for now a second-growth forest of poplar and alder that looks to be fifty or more years old surrounds the lake.[76] A statue

[75] Antonini, *La Lucania*, cited in Rodolico 1963, 312. Antonini went through four editions, 1745–1817, but the text probably refers to the situation in the 1740s; Saint-Non 1781–6, 3:147–50; Tenore 1827, 62 (he gave the figures as 150–400 *toises*, a measure roughly equal to 2 meters); Craven 1821, 350, 356; Petagna, Terrone, and Tenore 1827, 47–8 on Pollino; Casparini and Bruni 1845, 132–8 ("the wooded region of our kingdom extends from the vast lands next the sea up to the summits of the high mountains"); Lenormant 1883, 1:253, on the upper Bradano; Fittipaldi 1880, 27, on the upper Basento; Hooker 1927, 159–69, on Lagopesole; Filardi, n.d., 2:462, for the remark of Gioacchino Cutinelli-Rendina of Campomaggiore; Puglisi (1977, 93) gives several more examples of place names implying former forest.

[76] Cagli 1910, 55–9. The photographs are on pp. 47 and 52. ACSR, MAIC, DGA, I, busta 288, f. 4: manuscript of G. Labollita, "Monografia del bosco Monticchio." This has been published (Perugia: Martini and Boncompagnini, 1863) but I have not found a copy in Europe or America. Lear visited the lakes in 1850 and found "deep woods" all around. On the illegal cutting, ASB-P, Prefettura – Atti di Gabinetto, busta 25, Ispettore delle Finanze Caffelli to Prefetto, 24 June 1864; and Sottoprefetto di Melfi Musso to Prefetto, 20 July 1864. These letters implicate the inspector of forests in the scheme, and write of "enormous devastations" in the Monticchio forest. In November 1987 charcoal makers were at work on the shores of the lakes. In summer they attract a small tourist clientele, the successors to a parade of pilgrims that over the centuries came to visit the Abbey of San Michele di Monticchio overlooking the lakes.

to San Giovanni Gualberto, the patron saint of the Corpo Forestale d'Italia, contentedly admires the scene. Most of the rest of Lucania must distress the saint.[77]

Mountain deforestation proceeded throughout the first half of the twentieth century. One of the losers in Mussolini's famous "Battle for Wheat" was forest land, in Lucania as elsewhere. Between 1930 and 1950, an era of population growth and agricultural expansion, about 15 to 20 percent of remaining forests in Lucania disappeared. The land reform inaugurated by the Christian Democrats in 1948, while it had little effect on the minifundia areas of mountain Lucania, also hastened the destruction of what public forest remained at lower elevations. "Abusive" deforestation continued into the 1960s, by which time huge expenses for reforestation and torrent control drained public coffers.[78] No amount of public spending can make good the consequences of the deforestation of the nineteenth century.

The Alpujarra. In the Alpujarra considerable deforestation took place in the sixteenth century, with the resettlement of the barrancas by Christians unversed in the techniques of mountain agriculture. Nonetheless, large patches of forest remained into the nineteenth century, because pastoralism was only a minor part of life, because hogs, who like acorns and chestnuts, formed a large proportion of the livestock, and lastly because civic agreements checked the exploitation of common lands. But in the nineteenth century, population growth and the connection to distant markets, abetted by the desamortizaciones, ravaged the remaining forests of the Alpujarra (see Chapters 5 and 6). The effect was less dramatic and deleterious than the nineteenth-century deforestation in Lucania, but only because a disaster had occurred already three centuries previously.[79]

The deforestation of the Alpujarra formed a small chapter in the larger story of forest clearance in Andalusia and Spain. Eastern Andalusia lost a third of its public forests (and almost all forest was public

[77] Southern Italian mountain forests outside of Lucania suffered as well. The oak and beech of the Gargano Peninsula shrank by about a third in just a few years before 1825 according to the forestry expert Moreau de Jonnés (1826, 55–6). He reported that the Sila, in Calabria (see Chapter 6), had shrunk by a third since Strabo's time.

[78] Kayser 1961, 22; Lopinto 1972, 42. Trotter (1952) explains the widespread distribution of *Ilex aquifolium* on Lucanian slopes as a succession in lands recently deforested of oak and beech. Formica (1983, 96) explains that as late as the 1970s trees were felled to facilitate mechanized agriculture, with unfortunate results for soil erosion.

[79] And for geological reasons as well: the flysch of Lucania is especially given to erosion and landslides.

forest) between 1859 and 1910.[80] Spain's forests also shrank notably in the nineteenth century. In the woods of Berchules, in the Alpujarra, an observer put this down to "abusive cutting," presumably for firewood.[81] I have no statistics specific to the Alpujarra, but the testimony of visitors suggests that forest cover survived through to the end of the century in places where today it no longer exists.

Ansted, who worked near the Alpujarra during the lead mining boom of the 1830s, reported that although the low valleys and the ramblas were desolate, the lands around the high villages were "beautifully wooded," something an Englishman would not likely say about anything less than tall timber. Rubio, who visited Capileira in the 1870s, noted that one could not see the village from above until one came upon it, because of the thick woods of kermes oak. No such wood obscures the village today, although recently replanted pine forest covers a stretch above Capileira. Below Capileira, Rubio, and Marín in the 1890s, found "un bosque de enormes castaños," including one so large that, allegedly, a weaver, his loom, and his family lived in its hollowed out trunk (this tale is also told of a chestnut tree in Lanjaron). At Juviles, Rubio found centuries-old oak, chestnut, and walnut groves that "blocked the sun." Equally old stands near Trevélez lasted until the 1940s. In the 1920s around Padul, an old shepherd grazing his goats on bare rock told Almagro: "There is nothing else; when I was a lad this was an evergreen oak wood [*encinar*] and there was grass everywhere. The storms did not flood as they do now, and didn't cover the village with stones and mud. Every year we have less livestock." Almagro was an erosion expert and may have put words in the shepherd's mouth to illustrate his point. But he was convinced the deforestation in the late nineteenth century had brought on the hydrological problems of the Alpujarra that he investigated.[82]

On the north slope of the Sierra Nevada forest cover also disappeared before the nineteenth century. Widdrington, traveling in the 1820s, found on the Sierra between Guadix and Granada "the remains of a noble forest of oak; not a tree is now standing." Probably little forest remained on this flank of the massif after the mid-eighteenth century. Today extensive belts of pine girdle the slopes between 1,000

[80] Jiménez Blanco 1986, 1:369.

[81] AMAPA, legajo 372–1, [?] al Ministro de Fomento, 22 December 1877. Several other documents in this *carpeta* deal with abusive cutting.

[82] Ansted 1854, 161; Rubio 1881, 94, 204; Marín 1896; 200; Almagro 1932; 51; Aivar 1985; 129. The story of the weaver and the chestnut tree appears in the sixteenth-century works of Luis de Marmol and of Hurtado de Mendoza, and in the travels of Ibn Battuta from the fourteenth century.

and 1,500 meters, a result of forest service work that began on a major scale in 1954.[83]

Leaving aside (for the moment) recently replanted stands, scarcely any forest exists today in the Alpujarra or anywhere on the Sierra Nevada. Maquis is everywhere, colonizing abandoned land, but tall trees exist only singly or in small clumps. Some of them are relicts, showing what existed two or three centuries before: high above Válor, for instance, a series of lone oaks towers over the landscape. They are as large as any trees I have seen in the Mediterranean.

In the Alpujarra the nineteenth century was not quite the turning point in forest history that it was in the Pindus and Lucanian Apennines – or in the rest of Spain. More deforestation had taken place earlier (proportionally speaking), with the Christian resettlement and the Málaga wine boom. But the nineteenth century eliminated what tall forest had hitherto survived in the Alpujarra. Its ravages provided the coup de grace.

The Rif. The widespread deforestation of about 1600 to 1800 implied by pollen records probably extended into the nineteenth century, but no conclusive evidence exists one way or another. The species composition of Rifian landscapes was much the same 150 years ago, and so palynology reveals no clear changes. But in all probability tall trees have given way to shrubs of the same species under the growing pressure of peasant numbers. In the twentieth century deforestation has clearly continued, so that there will be very little left to use or abuse in the twenty-first.

Rifians themselves take the view that their forests once extended much further than they do now. Some told Coon that conifers once covered almost the entire region. If Rifians in the late 1920s preserved a memory of a well-forested Rif, it probably still existed into the nineteenth century. But the half-life of botanical folk memories is impossible to estimate. A later generation of Rifians told Maurer that fifty years previously (that would be in the 1920s) forest and maquis were both more extensive and more dense than they had become by the 1970s. The Beni Bouzra remember that their hills and valleys were forested before the Spanish came. They once used wood as a medium of exchange with their neighbors, from whom they got wool and livestock.[84]

[83] Widdrington 1834, 1:54 and 2:239; Arias Abellán 1981, 293.

[84] Coon 1931, 9; G. Maurer 1976, 20; Gaudio 1981, 112. Maurer's informants told him that the thicker vegetation made ambushes easier. G. Maurer (1968b, 20) reports that old men in several tribes noted that forests had once extended far further, especially in Beni Ammart.

The landscape itself bears out these memories and may help maintain them. Morocco preserves a large number of sacred groves, usually cemeteries or shrines, where people will not touch the vegetation and animals are kept out. Dozens of these exist in the Rif, with large and stately oleaster, holly oak, and cork oak. The contrast between these relict groves and the surrounding low scrub – sometimes dwarf examples of the same species – shows how effective goats and fuelwood gatherers can be in preventing forest growth. Mikesell, on the evidence of relict groves and soil and climate conditions of northern Morocco, suggests that forest in 1960 covered only one-tenth of the area it would cover without the interventions of people and goats. Just what proportion of the deforested nine-tenths lost its cover recently is open to question. Sauvage, a careful student of Moroccan forests, used relict stands to support his view that the "massive" deforestation had taken place within a century of his study – that is, between 1840 and 1940.[85]

Some Spanish and Moroccan statistics, although they exist only after 1930, imply very rapid deforestation in the twentieth century. The area in cork oak, a profit-making species that authorities watched carefully, declined from 150,000 hectares to 110,000 between 1935 and 1948. The area in cedar, about 16,000 hectares in 1942, had recently shrunk by half. By official counts, the total area of tall forest in the Spanish zone in 1955 came to about 137,000 hectares; by one estimate, the Rif lost 100,000 hectares of forest between 1956 and 1971 alone (that equals 5 percent of the land area of the Spanish Protectorate and almost all its forest).[86] No doubt the statistics leave much to be desired. But since the cultivated area of the Spanish zone doubled between 1936 and 1953, it is easy to accept tremendous rates of deforestation in modern times.[87]

Anecdotal evidence supports this view. The French merchant Roland Fréjus traveled across the eastern Rif in 1666. He noted at many points, even in the Nekor River valley, ancient trees, including oaks tall enough to shade a wide road. An eighteenth-century Spanish mariner described the hinterland behind Peñon de Vélez (a Spanish fort off the Rif coast) as having many trees, including cork oak, holly oak, and "alerces" ideal for boat building.[88] The locals, "Berberos Gomeranos," cut timber here

85 Fay 1972, 1:76; Mikesell 1960, 446–7; Sauvage 1941, 246–9.

86 Vázquez del Río 1935, 8; Roda y Jiménez and García Figueras 1950–5 1:50; Sánchez Cózar 1948, 6; Montalembert 1972, 68. On cedar: Sauvage 1941, 248; Sánchez Cózar 1942, 16.

87 Seddon 1981, 134; Roda Jiménez c. 1940s, 22.

88 Fréjus 1922, 139–42; Estrada 1768, 3:530–1. Leo Africanus and other sources also mention "alerce," which Mikesell (1961) suggests is sandarac (*Callitris articulata*). Chenier (1788, 18) reported that Peñon de Gomara and Peñon de Vélez were "surrounded by mountains and forests."

as their chief source of income. Segonzac visited the same spot 130 years later, and saw only cork oak, which the peasants burned in order to plant barley in its place. Virtually no timber remains on the coast today.[89]

In the late 1920s, Spanish army officers and other literate Europeans descended upon the Rif, providing detailed information about the interior for the first time since Leo Africanus in the early sixteenth century. In Beni Bou Frah, where no tall trees stood in 1928, large stumps betrayed the site of a recently felled forest, perhaps an indirect victim of the war.[90] In the western reaches of the Rif however, plenty of majestic timber remained to impress foreign visitors. French travelers reported that the cedar forest extended as far east as Targuist; a British diplomat and aerial photographs agreed. A Spanish forest map, from the 1930s, shows cedar covering about 50 square kilometers, southwest, south, and even east of Targuist. A recent map shows the domain of cedar much diminished, now confined almost exclusively to the southwest of Targuist. Today cedar covers perhaps 15 percent of its potential range in the Rif. Scrub oak covers most of the deforested land.[91] The greater part of this deforestation took place in the Spanish period (1927–56) and before; cedar area has not shrunk much since independence. But the forest is thinning out: wood stock (in cubic meters per hectare) declined by a third between 1953 and 1976, and continues to shrink under the pressure of (illegal) cutting. Middle-aged trees, the best for transport, are cut for timber, skewing the age structure (and weakening the regenerative capacity) of the cedar stands.[92]

The eminent botanists Emberger and Maire visited the central Rif in 1926, immediately after the conclusion of war. They went from Taza north to Aknoul and found all "forest formations currently very degraded by abusive exploitation, fire, and grazing." Aleppo pine existed only as single trees, sandarac and holly oak only as bushes. From Aknoul to Boured degraded holly oak gave way to pasture and cereals. Only the

89 Segonzac 1903, 64.

90 SHM, Monografías de Kabilas, 1–5, Beni Bu Frah. Pascon and van der Wusten (1983, 34, 233–5) suggest that in Beni Bou Frah matorral disappeared at a great rate in the twentieth century, but that tall forest had already vanished.

91 Célerier and Charton 1926, 26–8; the diplomat was A. Monck-Mason, "Tetuan-Melilla Tour," November 1934, PRO FO 636/14, ff. 1–5. Harris 1927, 46–7, for the aerial photos; the Spanish forest map is in the Biblioteca Nacional, Sección de Africa, Mapas, B.8–3. It is undated but looks to be from the late 1920s or early 1930s. The recent map is in Deil 1988. Burned cedar often gives way to *Q. ilex.* Sauvage 1941, 248–9.

92 Deil 1988, 18–19, 24–5, and Deil 1987, 249. Pons and Quézel (1985, 39) refer to the decline of high-altitude cedar forest in Morocco over the past thirty years as "spectacular."

valley of the Oued Asfalou, a tributary of the Oued Ouergha, had any trees, and these existed only below Boured. Poplars and tamarisk grew in the valley bottom, and holly oak, Aleppo pine, and sandarac on the valley sides. From Boured to Souk et Tnin, via the Oued Agouri and Oued Bouchta, they found only degraded holly oak, except for one tall relict stand. Between Souk et Tnin and Targuist they noted only degraded cork oak. In the mountains north of Targuist they found degraded cork and holly oak up to 1,700 meters, then cedar and holly oak above that. Above 1,800 meters stood pure cedar in dense stands. But they noted that woodcutters had invaded these stands and had cut a good deal. Goats too had arrived and had completely destroyed the undergrowth.[93] I have traced much the same route (in 1988) and found the vegetation only a little changed since 1926. The cedars in the mountains north of Targuist have almost disappeared; and where Emberger and Maire found degraded forest, today there is mostly scrub or unproductive fields. But in general the change over sixty years has been modest, suggesting that the bulk of deforestation took place earlier.

The area around Chechouan, in the western Rif, also lost its forest before the arrival of the Spanish. A photograph (probably from 1926) shows the two-horned mountain behind the town just as bare and treeless as it is today.[94] Probably the environs of major towns like Chechouan had lost their forest cover well before the twentieth century, but perhaps in some cases only in the nineteenth: Brooke describes Tetouan in the 1820s as "hemmed in" by "lofty dark wooded mountains."[95]

The evidence for the Rif is notably unsatisfactory, but the general picture seems to be as follows. Commercial agriculture (until the kif boom) and timber operations had only occasional effect on the area, so deforestation came principally through the needs of the local population for arable and fuel. Hence, deforestation progressed in rough proportion to population. The precise chronology of deforestation in the Rif is elusive, but most likely it proceeded gradually for centuries, as in the other peasant massifs, extending only to limited areas as population and settlement remained modest. It surged in the tenth to twelfth centuries, again in the seventeenth to eighteenth, and accelerated with population growth in the late nineteenth century, with war (1921–6) and then with intensified timber cutting in the oak and cedar zones (after 1926). Population growth since then has prevented any lapse in the rate of deforestation; and goats (together with aridity in the

[93] Emberger and Maire 1927, 1–5. Southeast of Targuist, in Beni Ouriarhel, "unregulated cutting" jeopardized the survival of the dyewood *tizra*. PRO FO 636/14, "Tetouan-Melilla Tour," November 1934.

[94] Harris 1927, 114bis. [95] Brooke 1831, 1:207.

east) have prevented any return of forest in denuded areas.[96] If, as Mikesell proposed, forest area in the former Spanish zone in 1960 came to only a tenth of its former extent, today it can come to at best a twentieth. In Morocco as a whole in the late 1970s, forest disappeared at a rate of 14,000 hectares per year. The rate is twice that today. A quarter of Morocco's forests – a million hectares – vanished between 1940 and 1982. And the forces behind this continuous assault show no signs of relenting any time soon in either the Rif or all Morocco.[97]

The massive deforestation of the Rif took place in the larger context of Morocco and the entire Maghreb. In the French zone of Morocco and in Algeria, commercial agriculture played a large role in destroying forests over areas wider than the needs of the local population would have done. The French colonization of Algeria involved huge deforestation: "le XIXe siècle a été un gaspilleur formidable de richesses forestière."[98] But learning from the experience of the nineteenth century in Algeria, the French set up forest conservation schemes in the Maghreb – often to the dismay of the local peasantry – that helped preserve remaining forest in the twentieth century. In the smaller and poorer Spanish zone of Morocco, both the destruction and the conservation encouraged by the colonial power came to much less. For the most part, Rifians made their own forest history, although they did not make it just as they pleased.

In general, throughout the upland areas of the Mediterranean lands forests came under intensified pressure after 1800, from the same combination of factors that had always affected them. The precise combination of clearing for agriculture, burning for pasture, timber cutting, and fuel gathering varied from place to place and time to time. And the rhythm of forest destruction varied somewhat too, reaching a crescendo earlier in the Alpujarra and the Pindus, and rather later in Lucania, and later still in the Taurus. The chronology of deforestation in the Rif is too vague to fit it confidently into this spectrum. Everywhere but the Taurus the ecological costs, and eventually the human costs, were high.

96 Pollen evidence (Reille 1977) shows no modern surge in deforestation. Perhaps the species composition of Rif woodlands is unchanged since 1800, but these species exist in shrub form and not as trees.

97 FAO Library (Rome) FAO/WP 4457 Alheritiere and Downes, 1981, 2; Fay 1984, 3 citing M. El Latifi, "Au secours de la forêt marocaine," *Al-Assas* 40–41 (1982). El Latifi reports that Morocco had 4.3 million hectares of forest in 1940 and 3.275 million in 1982. See also Fay 1987, 402–3.

98 Brunhes, quoted in Boudy 1928, 18 ("the nineteenth century has been a formidable waster of forest wealth").

Erosion

Where the cutting and burning of forests lead to productive fields and pastures, the landscape change results in a net economic gain for resident human populations. That net gain can turn into a loss, however, if fields and pastures lose their fertility. In level landscapes where people practice good husbandry, this need not occur. But on steeply sloped land, the utmost vigilance cannot prevent soil erosion and depletion once vegetation is cleared away, and anything less than the utmost vigilance permits erosion on a scale that quickly turns profitable land into poor land.

The relationship between vegetation and erosion is not entirely obvious and intuitive. It is normally vegetation cover, rather than root structure, that is most important in preventing erosion. Mosquito netting is as effective as grass. Modest reductions have little effect, but once a threshold of about 30 percent decline in vegetation cover is reached, then further devegetation leads to rapid erosion. It can also lead to soil depletion, a slightly different matter but with similarly unfortunate effects for agricultural productivity. Soil depletion is the loss not of particulate matter (soil erosion strictly speaking) but of specific chemical nutrients necessary to plant life. Direct exposure to the sun and rain, without the shield that vegetation provides, can bring this on. Furthermore, soil nutrients in part derive from the decomposition of dead leaves, grasses, and the like; devegetation breaks the chemical cycle on which future fertility depends.[99] Soil depletion routinely occurs in agricultural soils (where plant mass is harvested rather than left to decompose), unless fallow permits recovery. One of the pernicious but inconspicuous effects of population pressure (or market opportunities) on an agrarian landscape is the necessity (or the temptation) to reduce or eliminate fallow. I will concentrate on soil erosion rather than soil depletion in the pages that follow, because soil chemistry is hard to trace historically. But it may safely be presumed that the

[99] Thornes 1987. Thornes (1976, 11) reports that erosion rates in the Alpujarra vary with vegetation cover as follows: under thick matorral (maquis) soils erode at 3 times the rate of soils under dense forest; under degraded matorral, 12 times as fast, and bare earth erodes at 16 times the rate of forest soils. Cultivated land erodes at 100 times the rate of forest land in the Rif; see Fay 1972, 1:97–9. In southeastern Spain, land in almonds or vines erodes 25–30 times as fast as land in pine forest (López-Cadenas de Llano 1982, 25). *Pinus halepensis* formations hold soil about 57 times as well as the same slope without vegetation (*La mobilité* 1984, 316–21). Rackham (1982, 195) holds the maverick opinion that deforestation and erosion are unrelated in modern Greece. The professors of the University of Thessaloniki Forestry School (to my mind convincingly) take the more conventional view.

loss of vegetation cover that accelerated erosion also depleted soils, as did the population pressures that led to the elimination of fallow in the Rif, and no doubt elsewhere.[100]

Erosion transports soil from higher altitudes to lower ones, and in some settings may serve to concentrate soil where it can be of greatest use. Some geographers have contended that mountain soil erosion and deposition in lowland plains represent a net gain in Epirus and in southern Spain.[101] But in the Mediterranean the mountains generally lie close by the sea, so most eroded soils have not come to rest in broad lowland plains to the advantage of lowland farmers. Instead they have flowed out to sea, to the advantage of no one. Admittedly, if enough sediment piles up along the seacoast a new and often highly fertile coastal plain develops. But it is rash to consider this a net gain: it takes as much as 100 hectares of eroded fertile inland topsoil to create a single hectare of new fertile soil along a seacoast.[102] As we shall see, the coasts in many areas of the Mediterranean have moved out considerably because of deposition. Several fertile plains have grown – or been born – in the past two hundred years. But the cost to mountain soil fertility probably comes to several times the benefits conferred upon the coasts. To take the largest scale example in the Mediterranean world, the fertility of the Nile Valley and Delta has always come from the soils of the mountains of Ethiopia. Before the Aswan Dam, Egyptians ate at the expense of subsequent generations of Ethiopians.

This unhappy calculus of the costs of erosion grows more convincing if one subtracts – as one should – from any calculation of benefits the effects of the expansion of coastal swamps upon the incidence of malaria. Upland erosion and coastal deposition created marshes ideal for malarial mosquitoes. Before the twentieth century, many of the growing coastal plains – Antalya, Arta, and Metaponto, for example – remained but little cultivated on account of this summer scourge.

From the point of view of mountain people, and the stability of their way of life, the costs mattered, and not what benefits might accrue to others. In the Epirot Pindus, where the mountain populations were

[100] On the Rif fallow cycle: Fay 1972. In Epirus, Stuart reported a three-year cycle of first maize, second wheat, barley, or oats, and third fallow – where the farmer could afford it (PRO FO 195/935, "Report on Land Tenure in Epirus," 12 January 1870). In Kallarites an elderly villager told me that in the 1920s and 1930s they used a two-year cycle, maize and wheat, without a fallow.

[101] Hutchinson 1969, on Epirus; Sermet 1942, on Spain.

[102] This ratio varies sharply with many factors, chiefly the relief of the sea floor just offshore and the local currents. The 100:1 ratio comes from a University of Nanking study of the Hwang Ho in China, cited by López-Cadenas de Llano 1982, 23–4.

Greek and Vlach, and the owners of lowland properties often Turks (until 1912), soil erosion and deposition represented a transfer of resources from one ethnos and religion to another. In the Rif, erosion to the south, where most of the Rif is drained, carried sediments for deposition from Berber to Arab land. Elsewhere in the Mediterranean world, erosion transferred soil from Kurds and Armenians to Turks and Arabs, from Basques to Spaniards and Catalans, from Serbs to Rumanians and Greeks. But even in the western Taurus, the Lucanian Alps, and the Sierra Nevada, where neither ethnic nor religious differences separated highland from lowland peoples, erosion damaged mountain agriculture and grazing. If lowland folk profited from it (which I believe is more often false than true), mountain people found no solace in that.[103]

One further aspect of mountain soil erosion in the Mediterranean deserves explanation: the symbiotic relationship between grazing and erosion. Each promotes the other. Erosion reduces the rewards of agriculture, encouraging land abandonment, indeed encouraging shifting cultivation. To make abandoned land yield anything at all, mountain folk must loose their sheep and goats on it. As they do so, they prevent spontaneous vegetation from colonizing, thereby perpetuating conditions favorable to further soil erosion. Eventually – and it does not take long with steep slopes and friable soils – the land can support only the sparsest and hardiest vegetation; and this in turn can support only the hardiest of grazing animals: the goat. This symbiosis, in combination with fire, has maintained landscapes in which the goat and holly oak are the ecological dominants. This process, inaugurated by deforestation, vastly impoverished Mediterranean hill and mountain districts, sharply reducing their economic potential – to the point, in fact, where no economy more fruitful than extensive pastoralism is possible. In most places, the effect is irreversible.

The Taurus. The western Taurus has suffered comparatively little from soil erosion, because cultivation has played so little part in its history. Mountain peaks are often stony deserts, stripped of all soil and home only to a few spring flowers and fewer tenacious pines. Human beings have, however, not caused this, nor have they ever used these mountain tops. So the soil loss there is irrelevant to mountain people.

With few exceptions, the western Taurus is not a land of terraces,

[103] Maize, the lifeblood of mountain agriculture, is especially sensitive to topsoil depth and does poorly in eroded soils. Lowland crops, tree crops especially, are often better suited to thin soils. On ratios between yields and topsoil depths for maize: Pimentel et al. 1976.

unlike Pontic Anatolia or the Lebanon. Most cultivation has always taken place in fairly flat bowls and upland plains, as at Zerk or Yılanlı, and clearing and plowing have not destabilized soils appreciably. Wheat occupies some mild slopes around the villages of the Yılanlı Plateau, and these show signs of recent erosion. But the rivers that drain the mountains, the Aksu and the Köprü Su, run clear, and carry only minimal sediment loads. Gullying exists here and there, and a few fields have been abandoned, but nowhere has erosion yet truly threatened the mountain way of life. Mountain peasants remain sufficiently few, most of their fields on sufficiently level lands, that they have escaped the fate that has befallen most Mediterranean mountain folk. Their way of life remains sustainable – and will as long as they do not experience notable population growth.[104]

The erosion history of the western Taurus is unusual in the Anatolian as well as Mediterranean context. Anatolian rivers carry about 450 million metric tons of sediment to the sea every year. In many districts agricultural yields have suffered and good land has gone bad.[105] The hillsides of the Menderes Valley, which once supported the flourishing port city of Ephesus, long ago lost their soils, their productivity, while Ephesus became landlocked and malarial. The foothills of the Cilician Taurus have also eroded down to rock, and sediments have washed down to the plain of Adana, expanding it into the Mediterranean. In many locales in eastern Turkey, as cultivators have pushed nomads higher up mountain slopes, their overgrazing has promoted erosion. But not in the western Taurus.[106]

The Pindus. The flysch of the Pindus erodes rapidly when exposed to winter rains. The western flank of the Pindus, the Epirus side, has long had the most severe erosion problem in Greek lands. Several studies dating from the 1960s documented the extent of erosion. This was a time of rapid depopulation in the mountains and, consequently, land abandonment, so erosion rates then were probably higher than at most previous times, and certainly higher than they are now – in part because so little soil is left to erode away.[107]

Soil scientists of the 1960s compiled mountains of statistics. In the Pindus as a whole in 1964 sediment loads averaged about 20 million cubic

[104] Planhol 1958, 52–3. [105] Erinç 1978, 108.

[106] Firat 1965; Eggeling 1981; Uslu 1971, on Turkish erosion problems.

[107] On the relationship between erosion and the rapidity of rural exodus, *La mobilité* 1984, 327–30; Bousquet, Dufaure, and Pechoux 1983, 16. On erosion in Zagori: Hatzimpiros 1980. On Greece: Metaxas 1948 and 1953; Vouzaras and Karamitros 1981.

meters annually. In some watersheds it reached 75 cubic meters per hectare. By 1960 serious erosion afflicted 63 percent of the Arachthos watershed, and only 11 percent remained unaffected. An FAO team at work in the mid-1960s in northern Greece reported that erosion was complete over 3 percent of their project area, severe in 12 percent, advanced in 16 percent, moderate in 34 percent, and slight in 35 percent. The project covered a large segment of Macedonia and Thessaly, including fairly flat lands.[108] In the Kónitsa area, another team, this one of French and Greek rural sociologists, described erosion in the Kónitsa area as "brutal."[109] The pace of erosion has moderated in the Pindus since the 1960s, as rural exodus and land abandonment has slowed, although Greece as a whole now loses more soil each year than a generation ago. At 86 million cubic meters annually, soil loss is equivalent to a goodly sized island every year.[110]

How long such catastrophic erosion has gone on is impossible to ascertain. In the 1940s and 1950s erosion deeply concerned Greek agronomists, one of whom calculated that it cost the country 100,000 stremmata (about 10,000 hectares) of cropland per year. Counting only lands sloped 15 degrees or more, soil erosion nationally destroyed farmland equivalent to the area of the island of Chios every ten years.[111] In the Pindus, a survey made in 1949 cataloged the extent of erosion in the Arachthos watershed. It determined that 21 percent of the watershed had lost half its topsoil; 33 percent had lost all its topsoil and part of its subsoil; 30 percent had lost all topsoil and three-fourths of its subsoil; and 4 percent had lost all topsoil, all subsoil, and was completely useless land.[112] I have found no statistical data referring to earlier periods, but a casual inspection of the Greek landscape will convince anyone that erosion has undermined mountain agriculture for a long time.

[108] Puglisi 1965, 31; Margaropoulos 1961, 403–4; FAO Library (Rome) SF:54/GRE 45, 1968, 34.

[109] Mendras 1961, 13. Dimitriadis (1980, 200–1) also laments the erosion of the Aoos Basin around Kónitsa.

[110] This figure is the estimate of Professor Kotoulas of the University of Thessaloniki (personal communication, October 1987). This equals 647 cubic meters per square kilometer, or the equivalent of the island of Patmos to a depth of 2.5 meters. Some of these data appear in Kotoulas 1989, 76–7. Data from the early 1970s, less trustworthy I believe, imply rates of only 30 million cubic meters annually. Kentro Programmatismou Kai Oikonomikon Erevnon 1976, 56. According to N. Goulandris, of the Goulandris Natural History Museum, Greece now has about 900,000 hectares of badlands from erosion.

[111] Metaxas 1948, 19; Metaxas 1953, 8–9.

[112] Greece, Iperesia Syntonismou [C.A. Doxiadis] 1950, 30–1.

The Aliakmon River, which drains the eastern flank of the high Pindus, has contributed mightily to the creation of a 1,200-square-kilometer plain west of the Gulf of Thessaloniki. Alexander's Macedonian capital of Pella was on the sea in his day, and the small city of Edessa was near it in the fifth century B.C. By the fifth century A.D. the Aliakmon (with considerable help from the Axios and smaller Louidias rivers) had cut off Pella from the Gulf of Thessaloniki, creating a marshy lake that lasted until the 1930s, when an American company drained it. Today Pella is 30 kilometers from the shoreline, and Edessa about 60. Over the past century or so, the Aliakmon has carried an annual average of about 8.2 million cubic meters of sediment, a quarter of which came from the Pindus. Over its entire watershed, the Aliakmon has lost 1,123 cubic meters per square kilometer per year. In the mountains, the figure would be much higher than this average. Although erosion has probably accelerated in the last century, it clearly has been in progress, at least in fits and starts, for millennia.[113]

Scraps of anecdotal evidence support the logical inference that rapid erosion took place in the nineteenth century, when both the intensity and instability of land use was greater than it has recently become. In the 1860s the Aoos and Acheloos carried to the sea "immense quantities of mud."[114] It has long been evident that Lake Ioannina is slowly silting up. An Italian plane shot down over the lake in 1940 used to catch fishermen's nets, but for many years now sediment has covered it up. Today the lake shoreline moves outward at about 55 meters a year.[115] In an earlier epoch, Wace and Thompson noted widespread gully erosion menacing the village of Samarina around 1910. Stuart remarked on sheet erosion in the mountains around the Ioannina basin in 1869, connecting it to recent deforestation. Indirect evidence concerning yields, and the movement of the coastline at Arta, imply heavy erosion and deposition rates since the nineteenth century. All this stands to reason, given the forest history of the Epirot Pindus and the relationship between vegetation and erosion.[116]

The Lucanian Apennines. The flysch and argillaceous soils of Lucania and the torrential winter rains invite accelerated erosion wherever people and animals strip the plant cover and humus. Lucanian

113 Kotoulas 1984.

114 PRO FO 195/751, "General Report on Epirus," 1 December 1863 (by R. Stuart).

115 Professors Athanassiades and Kotoulas, personal communication, October 1987.

116 Stuart 1869, 291. On the relationship between soil and vegetation in northern Greece: Mavrommatis 1973.

7.4. Lucanian soils are especially prone to erosion. This slope, about 2 kilometers south of Pietrapertosa, is a typical example.

slopes are mild next to those of the Taurus or the Pindus, but in every other respect it is a land ideal for erosion. Its vegetation was not: it takes thirty thousand years to erode a centimeter of soil in beech or oak forest, and four thousand years in natural pasture.[117] To translate Lucania's geologic and climatic vulnerability to erosion into actual injury, this vegetation had to disappear. Humankind has obliged.[118]

Today the silt load of the Lucanian rivers ranks among the highest in the Mediterranean world. In the basin of the Bradano, rains wash down to the sea an annual average of 1,100 metric tons per square kilometer, down from 1,210 in the 1930s. The Sinni carries off about 1,400 metric tons of soil per square kilometer. In the smaller watershed of the Fiumara di Venosa the figure between 1957 and 1961 came to 1,450 metric tons per square kilometer, or a little more than a millimeter of topsoil per year over the whole basin. In comparison, the sediment

[117] Puglisi 1967, 11.

[118] Kayser 1961 is the classic treatment of erosion in Lucania. Radina (1976, 100) explains that deforestation is responsible for the majority of accelerated erosion in Lucania.

load per square kilometer of watershed for the Nile is 36 metric tons; for the Tigris-Euphrates, 55; for the Danube, 98; for the Mississippi, 153; for the Hwang Ho (by far the world's muddiest big river), 2,791 metric tons.[119]

The rates of erosion in Lucania, high enough to make impossible artifical lakes for irrigation or hydroelectric purposes, are only a small patch on what they must have been in the fairly recent past. In the century before 1960, erosion carried off soil to an average depth of 6 centimeters over the entire Bradano basin. Between 1600 and 1950 the southern shore of Lucania moved up to 1,500 meters out into the Ionian Sea. The era of most rapid deposition came between 1824 and 1924, when silt deposition covered 26,500 square kilometers of the seabed surrounding the Mezzogiorno. Since 1943 the shoreline has stabilized or indeed moved inland by small distances. The difference between then and now is that the rivers no longer feed the sea nearly as much mountain soil as before. In fact, the fertile coastal plain of Metaponto will prove only a fleeting boon if expensive interventions do not protect the coast from the wave action of the sea.[120]

Whatever happens to the coastline, the costs in the mountains have been high. By 1960, about 50,000 hectares of farmland in Lucania had fallen out of use because of gradual sheet erosion, and badlands covered an equal area. In southern Italy as a whole, erosion by 1967 had made uneconomic the cultivation of 1.5 million hectares, although peasants continued to farm much of this area.[121] The upper watersheds of the Basento, Agri, and Sinni are especially affected, but the whole Ionian versant of the Appennines has suffered greatly. By 1950, landslides affected one-sixth of the basins of the Sinni and Agri rivers. Imprudent land use, landslides, and accelerated erosion continue to afflict Lucania.[122] Geologic conditions spare the Bradano basin from serious landslides.

Erosion has always threatened the soils of Lucania, but only where farmers have cleared the soil has the threat been realized. Ancient Greek settlement exposed the low hills behind the coast and produced

[119] Neboit 1983, 85–90; D'Arrigo 1951; Ranieri 1961, 87. Merendi (1952, 7) claims that in southern Italy erosion since 1850 had carried off an average of 50 to 60 centimeters in depth of soil, more in the mountains, less in the lowlands. This is probably too high: it implies erosion rates five times the highest in effect in Lucania in recent years. Figures on rivers around the world are elaborated from *Encyclopaedia Britannica* 15:868.

[120] Cocco, De Pippo, and Pennette 1976.

[121] Rossi-Doria 1963, 97–8. Paranzino 1967, 301.

[122] Ippolito and Cotecchia 1954; Basilicata (Regione) 1986; Radina 1976. Erosion remains rapid in Lagonegro: Lazzari 1986, 110–13.

the first spate of erosion (see Chapter 3). Another came with the decline of ancient civilization and widespread land abandonment. But not until the population growth and expansion of agriculture that began in the sixteenth century and accelerated in the eighteenth did conditions once more become propitious for accelerated erosion. In the nineteenth century, with further poulation growth, with massive deforestation and "rapine" agriculture late in the century, and with emigration and land abandonment from 1880 to 1925, all circumstances (except the mild slopes) favored catastrophic erosion. Only in the Rif have conditions conspired so effectively to strip the soils from beneath the populations that depended on them.

The Alpujarra. The ubiquitous terraces of the Alpujarra, when properly maintained, served to stem soil erosion. For long stretches of time the Guadalfeo must have run clear. But terraces when neglected give way to erosion that acts with a pent-up force. So the erosional history of the Alpujarra was a very uneven one, with fits and starts, alternating periods of calm and fury. The same is likely true for other lands with widespread terracing, such as Lebanon, the Pontic mountains, the Balearic Islands, or the High Atlas. In the case of the Alpujarra it is fairly easy to pinpoint the periods of fury. The modern soil history of the Alpujarra has four key epochs, defined by trends in Iberian politics, the world economy, and local demography. The first came in the sixteenth century when newcomers unfamiliar with the local environment colonized the valley in the wake of the expulsion of the Moors. The second came in the late eighteenth and early nineteenth century with the extension of market agriculture and mining to the vicinity of the Alpujarra. The third, which overlapped with the second, was the desamortización and the expansion of cultivation late in the nineteenth century. The fourth, still in progress, is the depopulation and abandonment of the soil. Soil erosion surged in each of these epochs, impoverishing the Alpujarra, and requiring more and more labor and care on the part of Alpujarreños to sustain agriculture and maintain their lives. Recently most have given up the fight. A popular refrain put it succinctly:

A quien Dios quisó bien
En Granada le dió de comer.
A quien quisó mal
En la Alpujarra un bancal.[123]

[123] Sermet 1953, 156. "He whom God loved well, He gave to eat in Granada. To whom He loved poorly, He gave a terrace in the Alpujarra."

A quarter of Spain suffers from serious erosion. Nationally, about a billion metric tons of soil disappear every year. Granada and Almería have the worst problem in the country, being the only provinces in which more than half of the land area shows signs of erosion.[124] In the high Alpujarra in the early 1980s, annual erosion came to about 2,000 metric tons per square kilometer. Erosion in the Guadalfeo Basin as a whole, including the flatter lands near the coast, amounted to about half as much.[125] This is fairly rapid erosion, due in part to the effect of continued abandonment on land that for centuries only terraces had kept in place. In the late 1980s erosion had abated somewhat, to the point where it was only slightly above average for Andalusia. In places new plant cover has grown up in the wake of the abandonment of cultivation.[126] But accelerated erosion over centuries has left many slopes without any more soil to erode.

The transformation of the countryside after the expulsion of the Moors produced a "violencia erosiva" that washed away whole settlements.[127] The tree crops and terraces of the Moors had held the soil well, but the extensive wheat and sheep economy that replaced it did not. According to one view, the greater part of the valley's productive topsoil splashed down to the deltas of Motril and Adra, an irrecoverable loss for the Alpujarra.[128] The expansion of agriculture in the eighteenth and nineteenth century had serious effects as well. Erosion took place on a scale large enough to build out the deltas at Motril and Adra. Landslides began to plague the western part of the high Alpujarra more regularly. Soportújar, Carataunas, Bayarcal, and other villages felt the costs of local deforestation. In a memorable landslide in 1863, mud and debris obliterated fields and houses, buried bridges and mills, and cracked a church in Soportújar. Since 1863, these and other villages lived with the constant danger of landslides, much like the people of Lucania. Nowhere else in the Alpujarra, nor in southeastern Spain,

[124] FAO Library (Rome) Microfiche 42052 (López-Cadenas 1976, 4). Government figures reported in the newspaper *El Pais* (23 September 1991) indicate that 36% of Spain suffered from either severe or moderate erosion.

[125] Quirantes Puertas 1987, 147. Thornes (1976, 32) has lower figures for Mecina Bombarón, Ujíjar, and Berja, all between 370 and 575 metric tons per square kilometer per year. Scoging (1982) points out that erosion in the Alpujarra varies greatly with microlocal soil conditions.

[126] Spain, Junta de Andalucía 1987, 80–9.

[127] Bosque Maurel 1971, 82.

[128] Navarro Alcalá-Zamora 1981, 44. Quirantes Puertas 1987, 154, on erosion during Moorish times.

7.5. Tree crops offer minimal obstacles to accelerated erosion. Near Torviczón, in the Sierra Contraviesa, an almond grove loses ground to erosion and to encroaching maquis (February 1988).

have geologic and climatic conditions combined with anthropic landscape change to produce such unstable slopes.[129]

Serious as erosion was on these slopes, on those without terraces the soil loss must have been greater still in the nineteenth century. On the Sierra Contraviesa, where vines, and later figs and almonds, replaced spontaneous vegetation, the schist soils resisted terraces. A few clumps of holly oak testify to the former protection from the rains, but elsewhere, especially in the eastern part of the Sierra Contraviesa, erosion has carried off almost everything one could call soil. The barrancas have deepened, the ramblas widened, and the people have left. A few hectares of vines and fruit trees remain.[130]

Where the terraces have so long held erosion in check, they are now often decaying. The terraced valley above Válor and the steep walls of the barranca of Poqueira have become goat pasture. No one bothers to keep the stones in place. No one attends to the task of hauling soil

[129] Sermet 1953, 157; García Najera 1929, 5–11; Almagro 1932, 19–21.
[130] Quirantes Puertas 1987, 156–7.

Table 7.2. *Erosion rates in Rif watersheds*

River	Watershed Area (km^2)	Sediment Yield ($mt/km^2/yr$)
Loukos	1,820	4,300
Tlata	178	22,000
Nekor	780	7,900–28,500

Notes: (1) Alheritière (1981, 2) gives 10,000 to 12,000 metric tons per square kilometer per year ($mt/km^2/yr$) for the Nekor. (2) Lower figures for the Nekor and Tlata were published in the Rabat newspaper *Le Matin Economique*, 21 January 1988, p. 12; these appear to be official figures, widely suspected by foresters, hydrologists, and soil scientists (intentionally) to understate the problem.
Source: Tayaa 1985, 5; Tayaa and Brooks 1984, 27.

from below to maintain the higher terraces. Goats nibble away at the scrub guarding the edge of every terrace, and their hoofs now and then knock down a piece of terrace wall. Some of the old flat-stoned threshing floors, once the scene of all-night work and festivities, now carry a coating of mud and pebbles. Here, as in Epirus, the departure of the peasantry will complete the destruction of the land that its arrival began.[131]

The Rif. Today the Rif loses soil faster than any other area in the Mediterranean world. Geology, climate, and the human factor here are as favorable to erosion as in Lucania, and the slopes are much steeper. Soil scientists now measure soil loss in metric tons per hectare, rather than per square kilometer.

The Moroccan development agency for the western Rif, DERRO, estimated in 1965 that the annual loss to erosion in the Ouerrha basin was 4,500 metric tons per square kilometer, or to a depth of about 1.5 centimeters per year throughout the basin. In 1981 the Nekor watershed lost 10,000 to 12,000 metric tons of soil per square kilometer. In the early 1980s erosion throughout the Rif had grown still more serious (Table 7.2).

About three-fifths of the soil lost to Morocco through erosion is lost from the Rif. The tiny rivers of the north coast create giant patches of brown in the otherwise blue Mediterranean, like miniature Niles. The

[131] Ortega Alba (1981, 161) describes the Poqueira barranca. Between 1962 and 1972 38% or 41% of the cultivated land in the Alpujarra was abandoned; see Navarro Alcalá-Zamora 1981, 26–7, 45 (he gives different figures on these different pages). In a village on the north slope of the Sierra Nevada, half the arable went out of cultivation between 1954 and 1969; Luque Baena 1974, 35.

7.6. A scene from the Nekor River valley where erosion rates are among the highest in the world. Hillslope erosion produces the braided river bed, but irrigation and hard work continue to produce good crops.

Sebou River, which drains the south flank of the Rif as well as some of the Middle Atlas, carries a sediment load that averages 30 million metric tons, and some years reaches 100 million tons, equal to the load carried by the Mississippi at Baton Rouge; the watershed of the Sebou is about 4 percent of that of the Mississippi.[132] When erosion was a trifle slower, in the 1960s, DERRO calculated that every year between 1.5 and 2.0 percent of agricultural potential was lost to erosion; the rate is higher now.[133] Erosion is so fast that in most places it makes no sense for farmers to use fertilizer. Although soil formation proceeds fairly quickly in the Rif (a centimeter every 12 to 40 years), erosion carries it away at least ten times as fast as it is formed. The lowest of the rates in Table 7.2 implies erosion almost twice as serious as in China's Hwang Ho basin. The highest is almost ten times as great as the Hwang Ho's rate. This is a soil holocaust, the most dramatic case of overshoot in the Mediterranean in some time, and one of the most serious in the world.

[132] Tayaa 1985, 4. Other sources give 495 million tons as the sediment load of the Mississippi – at its mouth.

[133] FAO Library (Rome) SF: 18/MOR (1965) 1:14; Maioui (1984, 6) gives 2%.

7.7. The barren slopes behind this *ksar* in the Nekor River valley were once cultivated. Faint traces of old terraces remain.

How long such prodigious soil destruction stretches back into Rifian history is hard to say. No records of it exist before the 1960s (as far as I know). Geomorphologists estimate that the severe erosion of the Beni Bou Frah badlands dates to the 1920s.[134] But what this might mean for the Rif as a whole is hard to say. The strong currents off the northern shore sweep away the silt load of the rivers, complicating any attempt to estimate previous erosion by measuring undersea deposition, a method used in southern Italy. Geology, climate, and slope have not changed, at least not recently, and agricultural methods seem little changed since the time of Leo Africanus. What has changed is the scale on which agriculture is practiced, and the slopes to which it has recently climbed. When fewer Rifians struggled to scratch a living from their mountains, they practiced a long-fallow system on many of their fields. This meant that over two decades, only about a quarter of the rainy seasons would find a given slope bare and vulnerable. By the 1960s, with population pressure more intense, Rifians eliminated fallow in many cases, and

[134] Imeson et al. 1982, 49.

seeded slopes steeper than 45 degrees. They also "colonized" all high valleys where crops might grow.[135]

They have often done this in a hurry, without pausing to engage in the backbreaking work of building terraces. Outside experts have proposed countless schemes, all of them failures, technical fixes for sociopolitical problems.[136] Rifians, of necessity, practice true rapine agriculture, at least on the steep slopes. This gets them through in the short run, but undermines them in the not-so-long run. When presented with this analysis, which comes as no news to them, Rifians often have recourse to the comfort of people in dire straits, and say Allah will provide for the future. Nothing short of divine intervention can preserve what remains of Rifian soils and the Rifian way of life.

Consequences of Landscape Change

Deforestation and soil erosion have changed the way of life in many mountain areas and in some lowland zones as well. Agriculture has suffered with soil depletion and hydrological changes. Hunting and gathering have declined as forest habitat has shrunk. Fuel has grown more dear, at least until the arrival of kerosene and other substitutes for wood and charcoal. Eroded pastureland has yielded less in milk and meat, and often required the substitution of goats for sheep. All this hit mountain folk hard.

Changing mountain ecology touched those living in the lowland plains as well. In some cases it did so gently, providing new lands in coastal deltas. More often it did so roughly, through floods that destroyed crops, animals, bridges, and buildings; through deposition of beds of gravel and rubble on fields and pastures; through the formation or expansion of swamps that hosted the anopheles mosquito; through the siltation of navigable river channels and freshwater reservoirs, and, indirectly, through the expulsion from the mountains of hungry and occasionally desperate people. A few examples of the consequences of landscape changes in the past two hundred years will suffice. Most of these examples, such as the dwindling of game animals, ought also to be taken as further indication of the depletion of woodland in the nineteenth and early twentieth centuries.

[135] Fay 1972, 2:296–303. Heusch (1981, 422) says slopes up to 80 degrees are cultivated! This is hard to credit, but appears also in Tayaa and Brooks 1984, 24. G. Maurer 1968b, 24.

[136] Heusch 1981.

Table 7.3. *Price of fuelwood imported at Corfu, 1827–1859* (*£/passo*)

1827	.18
1830	.46
1835	.44
1840	.32
1845	.46
1850	.45
1855	.36
1858	.60
1859	.60

Note: Ordinarily "passo" in Italian means a "step." Here it must mean a measure of fuelwood, but how large I do not know. The 1850 figure is the average of 1849 (.44) and 1851 (.45), as the accounts are silent for 1850.
Source: PRO CO 136/1391–1423.

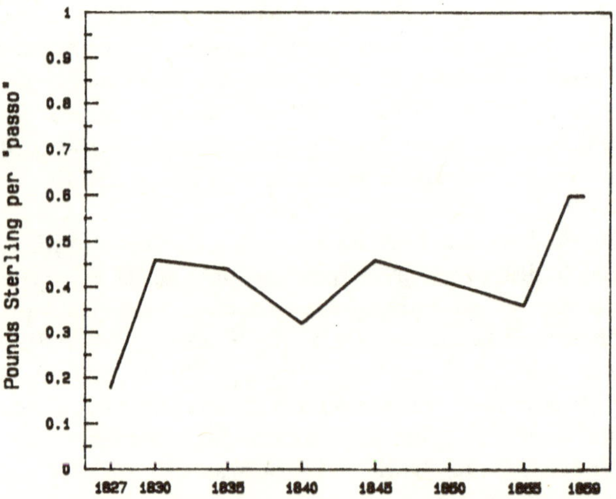

Graph for Table 7.3

Wood shortages

Mountain villages depended on wood for fuel and many other purposes. Every family routinely needed several cubic meters per year for cooking and heating. When wood grew scarce, women and children had to go further to find it, reducing the time available for other pursuits that might produce income. When it grew too scarce, villages had to move.

Systematic data on fuelwood prices are hard to find. Scattered data in Italian archives suggest an upward trend in Lucania from at least 1866.[137] In the country as a whole, fuelwood prices increased from 80 to 90 lire per cubic meter in 1903 to 120 to 130 in 1913, and the increase was higher in the mountains.[138] British archives reveal prices of imported fuelwood at Corfu, wood that invariably came from Epirus (Table 7.3). It shows an inconsistent but upward movement between 1827 and 1859. Additional prices from Ioannina (1863–71) show fuelwood becoming more expensive at a time when the cost of food and most everything else declined.[139] The Ioannina and Corfu data imply that wood grew scarce relative to demand in the nineteenth century, but they are not faultless guides to conditions of supply. Fuelwood prices indicate only what cities were prepared to pay. Even the most extensive and regular urban price series, therefore, could never prove the existence of a fuel crisis in mountain villages. Anecdotal evidence is more helfpul.

Reports of fuel shortages in the nineteenth century are common. Travelers in the Pindus complained of shortages and prices as early as 1813, and kept doing so,[140] while woodsmen found it took more and more work to make the same amount of charcoal.[141] A local doctor noted that villages in western Zagori had difficulty obtaining firewood in the 1880s.[142] In the 1860s a local gentleman observed to Ramage that fuelwood was running short in the Lucanian Apennines, and predicted a serious crisis in the near future. Other locals noted the extreme price of fuel in Potenza in 1903 and a shortage in the Sila by 1913.[143] In Lagonegro wood had grown so precious by 1879 that wine producers had replaced it with iron for funnels, strainers, and barrels![144]

In both the Pindus and Lucania, fuelwood remained a precious item

137 ACSR, MAIC, DGA, I, busta 319, ff. 2–4, on Pietrapertosa, for example. Other data for other mountain comune are in buste 257, 260–2.

138 Touring Club Italiano 1913 (?), 80. In north-central Italy fuelwood prices rose sharply in the 1790s. ASAEAG, Carte Bartolozzi, busta 185, "Legna e carbone, Misure e prezzi antichi e moderni." In the mountains of Lucca wood shortage pressed continually in the nineteenth century; see Sarti 1985, 36–7.

139 PRO FO 195/751, "General Report on Epirus," 1 December 1863, and 195/1002, "Commercial Report for 1871, Jannina."

140 Hughes 1820, 1:501, quoted earlier in this chapter; Holland 1819, 1:136; Boué 1854, 2:48.

141 Lazos (1935, 26–7) reports that whereas once a single stremma could yield 1,500 to 2,000 okes of charcoal, by the early twentieth century the average yield was 700 to 800. (An oke equals about 1.3 kilograms; a stremma equals about a tenth of an hectare.)

142 Lambridis 1887, 1:16.

143 Ramage 1868, 158; D'Alfonso 1913, 624–5; Franzoni 1903, 16.

144 ACSR, Arch. Parl. (Jacini), busta 3, fasc.: Lagonegro, ff. 13–33 (1879).

until well after World War II. Eventually alternative fuels came to compete with wood and charcoal, and their use declined dramatically. They underwent a small revival when petroleum derivatives jumped in price in the 1970s.[145] Today many peasant homes still rely on wood and charcoal, although they could easily switch to propane and kerosene if relative costs made it sensible. Options exist today that did not fifty years ago, so fuelwood shortage may be an irritation, but it is no longer a serious problem.

In a thousand other ways, wood shortages have impinged on mountain communities. Rifians could build only narrow homes by the 1940s because long, broad beams had become rare. Boatbuilding also became much more problematic. Proper wood for plows and other agricultural tools is now scarce.[146] Elsewhere railroad ties and telephone poles are now made of concrete – an expense that states rather than mountain villagers have had to pay. Each of these minor adjustments and limitations has meant little by itself. Their cumulative effect made mountain life more uncomfortable, and more costly, than it otherwise would have been.

Hunting and Gathering

The only thing more basic than fuel was food, and the loss of forests affected that as well. Throughout the Mediterranean mountains, hunting declined in the nineteenth century. It had never brought much to peasant tables, but if not replaced, what little it brought was a lot to lose. Large animals of many species died out, partly through campaigns of extermination, as in the case of the big cats,[147] but also through the destruction of forest habitat. The Lucanian Apennines supported herds of deer and wild boar as late as 1880.[148] Wild hogs, perhaps preserved by the Muslim proscription of pork, roamed the western Rif in the 1820s and the Taurus in the 1870s. These forest animals are gone now and, if absence of mention can be trusted, disappeared early in this century.[149]

[145] Volpini 1979, 89–92, on the modern use of fuelwood in Italy.

[146] Jiménez Salas 1946, 288; Pascon and van der Wusten 1983, 39.

[147] Big cats roamed the Taurus in the early nineteenth century (Brice 1978, 141); the Apeninnes (Filangieri 1980, 109); the Sierra Nevada as late as 1880 (Rubio 1881, 316–17); and the Rif in the 1840s and probably later (Coon 1951, 25; Merry del Val 1920, 23). Fréjus saw lions in the Rif's Nekor Valley in 1666; his editor in 1922 thought Fréjus must have been mistaken (Fréjus 1922, 138).

[148] Lenormant 1883, 1:225–8, 237–8.

[149] Harris found game in the western Rif before 1912, but Jiménez Salas mentions only hyenas and foxes by 1946. Hunger during the Rif War must have put a premium on

Big game hunting mattered much more to nobles than to peasants. King Otto, not the villagers of Lavrion, enjoyed hunting wild boar in the forests of southern Attica in the 1830s.[150] In Lucania, where the peasantry rarely had guns and a landed aristocracy did, hunting lasted until the late nineteenth century. Other forest animals, not hunted for food, disappeared at this same time, suggesting habitat destruction mattered as much as or more than hunting. Bears, deer, and wild boar all vanished from the Pindus,[151] and wolves have all but vanished almost everywhere. While these latter never provided food for mountain villages – quite the opposite in the case of sheep-eating wolves – their near-extinction shows how forest land has retreated in modern times in the Pindus, Apennines, and Rif. On the Taurus's western flank, and in southwest Anatolia generally, bears, panthers, foxes, and wolves roamed the forests in the 1870s, and "wild boar [was] everywhere."[152] Today forest animals are very rare.[153] In all likelihood, animal depopulation in the Taurus is mild compared with that of the other regions, because neither human population nor deforestation has changed habitats nearly as much as elsewhere. In the Sierra Nevada, although big cats have vanished, most of the other species present in the late nineteenth century are present today – none of them, however, strictly forest animals.[154] If deer or wild boar ever existed in the forests of the Sierra Nevada, they disappeared before the late nineteenth century.

The decline of hunting may have deprived villagers of a reserve food supply in dire moments, but it had little effect in normal times, because they rarely hunted large animals. Hares and fowl are the more frequent targets of the Mediterranean peasant out with his traps or gun, and these species are far from extinct. It is impossible to know whether they exist in greater or lesser number now than in the past. Both prosper in open country more than in closed forest, and so the clearing of recent

edible animals, and their numbers have probably never recovered. Harris 1921, 255; Jiménez Salas 1946, 288. See also Brooke 1931, 2:376; Gresa de Camps 1903, 76, on Rif wild fauna. E. Davis 1874, 302–4, on the Taurus.

150 Kontos 1929, 173–6.

151 PRO FO 195/751, "General Report on Epirus," 1 December 1863, lists these species now absent or extremely rare. Pouqueville (1822, 7) wrote that the Pindus was inhabited by troops of deer, roe, and wild boar.

152 E. Davis 1874, 302–4.

153 The only specimens of terrestrial wildlife I have encountered in the mountains of southern Turkey are tortoises, on Bozburun Dağ, near Zerk. Fellows (1838, 174) remarked on their size after an encounter only a few miles west of Bozburun.

154 Rubio (1881, 212, 316–17) mentions wolves, eagles, foxes, and mountain goats. I have never seen wolves in the Sierra Nevada, but have seen eagles, foxes, and mountain goats. Villagers say wolves still exist.

centuries may have worked to increase their numbers, adding marginally to the food supply of mountain villagers. Whatever the net effect of changes (however great) in fauna upon the welfare of mountain villagers, the difference must have been small.[155]

Gathering was another matter. Everywhere in the mountains gathering supplemented food production in important ways, often making the crucial difference between winter hunger and starvation. Chestnuts and acorns formed an important part of the peasant diet in Lucania and the Alpujarra (see Chapter 4). Other nuts and berries provided a reserve food supply in every mountain district. The sale of vallonea oak acorns (a dye source) helped Pindus villagers add to their meager incomes. Doing without these auxiliaries might make no difference in a year when soils yielded abundantly; but bad years were all too common, and mountain peasants needed a cushion of one sort or another to survive them. As forests shrank, the scope for gathering must have shrunk as well. Fuelwood and timber seekers made oak a special target, so the acorn supply must have dwindled. This may help account for the decline of wild hogs in Lucania and the Rif, and of herds of swine in Lucania and the Alpujarra. Where forest regenerated in the wake of oak destruction, it often returned as pine, a species with fewer uses. Villagers prized chestnut trees and would certainly have defended them where possible. But forest groves bought up in the alienations of public land must routinely have been cut or burned, as the new proprietors knew little and cared less about the trees' role in maintaining village life. Gathering mattered most in Italy and Spain, and nineteenth-century deforestation was greatest in Lucania. Consequently, it was there that the restriction of these auxiliary activities posed the greatest challenge to the mountain way of life.[156]

Gathering, whether of nuts or fuelwood, usually fell to women and children. They, rather than adult men, would have spent additional hours scouring the ground for these necessities. And insofar as deforestation upset hydrological regimes, desiccating springs and transforming streams into seasonal torrents, again women and children would be the ones who went further afield for water. They, perhaps, more than men, felt the immediate costs of landscape change. This difference

[155] Migratory birds are a favorite target of Mediterranean hunters. Of the five billion or so birds which fly twice annually between Europe and Africa, several hundred million do not make it. Peters and Lovejoy 1990, 363.

[156] Italy, Ministerio dell'Agricoltura (1976, 151) has statistics on acorn and chestnut production in Lucania, 1947–73. I have found no earlier data. These statistics show a sharp decline, a reflection of shrinking mountain population as well as reduced dependence on the forest.

may have mitigated the broader political effects of mountain landscape change, as men absorbed a diminished share of the costs, and their dissatisfactions translated more easily than women's into political action.

Agriculture

Deforestation and erosion, landscape effects often produced by the extension or merely the practice of agriculture, generally diminish its productivity. Forests moderate temperature and humidity, so deforestation invites wider fluctuations in air and surface temperatures, in winds, and in stream flows. A pernicious effect upon agriculture of local deforestation in the Mediterranean has been to promote seasonal extremes of drought and flood in watercourses. Seasonality is pronounced in the rainfall regime of most parts of the Mediterranean, but forest regularizes stream flow by holding much water back, rather like a sponge. Without the forest, runoff increases severalfold, depriving soils (and air) of moisture, and of course encouraging soil erosion. But even the less pernicious effects of local deforestation, such as the more open avenues to winds, can prove agriculturally damaging. Almonds, for instance, an important crop in the Alpujarra, withstand drought fairly well, but not high winds.[157]

Soil erosion menaces agriculture in obvious ways. Sheet erosion, "more insidious than spectacular,"[158] strips the topmost layers of soil and humus, which contain minerals and nutrients in far greater measure than do subsoils. Gully erosion reduces the extent of suitable arable.

The impact of deforestation and soil erosion upon mountain agriculture in the Mediterranean world is not easy to gauge. The pace of these landscape changes is usually slow enough that they do not strike villagers as important, although occasionally old men will say that the land used to be more fertile, or that a certain eroded and abandoned slope once yielded a crop. Statistical data on yields rarely exist before 1945, and data from recent decades reflect tumultuous changes – labor shortage, abandonment of poorer lands, increased use of chemical fertilizer and, in some cases, of machinery – making it impossible to judge the impact of changing landscape conditions. In the Alpujarra, for example, almost no one raises crops except under irrigation now, whereas fifty years

[157] Jiménez Blanco 1986 2:798–802.

[158] Kayser 1961, 38. He continues, explaining the agronomic meaning of erosion in the Lucanian context: "lorsque le sol, avec ses principes nutritifs, sa matière organique, sa structure, est entraîné, il laisse la place en général, non pas à une étendue stérile, mais à de nouvelles strates cultivables. Les 'qualités' des roches-mères, aisément labourables, limitent les dégâts. Mais, à partir de ce moment, le vrai sol ne se reconstitue plus."

Table 7.4. *Average yields in Potenza* circondario, *1879–1894*

	1879–83	1890–94
Wheat (hectolitres/ha)	8	5
Maize (hectolitres/ha)	15	4
Potatoes (quintals/ha)	30	4

Source: ACSR, PC, Zanardelli e la Basilicata, busta 7, cartella 5.

Table 7.5. *Land use in Potenza* circondario, *1879–1894 (ha)*

	1879–1883	1890–1894
Cultivated	8560	5889
Pasture	3550	3550
Sterile	5804	8475
Total	17914	17914

Source: See Table 7.4.

ago, more than two-thirds of the cereals were raised without it. Consequently yields per hectare are (presumably) much higher now than then, regardless of other variables – such is the importance of irrigation. The Rif, where these modern changes have scarcely made themselves felt as yet, and the Lucanian Apennines, where carefully gathered statistics exist for the late nineteenth century, offer the best glimpses into this question.

In Lucania, the Zanardelli inchiesta into agriculture took an interest in the trend in average yields. In the *circondario* (subprovincial district) of Potenza, an area of considerable land clearing after 1865, the figures show sharp declines between 1879 and 1894. The data for the three most important crops are presented in Table 7.4. Vineyard yields also declined dramatically in these years, probably a consequence of phylloxera, but those of some of the less important crops, barley and oats for instance, increased slightly. Their combined area, however, came to about one-quarter of that of wheat, far and away the most important crop in Potenza. The area sown to all other crops shrank in these years, and that described as "sterile" expanded (see Table 7.5). This strongly implies soil deterioration, precisely what one would expect of agriculture practiced in newly deforested lands like those of Potenza.

Reports from the circondario of Lagonegro indicate troubles in local agriculture and fuel supply traceable to deforestation. In 1879 one of the investigators on the Jacini inquiry wrote:

> Quasi tutto il suolo del Circondario era impiantato a boschi; da poco tempo in qua viene distrutta la maggiore parte dei medesimi, rendendola improduttiva. Con modo vandalico si è agito sin oggi per la distruzione dei boschi, molti dei quali vennero quotizzati. All'effimero prodotto di pochi anni si è sostuito il grave danno alla pastorizia. . . . Si comincia ad esperimentare l'effetto dell'errore commesso, preciso per la scarsezza della legna nei luoghi freddi.[159]

A local agricultural commission came to a similar conclusion a few years later, explaining that unsuitable agriculture had ruined Lagonegro's resources:

> Guardate le condizioni topografiche, climatologiche e telluriche del nostro circondario si deriva che le coltivazioni da noi usate sono rovinose, per lo meno nella zona montuosa. Con assiduità condannevole si sono dissodate le montagne, le colline, distruggendo antiche piante a pascoli cespugliosi. Sotto l'attuale reggime forestale sono scomparsi anche i boschi. A questo modo i terreni sono steriliti, agli animali mancano le pasture, agli uomini il combustibile, difetta il legname per le costruzioni e per gli usi agrarii, ogni terreno insomma è diventato grillaio.[160]

Further evidence concerning larger areas adds to the impression that mountain agriculture was suffering soil depletion and erosion and suggests the problem only grew worse after the early 1890s – until technological revolutions raised yields after 1950. Wheat yields on deforested

[159] ACSR, Arch. Parl. (Jacini), busta 3, fasc.: Lagonegro (10 December 1879). The author was probably Dottore V. A. Mauri. ("Almost all the soil of the district was in forests; recently the greater part of these were destroyed, rendering the soil unproductive. . . . Up until today the forests were vandalously destroyed, many of them having been parceled out and sold [in the land alienations]. Grave damage to pastoralism has followed upon an ephemeral yield of a few years. . . . The effect of the error committed is now beginning to be felt, specifically in the scarcity of firewood in cold places.")

[160] ACSR, MAIC, DGR, IV, busta 29, fasc.: Lagonegro, "Atti della prima sessione ordinaria del Comizio Agrario, 1885." The author is the president of the commission, named Barletta. ("Looking at the topographical, climatological and telluric [meaning seismic] conditions of our district, one concludes that our cultivation is ruinous, at least in the mountainous zone. The mountains and hills were plowed with assiduity worthy of condemnation, destroying old vegetation and scrub pasture. Under the current forestry regime, even the woods disappeared. In this manner the lands have become sterile, the animals lack pasture, people lack fuelwood, and also timber for construction and agricultural use, and all lands have become fit only for crickets.")

lands in Lucania dropped by about one-third between 1892 and 1910, while maize yields fell by about half between 1882 and 1903. Potato yields fell by about 40 percent between 1880 and 1903.[161] One small proprietor from Lagonegro, Donato d'Onofrio, summed the situation up about 1908, explaining that the former desmenial lands had all been abandoned because they had become completely sterile. He added that if he were a younger man (he was then fifty), he would go to America.[162] His judgment accorded with that of his more youthful fellows: roughly half the land cultivated in Lucania in 1878 was abandoned by 1903 and thousands of peasants tried their luck in America or Naples. They, with the help of speculators and timber merchants, had ruined the soils on which their livelihood chiefly depended, overshooting the capacity of their environment under existing technologies.

The Rif in more recent years has begun the same process of overshoot and ecological destruction, with unfortunate results for agricultural yields. Here there is as yet very little land abandonment, because population continues to grow, even with high rates of outmigration. Rifians do not use agricultural machinery or much in the way of chemical fertilizer. The scope of irrigation has increased somewhat over the past fifty years, improving yields here and there, but for the most part the history of average yields reflects the history of soil conditions. It is a disconcerting story (see Table 7.6).

The decline in cereal yields in the Rif corresponds to the era of most rapid population growth. Even in the best agricultural lands in the western Rif, the area around Chechouan and the Oued Laou, where irrigation is widespread, yields have declined from 11–17 metric quintals per hectare in the early 1950s to 10–12 in the mid-1960s, and 6–8 as of 1980.[163] The explanation for this diminished effectiveness of agriculture is not merely soil erosion and depletion, but also the extension of

[161] Wheat yields in deforested lands of Basilicata were only 3:1 or 4:1 in 1892, but dropped further to 2:1 or at best 3:1 by 1909. Carbone 1892, 177–8; Nitti 1909–10, 121–2. Nitti writes that yields on these lands had once been quite high. Of Basilicata maize yields, Spera (1903, 42) writes they declined from 500,000 to 140,000 hectoliters, while the area under maize fell by one-half. Potatoes fell from 500,000 metric quintals on 16,000 hectares to 150,000 quintals on 9,000 hectares between 1880 and 1903. By the 1970s average yields in Potenza province were higher than ever: 17 metric quintals per hectare for hard wheat, 13 for oats, 11 for maize and 35 for hybrid maize, 48 for potatoes. These were generally between one-half and one-quarter of the Italian average. Italy, Ministero dell'Agricoltura 1976, 199.

[162] Nitti 1909–10, 105. On land abandonment in the nineteenth century: Franchetti 1875, 61–3; on abandonment more recently: Formica 1975, 124–32; Formica 1983, 92–3.

[163] 1950s data from Spain, Dirección General de Estadística *Anuario estadístico* 1955, 114; mid-1960s from G. Maurer 1968b, 49; 1980 from Gaudio 1981, 101.

Table 7.6. *Rif cereal yields, 1934–1977 (metric quintals/hectare)*

Years averaged		Wheat	Maize	Barley	Source
1934–5	(2)	3.9	6.5	6.4	a
1936–40	(5)	6.2	4.3	6.2	b
1938–45	(8)	7.2	6.3	7.2	b
1950–1	(2)	8.0	9.0	8.3	b
1965 (estimate)		5.0	5.0	5.0	c
mid-1960s	(?)	3–5	—	3–5	d
1970s	(?)	4–5	—	—	e
1976–7	(2)	1.8	—	5.0	f

[a] Spain, ACEM, Delagación de Asuntos Indígenas, 1935a.
[b] Spain, Dirección General de Estadística, *Anuario estadístico*, 1941:52, 1946:47, and 1955: 114.
[c] FAO/DERRO, 1965:xi.
[d] G. Maurer 1968b, 49.
[e] Fay 1979, 87.
[f] Pascon and van der Wusten 1983, 180–2.

Note: 1934–51 data concern the administrative area known as the Rif within the Spanish Protectorate. The 1965 estimate concerns the DERRO area. The mid-1960s data cover the central Rif. The 1976–7 data concern two communities within the Beni Bou Frah lands. Given these difference, this table represents only a rough approximation of Rif cereal yield history.

cultivation into marginal (usually steeper) lands. Any yield is better than none for a Rifian with hungry children. Unfortunately, extracting a yield from unsuited lands often means no soil will survive to feed the Rifians' grandchildren. No amount of machinery or chemical fertilizer can make good this loss.

In the Rif, as elsewhere in the Mediterranean mountains, other crops besides cereals form an important part of sustenance. Unhappily, the effects of soil erosion are felt here as well. In Jbala, in the days of the Spanish Protectorate, raisins, a crucial winter food, yielded an average of 15 metric quintals per hectare and up to 40 in good years; by 1970 erosion and vine infections reduced that to only 10. Peasants accordingly gave up on vines. In the Rif as a whole, erosion by 1963 had ruined a third of existing vineyards, and another third appeared doomed. At Boured, at Chechouan, in the Nekor Valley, and elsewhere vines had become uneconomical on account of erosion.[164]

Every year Rifians have to work more and more land in order to get

[164] Fay 1972, 2:434; Fay 1979, 87. These data refer to the western part of the Rif and Jbala. Liger 1963, 107.

the same amount of food; and every year they need more food. Hence they clear land wherever possible, shorten or eliminate fallow entirely, and practice an unsustainable agriculture that is destroying their homeland.

In the Pindus the effect of soil erosion upon agriculture has likely been as great as anywhere in the Mediterranean mountains. In an inquiry concerning soil erosion, the Greek forest service compared yields in Agrapha from 1953 with those from the same village as reported by Leake 150 years earlier. Yields had declined sharply, by 95 percent in the vineyards, and by more than half in every other product except honey. Silk production had been abandoned so that land could go into maize, but then erosion had eaten away the maize lands and reduced pasture by four-fifths.[165] The fact that honey production increased about sevenfold in this century and a half strongly suggests that the composition of the local vegetation changed to maquis and garrigue.

No doubt some of the agricultural and vegetation change in Agrapha derives from factors other than soil erosion that went unrecognized by Greek forest service analysis. The disappearance of silk probably had as much or more to do with changing market conditions as with the need for more maize production. But equally surely, the Acheloos carried away the foundation of peasant life in Agrapha, depositing it in the swampy lowlands around Missolonghi and in the Gulf of Corinth.[166] Perhaps it is no accident that Agrapha is famous for its nineteenth-century brigands and klephts. At Agrapha, as in Lucania and the Rif, deforestation and erosion reduced the productivity of agriculture. Chemical fertilizers and agricultural machinery, which maintain yields in other settings, could not temper these gradual and unspectacular losses, and peasants (probably) suffered and (certainly) emigrated as never before.

Flocks

The productivity of pastoralism declined as well, aggravating the difficulties facing mountain folk. The retreat of forests toward the mountain peaks made swineherding increasingly difficult, affecting the Pindus, Apennines, and Sierra Nevada. Erosion and watershed disruption must have adversely affected the plant biomass of meadows and maquis,

[165] Metaxas 1953, 10–13.
[166] Missolonghi, incidentally, is where in 1824 Lord Byron died of malaria.

reducing potential stocking ratios.[167] Certainly the flocks and herds declined as peasants cleared and farmed every available slope. In Basilicata sheep numbers fell from 503,000 in 1822 to 220,000 in 1861. Thereafter they recovered somewhat, according to official data, although at least one careful observer doubted the reality of this recovery. Meanwhile the goat population fell from 102,000 to 62,000, and hogs from 126,000 in 1822 to 35,000 in 1875. After 1875 apparently numbers of sheep and goats increased again, as people left degraded land, but in Potenza and Lagonegro at least good pasture had grown scarce. The numbers of swine, dependent as they were upon acorns and chestnuts, never recovered.[168] At Agrapha, where Leake found 40,000 head of sheep and goats in 1805, there were 18,000 in 1953. In eastern Andalusia, including the Alpujarra, livestock numbers also declined after the 1860s, perhaps a result of pasture shortage after the desamortizaciones.[169] In every case it seems that the expansion of agriculture, or of private property, reduced the area available to flocks and in turn lowered their numbers. When people abandoned fields in the late nineteenth century, pastoralism made a small comeback, especially in Lucania and the Alpujarra.

Goats rather than sheep led this comeback, implying a change in landscape toward sparser and hardier vegetation. In Lucania, the ratio of sheep to goats in 1822 was 5 to 1; in 1861 it shrank to 3.5 to 1, by 1881 to 3.2 to 1, and by 1896 to 2.7 to 1. The ratio changed little thereafter, except during Mussolini's rule when goats were regarded as unfascist.[170] At Agrapha, wool production declined more than twice as

167 D'Amelio 1925, on erosion and declining pasture productivity in Lagonegro. The growing infertility of pasture (and arable) land even attracted the notice of the Italian automobile club: Touring Club Italiano 1913, 51–8; see also Rossi-Doria 1963, 97–8; Sprengel 1975

168 Morano 1981, 528, for the figures; Nitti 1909–10, 320, for the doubts. D'Aragona 1964, 80–1, for the figures for 1881 and 1908. Franciosa 1951, 61–7, for the decline of pastoralism in southern Italy as a whole. He counts almost 7 million sheep in 1800, but only 200,000 in 1950. Many Lucanian comune described further deterioration of pastoralism in the 1890s, attributed in part to exhausted pastures. See for example ACSR, PC, Zanardelli e La Basilicata, busta 6, fasc. 4, Memorandum di Comune di Acerenza a Zanardelli, 1902.

169 Metaxas 1953, 10–12; Jiménez Blanco 1986, 1:255–98.

170 Morano 1981, 528; for 1881, Spera 1903, 45; for 1896, ACSR, PC, Zanardelli e La Basilicata, busta 7, Relazione Sanjust, table C. The Jacini inchiesta permits calculation of the ratio for Lagonegro (2.1 to 1) and Potenza (3.7 to 1) provinces for 1881, Italy, Giunta per la Inchiesta Agraria . . . 1883, vol. 9, fasicolo 2, 26. In 1908 the ratio had fallen to 2.6 to 1. During Mussolini's rule goat numbers in Basilicata (or at least those reported to the authorities – a new goat tax probably led to underreporting) dropped by at least a third. Hence in 1930 the sheep-to-goat ratio had climbed to 4.2 to 1:

sharply as did the combined numbers of sheep and goats (1805–1953), suggesting the replacement of sheep by goats. In the vilayet of Ioannina, which included the Pindus and Epirus, Ottoman taxation revenues imply 5.7 sheep per goat in 1860 and only 1.5 in 1907, and this at a time when the Austrian market for Pindus sheep's wool remained strong. The ratio held quite steady through the early part of the twentieth century.[171] In eastern Andalusia goats also formed a larger share of the flocks in the swift recovery that came after 1905 and peaked around 1930.[172] In the Rif, estimates suggest a considerable increase of goats in proportion to sheep between 1929 and 1952. For the administrative district of the Rif only, goats outnumbered sheep by 2.6 to 1 in 1929, by 3.2 to 1 in 1934–5, and by 4.3 to 1 in 1952.[173] Now admittedly, livestock censuses are even less accurate than those of human populations, and when animal numbers are derived from taxation records the chance of wild inaccuracy is high. Precise information (regrettably lacking) on the evolution of animal weights and milk yields would clarify matters. But these population numbers and ratios, incomplete as they are, support the notion of landscape degradation and consequent decline in the productivity of pastoralism.

Watersheds

Part of the cost that deforestation and erosion imposes upon agriculture and pastoralism comes through disruption of watersheds. The desiccation of springs and the increasingly extreme pattern of drought and spate – two common hydrological disorders in the wake of de-

Viggiani 1946, 62. By 1960 it was down to 2.9 again: Italy, Istituto Centrale di Statistica 1961, 4:208–9 and 272–3. In 1971 the ratio stood at 2.8 to 1: Italy, Ministero dell' Agricoltura 1976, 181.

171 On the vilayet of Ioannina, the Ottoman data are reported by the British consul in PRO FO 195/741, "Statement of Receipts and Expenditure of the Province of Epirus of Lower Albania" (1861); PRO FO 294/23, f. 404, "Recettes du Vilayet de Janina," 1902; PRO FO 294/32, Chevalier de Camisin to H. Lamb, 23 December 1907. About a third of the revenue of the vilayet (in 1902) came from head taxes on sheep ("agnam") and goats ("achar"), so one may presume that the tax collectors did their level best to find them all. McCarthy (1982, 290) has numbers indicating a ratio of 1.49 to 1 in 1910–11. Exarchos (1982, 205) gives figures showing that the sheep-to-goat ratio was 1.44 to 1 in 1934 and 1.56 to 1 in 1914. The Ottoman tax on sheep went up by more than double between 1861 and 1902. I do not know about the tax on goats; if it did not keep pace, this might also help account for the rise in the number of goats vs. sheep in Epirus.

172 Jiménez Blanco 1986, 1:255–98.

173 Spain, ACEM, Intervenciones Militares 1929; Roda Jiménez 1941, 66; and Spain, Dirección General de Estadística, *Annuario Estadística* 1955, 70.

forestation – affected mountain people, although it is hard to tell how much. Foresters have often asserted that springs are not what they used to be, but normally they rely on deduction rather than evidence.[174] The most damaging of these hydrological problems, floods and siltation, afflicted the people of the plains. To a large extent, mountain folk exported the hydrological consequences of their environmental problems. These disorders are both evidence of the costs of the ecological changes taking place in the mountains and testimony to the existence of these changes. Nothing testifies to upstream soil erosion quite as solidly as the deposition of silt.

Debris and silt have damaged watersheds and alluvial agriculture throughout the Mediterranean. The villagers of Alpohori, at the foot of Mount Olytzika, found their fields "swamped by boulders" and scree, carried down the mountain by unimpeded torrents after the deforestation that came with war in and after 1912.[175] By 1953, mountain torrents had lowered the yield by 15 to 20 percent of about 200,000 hectares of Greek farmland, and destroyed another 350,000. Throughout the early twentieth century, Lucanian communities reported that hydrological disorders had caused lands to become unproductive or abandoned.[176] Most of this damage took place not in the mountains but in the plains.[177] Floods and muck prevented agriculture along the banks of the lower Basento in the late nineteenth century.[178] Even today the rivers of Lucania often change their course, and more often still dump millions of tons of gravel (an event locally called *misule*) on neighboring fields, notwithstanding the billions of lire invested in regularizing watersheds in Lucania.

The watershed of the Guadalfeo, which drains the Alpujarra, has perhaps suffered the greatest damage. Between 1790 and 1847 it changed its course three times, destroying fields on each occasion. Its Alpujarran

[174] Letsas 1953, 38–40; Kontos 1906, 207; Briganti 1916, 74–5. The deduction is however rather compelling, as water infiltration depends on forest. A hectare of beech forest holds about 3,000 to 5,000 cubic meters of water, of which 2,000 eventually evaporates while the rest either seeps into the earth or runs off. Without the forest, almost all is runoff. Pavan 1970, 19.

[175] Hammond 1967, 172.

[176] ASB-P, Prefettura – Atti Amministrativi, 1913–32, busta 115, "Rimboschimiento nel bacino del torrente Fiumicello" (1929). ASB-P, Prefettura – Atti Amministrativi, 1933–1953, busta 35/156, "Processo Verbale di Deliberazione del Consiglio Comunale di Terranova di Pollino," 26 January 1948. The Fiumicello is a tributary of the Sinni, near the villages of San Paolo Albanese and Cerosimo.

[177] Metaxas 1953, 18–19. Ramage 1868, 158, for gravel and stone deposition damage to agriculture near Cassano.

[178] Lenormant 1883, 1:334–5.

tributary, the Río Chico, was 2 meters wide in 1830, but carried enough water to drive flour mills. In the next thirty years its watershed was substantially deforested. Its first major flood occurred in 1860, and by 1908 the bed had widened to 200 meters (and this stream is only 5 kilometers long!) and the current could no longer power anything. One flour mill was buried in 20 meters of silt and debris, and the river regularly flooded crops around Bayacas and chewed away portions of the road between Orgiva and Albuñol. About 1830, Widdrington noticed widespread damage done by silt and mud in the Guadalfeo Basin. The destruction was sufficient, locals told him, that many of them had turned in desperation to robbery and brigandage: ecological bandits.[179]

The waters that strewed debris and silt across alluvial plains and coastal deltas caused more direct flood damage as well. The modern burst of mountain deforestation has exacerbated the seasonality of rivers in the Mediterranean world. Without the forest sponge, the seasonal imbalance of rainfall in the Mediterranean leads to pronounced cycles of flood and drought.[180] Watercourses dry up in summer, depriving lowland peoples. Winter's heavy rains bring floods, menacing life and property.

Only in the western Taurus do the rivers run strong in the summer and rarely flood in the winter. This is a measure of the effect of intact (or nearly so) montane forest upon hydrology but also a matter of geology; the porous limestone of the Taurus serves as a sponge too, releasing winter rains and snowmelt throughout the year. The florescence of the plain of Antalya depends upon this water as much as Egypt depends upon the Nile. The Pindus too is largely porous limestone, which helps mitigate the effects of deforestation there upon the seasonal water balance. But with slopes much barer than those of the western Taurus, the Pindus rivers show greater irregularity than the Aksu or the Köprü Su.

The Guadalfeo shows the greatest irregularity, and poses the greatest dangers, of any of these mountain watersheds. While the Lucanian and Rifian rivers dry up or carry only a trickle in August and often flood in winter, they are smaller than the Guadalfeo.[181] From the late sixteenth

179 Almagro 1932, 17–19, and photograph on p. 54; Carandell 1935, 56; Widdrington 1834, 50–1.

180 The seasonal water imbalances of Metsovo, Kipi, and Pramanda are the three highest in Greece. Papoulias 1973.

181 On Rifian floods: FAO Library (Rome), FAO/MOR/514 66/514 paper by W. Heymann (1973); Seigue 1985, 331. All the large rivers in Morocco rise in cedar zones, and the destruction of cedar for commercial purposes has long affected the regularity of flow and the purity of water in Moroccan rivers. Sauvage 1941, 244–5.

7.8. This *rambla,* or dry riverbed, serves as a highway except after a rain. Then its waters tumble into the Río Guadalfeo en route to the sea. A threshing floor is in the foreground.

century, when Castilians replaced Moors in the Alpujarra, to the 1930s when watershedwide torrrent control was introduced, the Guadalfeo regularly threatened all those along its banks. Soon after it destroyed an aged Moorish bridge in the sixteenth century, it acquired the name "mouth of the dragon" at Motril. It breathed its liquid fire often, destroying the sugarcane and olive groves around its mouth. Indeed the floods of the Guadalfeo after the sixteenth century prevented the full use of the vega of Motril until quite recently. The floodwaters of the Guadalfeo dragon have often inundated upland plains before reaching Motril. In the nineteenth century floods around Orgiva were especially common. In 1860 one destroyed the irrigated fields of Soportújar, leaving the village with only 64 hectares of arable, and most of that steeply sloped. Today, scientific torrent control has tamed the dragon somewhat, but the deforestation and watershed disruption of the sixteenth and nineteenth centuries cannot easily be undone. In the Alpujarra, floods with a return period of five hundred years (meaning floods of such magnitude can be expected on average once every five centuries) occurred several times in the nineteenth century, and three times in the twentieth.[182]

But certain points along the coast have also benefited from the upland deforestatioin and erosion. Where the undersea ledge is gently sloped, silt deposition has built out the deltas of the Granadine and Almerian rivers. This coast once resembled that of the Rif, steep and rocky, with little in the way of sea plain. Unfortunately for Morocco, strong currents carry off the tremendous silt loads of the Rif oueds, permitting no Motrils (although a small deltaic plain is growing at the mouth of the Oued Laou). The vega at Motril, home to sugar cane for centuries and now to citrus, has grown steadily since the sixteenth century. But it is at Adra, at the mouth of the Andarax, where the soils of the Alpujarra have truly colonized the sea. The Adra Delta was born, rather slowly, between the sixteenth and eighteenth centuries; old maps show its progress. But with the deforestation associated with the mining boom in the nineteenth century, the Andarax began to carry huge silt loads. The Adra Delta grew so fast that in midcentury enterprising locals bought plots of the Mediterranean (which the Spanish state alertly sold), knowing they would quickly emerge above sea level, eminently suitable for cultivation. Here erosion created a new landscape, a profitable one perfect for tropical crops that brought (and bring) good prices further north. Erosion in the Alpujarra brought forth a little Egypt, from which everyone but Alpujarrans benefited.

This bonanza came to an end between 1870 and 1910. The river was rerouted in an antimalarial campaign – malaria had recently become a serious menace; a port with moles was built (1908); and then, fittingly, a disastrous flood (1910) dumped gravel everywhere across the delta, inspiring a spate of emigration from Adra and environs to Oran and South America. Since 1910 the delta at Adra has stabilized and, with summer irrigation from subterranean nappes, produces quantities of

[182] Much of the Andalusian coast of Spain has felt the breath of dragons like the Guadalfeo. Málaga in 1907, for example, witnessed a devastating flood. In 1969 and 1973 flash floods devastated the Albuñol rambla and coast. The latter carried off soils to a depth varying between 5 and 42 centimeters. Almagro 1932, 8–12; Jiménez Blanco 1986, 1:340–1; Carrera Morales 1987, 35; Thornes 1976, 46–7.

fruits, sugar, and vegetables with the soil Alpujarreños have lost. Sermet, the geographer of Adra's seaward expansion, believed the price of erosion has been well worth the benefit of newly created and highly productive deltas at Adra and elsewhere. He is certainly right in cash terms, as the deltas produce for market and the Alpujarras produced for subsistence. He might even be right in calorie terms, because the thin soils of steep slopes have been concentrated in the now-rich deltas and perhaps produce more food. But mountain folk would not likely see it this way.[183]

The Lucanian rivers, all navigable to some distance inland in ancient times, had become seasonal torrents by the late nineteenth century. Their mouths, once surrounded by fields, pastures, and cities, became wastelands. The Metaponto coastal plain was empty when Swinburne (1777) and Craven (1818) crossed it. None of the river mouths or adjacent coastal plains had settlements when Ramage passed by in the 1860s. Silt deposition had built coastal swamps, perfect conditions for the anopheles mosquito, and provoked a sharp rise in the incidence of malaria. The Jacini inquiry (1879) discovered that Basilicata had 10,000 hectares under stagnant water as a result of recent deforestation.[184]

Malaria also grew more serious in lowland Epirus in the nineteenth century, partly as a result of the introduction of rice cultivation, but also because of deforestation, siltation, and the expansion of marshlands. The Arachthos and Louros rivers, rising in Zagori and Souli respectively, were navigable for small boats in the fifteenth century. The Louros carried traffic on its lower 30 kilometers as late as the 1830s, when its banks were still forested, and according to one report until shortly before 1912. But more recently these rivers too have become highly seasonal, and their lowland channels shallow, braided, and full of gravel. In the early twentieth century the Arachthos flooded so frequently that the villages in its valley arranged a system of watchmen, posted day and night after Saint Demetrius's day (mid-October), who rang churchbells when it was time to flee to higher ground. Arta, on the Arachthos, is about 4 or 5 kilometers further from the sea than it was in Ali Pasha's time. The entire Arta Plain has been laid down since classical times.

[183] Sermet 1934, 7–19; Sermet 1953, 154–5; Sermet 1973, 556–61. The rich coastal plains usually require constant refills of soil from the mountains. Torrent control, or the absence of upland soils, can thus imperil these plains. The one at Metaponto, for instance, is now shrinking. Others will too, unless protected – at great cost – so the wealth of the coasts may prove a passing phase.

[184] Carlyle [1962] 1985, 90; Ramage 1868, 160–2; Nitti 1909–10, 9; Azimonti 1909–10, 1:60; Sciacca 1977, 84; Ahlmann 1926, 106–10; Filangieri 1980, 202–3. Jacini inquiry noted in Clark 1984, 15.

Glyki, which Holland (1819) referred to as a port on the river Acheron, is now 12 or 15 kilometers from the coast and could scarcely harbor a rowboat. In recent years, with malaria control, torrent control, swamp drainage, and other mearures, these lowland landscapes have become habitable and in cases productive. The Plain of Arta, with irrigation, is highly fertile. But for a century or more the ecological disruption of the highlands created useless, indeed hostile landscapes on the coasts, and invalidated the best transport system available in what was then a land without wheels.[185]

On Anatolia's south coast sedimentation also affected river navigability and built malarial plains. Here deforestation and upland erosion had been modest until the late nineteenth century. But with the expansion of cultivation and the invigoration of the timber trade alluvial erosion and deposition troubled river mouths and adjacent lowlands. No doubt the settlement and cultivation of Roman times brought silt down to the coast. But the nineteenth century, especially around Adana at the foot of the eastern Taurus, appears to have witnessed the most severe hydrological problems.

Smaller ones developed at the foot of the western Taurus. Even the rather small Aksu and Köprü Su could accommodate vessels in ancient times, when they were known as the Cestrus and Eurymedon. According to Plutarch, 550 galleys fought in the Eurymedon when Kimon destroyed a Persian fleet, ad Livy wrote that 36 ships of the Rhodian navy put in here. In 1812 Captain Beaufort found the river mouth impassable. Further to the east, the Ceyhan River was navigable as late as the twelfth century, but by 1812 had silted up and, together with the Seyhan, built out the plain of Adana (the Çukurova) several kilometers into the sea. Since then the plain has continued to grow, probably at a faster pace. Between 1935 and 1952 large amounts of new land emerged from the sea on account of deposition by these rivers.[186] Not all the alluvium, of course, derives from mountain erosion. But the nineteenth-century forest clearances of the Cilician Taurus can only have hastened the silting up of the waterways and the expansion of the plain. Elsewhere this coast is too steep to permit silt plains, except at Silifke, where the Göksü has built a small delta, and at Antalya, where the plain is not alluvial.

[185] On malaria: Bruce-Chwatt and Zulueta 1980, 33–4; on navigability see Nicol 1984, 229; Greece, Iperesia Syntonismou [C. A. Doxiadis] 1950, 24; and AMAE-P, Mémoires et Documents, Turquie, 22:398, Boislecomte to Rigny, 4 April 1834; on Glyki, Holland 1819, 2:234; on Arta's position, Holland 1819, 1:111; on the Arta Plain, Bintliff 1977, 50. Thessaloniki acquired a malarial swamp hinterland through silt deposition over the centuries: Kotoulas 1984.

[186] Plutarch cited in Beaufort 1818, 143–4, 297–9; Russell 1954, 387–9.

Malaria long reigned all along this coast, but it probably grew more prevalent in the nineteenth century. In ancient and Seljuk times flourishing cities imply that malaria had at most a slight presence; but by the late nineteenth century the Antalya Plain had become "fearfully unhealthy" on account of "deadly malaria," and the cities shrunken remnants of their former glory. This condition changed only with the advent of the Turkish republic which curtailed the cultivation of rice in the coastal plain, specifically for health reasons.[187] The evidence suggests that malaria minimized settlement on the Antalya Plain from the seventeenth to the twentieth century. Disruption of lowland watersheds from upstream erosion could have played only a small role at Antalya, however. Travelers referred to the plain as empty in summer, despite its fertility, but did not mention swamps. Perhaps the history of rice, rather than the history of erosion, holds the clues, at least for the coastal district east of the city of Antalya. At Silifke, Adana, and other smaller coastal plains of Anatolia hydrological disorders promoted malaria deadly enough to keep them almost empty from early Ottoman times until the twentieth century.[188] In recent decades these plains, like those of the Granada and Almería coast, have become the seat of highly profitable subtropical agriculture. Cotton, bananas, and citrus all flourish, now that malaria has been banished to insignificance and modern agronomy introduced. Population has surpassed all previous levels. But it is well to remember, amid this boom, that the ecological disruptions of upland watersheds helped to render these coastal plains useless for generations.

Elsewhere around the inland sea, deforestation and erosion have created new swamps as well as deltas. The happy consequences at Adra – and even there malarial swamps formed in the early nineteenth century[189] – are unusual. More common were the restriction of navigability and the entrenchment of malaria, as at Adana, Arta, and Metaponto. This of course had proceeded for centuries before the modern landscape changes. The cases of Ephesus and the Büyük Menderes, and of Pella and the Aliakmon, both mentioned earlier, have many analogues around the Mediterranean. The Ebro, navigable to Logrono in ancient times, by 1700 admitted ships only as far as Tortosa. Seagoing vessels plied the Guadalquivir as far upstream as Córdoba in 1350, which helps explain the city's prominence in Moorish times.[190] After 1717 silt so clogged the river that the Spanish had to replace Seville by Cadiz as the

[187] E. Davis 1874, 202. Planhol 1958, 105, 107. A local told Davis (p. 204) the "mosquitoes and heat were terrible" in the summer in Antalya. Davis mentions malaria in the next sentence, but did not know of the connection between malaria and mosquitoes. No one did at that time.

[188] Gerber 1987, 86–7. [189] Ruz Márquez 1981, 64. [190] Regnault 1904, 39.

port of trade with the Indies. In the nineteenth century these old problems deepened wherever rivers drew their water from mountains.

Throughout the Mediterranean, even today old hydrological disruptions in mountain watersheds pose very serious problems. In Epirus silt from the Pindus is rapidly filling up the lake behind the hydroelectric dam on the Acheloos, reducing its generating capacity and drastically shortening the useful life of the dam. In the Rif every barrage built in the last twenty years serves as a sediment trap. Most of the small artificial lakes now have growing islands in them, and will be marshes in a decade unless expensive interventions take place. In Lucania none of the rivers have dams, despite the need for irrigation water and electricity, because engineers recognize that silt loads would shortly invalidate their work.

Interventions

Addressing these hydrological problems at their roots requires forest preservation and reforestation, as well as careful agronomy. The forest services in Mediterranean countries have generally not been up to the task, although fully aware of the consequences, as had been in their day Democritus, Theophrastus, Seneca, and others. In Italy and Spain scientists analyzed these problems as much as ninety years ago – indeed in the Kingdom of Naples two hundred years ago.[191] Subsequent generations of foresters, soil scientists, and hydrologists have repeated the same refrains. The difficulty is that healing these mountain landscapes would be enormously expensive and would require getting thousands of people to change their ways. Rare attempts often provoked resistance.[192] Furthermore officials often had priorities reflecting their great social distance from the peasantry: in the late 1890s the Italian Ministry of Agriculture conducted field trials to try to improve breeds of hunting dogs.[193] In the 1920s, the reforestation budget at Calvello was spent

[191] FAO Library (Rome), microfiche 85:53483 (Monaco 1985), 4. In 1812 one Giuseppe Lambert, presumably Joseph Lambert of Murat's administration, gave an address on the damages of deforestation: ASAEAG, busta 63, ms. 439, "Sul disboscamento eccessivo e dannoso dei monti . . . ," 29 January 1812. A good Catholic, Lambert regretted that the true religion did not include safeguards for trees, as did that of the ancient Gauls.

[192] In 1875 Potentini routed the local Guardie Forestali who had objected to the illegal presence of five hundred goats in state forest. ACSR, MAIC, DGA, VI, 150/2, Potenza, "Ribellione commessa dai cittadini," 24 May 1875.

[193] ACSR, MAIC, DGA, VI, busta 255, f. 7, "Primo Field Trial Italiano (miglioramento delle razze canine da caccia)."

honoring each of the fallen from World War I with a personal tree, while serious hydrological problems went unaddressed.[194] Now that ministries are a trifle more interested, and the mountain economy is dwindling on the European shores of the Mediterranean, reforestation, torrent control, and soil conservation are easier than ever before. Italy and Spain have spent large sums and made some improvements, but at recent rates of reforestation it will take Calabria three hundred years to achieve soil stability.[195] Greece has spent much less and has less to show for it. The costs are such that forest services must in effect practice triage, treating only those watersheds subject to the cheapest cures, while letting the most serious and intractable cases deteriorate. In any event, there is nothing much one can do about massive erosion once the soil is deposited on the sea floor.

In the Rif, where large populations still live from mountain agriculture, addressing the ecological problems is not only prohibitively expensive but politically impossible. Foresters know that barrages are impractical and reforestation necessary, but the government, not popular to begin with in the Rif, cannot afford to alienate the peasants whose lands would be affected: better to permit the destruction of niches for future generations and hope in the meantime that something will happen so that they will not be needed. This is in effect the policy of the Moroccan state, as it was in effect the policy of the Greek, Italian, and Spanish states until recently. Whether emigration or something else will intervene so that the mountain niche can safely be sacrificed in Morocco, as happened (painfully) in the lands of the Christian Mediterranean, remains to be seen. In the western Taurus, happily, conditions do not yet pose this dilemma.[196]

194 ASB-P, Prefettura – Atti di Gabinetto, 1913–32, busta 115, Calvello podestá to prefetto, 12 February 1930.

195 Pavan 1970, 28.

196 On the history of forest policy, soil conservation, and reforestation in Greece: Kentro Pragrammatismou Kai Oikonomikon Erevnon 1976, 236–9; Prentice 1956; Kontos 1929 and 1932, 83–97; Andreades 1905; Oikonomopoulos 1966; Letsas 1953, 66–103, on the Salonika area; Puglisi 1965; and Papastavrou and Makris, 1985–6, 1:27–51 and 2:16–42.

On Turkey: Özdönmez 1982, 213–15; Saatçioğlu 1952; Bricogne 1876.

On southern Italy: Trifone 1957; Pavan 1970; Radina 1976; Basilicata (Regione), Ufficio Geologico Regionale 1986; Sacchi 1950; Puglisi 1967; and the special issue of *Monti e boschi* 3 (1952): 483–571. Information on local efforts in various villages of Lucania is in ACSR, MAIC, DGA, VI, buste 7, and 272, (Pollino); ACSR, PC, Zanardelli e La Basilicata, busta 7, cartella 6, has the text of the special reforestation law of 1904; ASB-P, Prefettura – Atti di Gabinetto, 1861–1934, cartella 54bis/39; ASB-P, Prefettura – Atti Amministrativi, 1913–32, cartella 115.

7.9. Planted pines above this wheat field between Boured and Bab Sidi Kebat now slow soil loss. Severe gully erosion has inspired one of the few reforestation projects in the Rif.

7.10. Another reforestation project in the Rif, this one southwest of Bab Sidi Kebat. In this case there is no soil left to protect: the intervention came too late.

Conclusion

Although incomplete and imperfect, the evidence for massive deforestation and soil erosion in the mountains of the Mediterranean between 1800 and 1950 is compelling. The timing and pace differed from place to place. The costs of landscape change varied according to local geology and economies. In places it worked to the advantage of lowland and coastal peoples, although more often it did not. Nowhere did it work to the advantage of mountain people. Soil, fuel, timber, even pasture grew more scarce or distant, and hard lives became harder. Pavari's 1925 description of Lucania can serve for many districts:

> When Basilicata could truly be called Lucania, that is when all the mountains were covered with dense forests of oaks below and beech above, these disastrous things did not happen. But the deforestation, unchecked in the final decades of the last century, above all as a consequence of the alienation of desmenial and church lands, increasingly tore and destroyed the forest clothing, and put the denuded land at the mercy of the agents of disintegration. This ruin was intensified by rapine agriculture and the complete lack of any agrarian water control. Populous regions flourishing with pastoral and forest industries, like the basin of the Camastra for example, thus became depopulated because of the dreadful death of the mountain, due to the deforestation, the plowing, and then the abandonment of cultivation; the people emigrated en masse and went to search in America for the bread their own land denied them.[197]

Deforestation and erosion helped close the mountain niche. It also narrowed the niche available to all rural people for a while. They had

On Spain: FAO Library (Rome), microfiche 76:42052, paper by López Cadenas de Llano, "Informe Nacional España," 1976, 7–12; Bauer Manderscheid 1980; Jiménez Blanco 1986, 1:347–78; O. Elorrieta y Artaza 1948, 35–130. The situation in the province of Granada is touched upon in AMAPA, legajos 353/22 and 9/11.

On the Rif: Plit 1983; Fay 1972, 2:512–19; Metro 1958, 26–9; Heusch 1981; FAO Library (Rome), anonymous working paper FAO/DP/MOR/73/016 (1976).

197 Pavari 1925, 3. "Quando la Basilicata poteva veramente chiamarsi Lucania, quando cioè tutta la montagna era ricoperta di dense selve di querce e cerri in basso, di faggio in alto, questi disastroso fenomeni non so verificanavo. Ma il disboscamento, sfrenatosi negli ultimi decenni del secolo scorso, sopratutto in conseguenza della quotazione dei beni demaniali ed ecclesiastici, lacerò e distrusse con progressione crescenta la veste selvosa e, denudato il terreno, lo dette in preda agli agenti disgregatori; questa rovina fu intensificata dalle colture di rapina, dalla mancanza assoluta di ogni sistemazione idraulica dei terreni agrari.

Regioni giá poplose a fiorenti per le industrie armentizie e silvane, come ad esempio il bacino della Camastra, divenesso poi spopolate a cause della spaventosa morte della montagna dovuta al disboscamento, ai dissodamenti ed al successivo abbandono delle colture; le gente emigrò in massa ed andó a cercare in America il pane che la sua terra le rifiutava."

long found life least difficult between the tree line and the malarial stratum. The tree line, at about 1,800 or 2,000 meters, forms the upward limit for human settlement in the Mediterranean. Above it, only the temporary summer dwellings of herders exist, and agriculture is all but impossible. Sea level formed the lower limit open to human habitation – until malaria. Malaria decimated the coasts and lowlands more thoroughly than the most energetic Barbary or Genoese sea raiders. With the firm establishment of malaria in the lowlands (which happened at varying times but in many places in the eighteenth and nineteenth century), the lower limit on permanent settlement moved up. The rise of malaria, especially in Lucania and Epirus, but also in Anatolia, made lowlands into deserts. Only the foolhardy and the desperate ventured down except for the winter months. As malaria rendered these districts uninhabitable, more people crowded into the hills and mountains, concentrating population at elevations between 500 and 1,800 meters (in most cases between 500 and 1,200). Disrupting the upland landscape at faster rates because of their growing numbers, mountain folk accelerated the rise of malaria, turning places like the Ionian coast of Lucania into a killing ground by 1870. Survivors usually carried the infection for life, regularly falling ill with recurrent bouts: Lucanian peasants (before quinine) used to say that earthquakes pass, plagues pass, but malaria never passes.

As malaria ravaged the lowlands, deforestation in the high mountains, often the work of shepherds, made the highest villages increasingly uninviting. At high altitudes large amounts of fuelwood are indispensable, and traveling long distances to get it usually involves especially difficult terrain. Mountain folk thus found themselves squeezed between a rock and hard place that were slowly moving closer and closer together: malaria spread up the valleys, and the tree line descended where shepherds used or burned the forest in their high summer camps. Furthermore, the hard place kept getting harder, as malaria intensified from a sporadic risk into a true scourge. It is no wonder that mountain people developed a high threshold for risk, and often found attractive the gambles represented by migrant labor, brigandage, or emigration.

8

CONCLUSION

> One generation passeth away and another cometh, but the Earth abideth forever.
>
> Ecclesiastes

> If you plan for one year, plant rice.
> If you plan for ten, plant trees.
> If you plan for a hundred years, educate mankind.
> Kuan-Tzu [Chuang-dze], fourth century B.C.

When I began this book (some years ago now), I intended to argue and document the proposition that the physical landscapes of the Mediterranean mountains are often recent creations. I hope I have persuaded the reader of that. But as the book progressed it acquired a broader purpose: to explain how the physical and social landscapes of this mountain world have come to be, and how they have affected one another. In other words, while not setting out to do so, and shortsighted enough not to see from the start where things would lead, I have wrestled with one of the oldest questions in geography and history: the relationship between humankind and nature. In the process of being pinned to the mat, I have learned some interesting things about my chosen mountains and about other mountains in the Mediterranean world. All this has provoked some musings about other parts of the world (I will keep them brief).

The Mediterranean landscape bears the scars of ancient wounds. Many of its lowlands have been intensively farmed for eight thousand years, have been subject to all manner of disruption, yet still give good yields. The lowlands have held up well in comparison, say, to the American Middle West where a little over a century of farming has cost an alarming proportion of soil.

But in Mediterranean mountain areas, the severity and number of recent wounds is impressive and disquieting. The imprint of human activity remained bounded until the ecological pressures generated by population growth and by market integration and concentrated demand increased. Thereafter, the degradation of vegetation and soilscapes declined sharply, and continues to do so today wherever these pressures remain. Equally impressive is the degree to which mountain life rested on much the same foundations throughout the Mediterranean world. To be sure, the shepherd massifs differed from the peasant massifs. But everywhere mountain folk combined cultivation, herding, gathering, and craft work in different proportions but similar equations to create lives sharply different from those led by lowlanders. The role of environment in shaping history and culture must surely be much stronger in the mountains than in the plains.[1] Of course the same claim can be made for the desert and for other marginal zones. But the reverse cannot: history and culture have played a much larger role in shaping mountain environments than they have desert ones.

Impressive too is the irregularity of the rhythm of ecological history in the mountains. Pulses of rapid transfiguring change seem to punctuate periods of relative calm. The arrival of maize in many cases inaugurated fundamental changes. So did the replacement of Moors by Christians in the Alpujarra, or the settling of nomads in the Taurus. These pulses of change seem always to be irreversible: even should the maize, the Christians, or the settlements disappear, old equilibriums will never reassert themselves. This suggests ecological lability – the pronounced inclination of an ecosystem to change irreversibly. Put another way, it implies the potential instability of stable systems.[2] Mountain landscapes do not show fluctuations within a stable equilibrium; they do not show cyclical stability, resilience, or elasticity, except perhaps over geologic time. Whether wilderness or anthropic creations, they seem to have only a fragile stability, easily upset by unintentional human action.

In this context, it seems clear that peasant settlement in the Mediterranean mountains has been (or someday will have been in the Rif) a brief phase. Geographically, the ideal zone for settlement lay above the reach of malaria and below the timberline; chronologically, it lay between the arrival of maize and the disappearance of forest and soils. The florescence of mountain communities existed only within these spatial and temporal confines. Various adaptations – labor migration, silk, kif – could extend the life of mountain communities, but all struggled beneath an ecological sword of Damocles. Villagers might break

[1] Rhoades (1979) argues this for Switzerland and Nepal.

[2] Gigon (1983) explains this concept in theoretical terms.

the thread themselves, but even if they took care not to, outside forces might inadvertently do it as well.

Again, the same case might be made more widely. All adaptations to environment are temporary, because they either undermine themselves or are at risk to changes to the environment coming from the outside. Yet crucial distinctions exist among various ways of life and various natural environments. As a rule, agricultural and pastoral societies corrode their resource base more quickly than do hunting and gathering societies. Taking the very long view, here is geographer Karl Butzer:

> Deforestation, grazing, and cultivation create fragile soilscapes that are susceptible to rainsplash, accelerated runoff, reduced infiltration and aeration, periodic dehydration, erosion and declining fertility. High human population densities are possible, far exceeding those in foraging societies in which people are but a small segment of the consumer community. But agricultural and livestock productivity will fluctuate far more dramatically than that for game and vegetable foods in an intact ecosystem and will eventually decline without massive inputs of technology, labor, and capital. Agricultural ecosystems consequently represent short-term maximization strategies, unless carefully tailored through generations of trial and error to approximate a homeostatic equilibrium. The geo-archaeological record demonstrates that homeostatic equilibrium has rarely been maintained over the long run.[3]

Taking this very long view, mountain agriculture represents a shorter-term maximization strategy than most any other agriculture, because (other things being equal) it invites much faster erosion. Furthermore, because mountain life depended on so many auxiliary activities, some of which could be undercut by politics and economics in the plains and overseas, any sort of equilibrium that might preserve soils and vegetation was quite impossible to maintain for any length of time.

In the Mediterranean mountains various mixtures of internal and external forces led to environmental degradation. In the case of the Alpujarra, more clearly than any other, repeated lightning strikes came in the form of commercial connections to the outside world. In the Rif, population growth and subsistence agriculture did most of the damage, and the kif market is now merely completing the job. Whatever the combination of internal and external forces, nowhere did peasants and herders practice economies that were indefinitely sustainable except at extremely low population densities. Nowhere did peasants and herders devise systems of human ecology in durable harmony or balance with

[3] Butzer 1982, 155.

the natural world. Only low population preserved the forests of the Taurus up through the past century. The ecological angels one sometimes reads about in accounts of pre-Columbian America or other precapitalist societies did not inhabit the Mediterranean mountains.[4]

One interesting conclusion that I have not been able to substantiate, but which I still believe, is that the political upheavals of the nineteenth and twentieth centuries in southern Europe, Anatolia, and the Maghreb owe part of their vigor to the corrosion of mountain environments. Perhaps careful inquiry into the geographic origins of those who enlisted in radical movements would provide an answer. There is none to be found in voting patterns. Peasants did not vote until quite recently in the Mediterranean world, and still do not in several countries. And where they do, politics revolves around patron–client relations much more than political belief, especially in rural areas. One votes for the man who delivers a government pension or a loan from the agricultural bank. Equally unrevealing is the geography of political violence: sizable popular insurgencies existed in all these mountains except the Taurus, but that may reflect nothing more than sensible tactics on the part of insurgents. Rebels who did not take advantage of mountain terrain did not last long. Although I lack the evidence to support this idea, it is at least consistent with what I have learned about these mountains, and a logical inference.

In hopes of providing some depth of understanding, I have with rare exception concentrated on only five of the many mountain chains in the Mediterranean world. Might environmental history have been different elsewhere? In detail certainly, but generally not in broad outline. A quick circuit of the Mediterranean will suffice.

In the Syrian anti-Lebanon pollen diagrams suggest a familiar vegetation and settlement history. Oaks and cedars dominated until about 220 years ago, when maize arrived and trees began to disappear. Today no cedar is left.[5] This history suggests a transition from shepherd to peasant massif in the eighteenth century. More broadly in Syria and Iraq, although lowland forest vanished in the Bronze Age, montane forests lasted until the nineteenth century A.D.[6] In the Pontic ranges of northern Anatolia, a seemingly durable system of human ecology broke down in the twentieth century, under the twin pressures of population growth and economic collapse, producing deforestation, erosion, declining yields, and emigration. Villages such as Anna and Gölve in

[4] Recent examples include Sale 1990; Merchant 1989.

[5] Bottema 1977. [6] Rowton 1967.

Giresun Province were surrounded by forests as late as 1900, but suffered a gradual "ecological tragedy," which sent hundreds of refugees to the shantytowns of Istanbul.[7]

Even the Alps, settled for many centuries, suffered severe disruptions in modern times. The Po Delta has developed mostly since 1750, a result primarily of accelerated erosion in the southern Alps.[8] In Corsica and Provence mountain deforestation accelerated before the nineteenth century. The Genoese played a large role in destroying Corsican forests, especially in the eighteenth century. In Provence and the Alpes Maritimes, substantial deforestation took place with the medieval expansion of population, although much more came in the eighteenth century and after.[9] In both these cases, involvement in exchange economies hastened the demise of forests.

In the Spanish central Pyrenees, a bit more remote from economic centers, population changes played the major role in vegetation and soil history. Medieval fluctuations – an expansion in the thirteenth century and a sharp decline in the fourteenth – destabilized landscapes, but by far the greatest changes came in the nineteenth century. Here the establishment of the potato permitted denser population than ever before, with unfortunate results for forests and pastures.[10]

In western Algeria (Oranie), the nineteenth and twentieth centuries witnessed large-scale forest clearance and massive soil erosion. French colonial settlement after 1850 drove Arabs and Berbers into the higher reaches of the Tell, where their cultivation quickly led to ecological disasters. Shortly after clearing fields they found yields in decline and abandoned their land to try again elsewhere.[11] Here, as in southern Italy and all of Spain, political forces promoted sudden shifts in land use patterns, with devastating consequences. Algeria, far more than Morocco or Tunisia, felt the environmental consequences of colonial power. But even Tunisia's mountains, really the easternmost foothills of the Atlas, have not escaped the modern trend of deforestation and soil erosion. Forest cover shrank by a third between 1910 and 1970, and what remained was in poor shape.[12]

[7] Karpat 1976, 53–5, 73–4, 256, n. 43. See also Tunçdilek 1951, 102–3.

[8] Vita-Finzi 1975, 140. [9] Clavé 1864, 362–77; Sclafert 1959; Douguedroit 1976.

[10] Chocarro, Fanlo, Fillat, and Marin 1990. Sánchez Sánchez and Rodríguez Rodríguez (1989) discuss mountain life generally in Spain. The decline of the Alpujarra has many analogues.

[11] Benchetrit 1954, 164–5; Sari 1977.

[12] Bensalem 1977. In the 1970s, fuelwood removals, mostly for charcoal, amounted to twice the annual growth. FAO Library (Rome), anonymous working paper, SCM 85/3 (1985).

Throughout the Mediterranean mountains, similar forces have produced similar results, a reduction in the capacity of landscapes to support human life. And of course this tale of woe is still in progress. The sad fate of abandoned terraces in the Alpujarra is repeated in the High Atlas, in North Yemen (not, strictly speaking, part of the Mediterranean world), and in parts of the Taurus. The loss of soil, and of carefully carved-out, highly productive fields, will be permanent.[13]

The sources of environmental degradation and instability that produced the history of the Mediterranean mountains are abroad in the world today. Parallels are especially apt in other mountain areas, where deforestation and erosion exact their greatest costs. Peasant mountains, like the south flank of the Himalayas and the northern Andes, show the effects of population growth on forests. In Peru and Bolivia, soil erosion and depletion have brought on declining yields, forcing peasants to overgraze in order to get sufficient animal fertilizer. The southern and central Andes attest to the impact of concentrated demand, in this case for minerals.[14]

In late 1989 a mudslide in Thailand destroyed several villages and killed scores of people. Years of logging had removed tree cover, opening the soil to erosion. Soil trickled away and then, with a big rain, came rumbling down, sweeping boulders and logs before it, like a runaway glacier. In November of 1885 the same thing happened in Campomaggiore Vecchio – it was then just Campomaggiore – in Lucania. The frequency of *frane* increased throughout southern Italy in the late nineteenth century on account of the massive deforestation. For the same reason, the frequency of landslides and floods is now increasing in southeast Asia, on the south face of the Himalayas, in Central America and the Andes. Two-thirds of Bangladesh spent several days underwater in 1989, in the worst flood in recorded memory. The denudation of the Brahmaputra watershed is almost surely responsible. In this and in many other ways, the sad environmental history of the Mediterranean mountains is now repeating itself a century later in the mountain systems of the tropics. History repeats itself, wrote Marx, the first time as tragedy, the second time as farce. In this case, the second time is tragedy as well, and on a larger scale.

Almost every mountain landscape now feels pressure either from settlement and subsistence economies or from the reach of concen-

[13] Vogel 1988; Tunçdilek 1951, 102. Tunçdilek assumes, wrongly I believe, that significant abandonment of terraces had been characteristic of the Taurus for centuries.

[14] Bahre 1979; Gulliet 1984.

8.1. These ruins are the village of Campomaggiore Vecchio, destroyed by a mudslide in 1885. In Lucania, nineteenth-century deforestation increased the frequency of *frane*.

trated demand. Anthropic erosion in the Drakensberg (South Africa) results chiefly from overcrowding that derives from state policy. In the Ethiopian highlands, the peasant economy has produced catastrophic erosion in recent decades. In El Salvador deforestation and erosion come from both population growth and from the effects of distant demand, which have led to the creation of large plantations devoted to export crops, driving subsistence producers onto steeper land. Mountain landscapes have low thresholds for both carrying capacity and fluctuation in land use, and around the world, especially the tropical world, these thresholds are routinely exceeded.[15]

What social and political consequences will flow from mountain overshoot around the world is a hard question. No doubt results will vary greatly from place to place, as they have in the Mediterranean. In some cases emigration may leave behind empty villages but preserve social peace. But there are no Americas for the displaced mountain folk of

[15] Eckholm (1975) emphasizes the similarities in environmental problems facing mountain areas.

Ethiopia, Peru, and Nepal. Nor are there any empty coastal lowlands. This leaves them with cities and, in some cases, tropical forests, and considerable problems await immigrants in both settings, some of which their arrival will exacerbate. Perhaps some mountain people can hit upon adaptations that will permit them to survive in greater number without destroying their landscapes. They need something they can sell that is valuable, lightweight for transport, and hard to produce except in special conditions, like Alpujarran silk in Moorish days. If they do not find it, they will have to find their niches elsewhere. Should enough fail, they will surely project their misfortunes and discontents onto national and international politics.[16]

Mountains, like desert edges, fray from ecological strain much more quickly than do other environments. They also repair themselves much more slowly, if at all, than gentler lands. Thus the disturbing trends in mountain environmental history are not likely a portent of things to come for the globe as a whole. More likely, ecological strain will continue to be distributed very unevenly, ravaging some zones and sparing others. This was certainly the pattern of the Mediterranean world. Even were the sources of strain evenly spread, the effects would be felt more severely in the fragile environments. And of course the sources of strain, rapid population growth and the laser of concentrated demand, are not distributed evenly. If history is any guide, the fortunes and misfortunes brought on by environmental change will be visited upon humankind quite unequally, adding to the gulfs dividing the economically secure from the insecure. History may well prove to be a poor guide, however, since the pace, scale, and scope of environmental change in the contemporary world is unprecedented.

[16] One scenario likely to be played out again and again is the emergence of ecological refugees turning to lawlessness, of which I have mentioned examples in the Mediterranean. In China, Van Slyke (1988, 77–80) notes that in Huai, east of the Grand Canal, hydrological disruptions often produced ecological bandits. Murphey (1983, 114) reports that Chinese authorities often burned off hillsides to destroy bandit refuges. Such destruction produced further ecological refugees and bandits. Mountain people in China were generally outside the networks of deference and clientilism that underlay Chinese politics, which barred their participation (or co-optation). Their sociopolitical situation, chiefly a function of their geographical situation, inclined them therefore toward illegality, revolution, and violence. Fairbank 1987, 42–3.

BIBLIOGRAPHY

Archival Sources and Unpublished Manuscripts

I. Greece

A. Society for Epirot Studies [Etaireia ton Epeirotikon Meleton], Ioannina.
   1. Lamour, Philippe. "Diefthetisis tis periochis Epeirou" (1957), typescript.

B. Epirus Forestry Service Office [Epitheorisi Dason Epeirou], Ioannina.
   1. "Periferiako programma dasikis anaptixeos, 1980–1985" (1980), typescript.

C. British School of Archaeology, Athens.
   1. Notebooks of S. S. Clarke, 1923–4, 3 vols.
   Summer 1923, Cyclades, Epirus, S. Greece, Thessaly.
   Winter 1923–4, Epirus, S. Albania.
   Spring 1923, Central Greece, Epirus, S. Albania.

II. Italy

A. Archivio Storico della Accademia Economico-Agrario dei Georgofili, Florence: buste 57, 63, 69–70, 74, 92, 94, 98, 197, 181–2.

B. Archivio Centrale di Stato, Rome.
   1. Archivi Parlamentari:
      a. Giunta Parlamentare d'inchiesta sulle condizioni dei contadini nelle Province meridionali e nella Sicilia (1904–10): buste 1–5.
      b. Giunta Parlamentare per l'inchiesta agraria sulle condizioni della classe agricola 1887 (Jacini): buste 3–4, on Basilicata and Calabria.
   2. Presidenza del Consiglio, fondo Zanardelli e La Basilicata: buste 1, 2, 4, 6, 7.
   3. Ministero dell'Agricoltura, Industria e Comercio.
   Direzione Generale dell'Agricoltura:
      Verso I, buste 150, 219, 257–62, 287–8, 319, 350–1.
      Verso IV, buste 29, 34, 97.
      Verso VI, buste 7, 39, 120, 129, 153, 163, 207, 255, 270, 272, 359.

C. Archivio di Stato di Basilicata, Potenza.
   1. Prefettura – Atti Amministrativi 1933–53: cartelle 29, 35, 63, 356, 1180, 1181.

2. Prefettura – Atti Amministrativi, 1913–32: cartelle 115, 116, 958–60, 1665.
3. Prefettura – Atti di Gabinetto, 1861–1934: cartelle 11, 25–6, 54bis, 289, 401, 438–9; 2° versamento, 1° elenco (1926–40), buste 44, 46, 49.
4. Sottoprefettura di Lagonegro: fasc. 49, 52.
5. Intendenza di Basilicata, 1806–60: busta 1338.
6. Ispettorato Provinciale dell'Agricoltura, fasc. 4.

D. Biblioteca Provinciale di Potenza, Sezione Lucana:
1. x. 18 anonymous 1833 pamphlet.
2. x. 29 "Quadro della popolazione . . . 1853"

E. Biblioteca di Studi Meridionali "Giustino Fortunato," Rome.
Isnardi, Giuseppe, "Cenni di geografia e di storia economica della Calabria" (1955), typescript.

F. Biblioteca Nazionale (Naples), Ms. XIV–II–39, Relazione Gaudioso (1736).

III. Food and Agriculture Organization Library, Rome.

A. Anonymous Working Papers.
1. AG: DP/MOR/71/536. Rapport technique no. 1, "Lutte contre l'érosion et conservation des sols, Maroc. Analyse des problèmes à l'échelle nationale" (1975).
2. DP/MOR/73/016. Rapport technique no. 1, "Aménagement et amélioration des parcours forestiers. Maroc. Etude des données de base" (1976).
3. SCM/65. Joint Sub-Commission on Mediterranean Forestry Problems, 6th Session. "General Land-Use Situation in the Mediterranean Region and Its Trends – Role of the Forest" (1958).
4. SCM/85/3. AFC/EFC/NEFC Committee on Mediterranean Forestry Questions, 12th Session. "Forest Management and Utilization" (1985).
5. SCM 85/4. AFC/EFC/NEFC Committee on Mediterranean Forestry Questions, 12th Session. "Prevention and Control of Forest Fires" (1985).
6. SF:18/MOR. "Le développement économique rural du Rif occidental," 2 vols. (1965).
7. SF: 54/GRE/45 Pre-Investment Survey of Selected Forest Areas. Greece. 5 vols. (1968).

B. Authored Papers.
1. Alheritière, D., and R. G. Downes, "Aspects institutionnels de la conservations des sols et de la lutte contre l'érosion dans le royaume du Maroc." FAO/WP/4457 (1981).
2. Cano Pedrajas, Laureano. "Regeneration of the Northern Slope of the Sierra Nevada (Granada)," FAO/CE/OIT WP 1/R. 21 (1980).
3. Duchaufour, Ph. "Rapport de mission en Grèce," SF/FAO/GRE 20/230 (1969).

4. Heymann, W. "Projet de rapport final de mission de l'expert W. Heymann FAO/DERRO à la Province de Nador," FAO/MOR/514 66/514 (1973).
5. López Cadenas de Llano, Filiberto. "Informe Nacional España," FAO library microfiche 76:42052 (1976).
6. Monaco, Giuseppe. "Italie. Régione Calabria. 14e session. Groupe de travail pour l'aménagement des bassins de montagne. Rapport. Aspects économiques, sociaux et institutionnels," FAO library microfiche 85:53483 (1985).
7. Quinlivan, B. J. "Forestry Development and Reforestation in Greece. Pasture Development Report No. 1," FO/GRE/78/003 (1982).

IV. Spain

A. Biblioteca Nacional, Madrid.
   1. Sección de Manuscritos: ms. 2,785, 12,934.
   2. Sección de Africa.
      a. "Censo aproximado de Gomara" (1929), typescript.
      b. Intervención Territorial del Rif. "Resumen Semestral" (1952), typescript.
      c. Mapas, B. 8-3. "Croquis de las manchas de arbolado (Rif)" (n.d.).
   3. Sección de Africa, Colección García Figueras.
      a. Jagerschmidt, Charles. "Reconocimiento de la costa norte de Marruecos" (1853), typescript.
      b. ACEM. Delegación de Economía. Servicio de Montes. "Memoria de los trabajos realizados en el año 1943," typescript.
      c. ACEM. Delegación de Economía. Servicio de Montes. "Memoria de los trabajos realizados en el año 1945," typescript.
      d. Intervenciones Militares del Rif. "Estadísticas 1929."
B. Archivo Histórico Nacional, Madrid.
   1. Estado, legajo 8357.
   2. Estado, signatura 560–2, "Mapas de Granada y Almería," c. 1860. Estado, signatura 771, "Taha de Luchar en la Alpujarra vaja."
   3. Fondos Modernos, Obras Públicas, legajos 172–80.
C. Archivo del Ministerio de Agricultura, Pesca y Alimentación, Madrid: legajos 6, 9, 10, 19, 37, 56, 226, 257–64, 353, 355–356, 372.
D. Biblioteca de la Real Academia de la Historia, Madrid: mss. 9-5.724, 9-6.258, 9-6.223, 9-6.224.
E. Archivo del Ministerio de Asuntos Etranjeros, Madrid: ms. 67.
F. Servicio Histórico Militar, Madrid. Monografías de Kabilas, vols. 1–3.
G. Instituto Olóriz de Investigaciones Anatómicas, University of Granada. Murillo, N., and L. Roca. "Estudio antropológico y demográfico de la Alpujarra a finales del siglo xix," typescript.

V. France

Archives du Ministère des Affaires Etrangères, Paris.
Série: Mémoires et Documents – Fond divers. Turquie, vols. 8, 13, 22, 23, 33, 35, 41, 53, 101, 116, 119, 120.
Série: Mémoires et Documents. Maroc, vols. 9, 10.
Série: Correspondance Commerciale. Smyrne, vol. 43.

VI. Great Britain

A. Public Record Office, London.
 1. Foreign Office Papers. 195/97, 137, 715, 741, 751, 771, 801, 935, 1002, 1065, 1150, 1351.
 196/2, 3.
 294/18, 23, 32.
 424/20.
 636/1,13–14.
 925/3090, 3098, 3373, 3636.
 2. Colonial Office Papers: 136/1261, 1391–1427.
B. British Library, Additional Manuscripts, London: mss. 34, 922; 61, 742.

Books, Articles, Papers, Theses

A Handbook for Travellers in the Ionian Islands, Greece, Turkey, Asia Minor and Constantinople. 1845. London: John Murray.

A Handbook for Travellers in Turkey. 1854. 3d ed. London: John Murray.

Abadan Unat, N., ed. 1976. *Turkish Workers in Europe, 1960–75.* Leiden: E. J. Brill.

Abd el-Krim et la république du Rif. Actes du colloque internationale d'études historiques et sociologiques, 18–20 janvier 1973. 1976. Paris: Maspero.

Abélanet, Bernard. 1969. "Les problèmes économiques du vignoble d'Almeria." *Revue géographique des Pyrénées et du Sud-Ouest* 40:143–56.

Abenia Taure, Ignacio de. 1859. *Memorias sobre el Rif.* Zaragoza: Gallifa.

Actes du Symposium International d'Histoire Forestière. 1982. 2 vols. Nancy: Ecole Nationale du Génie Rural.

Agard, J. 1957. "Les gîtes minéraux et l'industrie minière de la zone nord du Maroc." *Notes Marocaines,* 9–10:13–22.

Agnello, Giuseppe. 1975–6. "Provvidenze borboniche per la tutela del patrimonio boschivo." *Archivio storico siracusano* 4:137–63.

Aguirre, Ruperto. 1858. *Espedición al Riff.* Madrid: Ducayal.

Ahlmann, H. W. 1925, 1926. "Etudes de géographie humaine sur l'Italie subtropicale." *Geografiska Annaler* (Stockholm) 7:257–322; 8:74–124.

Aivar, José. 1985. *Sierra Nevada y Alpujarra Alta.* Granada: Editorial Everest.

Akan, Recep. 1950. "Eğridir işletmesi ormanları ve iktisadî önemi." *Orman ve Av* 22:57–59, 80–82, 120–2, 141–4.

Akman, Y., M. Barbero, and P. Quézel. 1978. "Contribution à l'étude de la végétation forestière de l'Anatolie méditeranéenne." *Phytocoenologia* 5:1–79, 189–346.

Alarcón, Pedro Antonio de. 1874. *La Alpujarra. Sesenta leguas a caballo.* Madrid: Guijarro.

Alberi, Eugenio, ed. 1858. *L'Italia nel secolo decimosesto ossia le relazioni degli ambasciatori veneti.* Vol. 2. Florence: Società Editrice Fiorentina.

Aldaya, F. 1968. "Los mantos alpujarrides al sur de Sierra Nevada." *Acta geológica hispánica* 4:126–30.

Algranati, G. 1929. *Basilicata e Calabria.* Turin: UTET.

Almagro Sanmartín, José. 1932. *Torrentes y pantanos en Sierra Nevada.* Madrid.

Alvarez Sereix, Rafael. 1883. *La desamortización forestal.* Madrid: Moreno y Rojos.

Anagnostopoulos, C. I. 1983. "Apographi kafsoxilon." *Dasiki erevna* 4:67–91.

Anastasiadou, Sophia. 1955. *Pindos.* Athens.

Andreades, A. 1905. "Les forêts Grecques." *Le mouvement économique* (Bucharest): 1–20.

Andrews, Peter Alford, with the assistance of Rüdiger Benninghaus. 1989. *Ethnic Groups in the Republic of Turkey.* Wiesbaden: Ludwig Reichert Verlag.

Andrisano, Teodoro, and Stefano Quartulli. 1986. "La carbonizzazione nel Mezzogiorno." *Monti e boschi* 37:13–19.

Angel, J. L. 1972. "Ecology and Population in the Eastern Mediterranean." *World Archaeology* 4:88–105.

Anonymous. 1827. *Narrative of an Excursion from Corfu to Smyrna.* London: Black, Young and Young.

Anonymous. 1887. "A través del Riff." *Revista de geografía comercial,* 31 January 1887.

Anonymous. 1900. *La tenuta di Monticchio in Basilicata.* Rome: A. Linari.

Anonymous. 1953. "Hidrografía del Norte del Marruecos." *El agua* (Barcelona):52–60.

Anonymous. 1964. *Land Reform – Apulia, Lucania, and Molise.* Bari: Laterza.

Anonymous. 1971. "La questione di Pollino." *Basilicata* 15, no. 7:35–37.

Anoyatis-Pele, D. 1988. "Intervention de transport terrestre en Grèce dans l'économie au XVIIIe siécle." In Ethniko Kentro Koinonikon Erevnon and Kentro Neoellinikon Erevnon, *O agrotikos kosmos ston Mesoyeiako choro,* 627–35. Athens.

Ansted, David T. 1854. *Scenery, Science and Art.* London: J. Van Voorst.

Antalya 1967. İl Yilliği. 1969? Ankara: Ajans – Türk matbaacılık Sanayii' nin tipo ve ofset Servi slerinde basilmiştir.

Antoniades-Bibicou, Hélène. 1965. "Villages désertés en Grèce: un bilan provisoire." In Ecole Pratique des Hautes Etudes-VIe section, *Villages désertés et histoire économique, XIe–XVIIIe siècle,* 343–417. Paris: S.E.V.P.E.N.

Antonucci, E., and U. Trillò. 1931. *Provenienza e destinazione delle correnti dell' emigrazione italiana per l'estero dal 1876 al 1930.* Rome: Istituto Poligrafico dello Stato.

Apodimoi Ellines. 1972. Athens: Ethnikon Kentron Koinonikon Erevnon.

Arapoglou, Michel. n.d. "Espace et société zagorienne." In *Jannina,* 85–96. N.p.

Aravantinos, Panayiotis. 1856. *Chronographia tis Epeirou.* 2 vols. Athens: Vlastou.

Ardos, Mehmet. 1969. "Problèmes géomorphologiques du versant sud du Taurus Central (Turquie Méridionale)." *Méditerranée,* 1st ser., 10:233–56.

Ardos, Mehmet. 1979. *Problèmes géomorphologiques du Taurus central et de sa bordure méditerranéenne (Turquie).* Istanbul: Enstitüsü Yayın, no. 107.

Arias Abellán, Jesús. 1981. "La repoblación forestal en la vertiente norte de Sierra Nevada." *Cuadernos geográficos* 11:283–305.

Arias Abellán, Jesús. 1984. *Propriedad y usos de la tierra en el Marquesado del Cenete.* Granada: Universidad de Granada.

Arlacchi, Pino. 1980. *Mafia, contadini e latifondo nella Calabria tradizionale.* Bologna: Mulino.

Arlacchi, Pino. 1982. "Perché si emigrava dalla società contadina e non dal latifondo?" In Pietro Borzomati, ed., *L'Emigrazione calabrese dall'unità ad oggi. Atti del II Convegno di studio,* 157–70. Rome: Centro Studi Emigrazione.

Arlacchi, Pino. 1983. *Mafia, Peasants and Great Estates.* Cambridge: Cambridge University Press.

Arlach, H. de T. D'. 1856. *Le maroc et le Riff en 1856.* Paris: Ledoyen.

Arnaud, Louis. 1952. *Au temps des melhalla; ou le Maroc de 1860 à 1912.* Casablanca: Editions Atlantides.

Aronica, Ignazio. 1961. "Bonifica dei territori montani nel Mezzogiorno (sintesi)." *Informazioni SVIMEZ* 14, no. 27:39–40, 912–15.

Arribas Palau, Mariano. 1972. "Datos sobre el comercio entre España y Marruecos en tiempo de Mawlāy al-Yazîd." *Hesperis Tamuda* 13:95–138.

Arrigo, A. di. 1950. "Aspetti fisiografici dell'erosione del suolo nel mezzogiorno." *Monti e boschi* 1:501–13.

Arriola Calleja, T. 1955. "La erosión del suelo español a través de la legislación forestal." *Montes* 61:19–26.

Artola, Miguel, A. M. Bernal, and I. Contreras. 1978. *El latifundio.* Madrid: Servicio de Publicaciones Agrarias.

Arundell, F. V. J. 1834. *Discoveries in Asia Minor.* 2 vols. London: Bentley.

Aschmann, Homer. 1973. "Man's Impact on Several Regions with Mediterranean Climates." In Francesco di Castri and Harold Mooney, eds., *Mediterranean Type Ecosystems: Origin and Structure,* 363–72. Berlin: Springer-Verlag.

Astarita, Tommaso. 1992. *The Continuity of Feudal Power: The Caracciolo di Brienza in Spanish Naples.* New York: Cambridge University Press.

Asteris, K. I. 1963. "Paratirisis epi ton afxitikon kai loipon tinon xilometrikon stoicheion mavris pefkis Zagoriou Epeirou." *To dasos* 32:49–63.

Attenborough, David. 1987. *The First Eden: The Mediterranean World and Man.* Boston: Little, Brown and Company.

Ayache, Germain. 1981. *Les origines de la guerre du Rif.* Rabat: Société Marocaine des Editeurs Réunis.

Ayache, Germain. 1983. *Etudes d'histoire marocaine.* Rabat: Société Marocaine des Editeurs Réunis.

Azimonti, Eugenio, ed. 1909–10. *Inchiesta parlamentare sulle condizioni dei contadini nelle provincie meridionali e nella Sicilia. Vol. 5: Basilicata e Calabria.* pts. 1–3. Rome: Tipografia Nazionale.

Azimonti, Eugenio. 1929. *La colonizzazione in Basilicata.* Rome: Tipografia del Senato.

Bahre, Conrad J. 1979. *Destruction of the Natural Vegetation of North-Central Chile.* Berkeley: University of California Press.

Bairacli Levy, Juliette de. 1954(?). *Spanish Mountain Life: The Sierra Nevada.* London: Faber and Faber.

Baker, James A. 1877. *Turkey in Europe.* London: Cassell, Petter & Galpin.

Baker, Lt.-Col. [no first name]. 1837. "Memoir of the Northern Frontier of Greece." *Journal of the Royal Geographical Society* 7:29.

Baker, P. T. 1978. *The Biology of High Altitude Peoples.* Cambridge: Cambridge University Press.

al-Bakri [Bakri, Abu Ubayd Abd Allah ibn Abd al-Aziz]. 1911. *Description de l'Afrique septentrionale.* Algiers: A. Jourdan.

Baldacci, A. 1893. "Ricordo di un viaggio botanico fra Prevesa e Janina." *Bollettino della Società botanica italiana* 2:84–8.

Baldacci, Antonio. 1896. "Prodotti vegetali che se usano nell'Albania e nell'Epiro." *Atti della Reale accademia dei georgofili* 19:58–82.

Baldacci, Antonio. 1896, 1897. "Itinerari albanesi." I–III. *Memorie della Società geografica italiana* 6:45–79, 378–409; and 7:15–44.

Baldacci, Antonio. 1917. *Itinerari albanesi (1892–1902)* Rome: Società Geografica Italiana.

Baldacci, Antonio. 1942. "Contributi alla conoscenza forestale dell'Albania Dardanica (Kossovo)." In *Le Terre Albanesi redente,* 219–66. Rome: Reale Accademia d'Italia.

Ballouche, A., et al. 1986. "Holocene Environments of Coastal and Continental Morocco." In F. López-Vera, ed., *Quatenary Climate in Western Mediterranean,* 517–32. Madrid: Universidad Autónoma de Madrid.

Banco de Granada. 1970. *Nuevas posibilidades del campo granadino.* Granada: Gráficas del Sur.

Barachette, R. 1973. *Lutte contre l'érosion et conservation des sols. Activités pre-projet.* Rabat: Ministère de l'Agriculture. Rome: FAO.

Barathon, J. J. 1989. "Bassins et littoraux du Rif Oriental (Maroc): Evolution morphoclimatique et tectonique depuis le Néogene supérieur." *Etudes méditerranéennes* (Poitiers) fascicule 13, 1–531.

Barbero, M., and P. Quézel. 1983. "La végétation de la Grèce et l'action de l'homme." *Méditerranée,* 3d ser., 47:65–71.

Barbieri, G. 1955. "Osservazioni geografico-statistiche sulla transumanza in Italia." *Rivista geografica italiana* 62:15–30.

Barbieri, Raffaele. 1953. "I sistemi colturali nel Mezzogiorno in relazione alle caratteristiche dell'ambiente." In *Problemi dell'agricoltura meridionale,* 579–610. Naples: Cassa per il Mezzogiorno.

Barkan, Ö. L. 1958. "Essai sur les données statistiques des registres de recensement dans l'Empire Ottoman au XVe et XVIe siècles." *Journal of the Economic and Social History of the Orient* 1:9–36.

Barletta, Pasquale. 1870. *Statistica di tutte le contrade della Sila . . .* Naples: Stamperia Governativa.

Baroni, E. 1955. *Guida botanica d'Italia.* Bologna: Cappelli.

Basilicata (Regione). Ufficio Statistica. 1983(?). *La Basilicata che cambia.* Potenza: Armento.

Basilicata (Regione). Ufficio Geologico Regionale. 1986. *Programmi ed interventi nel settore della difesa del suolo: l'assette idrogeologico nella Communità Montana del Lagonegrese. Atti del Convegno.* Matera: EDAD.

Bassett, Henri. 1926. "Vie sociale." *Bulletin de l'enseignement public au Maroc* 71:56–62.

Bates, Daniel G. 1973. *Nomads and Farmers: A Study of the Yörük of Southeastern Turkey.* Ann Arbor: University of Michigan Press.

Bates, Daniel G. 1980, "Yoruk Settlement in Southeast Turkey." In Philip C. Salzman, ed., *When Nomads Settle,* 124–39. New York: Praeger.

Bauer Manderscheid, Erich. 1980. *Los montes de España en la historia* Madrid: Ministerio de Agricultura.

Bean, George. 1968. *Turkey's Southern Shore.* London: Ernest Benn.

Beaufort, Francis. 1818, *Karamania.* London: R. Hunter.

Beaujour, Felix. 1800. *A View of the Commerce of Greece.* London: James Wallis.

Bechmann, Roland. 1990, *Trees and Man: The Forest in the Middle Ages.* New York: Paragon House.

Becker, Jerónimo. 1909. *El Rif.* Madrid: Patronato de Huérfanos de Administración Militar.

Beckinsale, Monica, and Robert Beckinsale. 1975. *Southern Europe: A Systematic Geographical Study.* New York: Holmes & Meier.

Beloch, Karl Julius. 1937. *Bevölkerungsgeschichte Italiens. I. Grundlagen Die Bevölkerung Sizilien und des Königreichs Neapel.* Berlin: De Gruyter.

Benchetritt, M. 1954. "L'érosion accélérée dans les chaînes telliennes de l'Oranie." *Revue de géomorphologie dynamique* 4:144–67.

Benedicto Pérez, Fernando. 1949. "Trabajadores rifeños en Argelia." In *Conferencias desarolladas en la Academia de Inventores 1948,* 5–17. Tetouan: ACEM.

Benekos, Petros. 1974. *Chouliaradon apanta.* Ioannina: n.p.

Bensalem, B. 1977. "Examples of Soil and Water Conservation Practices in North African Countries, Algeria, Morocco, and Tunisia." *F.A.O. Soils Bulletin* 33:151–60.

Bent, J. Theodore. 1890. "Explorations in Cilicia Tracheia." *Proceedings of the Royal Geographical Society,* n.s. 12:445–63.

Berl, Alfred. 1909. "Une course en Epire." *L'Hellenisme* 6, no. 1:18–33.

Bernal, Antonio, and Michel Drain. 1985. "Progreso y crisis de la agricultura andaluza en el siglo XIX." In R. Garrabou and J. Sanz Fernández, eds., *Historia agraria de la España contemporánea. II. Expansión y crisis,* 412–42. Barcelona: Crítica.

Bertagnolli, C. 1888. *Delle trasformazioni vecchie e recenti dell'agricoltura nostra e forestiera.* Bologna: Fava & Gaagnani.

Best, J. J. 1842, *Excursions in Albania.* London: Allen.

Beug, H. J. 1975. "Changes of Climate and Vegetation Belts in the Mountains of Mediterranean Europe during the Holocene." *Biuletyn geologiczny* 19: 101–10.

Biancheri, B. 1956. "Il fenomeno erosivo nel Comune di Matera." *Nuova agricultura Lucana*: 6–9.

Bianconi, F. 1877, *Ethnographie et statistique de la Turquie d'Europe et de la Grèce.* Paris: Lasailly.

Biblioteca dei Comuni Italiani. 1853. *Annuario economico-statistica dell'Italia per l'anno 1853.* Turin: Ferrero e Franco.

Biger, Gideon, and Nili Liphschitz. 1991. "Regional Dendrohistory and Timber Analysis: The Use of Wood in Buildings of the Nineteenth-Century Jaffa." *Mediterranean History Review* 6:86–104.

Biger, Gideon, and Nili Liphschitz. (forthcoming). "Foreign Tree Species as Construction Timber in Nineteenth-Century Palestine." *Forest and Conservation History.*

Bilbao, Luís María. 1984. "L'expansion de la culture du maïs et le déplacement des centres de gravité économique dans le Pays Basque-Espagnol." In A. Guarducci, ed., *Agricoltura e trasformazione dell'ambiente: secoli XIII–XVIII*, 557–620. Florence: Istituto Internazionale di Storia Economica.

Bintliff, John L. 1977. *Natural Environment and Human Settlement in Prehistoric Greece.* Oxford: British Archaeological Reports, Supplementary Series, no. 28.

Bintliff, John L. 1982. "Palaeoclimatic Modelling of Environmental Changes in the East Mediterranean Region since the Last Glaciation." In J. L Bintliff and W. van Zeist, eds., *Palaeoclimates, Palaeoenvironments, and Human Communities in the East Mediterranean Region*, 485–527. Oxford: British Archaeological Reports, International Series, no. 133.

Bintliff, John L., and W. van Zeist, eds. 1982. *Palaeoclimates, Palaeoenvironments, and Human Communities in the East Mediterranean Region.* Oxford: British Archaeological Reports, International Series, no. 133.

Bintliff, John L., Donald A. Davidson, and Eric G. Grant, eds. 1988. *Conceptual Issues in Environmental Archaeology.* Edinburgh: Edinburgh University Press.

Bisson, J.-C., J.-C. Brulé, R. Escallier, G. Maurer, J.-M. Miossec, G. Mutin, P. Signales, J.-F. Troin. 1985. *Le Maghreb: Hommes et espaces.* Paris: Armand Colin.

Black-Michaud, Jacob. 1975. *Cohesive Force: Feud in the Mediterranean and the Middle East.* New York: St. Martin's.

Blancard, Jules. 1846a. "Souli." *Revue de l'Orient* 11:229–35.

Blancard, Jules. 1846b. "Zagori." *Revue de l'Orient* 11:62–9.

Blancard, Jules. 1882. *L'Épire et la Thessalie.* Paris: Firmin-Didot.

Blanco Izaga, Emilio. 1930. *La vivienda rifeña.* 2 vols. Ceuta: Revista Africa.

Blanco Izaga, Emilio. 1939. *El Rif: La ley Rifeña.* Ceuta: Imperio.

Blázquez, Antonio. 1923. "La leyenda de los bosques devastados." *Revista de geografía colonial y mercantil* 20:406–10.

Blok, Anton. 1972. "The Peasant and the Brigand: Social Banditry Reconsidered." *Comparative Studies in Society and History* 14:499–503.

Blumenthal, Ekkehard. 1963. "Die altegreichische Siedlungkolonisation im Mittelmeerraum unter besonderer Berucksichtung der Südküste Kleinasiens." *Tübinger Geographische Studien* 10:1–182.

Bocciante, Maria. 1922. "Sulla distribuzione geografica dei calanchi in Italia." *L'universo* 3:585–605.

Boehm, Christopher. 1984. *Blood Revenge: The Anthropology of Feuding in Montenegro and Other Tribal Societies.* Lawrence: University Press of Kansas.

Bois, Charles. 1949. "Années de disettes, années d'abondance, sécheresse et pluies au Maroc." *Revue pour l'étude des calamités* 26–7:33–71.

Boletín de la Sociedad Geográfica Nacional 74 (1934): 339–40.

Bonin, G., and J. Gamisans. 1976. "Contribution à l'étude des forêts de l'étage supraméditerranéen de l'Italie méridionale." *Documents de phytosociologie* 19–20:73–86.

Borzomati, Pietro, ed. 1982. *L'Emigrazione calabrese dall'unità ad oggi. Atti del II Convegno di studio.* Rome: Centro Studi Emigrazione.

Bosco, Umberto, et al. 1965. *Basilicata.* Milan: Electa.

Boserup, Esther. 1965. *Conditions of Agricultural Growth.* Chicago: Aldine.

Bosque Maurel, Joaquín. 1969. "Tradición y modernidad en las Alpujarras granadinos." In *Aportación española al XXI Congreso Geográfico,* 165–83. Madrid: Instituto Geográfico y Catastral.

Bosque Maurel, Joaquín. 1971. *Granada, la tierra y sus hombres.* Granada: Organización Sindical.

Bosque Maurel, Joaquín. 1972. *Sierra Nevada.* Granada: Obra Cultural de la Caja de Ahorros de Granada.

Bosque Maurel, Joaquín. 1979. *Andalucía. Estudios de geografía agraria.* Granada: Aljibe. 1979.

Bottema, Sietse. 1974. "Late Quaternary Vegetation History of Northwestern Greece." Ph.D. diss. Groningen, Netherlands.

Bottema, Sietse. 1977. "A Pollen Diagram from the Syrian Anti-Lebanon." *Palaeorient* 3:259–68.

Bottema, Sietse. 1982. "Palynological Investigations in Greece with Special Reference to Pollen as an Indicator of Human Activity." *Palaeohistoria* 24: 257–89.

Bottema, Sietse, and H. Woldring. 1984. "Late Quatenary Vegetation and Climate of Southwestern Turkey II." *Palaeohistoria* 26:123–49.

Boudy, P. 1928. *L'arbre et les forêts du Maroc.* Rabat: Elgaly & Jaquet.

Boudy, P. 1948–54. *Economie forestière Nord-Africaine.* 4 vols. Paris: Larose.

Boué, Ami. 1840. *La Turquie d'Europe.* 4 vols. Paris: Bertrand.

Boué, Ami. 1854. *Recueil d'itinéraires dans la Turquie d'Europe.* 2 vols. Vienna: Braumüller.

Bourelly, Giuseppe. 1987. *Il brigantaggio dal 1860 al 1865.* Venosa: Edizioni Osanna.

Bousquet, B. 1974. "La Grèce occidentale: Interpretation géomorphologique de l'Epire, de l'Acarnania et des Iles Ioniennes." 2 vols. Ph.D. diss. Université de Paris IV.

Bousquet, B., J. J. Dufaure, and P.-Y. Pechoux. 1983. "Temps historiques et évolutions des paysages égéens." *Méditerrannée,* 3d ser., 48:3–25.

Boutilly, V. 1931. "Généralités sur les forêts de l'Algérie." In Exposition Coloniale

Internationale, *Congrès de la Production forestière coloniale et Nord Africaine*, 3–12. Paris: Quinzaine Nationale de la Production Agricole d'Outre Mer.

Bowen, George F. 1852. *Mt. Athos, Thessaly and Epirus.* London: Rivington.

Brandes, Stanley H. 1972. "El impacto de la emigración en un pueblo de la Sierra de Beja." *Ethnica*, no. 4:9–26.

Braudel, Fernand. 1976. *The Mediterranean and the Mediterranean World in the Age of Philip II.* 2 vols. Translated by Siân Reynolds. New York: Harper & Row.

Brenan, Gerald. [1957] 1988. *South from Granada.* Cambridge: Cambridge University Press.

Brenan, Gerald. 1974. *Personal Record 1920–1972.* London: Jonathan Cape.

Brice, William C., ed. 1978a. *Environmental History of the Near and Middle East.* London: Academic Press.

Brice, William C. 1978b. "The Desiccation of Anatolia." In William C. Brice, ed., *Environmental History of the Near and Middle East*, 141–7. London: Academic Press.

Bricogne, A. 1876. "La mission forestière en Turquie." *Revue des eaux et forêts* 15:361–86.

Briganti, Gaetano. 1916. "Rapporti fra boschi e pascoli nel Mezzogiorno." In *Atti del III Congresso forestale italiano e I Congresso per l'irrigazione 1914*, 73–90. Portici: Torre.

Brinkmann, Roland. 1976. *Geology of Turkey.* Stuttgart: Enke.

Bronzini, Giovanni B. 1964. *Vita tradizionale in Basilicata.* Matera: Montemurro.

Brooke, Arthur de Capell. 1831. *Sketches in Spain and Morocco.* 2 vols. London: Henry Colburn & Richard Bentley.

Bruce-Chwatt, L. J., and Julian De Zulueta. 1980. *The Rise and Fall of Malaria in Europe.* Oxford: Oxford University Press.

Brückner, H. 1982. "Ausmass von Erosion und Akkumlation in Verlauf des Quartärs in der Basilicata (Süditalien)." *Zeitschrift für Geomorphologie*, suppl. 43:121–37.

Brunt, P. A. 1971, *Italian Manpower, 225 B.C. – A.D. 14.* Oxford: At the Clarendon Press.

Buffault, Paul. 1941. "Le déboisement de l'Afrique du Nord." *Revue des eaux et forêts* 79:471–87.

Bujatti, Giovanni. 1953. "Pastorizia transumante a pastorizia stanziale nella montagne appenniniche." *Monti e boschi* 4:24–7.

Butzer, Karl W. 1974. "Accelerated Soil Erosion: A Problem of Man–Land Relationships." In Ian R. Manners and Marvin W. Mikesell, eds., *Perspectives on Environment*, 57–79. Washington, D.C.: Association of American Geographers.

Butzer, Karl W. 1982. *Archaeology as Human Ecology.* New York: Cambridge University Press.

Butzer, Karl W., J. F. Mateu, E. Butzer, and P. Kraus. 1985. "Irrigation Agrosystems in Eastern Spain: Roman or Islamic Origins?" *Annals of the Association of American Geographers* 75, no. 4:479–509.

Caballero y Segares, A. 1940. *Flora analítica de España.* Madrid. Sociedad anónima española de traductores y autores.

Cabrera y Latorre, Angel. 1924. *Magreb el Aksa. Recuerdos de cuatro viajes por Yebala y por el Rif.* Madrid: Voluntad.

Cabrillana, Nicolas. 1965. "Villages désertés en Espagne." In Ecole Pratique des Hautes Etudes – VIe Section, *Villages désertés et histoire économique XIe–XVIIIe siècle,* 461–512. Paris: S.E.V.P.E.N.

Cafiero, Salvatore, and Guido de Rossi. 1966. "La spopolamento della montagna meridionale." In *L'esodo rurale e lo spopolamento della montagna nella società contemporanea. Atti del Convegno Italo-Svizzero,* 79–92. Milan: Vita e Pensiero.

Cagli, Cesare. 1910, *La Basilicata ed il problema dell' immigrazione e della colonizzazione interna.* Rome: Colombo.

Cahen, C. 1968. *Pre-Ottoman Turkey: A General Survey of the Material and Spiritual Culture and History, c. 1071–1330.* London: Segwick & Jackson.

Caillé, Jacques. n.d. *Charles Jagerschmidt, Chargé d'Affaires de France au Maroc (1820–1894).* Publications de l'Institut des Hautes-Etudes Marocaines, vol. 51. Paris: Larose.

Calatrava Requeña, Javier. 1984. "Análisis de la potencialidad del turismo rural como elemento generador de rentas complementarias en zonas en depresión socioeconómica: el caso de las Alpujarras granadinas." In *Colloquio Hispano-Francés sobre espacios rurales,* 2:305–28. Madrid: Instituto de Estudios Agrarios, Pesqueros y Alimentarios.

Caldora, Umberto. 1960. "La statistica murattiana del Regno di Napoli: le relazioni sulla Calabria." *Quaderni per la geografia umana per la Sicilia e la Calabria* 5:9–113.

Caldora, Umberto. 1962. "Per la storia del brigantaggio in Basilicata durante il periodo napoleonico." *Archivio storico per la Calabria e la Lucania* 21:393–405.

Calvello, Comune di. 1982. *Calvello: Storia-Arte-Tradizione.* Potenza: Zaferone & Di Bello.

Camariano, Ariadna. 1984. *Contributions à l'histoire des relations gréco-roumanes: L'Epire et les pays roumains.* Jannina: Association d'Etudes Epirotes.

Campbell, John K. 1964. *Honour, Family and Patronage: A Study of Institutions and Moral Values in a Greek Mountain Community.* Oxford: At the Clarendon Press.

Capaz y Montes, Fernando. 1928. *Geografía de Gomara.* Tetouan: Alta Comisaria de España en Marruecos.

Capone, Dario. 1970. "La conservazione del suolo in Calabria." *Regione calabrese* 1, no. 1:61–9.

Cara Barrionuevo, Lorenzo. 1986. *Arqueología de la Baja Alpujarra.* Almería: Instituto de Estudios Almerienses.

Carandell, Juan. 1934. "El habitat de Sierra Nevada." *Boletín de la Sociedad geográfica nacional* 74:644–99.

Carandell, Juan. 1935. "Las condiciones del modelado erosivo en la vertiente mediterránea de la Cordillera Bética." *Boletín de la Real sociedad española de historia natural* 35:39–62.

Carbone, Giovanni. 1892. "Il disboscamento in Basilicata," *Cronaca lucana* 23:177–8.

Carlyle, Margaret. [1962] 1985. *The Awakening of Southern Italy.* Westport, Conn.: Greenwood.

Caro Baroja, J. 1957. *Los moriscos del Reino de Granada.* Madrid: Instituto de Estudios Políticos.

Carrascosa, Miguel. 1960. . . . *A las puertas de la Alpujarra.* Granada: F. Román.

Carrera Morales, José Ángel. 1987. "Lucha contra la erosión: la experiencia española." In Spain, Instituto para la Conservación de la Naturaleza, *Panorama de la actualidad forestal en España,* 33–43. Madrid: Ministerio de Agricultura.

Carrier, E. H. 1932. *Water and Grass: A Study in the Pastoral Economy of Southern Europe.* London: Christopher's.

Carter, Francis W., ed. 1977. *An Historical Geography of the Balkans.* London: Academic Press.

Carvajal Gutierrez, María Carmen. 1986. *Población y emigración en la provincia de Granada en el siglo XX.* Granada: Diputación Provincial de Granada.

Carvajal Gutierrez, María Carmen. 1973. "La emigración al extranjero en la provincia de Granada." *Cuadernos geográficos* 3:25–54.

Casparini, Guglielmo, and Achille Bruni. 1845. *Breve ragguaglio dell'agricoltura e pastorizia del Regno di Napoli di qua del Faro.* Naples: Filiatre-Salezio.

Cassa per il Mezzogiorno. 1955. *Atti del II convegno tecnico (Cosenza 25–26 settembre 1954). I problemi della montagna nell'Italia meridionale.* Rome: Istituto Poligrafico dello Stato.

Castillo Requeña, José Manuel. 1981. "Mecanismos de la precipitación en Sierra Nevada." *Cuadernos geográficos* 11:127–52.

Catton, William. 1980. *Overshoot: The Ecological Basis for Revolutionary Change.* Urbana: University of Illinois Press.

Cazorla Pérez, J. 1965. *Estructura socio-económica de Andalucía oriental.* Granada: Caja de Ahorros de Granada.

Ceballos, Luís, and M. Martín Bolaños. 1928. "El Pinsapo y el abeto de Marruecos." *Servicio forestal de investigaciones y experiencias* 1:47–101.

Ceballos y Fernández de Córdoba, Luís. 1959. *Pasado y presente del bosque en la region mediterránea.* Madrid: Academia de Ciencias Exactas, Físicas y Naturales.

Célerier, J., and A. Charton. 1926. "Le milieu physique." *Bulletin de l'Enseignement public au Maroc* 71:9–32.

Centre de la Méditerranée Moderne et Contemporaine. 1973. *L'ankylose de l'économie méditerranée au XVIIeme et au début du XIXeme siècle: Le rôle de l'agriculture.* Nice.

Centre de la Méditerranée Moderne et Contemporaine. 1977. *Typologie des crises dans les pays méditerranées: XVIe–XXe siècles.* Nice.

Centre de la Méditerranée Moderne et Contemporaine. 1979. *Les migrations dans les pays méditerranées au XVIIeme et au début du XIXeme* siècle. Nice.

Centre National de la Recherche Scientifique. 1977. *L'élévage en Méditerranée occidentale.* Paris: Editions du CNRS.

Cerón Martínez, Salvador. 1879. *Estudio sobre materiales y efectos usados en la Marina.* Cadiz.

Chapman, R. W. 1978. "The Evidence for Prehistoric Water Control in South-East Spain." *Journal of Arid Environments* 1:261–74.

Chasiotos, Dimitrios. 1887. *Diatrivai kai Ipomnimata peri Epeirou.* Athens: Perri.

Cheetham, Nicholas. 1981. *Medieval Greece.* New Haven: Yale University Press.

Chenier, M. 1788. *The Present State of Morocco.* London: Robinson.
Cherubini, G. 1981. "La civiltà del castagno in Italia alla fine del medievo." *Archeologia medievale* 8:247–80.
Chiliadakis, Stelios. 1940. *Taxidia stin Epeirou.* Athens.
Chirol, M. Valentine. 1881. *'Twixt Greek and Turk.* Edinburgh: Blackwood.
Chocarro, Christina, R. Fanlo, F. Fillat, and P. Marin. 1990. "Historical Evolution of Natural Resource Use in the Central Pyrenees of Spain." *Mountain Research and Development* 10:257–65.
Chouliarakis, Michail. 1973. *Geografiki kai plithismiaki exelixeis tis Ellados, 1821–1971.* Athens: Ethniko Kentro Koinonikon Erevnon.
Chouliarakis, Michail. 1988. *Exelixeis tou plithismou ton agrotikon periochon tis Ellados, 1920–1981.* Athens: Ethniko Kentro Koinonikon Erevnon.
Chouliarakis, Michail, E. Makris, E. Gritsopoulos, M. Gevetsi, and A. Aghiopetris. 1972. *Statistiki meletai, 1821–1971. I statistiki kata ta 150 eti apo tis palingenesias tis Ellados.* Athens: Ethniko Kentro Koinonikon Erevnon.
Christovasilis, Chr. 1905. *I Epeiros yeografikos kai ethnologikos.* Athens: Konstantinidou.
Cinel, Dino. 1982. "The Seasonal Emigrations of Italians in the Nineteenth Century: From Internal to International Destinations." *Journal of Ethnic Studies* 10:43–68.
Clark, Martin. 1984. *Modern Italy, 1871–1982.* London: Longmans.
Clavé, J. 1864. "Etudes Forestières. Les forêts de la Corse." *Revue des deux mondes* 51:353–80.
Clavé, J. 1866. "L'Alienation des forêts de l'Etat." *Revue des deux mondes* 53: 197–214.
Clavé, J. 1878. "L'exposition forestière. Les bois étrangères." *Revue des deux mondes*: 809–40.
Clemente Rubio, Simón de Roxas. 1807. *Ensayo sobre las variedades de la vid común que vegetan en Andalucía.* Madrid: Villalpando.
Cocco, E. T. De Pippo, and M. Pennette. 1976. "L'uso della fotografia aerea e del calcolatore elettronico nello studio degli spostamenti delle linee di costa: L'evoluzione del littorale Alto Ionico (golfo di Taranto) negli ultimi 30 anni." *Bolletino della Società geologica italiana* 95:275–312.
Cockerell, C. R. 1903. *Travels in Southern Europe and the Levant, 1810–1817.* London: Longmans, Greene.
Cohen, Aron. 1987. *El Marquesado del Zenete. Tierra de minas.* Granada: Diputación Provincial de Granada.
Cohen, Mark Nathan. 1989. *Health and the Rise of Civilization.* New Haven: Yale University Press.
Colamonico, Carmelo. 1961. "Carta agricolo-forestale della Calabria." *Almanacco calabrese*:141–8.
Colangelo, Angelo R. 1977. "Cento anni di emigrazione." In Nino Calice, ed., *Basilicata tra passato e presente,* 11–42. Milan: Teti.
Congreso de Historia rural (siglos XV–XIX). Actas del Coloquio de Octubre 1981. 1984. Madrid: Universidad Complutense.

Congresso (XXIII) Nazionale delle Bonifiche (Roma 1967). 1967. *Atti del XXIII Congresso. La Protezione del suolo e la regolazione delle acque.* Bologna: Il Mulino.

Consejo Económico Sindical de la Zona de las Alpujarras. 1970. *Pleno III (1969–70).* Granada.

Cook, M. A. 1972. *Population Pressures in Rural Anatolia, 1450–1600.* London: Oxford University Press.

Cooke, Ronald, and Richard Reeves. 1976. *Arroyos and Environmental Change in the American South-West.* Oxford: At the Clarendon Press.

Coon, Carleton S. 1931. *Tribes of the Rif.* Cambridge: Peabody Museum of Harvard University.

Coon, Carleton S. 1951. *Caravan: The Story of the Middle East.* New York: Henry Holt.

Coppola, Pasquale. 1974. "Verso la distruzione del territorio? Il caso della Basilicata." In Ugo Leone, ed., *Ambiente e sviluppo nel Mezzogiorno.* Naples: Edizioni Scientifiche Italiane.

Cornelisen, Ann. 1969. *Torregreca: A World in Southern Italy.* London: Macmillan.

Corvol, Andrée. 1987. *L'homme aux bois: Histoire des relations de l'homme et de la forêt (XVIIe–XXe siècles).* Paris: Fayard.

Costanza, Mariateresa. 1985. "Il bosco di Policoro del settecento alla riforma fondiara." *Monti e boschi* 36, no. 3:12–18.

Cote, Marc. 1986. "Les migrations internationales dans le Bassin Méditerranéen." *Méditerranée,* 3d ser., 59:103–6.

Cotecchia, Vincenzo, et al. 1983. *Tour in Apulia and Lucania. Geological and Hydrological Aspects.* Bari: Istituto de Geologia, Università di Bari.

Cotronei, Bruno. 1962. *Studio preliminare per le previsioni di sviluppo demografico della Basilicata.* Rome: Nicoletti & Terenzi.

Craven, Keppel. 1821. *A Tour through the Southern Provinces of the Kingdom of Naples.* London: Rodwell and Martin.

Crea, Giuseppe. 1956. "L'Aspromonte ed i suoi boschi." *Monti e boschi* 7:69–81.

Creuzé de Lesser, M. 1806. *Voyage en Italie et en Sicilie.* Paris: L'Ainé.

Cribb, Roger. 1991. *Nomads in Archaeology.* Cambridge. Cambridge University Press.

Crosby, Alfred W. 1986. *Ecological Imperialism: The Biological Expansion of Europe, 900–1900.* New York: Cambridge University Press.

Cuadrado Iglesias, Manuel. 1980. *Aprovechamiento en común de pastos y leñas.* Madrid: Ministerio de Agricultura, Pesca, y Alimentación.

Cubillo, Ignacio del. 1948. "Actividades maritimas de la zona de protectorado de España en Marruecos." In *Conferencias desarolladas en la Academia de Interventores durante 1947,* 87–104. Tetouan: Majzen.

Cuinet, Vital. 1890–5. *La Turquie d'Asie: Géographie administrative, statistique, descriptive et raisonée de chaque province de l'Asie Mineure.* 4 vols. Paris: Ernest Leroux.

D'Alfonso, N. R. 1913. "Il diboscamento in Calabria." *Nuova antologia di lettere, scienze ed arti.* (Sept–Oct.): 622–31.

D'Amelio, R. 1925. "I pascoli nel Lagonegro." *Terra lucana* 5:163–7.

D'Aragona, Gabriele G. 1964. "Economia agraria della Basilicata nel cinquantennio 1860–1914." *Annali del Mezzogiorno* 4:55–84.

D'Arrigo, Agatino. 1951. "Problema della difesa del suolo nel Mezzogiorno." *L'Universo* 36:1–14.

Da Molin, Giovanna. 1979. *La popolazione del regno di Napoli a metà quattrocento.* Bari: Adriatica.

Daget, Philippe, ed. 1980. "Atlas d'aréologie périméditerranéenne." *Naturalia monspeliensia,* numéro hors série.

Damianakos, Stathis. 1985. "Banditisme et imaginaire pastoral en Grèce." *Etudes rurales* 97–8:219–40.

Damiano, Nicola, and Mario Messina. 1975. "La emigrazione in Basilicata." *Sviluppo* 1 no. 2:48–67.

Daoutopoulos, George A. 1982. "The Changing Demographic Characteristics of the Rural Population of Greece." M. A. thesis, Michigan State University.

Darby, H. C. 1956. "The Clearing of the Woodland in Europe." In W. L. Thomas, ed., *Man's Role in Changing the Face of the Earth,* 1:183–216. Chicago: University of Chicago Press.

Dasmann, Raymond F. 1965. *The Destruction of California.* New York: Macmillan.

Datos estadísticos del Territorio de Chauen 1947. N.d. N.p. In Biblioteca Nacional, Madrid, Sección de Africa.

Davidson, Donald. 1980. "Erosion in Greece during the First and Second Millennia B.C." In R. A. Cullingford, D. A. Davidson, and J. Lewin, *Timescales in Geomorphology,* 143–58. New York: John Wiley.

Davillier, Jean Charles. 1876. *Spain.* New York: Scribner, Welford & Armstrong.

Davis, E. J. 1874. *Anatolica; Or, The Journal of a Visit to Some of the Ancient Ruined Cities of Caria, Phrygia, Lycia and Pisidia.* London: Grant.

Davis, J. 1973. *Land and Family in Pisticci.* London: Athlone.

Davis, P. H., ed. 1965–88. *Flora of Turkey and the East Aegean Islands.* 10 vols. Edinburgh: Edinburgh University Press.

Dāwūd, Muhammad. 1959. *Tārikh Tiṭwān.* 10 vols. Tetouan: Maktabat al-Nāsir.

De Grazia, Paolo. 1921. "La diminuazione della popolazione in Basilicata." *Bollettino della Reale società geografica italiana* 58:411–40 and 525–53.

De Gubernatis, Enrico. 1872. "L'Epiro." *Bollettino della Società geografica italiana* 8:1–25.

De Gubernatis, Enrico. 1884. *Cenni sull'Epiro e sulla sua nuova carta.* Rome: Capaccini e Ripamonti.

De Jesus, Prentiss. 1980. *The Development of Prehistoric Mining and Metallurgy in Anatolia.* 2 vols. Oxford: British Archaeological Reports, International Series, no. 74.

De Leo, G. 1917. *Utilità dei boschi. Festa nazionale degli alberi.* Potenza: La Perserverenza.

De Leo, Pietro, ed. 1988. *Minoranze etniche in Calabria e in Basilicata.* Terreni: Di Mauro.

De Leo, Pietro, Gaetano La Terza, Enzo Roseto, Mario Tommaselli, Giuliano Cesca, M. Codogno, Simona Fascetti, D. Puntillo, and Mara Cagnin. 1984. *Il Pollino: storia, arte, costume.* Rome: Editalia.

De Mas, Paolo. 1978. *Marges marocaines: Limites de la cooperation au développement dans une région périphérique:Le cas du Rif.* The Hague: NUFFIC/IMWOO/Project Remplod.

De Meo, Giuseppe. 1962. *Saggi di statistica economica e demografica sull'Italia meridionale nei secoli XVII e XVIII.* Rome: Istituto di statistica economica dell'Università di Roma.

Debierre, M. 1931. "L'exploitation et la production des forêts de Tunisie." In Exposition Coloniale Internationale, *Congrès de la production forestale coloniale et Nord Africaine,* 13–25. Paris: Quinzaine Nationale de la Production Agricole.

Deil, Ulrich. 1987. "La végétation actuelle et l'occupation des terres dans la région du Jebel Arz (Haut Rif Central, Maroc)." In *Hommage à Gérard Maurer,* 241–55. Poitiers. Université de Poitiers.

Deil, Ulrich. 1988. "La distribution actuelle et potentielle du cèdre dans le haut Rif central." *Revue de géographie du Maroc* 12:17–32.

Delibrias, G., L. Faugeres, and R. Neboit. 1978. "Evolution des paysages sur les rives nord-méditerranéennes, au cours du Post-Glaciaire (Italie du Sud, Macédoine Grecque)." In Société Géologique de France, *VI Réunion Annuelle des Sciences de la Terre,* 141. Paris: Société Géologique de France.

Delille, Gérard. 1977a. *Agricoltura e demografia nel Regno di Napoli nei secoli XVIII e XIX.* Naples: Guida.

Delille, Gérard. 1977b. "Crises et productivité agricole: L'exemple du Royaume de Naples." In Centre de la Méditerranée Moderne et Contemporaine. *Typologie des crises dans les pays méditerrannéens: XVIe–XXe siècles,* 115–36. Nice.

Despois, Jean. 1956. "La culture en terrasses en Afrique du Nord." *Annales: Economies, sociétés, civilisations* 1:42–50.

Despois, Jean. 1961. "Development of Land Use in Northern Africa, with References to Spain." In L. D. Stamp, ed., *History of Land Use in Arid Regions,* 219–37. Paris: UNESCO, Arid Zone Research, vol. 17.

Desprairies, Alain. 1979. *Etude sedimentologique de formations à caractère flysch et molasse, Macédoine, Epire (Grèce).* Paris: Société géologique de France.

Di Castri, Francesco and Harold Mooney, eds. 1973. *Mediterranean Type Ecosystems: Origin and Structure.* New York: Springer-Verlag.

Di Comite, Luigi, ed. 1982. *Aspetti della situazione demografica di alcuni paesi europei del bacino mediterraneo.* Bari: Cacucci.

Di Gianfrancesco, Mario. 1979. *La rivoluzione dei trasporti in Italia nell'età risorgimentale. L'unificazione del mercato e la crisis del Mezzogiorno.* L'Aquila: Japadre.

Di Stefano, Mario. 1969. *Il patrimonio boschivo ed i territori montani in Italia e in Campania.* Naples: Giannini.

Diamandi, Vasile. 1906. *Renseignements statistiques sur la population roumaine de la peninsule des Balkans.* Paris: Cornély.

Díaz Lobón, Eduardo. 1982. *Granada durante la crisis del antiguo régimen (1814–1820).* Granada: Excellentíssima Diputación Provincial.

Dickinson, Robert E. 1955. *The Population Problems of Southern Italy.* Syracuse, N.Y.: Syracuse University Press.

Didier, Ch. 1845. "L'Alpuxarra et Sierra Nevada." *Revue des deux mondes* 11: 487–518 and 812–41.

Dimitras, Elie. 1971. *Enquêtes sociologiques sur les emigrants grecs.* 3 vols. Athens: Centre National de Recherches Sociales.

Dimitriadis, Evangelos. 1980. *Koinonikos schimatismos kai Poleologikos choros: eikosi oikismoi tis eparchias Konitsas Epeirou.* Thessaloniki: University of Thessaloniki, Scientific Annals of the Architecture Department of the Polytechnic School.

Dodwell, Edward. 1819. *Classical and Topographical Tour through Greece during the Years 1801–1805 and 1806.* London: Rodwell & Martin.

Domenech Lafuente, Angel. 1940. *Apuntes sobre geografía de la Zona Norte del Protectorado de España en Marruecos.* Ceuta: Imperio.

Douglas, Norman. 1915. *Old Calabria.* London: M. Secker.

Douglass, William A. 1984. *Emigration in a South Italian Town: An Anthropological History.* New Brunswick, N.J.: Rutgers University Press.

Douguedroit, Annick. 1976. *Les paysages forestiers de Haute-Provence et des Alpes-Maritimes. Géographie, écologie, histoire.* Aix-en-Provence: EDISUD.

Dourojeanni, Marc. 1990. "The Environmental Impact of Coca Cultivation and Cocaine Production in the Peruvian Amazon." Unpublished paper.

Doxiadis, C. A. 1946. *The Sacrifices of Greece in the Second World War.* Athens: Ministry of Reconstruction.

Drusilli, Bruno. 1977. "Terre incolte o insufficientemente coltivate: problemi e prospettive nell'Appennino Bolognese." *Monti e boschi* 28, no. 3:21–6.

Dumont, Albert. 1873. *Le Balkan et l'Adriatique.* Paris: Didier.

Dumont, J. F., M. Gutnic, O. Monod, and A. Poisson. 1979. *Géologie des Taurides Occidentales (Turquie).* Paris: Société Géologique de France.

Dunn, Archibald J. 1905. *Turkey and Its Future.* London: Effingham Wilson.

Dupont, J. 1907. *Géographie de l'Empire Ottoman.* Paris: Poussielgue.

Durante, Raffaele. 1917. *Relazione sui Demani di Lagonegro.* Potenza: Editrice.

Eckholm, Erik. 1975. "The Deterioration of Mountain Environments." *Science* 189:764–70.

Eggeling, J. W. 1981. "Nomades et sédentaires en Anatolie occidentale." *Méditerranée,* 3d ser., 41 no. 1:35–8.

Eisma, Doeke. 1964. "Stream Deposition in the Mediterranean Area in Historical Times." *Nature* 203:1061.

Eisma, Doeke. 1978. "Stream Deposition and Erosion by the Eastern Shore of the Aegean." In William G. Brice, ed., *Environmental History of the Near and Middle East,* 67–81. London: Academic Press.

Ellen, Roy. 1982. *Environment, Subsistence and System: The Ecology of Small-Scale Social Formations.* Cambridge: Cambridge University Press.

Elorrieta y Artaza, José. 1949. *El castaño en España.* Madrid: Instituto Forestal.

Elorrieta y Artaza, José. 1964. *Ensayo de los pinos Ponderosa y Jeffreyi en la vertiente sur de Sierra Nevada (Granada) Años 1922 a 1962.* Madrid: Ministerio de Agricultura, Instituto Forestal.

Elorrieta y Artaza, Octavio. 1948. *Economia forestal. Las tierras incultas y los montes en la Política Económica de España.* Madrid: Diana.

Emberger, Louis. 1929. *Spicilegium rifanum.* Rabat: Mémoires de la Société des Sciences Naturelles du Maroc, no. 17.

Emberger, Louis, and René Maire. 1927. *Spicilegium rifanum.* Mémoires de la

Société des Sciences Naturelles du Maroc, no. 17. Rabat: Archives Scientifique du Protectorat Français.
Enciclopedia universal ilustrada. 1958. Madrid: Espasa-Calpe.
Enisleidis, Christos. 1951. *I Pindos kai ta choria tis.* Athens: Leontiadis.
Erinç, Sirri. 1950, "Climatic Types and the Variation of Moisture Regions in Turkey." *Geographical Review* 40:224–35.
Erinç, Sirri. 1978. "Changes in the Physical Environment in Turkey since the End of the Last Glacial." In William G. Brice, ed., *Environmental History of the Near and Middle East,* 87–110. London: Academic Press.
Espinosa Fernández, F. 1976. "Cartografía de la vegetación de Sierra Nevada." Ph.D. diss., Universidad de Granada, Departamiento de Botánica y Ecología Vegetal.
Estourmel, Joseph-Marie Creton, Comte d'. 1844. *Journal d'un voyage en Orient.* 2 vols. Paris: Crapelet.
Estrada, Juan Antonio. 1768. *Población general de España, sus reynos y provincias, ciudades, villas y pueblos.* 3 vols. Madrid: A. Ramirez.
Ethniko Kentro Koinonikon Erevnon and Kentro Neoellinikon Erevnon. 1988. *O agrotikos kosmos ston Mesoyeiako choro.* Athens.
Evliya [Çelebi]. 1896–1938. *Seyahatnamesi.* 10 vols. Istanbul: Iqdām Matbaası.
Exarchos, Ioannis. 1982. "I ktinotrofia stin Epeirou." *Epeirotiko imerologio* 4:199–228.
Exposition Coloniale Internationale. 1931. *Congrès de la production forestière coloniale et nord africaine.* Paris: Quinzaine Nationale de la Production Agricole.
Fadlouliah, Abdellatif. 1987, "L'évolution récente de la population dans le haut Rif central." In *Hommage à Gérard Maurer,* 463–82, Poitiers: Université de Poitiers.
Fairbank, John K. 1987. *The Great Chinese Revolution, 1800–1985.* New York: Harper & Row.
Faltaits, K. 1930. *Oi Epeirotai pou xenitevontai.* Athens: Sofichnopoulos.
Fanitsios, Vasilis. 1959. *To dasiko zitima st'Agrapha, sto Zagor' kai s'ola ta vounisixmas choria.* Athens.
Faroqhi, Suraiya. 1990. "Towns, Agriculture and the State in Sixteenth-Century Ottoman Anatolia." *Journal of the Social and Economic History of the Orient* 33:125–56.
Fay, Gérard. 1972. "Recherches sur l'organisation de la vie rurale et sur les conditions de la production dans la basse montagne rifaine." (2 vols. Thèse, Université de Paris VI.
Fay, Gérard. 1979. "L'évolution d'une paysannerie montagnarde: Les Jbalas sud-rifains." *Méditerranée,* 3d ser., 35, nos. 1–2:81–91.
Fay, Gérard. 1984. "Un projet agro-sylvo-pastorale pour le Rif Occidental (Tanghaya)." *Revue de géographie du Maroc* 8:3–22.
Fay, Gérard. 1987. "L'aménagement des espaces ruraux du Maroc: Un impératif et une urgence." In *Hommage à Gérard Maurer,* 391–414. Poitiers: Université de Poitiers.
Fellows, Charles. 1838. *Travels and Researches in Asia Minor.* London: John Murray.
Fernández de Castro y Pedrera, Rafael. 1911. *El Rif: los territorios de Guelaia y Quebdana.* Malaga: Zambrana Hermanos.

Fernández de Castro y Pedrera, Rafael. 1945. *Melilla prehispánica.* Madrid: Instituto de Estudios Políticos.

Fernández Liencres y Herrera, Francisco. 1888. *Causas de la decadencia de la agricultura en la provincia de Granada y medios para remediarlo.* Granada: Ventura.

Ferrer, Manuel E., Enrique López, and Joaquín Bosque Maurel. 1971. *Sierra Nevada.* Granada: Anel.

Festa nazionale della montagna per L'Italia meridionale. 1968. Salerno: Ispettorato Forestale di Potenza.

Filangieri, Angerio. 1980. *Territoria e popolazione nell'Italia meridionale.* Milan: Angeli.

Filardi, Giuseppe, ed. n.d. *Campomaggiore. Storia di un paese lucano.* 2 vols. Matera: BMG.

Firat, Fehim. 1965. "Forest and Erosion Problems in Turkey." *Istanbul Üniversitesi. Orman Fakültesi Dergisi* 15, no. 1:29–46.

Fittipaldi, E. 1880. *Potenza e l'Alto bacino del fiume Basento.* Potenza: Mogaldi.

Floristan, A., and J. Bosque Maurel. 1957. "Movimientos migratorios de población en la provincia de Granada." *Estudios geográficos* 18:361–402.

Florschutz, F., J. Menéndez Amor, and T. A. Wijmstra. 1971. "Palynology of a Thick Quatenary Succession in Southern Spain." *Palaeogeography, Palaeoclimate, Palaeoecology* 10:233–64.

Foerster, Robert F. 1924. *The Italian Emigration of Our Times.* Cambridge, Mass.: Harvard University Press.

Fogg, Walter. 1939. "Tribal Markets in Spanish Morocco." *Journal of the Royal African Society* 39:322–6.

Fogg, Walter. 1940. "Villages, Tribal Markets and Towns: Some Considerations Concerning Urban Development in the Spanish and International Zones of Morocco." *Sociological Review* 32:85–107.

Font y Quert, P. 1927. "Els Cedres del Rif." *Ciència. Revista catalana de ciència i tecnologia* 2:521–30.

Forbes, Rosita. 1924. *Sultan of the Mountains: The Life Story of Raisuli.* New York: Henry Holt.

Ford, Richard. [1845] 1966. *A Handbook for Travellers in Spain.* 2 vols. Carbondale: Southern Illinois University Press.

Formica, Carmelo. 1975. "Esodo agricolo e trasformazioni dell'agricoltura nel Mezzogiorno." *Annali del Mezzogiorno* 15:93–141.

Formica, Carmelo. 1979. *Lo spazio rurale nel Mezzogiorno: Esodo, desertificazione e riorganizzazione.* Naples: Edizioni scientifiche italiane.

Formica, Carmelo. 1983. "Trasformazione dell'ambiente e paesaggio agrario nel Mezzogiorno." *Nord e Sud* 30:89–106.

Foti, Giuseppina, Francesca Paolino, and Francesco Suraci. 1983. *Il disegno dell'ambiente: Analisi della struttura urbana agricola e storico-ambientale della Comunità montana Versante jonico meridionale.* Reggio Calabria: Laruffa.

Foucauld, Charles Eugène. 1888. *Reconnaissance au Maroc.* 2 vols. Paris: Challamel.

Franchetti, Leopoldo. 1875. *Le condizioni economiche ed amministrative delle provincie napoletane.* Florence: Gazzetta d'Italia.

Franciosa, Luchino. 1926. "Note sociali e di economia agraria sulla Basilicata." *L'Italia agricola* 2:58–80.

Franciosa, Luchino. 1927a. *Agricoltura e movimento demografico nelle Puglie e nella Basilicata.* Turin: Pozzo.

Franciosa, Luchino. 1927b. *Problemi e questioni della Basilicata, 1923–25.* Rome: Ricciotti.

Franciosa, Luchino. 1930a. *Basilicata. Rapporti fra proprietà, impresa e mano d'opera nell'agricoltura italiana.* Rome: Istituto Nazionale di Economia Agraria.

Franciosa, Luchino. 1930b. "Il progresso demografico del Mezzogiorno nell'ultimo cinquantennio." In *Congresso (XI) geografico italiano. Atti II,* 235–42. Naples: Società geografica italiana.

Franciosa, Luchino. 1930c. "Le migrazioni interne nel Mezzogiorno." In *Congresso (XI) geografico italiano. Atti III,* 68–73. Naples: Società geografica italiana.

Franciosa, Luchino. 1942. *La casa rurale nella Lucania.* Florence: Consiglio Nazionale delle Ricerche.

Franciosa, Luchino. 1948. "Distribuzione planimetrica e altimetrica della popolazione nella Basilicata." *Bollettino della Società geografica italiana* 8, nos. 5–6:269–83.

Franciosa, Luchino. 1951. "La transumanza nell'Appennino centro-meridionale." *Memorie di geografia economica* 4:5–99.

Franzoni, Ausonio. 1903. *L'emigrazione in Basilicata.* Brescia: Unione Tipo-Lit Bresciana.

Fréjus, Roland. 1922. *Relation d'un voyage fait dans la Mauritanie en Afrique . . . en 1666.* In H. de Castries, ed., *Sources Inédites de l'Histoire du Maroc,* 2d ser. 1: 118–88. Paris: Leroux.

French, M. H. 1970. *Observations on the Goat.* Rome: FAO Agricultural Studies, no. 80.

Frischauf, Johannes. 1893. *Eine Wallfahrt nach Dodona.* Graz: Leuschner & Lubensky.

Gabaccia, Donna. 1988. *Militants and Migrants: Rural Sicilians Become American Workers.* New Brunswick, N.J.: Rutgers University Press.

Galanti, Giuseppe Maria. 1788–94. *Nuova descrizione storica geografica delle Sicilie.* 4 vols. Naples: Letterario.

Galasso, Giuseppe. 1958. "Popolazione meridionale del 1861 al 1951." *Nord e Sud* 5:48–95.

Galasso, Giuseppe. 1982. *L'altra Europa. Per un'antropologia storica del Mezzogiorno d'Italia.* Milan: Mondadori.

Gallego Roca, Francisco Javier. 1987. *Morfología urbana de las poblaciones del reino de Granada a través del Catastro del Marqués de la Ensenada.* Granada: Diputación Provincial.

Gambi, G. 1955. "Equilibrio agro-silvo-pastorale e spopolamento montano." *Monti e boschi* 6:275–76.

García Barbancho, D. 1967. *Las migraciones interiores españolas. Estudio cuantativo desde 1900.* Madrid: Escuela Nacional de Administración Pública.

García Cicuendez, Pío. 1937. *Ganadería y pastizales.* Tetouan: Hispania.

García Cuenca, Tomás. 1948. "La ganadería en Marruecos." In *Conferencias desarolladas en la Academia de Interventores durante 1947*, 53–84. Tetouan: Majzen.

García Fernández, Paulino. 1985. *Población de los actuales terminos municipales, 1900–1981.* Madrid: Instituto Nacional de Estadística.

García Figueras, Tomás. 1957. *España y su protectorado en Marruecos (1912–1956).* Madrid: Consejo Superior de Investigaciones Científicas.

García Figueras, Tomás, and R. de Roda Jiménez. 1950–5. *Economía social de Marruecos.* 3 vols. Madrid: Instituto de Estudios Africanos.

García Maceira, Antonio. 1916. *Los montes y la emigración.* Madrid: Imprenta Alemana.

García Manrique, Eusebio. 1982. "Las actividades humanas como favorecidas de la erosión." In Junta de Andalucía, Dirección General de Medio Ambiente, *La erosión de los suelos de Andalucía*, 45–55. Seville.

García Martín, Pedro, and José María Sánchez Benito. 1986. *Contribución a la historia de la transhumancia en España.* Madrid: Ministerio de Agricultura, Pesca y Alimentación.

García Najera, José María. 1929. "Estudio-proyecto de corrección y fijación de los terrenos en movimiento sobre que asientan los pueblos de Carataunas y Soportujar (Granada)." In Instituto Forestal de Investigaciones y Experiencias, *Trabajos de las secciones: Hidraúlica torrencial, Combustibles vegetales, Flora y mapa forestal, Resinas* II, 3:5–36.

García Ruíz, José M., and Teodoro Lasanta-Martínez. 1990. "Land Use Changes in the Spanish Pyrenees." *Mountain Research and Development* 10:267–79.

García Viana, J. M. 1943. *Repoblación forestal.* Tetouan: Alta Comisaria de España en Marruecos.

Garrabou, Ramon, Carlos Barciela, and J. I. Jiménez Blanco, eds. 1986. *Historia agraria de la España contemporánea. III. El fin de la agricultura tradicional (1900–1960).* Barcelona: Crítica.

Garrabou, Ramon, and Jesús Sanz, eds. 1985. *Historia agraria de la España contemporánea. II. Expansión y crisis, 1850–1900.* Barcelona: Crítica.

Garrad, K. 1956. "La indústria sedera granadina en el siglo XVI y su conexión con el levantamiento de las Alpujarras (1568–71)." *Miscelánea de estudios arabes y hebraicos* 5:73–98.

Garzón Pareja, Manuel. 1972. *El arte de la seda en Granada.* Granada: Gráficas del Sur.

Garzón Pareja, Manuel. 1984. "Repoblación y agricultura en Granada." In *Congreso de Historia Rural. Siglos XV al XIX. Actas del Coloquio de Octubre 1981*, 561–72. Madrid: Universidad Complutense.

Gatta, Costantino. [1732] 1966. *Memorie topografiche-storiche della provincia di Lucania.* [Naples: Muzio.] Bologna: Forni.

Gaudio, Attilio. 1981. *Maroc du nord.* Paris: Nouvelles Editions Latines.

Gaudioso, Francesco. 1987. *Calabria ribelle.* Milan: Franco Angelli.

Gavioli, O. 1932. "Contributo alla flora del M. Pollino." *Archivio botanico* 8:32–40.

Gavioli, O. 1934. "Limiti altimetrici delle formazioni vegetali in alcuni gruppi dell'Appennino Lucano." *Nuovo giornale botanico italiano* 41:558–673.

Gavioli, O. 1936. "Ricerche sulla distribuzione altimetrica della vegetazione in Italia. III. Limiti altimetrici delle formazioni vegetali nel gruppo del Pollino." *Nuovo giornale botanico italiano* 43:636–706.

Gaviolo, O. 1947. "Synopsis florae Lucanae." *Nuovo giornale botanico italiano* 54:10–278.

Gazzo, Yves. 1979. *Afrique du Nord: D'hier à demain.* Paris: Economica.

Geertz, Clifford. 1963. *Agricultural Involution.* Berkeley: University of California Press.

Gell, William. 1827. *The Itinerary of Greece.* London: J. Rodwell.

Gerber, Haim. 1987. *The Social Origins of the Modern Middle East.* Boulder, Colo.: Lynne Riemer.

Germano, Elisabetta. 1909. *La Basilicata. Monografia geografica.* Florence: Susini.

Gharbaoui, Ahmed. 1986. "L'homme et la montagne dans la dorsale calcaire du Rif." *Revue de l'Occident Musulmane et de la Méditerranée* 41–42:197–207.

Ghirelli, Angelo. 1926. *El norte de Marruecos.* Melilla: Artes Gráficas–Postal Expres.

Giandotti, Mario. 1916. *Boschi e acque.* Rome: Genio Civile.

Gibbon, Edward. 1909. *The History of the Decline and Fall of the Roman Empire.* 7 vols. London: Methuen.

Gigon, Andreas. 1983. "Typology and Principles of Ecological Stability and Instability." *Mountain Research and Development* 3:95–102.

Gilliéron, Alfred. 1877. *Grèce et Turquie.* Paris: Sandoz & Fischbacher.

Giménez-Serrano, José. 1846. *Manual del artista y del viagero en Granada.* Granada: Puehol.

Giordano, Guglielmo. 1962. "Montagne e foreste del Libano." *Monti e boschi* 13:105–22.

Giusti, Ugo. 1943. *Caratteristiche ambientali italiane, agrarie, sociali, demografiche, 1815–1942.* Rome: Istituto Nazionale di Economia Agraria.

Giustiniani, Lorenzo. 1797–1805. *Dizionario geografico-ragionato del Regno di Napoli.* 13 vols. Naples: Manfredi.

Glesinger, Egon. 1960. "The Mediterranean Project." *Scientific American* 203: 86–103

Godoy López, L. 1979. *La ganadería andaluza.* Granada: Universidad de Granada.

Goicoechea Acosta, Margarita. 1982. "Aproximación al estudio de la influencia del clima en la erosión. El caso de la Alpujarra Alta granadina." *Cuadernos geográficos* 11:153–81.

Gómez-Campo, C. ed. 1985. *Plant Conservation in the Mediterranean Area.* The Hague: W. Junk.

Gómez-Campo, C., and J. Malato-Beliz. 1985. "The Iberian Peninsula." In C. Gómez-Campo, ed., *Plant Conservation in the Mediterranean Area,* 47–70. The Hague: W. Junk.

Gómez Oliver, Miguel. 1985. *La desamortización de Madoz en la provincia de Granada.* Granada: Centro de Estudios Históricos de Granada y su Reino.

González Montoya, J. 1821. *Paseo estadístico por las costas de Andalucía, desde Sevilla a Granada, en el verano de 1820.* Madrid: León Almarita.

González-Rothvoss, Mariano. 1953. "Influencia de la emigración en el crecimiento de la población española en los ultimas cien años, 1850–1950." *Revista Internacional de Sociología* 41:61–84.

González Vázquez, Ezequiel. 1945. *El Marruecos forestal.* Madrid: Técnicas.

Gould, A. G. 1973. "Pashas and Brigands: Ottoman Provincial Reform and Its Impact on the Nomadic Tribes of Southern Anatolia, 1840–1885." Ph.D. diss. UCLA.

Gozalbes Cravioto, Enrique. 1982. *Atlas arqueológico del Marruecos mediterráneo.* Granada: Arte.

Granada. Consejo Económico Sindical. 1961. *Las Alpujarras. Ponencias provisionales.* Granada: Organización Sindical Granada.

Granada. Gobierno Civil. 1985. *Granada en el Mercado Común.* Granada.

Grant, Michael, and Rachel Kitzinger, eds. 1988. *Civilization of the Ancient Mediterranean.* 3 vols. New York: Scribner's.

Greece. Ethniki Statistiki Iperesia. 1955. *Plithismos tis Ellados . . . 1951 – Population de la Gréce . . . 1951.* Athens: Ethnikon.

Greece. Ethniki Statistiki Iperesia. 1958. *Apotelesmata Yeorgikis Apografis Tis Ellados 1950 – Résultats du recensement de l'agriculture de la Grèce 1956.* Athens.

Greece. Ethniki Statistiki Iperesia. 1962. *Plithismos tis Ellados . . . 1961 – Population de la Grèce . . . 1961.* Athens: Ethnikon.

Greece. Ethniki Statistiki Iperesia. 1963. *Apotelesmata tis apografis plithismou-katoikou tis 19 martiou 1961.* Athens.

Greece. Ethniki Statistiki Iperesia. 1966. *Apotelesmata tis Apografis Yeorgias-Ktinotrofias tis 19 Martiou 1961 – Résultats du recensement de l'agriculture-élévage effectué le 19 Mars 1961.* 2 vols. Athens.

Greece. Ethniki Statistiki Iperesia. 1972. *Plithismos tis Ellados . . . 14 Martiou 1971 – Populaton de la Grèce . . . du 14 mars 1971.* Athens.

Greece. Ethniki Statistiki Iperesia. 1978. *Apotelesmata tis apografis yeorgias kai ktinotrofias tis 14 martiou 1971 – Results of the Agriculture-Livestock Census of March 14, 1971.* Athens.

Greece. Ethniki Statistiki Iperesia. 1982. *Plithismos tis Ellados . . . 5 Apriliou 1981 – Population de la Grèce . . . 5 avril 1981.* Athens.

Greece. Ethniki Statistiki Iperesia. 1986. *Statistical Yearbook of Greece.* Athens.

Greece. Iperesia Syntonismou [C. A. Doxiadis]. 1950. *Development of the Louros-Arachthos Valleys.* Athens.

Greece. Ipourgeion Esoterikon. 1884. *Pinakes tou eparchiou Epeirou kai Thessalias kata tin apografin tou 1881.* Athens: Perri.

Greece. Ipourgeion Stratiotikon. 1880. *Odoiporiká Epeirou kai Thessalias.* Athens: Ethnikou Typografeiou.

Greece. National Statistics Office. 1962. *Report on the Exploratory Survey into Motivations and Circumstances of Rural Migration.* Athens.

Greece. Stratiotikou. 1878. *Pinakes odoiporikoi Makedonias, Thessalias kai Epeirou.* Athens.

Greece. Yeniki Statistiki Iperesia. 1928. *Apografi tou plithismou tis Ellados . . . 1920 – Recensement de la population de la Grèce . . . 1920.* Athens: Ethnikon.

Greece. Yeniki Statistiki Iperesia. 1929. *Plithismos tis Ellados kata tin apografi tis 15–16 Maiou 1928.* Athens: Ethnikon.

Greece. Yeniki Statistiki Iperesia. 1946. *Plithismos tis Ellados . . . 1940 – Population de la Grèce . . . 1940.* Athens: Ethnikon.

Greg, W. R. 1833. *Sketches in Greece and Turkey.* London: Ridgeway.

Gresa de Camps, Santiago. 1903. *Las Kabilas de Bocoya, Beniburiaga y Tlemsámana.* Barcelona: F. Sánchez.

Grimaldi, Domenico. 1770. *Saggio di economia campestre per la Calabria Ulteriore.* Naples: Orsini.

Grisebach, A. von. 1841. *Reise durch Rumelien und nach Brussa im jahre 1839.* 2 vols. Göttingen: Vandenhoeck & Ruprecht.

Grispos, Panos. 1973. *Dasiki historia tis neoteras Hellados: apo tou IE aionos mechri tou 1971.* Athens: Iperesias Dasikon.

Grmek, Mirko. 1989. *Diseases in the Ancient Greek World.* Baltimore: Johns Hopkins University Press.

Grohmann-Kerouach, Brigitte. 1971. *Der Siedlungsraum der Ait Ouriaghel im östlichen Rif. Kulturgeographie eines Ruckzugsgebeites.* Heidelberg: Geographischen Instituts der Universität Heidelberg.

Gronhaug, Reidar. 1974. *Micro-Macro Relations: Social Organization in Antalya, Southern Turkey.* Bergen: Department of Anthropology, University of Bergen.

Groome, Helen J. 1990. *Historia de la politíca forestal en el estado español.* Madrid: Comunidad de Madrid.

Grove, Jean. 1988. *The Little Ice Age.* London: Routledge.

Guarducci, A. ed. 1984. *Agricoltura e trasformazione dell'ambiente: secoli XIII–XVIII.* Florence: Istituto Internazionale di Storia Economica "F. Datini."

Gucci, Eugenio. 1960. "Osservazioni sulla conservazione del bosco e sulle possibilità di nuovi rimboschimenti nel Lagonegrese." *Monti e boschi* 11: 66–71.

Guerraoui, Driss. 1986. *Agriculture et développement au Maroc.* Paris: Publisud.

Guillaume de Vaudoncourt, Frédéric. 1816. *Memoirs on the Ionian Islands Considered in a Commercial, Political, and Military Point of View . . .* London: Baldwin, Craddock & Joy.

Guillou, André. 1965. "La Lucanie Byzantine." *Byzantion* 25:119–49.

Gulliet, David W. 1984. "Agro-pastoral Land Use and the Tragedy of the Commons in the Central Andes." In P. D. Beaver and B. L. Purrington, eds., *Cultural Adaptations to Mountain Environments,* 12–23. Athens: University of Georgia Press.

Güngör, Kemal. 1941. *Cenubî Anadolu yürüklerinin etno – antropolojik tetkiki.* Ankara: İdeal Başımevi.

Gürdal, M. 1976. "Antalya Yürükleri." *Türk Etnografya Dergisi* 15:65–72.

Hadjigeorgiou, Th. 1958. *I apodimia ton Epeiroton.* Athens: Epeirotiki Bibliothiki.

Hahn, Johann Georg von. 1854. *Albanesische Studien.* Jena: Mauke.

Hamilton, William John. 1842. *Researches in Asia Minor, Pontus and Armenia.* 2 vols. London: J. Murray.

Hammond, Nicholas G. L. 1967. *Epirus: The Geography, the Ancient Remains, the*

History and the Topography of Epirus and Adjacent Areas. Oxford: At the Clarendon Press.

Hammond, Nicholas G. L. 1983. *Venture into Greece: With the Guerrillas, 1943–44.* London: W. Kimber.

Hardin, Garrett. 1968. "The Tragedy of the Commons." *Science* 162:1243–8.

Hardin, Garrett. 1985. *Filters against Folly.* New York: Viking.

Harris, Walter B. 1888. "My Ride to Sheshwan." *Blackwood's Magazine* (December):786–92.

Harris, Walter B. 1921. *Morocco That Was.* London: William Blackwood & Sons.

Harris, Walter B. 1927. *France, Spain and the Rif.* New York: Longmans, Green & Co.

Hart, David Montgomery. 1976a. *The Aith Waryaghar of the Moroccan Rif.* Tucson: University of Arizona Press.

Hart, David Montgomery. 1976b. "De 'Ripublic' à 'République' les institutions socio-politiques rifaines et les reformes d'Abd el-Krim." In *Abd el-Krim et la république du Rif,* 33–45. Paris: Maspero.

Hart, David Montgomery. 1987. *Banditry in Islam.* London: Menas.

Hatzimpiros, Kimon. 1980. "Edaphos – provlimata diavrosis." In Kentro Erevnon Zagoriou, *Praktika oikologikis synantisis,* 30–3. Ioannina (?).

Heichelheim, Fritz. 1956. "Effects of Classical Antiquity on the Land." In W. L. Thomas, ed., *Man's Role in Changing the Face of the Earth,* 1:165–82. Chicago: University of Chicago Press.

Heilmann, P. G. F. 1972. *On the Formation of Red Soils in the Lower Crati Basin [S. Italy].* Enschede, Netherlands: International Institute for Aerial Survey and Earth Sciences.

Helle von Samo, A. Ritter zur. 1877. *Die Völker des osmanischen Reiches.* Vienna: Carl Gerold's Sohn.

Helling, Barbara, and George Helling. 1958. *Rural Turkey: A New Socio-Statistical Appraisal.* Istanbul: Istanbul Üniversitesi Yayını 795.

Hémardinquer, Jean-Jacques. 1973. "Les débuts du Maïs en Méditerrannée (Premier aperçu)." In *Histoire économique du monde méditerranéen: Mélanges en l'honneur de F. Braudel,* 227–34. Toulouse: privately published.

Herr, Richard. 1988. *Rural Change and Royal Finances in Spain at the End of the Old Regime.* Berkeley: University of California Press.

Heusch, B. 1981. "Sociological Constraints in Soil Conservation: A Case Study." In R. P. C. Morgan, *Soil Conservation: Problems and Prospects,* 419–24. New York: John Wiley & Sons.

Higgs, Eric. 1978. "Environmental Changes in Northern Greece." In William G. Brice, ed., *Environmental History of the Near and Middle East,* 41–9. London: Academic Press.

Histoire économique du monde méditerranéen. Mélanges en l'honneur de F. Braudel. 1973. Toulouse: privately published.

Historia de Andalucía. 1982. 9 vols. Madrid: CUPSA.

Hobhouse, J. C. 1813. *A Journey through Albania and Other Provinces of Turkey . . .* 2 vols. London: Cawthorn.

Hobsbawm, E. J., 1965, *Primitive Rebels.* New York: Norton.

Hobsbawm, E. J. 1969. *Bandits.* London: Weidenfeld and Nicolson.

Høeg, Carsten. 1925–6. *Les Saracatsans. Une tribue nomade grecque.* 2 vols. Copenhagen: Pio-Branner.

Hofer, Philip. 1967. *Edward Lear as a Landscape Draughtsman.* Cambridge: Harvard University Press.

Holland, Henry. 1819, *Travels in the Ionian Isles, Albania, Thessaly, Macedonia, etc. during the years 1812 and 1813.* 2 vols. London: Longman, Hurst, Rees, Orme and Brown.

Hommage à Gérard Maurer. 1987. Poitiers: Université de Poitiers.

Hooker, Katharine. 1927. *Through the Heel of Italy.* New York: Henke.

Horwitz, Ellie. 1976. "Forestry in Greece." *Journal of Forestry* 74 (November):780.

Houssaye, Henri. 1879. "La Grèce et les provinces grecques de la Turquie." *Revue des deux mondes*: 840–57.

Houston, J. M. 1967. *The Western Mediterranean World.* New York: Praeger.

Hughes, J. Donald. 1975 *Ecology in Ancient Civilizations.* Albuquerque: University of New Mexico Press.

Hughes, J. Donald. 1982. "Deforestation, Erosion, and Forest Management in Ancient Greece and Rome." *Journal of Forest History* 26:60–75.

Hughes, J. Donald. 1988. "Land and Sea." In Michael Grant and Rachel Kitzinger, eds., *Civilization of the Ancient Mediterranean,* 1:89–133. New York: Scribner's.

Hughes, Thomas Smart. 1820. *Travels in Sicily, Greece, and Albania.* 2 vols. London: Colburn & Bentley.

Hugonie, G. 1976. "L'évolution récente de l'utilisation des sols montagnards en sicile septentrionale." *Méditerranée,* 2d ser., 24, no. 1:3–17.

Hurtado de Mendoza, Diego. [1573] 1970. *Guerra de Granada,* edited by Bernardo Blanco-González. Madrid: Clásicos-Castilia.

Hutchinson, George Evelyn. 1978. *An Introduction to Population Ecology.* New Haven: Yale University Press.

Hutchinson, J. 1969. "Erosion and Land Use: The Influence of Agriculture on the Epirus Region of Greece." *Agriculture History Review* 17:85–90.

Hütteroth, W. D. 1968. *Landliche Siedlungen in südlichen Inneranatolien.* Göttingen: Geographischen Instituts der Universität Göttingen.

Ibarruri, Dolores. 1984. *They Shall Not Pass.* New York: International Publishers.

Ibn Battuta. 1953. *Travels in Africa and Asia,* translated by H. A. R. Gibb. London: Routledge & Kegan Paul.

Ibrahim, Manzour Efendi. 1928. *Mémoires sur la Grèce et l'Albanie pendant le gouvernement l'Ali-Pacha.* Paris: J. N. Barba.

Iglesías Casado, Antonio. 1982. "El papel del estado ante el problema de la erosión." In Junta de Andalucía, Dirección General de Medio Ambiente, *La erosión de los suelos de Andalucía,* 69–75. Seville.

Imber, C. H. 1980. "The Navy of Suleyman the Magnificent." *Archivum Ottomanicum* 6:211–82.

Imeson, A. C., et al. 1982. "The Relationship of Soil Physical and Chemical Properties to the Development of Badlands in Morocco." In R. B. Bryan and A. Yair, eds., *Badland Geomorphology and Piping*, 47–70. Norwich: Geobooks.

Inalcik, Halil. 1960. "Bursa and the Commerce of the Levant." *Journal of the Economic and Social History of the Orient* 3:131–47.

Inalcik, Halil. 1973. *The Ottoman Empire: The Classical Age 1300–1600*. London: Weidenfeld & Nicholson.

Inchiesta Zanardelli sulla Basilicata. 1976. Edited by Paola Corti. Turin: Einaudi.

Institut Agronomique Méditerranéen de Montpellier, Centre International des Hautes Etudes Agronomiques Méditerranéennes. 1979. *Analyse comparative des structures agricoles au niveau régional de l'Espagne, la France, la Grèce, l'Italie et le Portugal.* 2 vols. Montpelier.

Institut Français de Cooperation Technique et Faculté d'Agronomie de Karadj (Iran). 1961. *Colloque sur la conservation et la restauration des sols (Teheran, 1960). Compte rendu générale.* Coutances: Bellée.

Institut Français du Pétrole (Rueil-Malmaison). 1966. *Etude géologique de l'Epire.* Paris: Editions Technip.

Ippolito, F., and V. Cotecchia. 1954. "Le frane ed i dissesti nelle medie valli dell'Agri e del Sinni." *Geotecnica* 1:45–65.

Isambert, Emile. 1878a. *Itinéraire descriptif, historique et archéologique de l'Orient*. 3 vols. Paris: Hachette.

Isambert, Emile. 1878b. *Odoiporiká. Makedonias, Epeirou, kai Thessalias.* Athens: Miliaraki.

Isnardi, Giuseppe. 1954. "Calabria: natura e uomini." *Almanacco calabrese*, 161–71.

Isparta 1967. Il Yıllığı. 1969? Ankara: Ajans-Türk Matbaacılık.

Issawi, Charles, ed. 1980. *The Economic History of Turkey, 1800–1914*. Chicago: University of Chicago Press.

Istanbullu, T. 1982. "Forest Ownership in Relation with Land Use Policy in Turkey." In *Actes du Symposium Internationale d'Histoire Forestière*, 2:87–90. Nancy: Ecole Nationale du Génie Rural.

Italy. Commissione Interministeriale per lo Studio della Sistemazione Idraulica e della Difesa del Suolo. 1970. *Atti della commissione. I. Relazione conclusiva.* Rome: Chiarini.

Italy. Giunta per l'Inchiesta Agraria e sulle Condizioni della Classe Agricola [F. Jacini]. 1881–6. *Atti della Giunta.* 15 vols. Rome: Forzani.

Italy. Istituto Centrale di Statistica. 1934. *Catasto agrario 1929. f. 77: Potenza.* Rome: Istituto Poligrafico dello Stato.

Italy. Istituto Centrale di Statistica. 1963–70. *Primo censimento generale dell'agricoltura (1961)*. 7 vols. Rome: ABETE.

Italy. Istituto Centrale di Statistica. 1967. *Popolazione residente e presente dei comuni ai censimenti dal 1861–1961*. Rome.

Italy. Istituto Centrale di Statistica. 1983–6. *12° censimento generale della popolazione, 25 ottubre 1981*. Rome.

Italy. Istituto Centrale di Statistica. 1987. *Annuario statistico italiano 1987*. Rome.

Italy. Istituto Centrale di Statistica. 1989. *La superficie forestale nelle comunità montane al 31 dicembre 1986.* ISTAT collana d'informazione, no. 5. Rome.

Italy. Istituto Nazionale di Economia Agraria. 1931–40. *Monografie di famiglie agricole.* 17 vols. Rome.

Italy. Istituto Nazionale di Economia Agraria. 1932. *Inchiesta sulla piccola proprietà coltavatrice formatasi nel dopoguerra. VIII. Basilicata.* Rome: Operaia Romana.

Italy. Istituto Nazionale di Economia Montana. 1978a. *Le aziende agricole e forestali nelle communità montane. III. Appennino meridionale.* Bologna: Pàtron.

Italy. Istituto Nazionale di Economia Montana. 1978b. *Il ruolo delle communità montane per la difesa del suolo e l'attività forestale.* Rome.

Italy. Ministero dell'Agricoltura e delle Foreste. 1976. *Carta della montagna. II. Monografie Regionali. Basilicata.* Rome.

Italy. Ministero di Agricoltura, Industria e Commercio. 1907. *Le correnti di migrazione interna in Italia durante il 1905.* Rome.

Italy. Ministero dei Lavori Pubblici. Provveditorio Regionale alle OO.PP. per La Basilicata. 1963 (?). *Indagini preliminari per lo studio del piano territoriale di coordinamento della Basilicata. II. Documentazione statistiche.* Naples: L'Arte tipografica.

Izquierdo, Francisco. 1969. *El apócrifo de la Alpujarra Alta.* Madrid: Azu.

Jahandiez, E., and R. Maire. 1931–41. *Catalogue des plantes du Maroc.* 4 vols. Algiers: Minerva.

Jedlowski, Eduardo. 1956. "Aspetti dell'economia montana del Gruppo del Pollino." In *Edizione speciale in occasione della VII festa nazionale della montagna per le regioni dell'Italia meridionale,* 3–20. Rome: San Giuseppe.

Jedlowski, Eduardo. 1959a. "Panorama forestale del monte Pollino." *Monti e boschi* 10:499–513.

Jedlowski, Eduardo. 1959b. "Il quadro economico di un comune montano tipico della Lucania." *Monti e boschi* 10:110–21, 147–55.

Jeréz Perchet, Augusto. 1875. *Impresiones de viaje. Andalucía, El Riff, Valencia.* Madrid: Bailly-Bailliere.

Jiménez Blanco, José Ignacio. 1986. "La producción agraria de Andalucía Oriental, 1874–1914." 2 vols. Tesis, Universidad de Madrid Complutense.

Jiménez Ortoneda, Jesús. 1930. *Estudio de la Región del Rif.* Toledo: Colegio de María Cristina.

Jiménez Salas, José A. 1946, 1949. "Suelos de la zona norte del Protectorado de España en Marruecos." *Anales del Instituto Español de Edafología, Ecología y Fisología Vegetal* 5:277–305 and 8:3–31.

Jones, E. L. 1981. *The European Miracle.* Cambridge: Cambridge University Press.

Jones, W. H. S. 1909. *Malaria and Greek History.* Manchester: Manchester University Press.

Jordana, J. 1870. "Desamortización forestal." *Revista forestal* 3:256–69.

Jordana y Morera, José, et al. 1882. *La producción agrícola y forestal de la Argelia.* Madrid: El Correo.

Junta de Andalucía. 1987. *Características socioeconómicas de las comarcas andaluzas.* Seville.

Junta de Andalucía. Dirección General de Medio Ambiente. 1982. *La erosión de los suelos de Andalucía.* Seville.

Jutglar, Antoni, et al. 1968. *La immigració a Catalunya.* Barcelona: Edició de Materials.

Kaftantzoglou, Roxani, and Matina Naoumi. 1988. "Morfes oikoyeneias sto Syrrako." In Ethniko Kentro Koinonikon Erevnon and Kentro Neoellinikon Erevnon, *O agrotikos kosmos ston Mesoyeiako choro,* 129–30. Athens.

Kandrelis, Sotirios. 1980. "Dasos, perivallon, anthropos." In Kentro Erevnon Zagoriou, *Praktika oikologikis synantisis,* 66–76. Ioannina (?).

Karal, Enver Ziya. 1943. *Osmanli Imparatorluğunda Ilk Nüfus Sayımı 1831.* Ankara: T. C. Başvekalet Istatistik Umum Müdürlügü.

Karpat, Kemal H. 1976. *The Gecekondu.* New York: Cambridge University Press.

Karpat, Kemal. H. 1978. "Ottoman Population Records and the Census of 1881/2–1893." *International Journal of Middle East Studies* 9:237–74.

Karpat, Kemal H. 1985. *Ottoman Population, 1830–1914: Demographic and Social Characteristics.* Madison: University of Wisconsin Press.

Kasapliğil, Baki. 1952. "The Forest Vegetation in the Mediterranean Regions of Turkey." *Istanbul Üniversitesi. Orman Fakültesi Dergisi* 2, no. 2:57–65.

Kassandreas, P. 1966. "Axiopoisis tou apolamvanomen xilodous ongou ek ton dasos tis pindou." *Agrotiki Oikonomia:* 49–58.

Kavadias, Georges B. 1965. *Pasteurs-nomades méditerranéens: Les Saracatsans de Grèce.* Paris: Gauthier-Villars.

Kayser, Bernard. 1961. *Recherches sur les sols et l'érosion en Italie Meridionale.* Paris: Société d'Edition d'Enseignement Supérieur.

Kayser, Bernard. 1963. "L'érosion par franes en Lucanie." *Méditerranée,* lst ser., 4, no. 1:93–100.

Kayser, Bernard, P.-Y. Pechoux, and Michel Sivignon. 1971. *Exode rural et attraction urbaine en Grèce.* Athens: Centre National de Recherches Sociales.

Kemal, Yashar, 1978. *Memed My Hawk.* New York: Pocket Books.

Kenbib, M. 1984. "Contrebande d'armes et 'anarchie' dans le maroc précolonial (1844–1912)." *Dar el-Niaba,* no. 4:8–16.

Kentro Erevnon Zagoriou. 1980. *Praktika oikologikis synantisis.* Ioannina (?).

Kentro Programmatismou kai Oikonomikon Erevnon. 1976. *Programma Anaptixeos, 1976–1980. Tomeas dason.* Athens.

Kiepert, H. 1878. "Zur Ethnographie von Epirus." *Zeitschrift der Gesellschaft für Erdkunde.* Berlin: Reimer.

King, Russell. 1973. *Land Reform: The Italian Experience.* London: Butterworth.

Koeniguer, Jean-Claude. 1986. "Les changements climatiques au cours de l'histoire des derniers millénaires dans les régions de la Méditerranée occidentale." In F. López-Vera, ed., *Quaternary Climate in Western Mediterranean,* 549–63. Madrid: Universidad Autónoma de Madrid.

Kofinas, G. 1948. "Ta dasi tis Archaias Ellados." *To dasos* 2:3–8.

Kofiniotis, A. E. 1937. *Dasi kai cheimarroi.* Athens: Ipourgeiou Yeorgias.

Kolars, John F. 1963. *Tradition, Season, and Change in a Turkish Village.* Chicago: University of Chicago Department of Geography Research Paper, no. 82.

Kolars, John F. 1966. "Locational Aspects of Cultural Ecology: The Case of the Goat in Non-Western Agriculture." *Geographical Review* 56:577–84.

Koliopoulos, John. 1987. *Brigands with A Cause: Brigandage and Irredentism in Modern Greece, 1821–1922.* Oxford: At the Clarendon Press.

Koller, P. Angel. 1952. *Los Bereberes Marroquies. Estudio etnográfico.* Tetouan: Marroquí.

Komlos, John. 1989. *Nutrition and Economic Development in the Eighteenth-Century Hapsburg Monarchy.* Princeton, N.J.: Princeton University Press.

Kontos, Petros. 1906. *Dasi kai politismos.* Athens: Leonis.

Kontos, Petros. 1909. *Epidrasis epi ton Ellinikon dasikin vlastisin tou Ellenikou klimatos.* Athens: Leonis.

Kontos, Petros. 1929. *Dasiki Elliniki istoria.* Athens.

Kontos, Petros. 1932. *Dasi kai ktinitrofia eis tin Ellada apo oikonomikis kai politikis apopseos.* Thessaloniki: Triandafilos.

Kotoulas, Dimitrios. 1984. "Bodenabtrag und -ablagerungen in Greichenland am Beispiel des Gebirgslandes und der Ebene von Thessaloniki." Paper presented at the International Congress Interpraevent 1984, Villach.

Kotoulas, Dimitrios. 1989. "Synepeies apo tin Katastrofi ton dason." In Mouseio Goulardri Fisikis Istorias, *Ellinika dasi,* 75–81. Athens: Mouseio Goulandri Fisikis Istorias.

Krystallis, Kostas. n.d. *Apanta* Athens: Avlos.

Kuhnholtz-Lordat, G. 1939. *La Terre incendiée: Essai d'agronomie comparée.* Nîmes: Editions de la Maison Carrée.

Kürschner, Harald. 1984. *Der östlicher Orta Toroslar (Mittlerer Taurus) und angrenzende Gebiete. Eine formationskundliche Darstellung der Vegetation Südost-Anatoliens.* Wiesbaden: Ludwig Reichert Verlag.

Kyrgiannis, Miltos. 1984. *To pasaliki Ioanninov stin epochi tou Ali-pasa-tepeleni (1788–1822).* Athens: Strouboukis.

La Marca, Ferdinando. 1916. "Le industrie forestali nel Mezzogiorno d'Italia." In *Atti del III Congresso forestale italiano e I Congresso per l'irrigazione 1914,* 383–428. Portici: Torre.

La Martinière, H. M. P., and N. Lacroix, eds. 1894. *Documents pour servir à l'histoire du Nord-Ouest africain. I. Rif et Jbala.* Algiers: Gouvernement Général.

La Poer Beresford, G. de. 1855. *Scenes of Southern Albania.* London: Day & Sons.

Labollita, Giocchino. 1852. *Osservazioni generali sulle cause funeste della visibile perdita dei boschi dei Comuni, e di altri corpi morali di Regno delle due Sicilie.* Naples: De Marco.

Laborde, Alexander. 1809. *A View of Spain.* 5 vols. London: Longman, Hurst, Rees & Orme.

Laborde Vallverdu, Agustín. 1978. *Poqueira, bandera blanca.* Granada: Excellentíssima Diputación Provincial.

Lacava, M. 1890. *Viabilità in Basilicata.* Potenza: Garamone.

Lambridis, Ioannis. 1870. *Zagoriaka.* Athens: Avgis.

Lambridis, Ioannis. 1884. *Dokimion istorikis tinos perilipseos. Artis kai Prevezis.* Athens: Kallous.

Lambridis, Ioannis. 1887–90. *Epeirotika meletimata.* 10 vols. Athens: Varvarigos.

Larras, N. 1906. "La population du Maroc." *La géographie* 13:337–48.

Larrere, R., et al. 1982. "Les reboisements en montagne depuis l'Empire." In *Actes du Symposium Internationale d'Histoire Forestière*, 1:254–78. Nancy: Ecole National de Génie Rurale.

Latham, J. D. 1965. "The Reconstruction and Expansion of Tetuan: The Period of Andalusian Immigration," In George Makdisi, ed., *Arabic and Islamic Studies in Honor of Hamilton A. R. Gibb*, 387–408. Cambridge: Cambridge University Press.

Laure, Guido. 1953. "I boschi siciliani nella preistora, nella storia, nell'attualità." *Monti e boschi* 4:253–61.

Lawless, Richard I. 1977. "The Economy and Landscapes of Thessaly during Ottoman Rule." In Francis W. Carter, ed., *An Historical Geography of the Balkans*, 501–34. London: Academic Press.

Lazaridis, Kostas. 1972. *I Atomiki idioktisia sto Zagori*. Ioannina.

Lazos, Nikos. 1935. *Ta dasi tis Thessalias kai Epeirou*. Athens: Rodi.

Lazzari, Silvestro. 1986. "Criteri di interventi e linee operative per la conservazione e le difesa del suolo nella communità montana del Lagonegrese." In Basilicata (Regione), Ufficio Geologico Regionale, *Programmi ed interventi nel settore della difesa del suolo: l'assette idrogeologico nella Communità montana del Lagonegrese. Atti del Convegno*, 95–122. Matera: EDAD.

Le Houérou, Henri N. 1974. "Fire and Vegetation in the Mediterranean Basin." *Annual Proceedings of the Tall Timbers Fire Ecology Conference* (Tallahassee Tall Timbers Research Station) 13:237–77.

Le Tourneau, Roger. 1949. *Fès avant le protectorat*. Casablanca: Société marocain de librairie et d'édition.

League of Nations. 1925. *Deux rapports sur le paludisme en Asie Mineure, en Syrie et en Palestine*. Geneva.

Leake, William Martin. 1824. *Journal of a Tour in Asia Minor*. London: John Murray.

Leake, William Martin. 1835. *Travels in Northern Greece*. 4 vols. London: J. Rodwell.

Lear, Edward. 1852a. *Journals of a Landscape Painter in Albania, Illyria, etc.* London: Richard Bentley.

Lear, Edward. 1852b. *Journals of a Landscape Painter in Southern Calabria, etc.* London: Richard Bentley.

Lehasca, Attanasio. 1843. *Cenno storico dei servizi militari prestati nel regno delle due Sicilie dai Greci, Epiroti, Albanesi, e Macedoni in epoche diverse*. Corfu.

Lenormant, François. 1883. *A travers l'Apulie et la Lucane*. 2 vols. Paris: A. Lévy.

Leo Africanus [Al-Hassan ibn-Mohammed al-Wezaz al-Fasi]. 1896. *The History and Description of Africa*. 3 vols. London: Hakluyt.

Leo Africanus. 1956. *Description de l'Afrique*. Translated by A. Epaulard. Paris: Libraire d'Amérique et d'Orient.

Letsas, Alexandros. 1953. *I stavroforia tou prasinou*. Thessaloniki: Makedoniki Bilbliothiki.

Levi, Carlo. 1963. *Christ Stopped at Eboli*. Translated by Frances Frenaye. New York: Farrar, Straus & Giroux.

Lévi Provençal, Evariste. 1918. "Pratiques agricoles et fêtes saisonnières des tribus Djebalah de la vallée moyenne de l'Ouargah." *Archives Berberes* 4:83–108.

Liger, P. 1963. "Le vignoble rifain." *Revue de géographie du Maroc* 3–4:103–9.

Lindner, Rudi P. 1983. *Nomads and Ottomans in Medieval Anatolia.* Bloomington: Indiana University Uralic and Altaic Studies, no. 144.

Liszewski, Stanislaw, and Andrzej Suliborski. 1977. "Colonisation et économie en haute montagne: Trevelez." *Méditerranée,* 2d ser., 28, no. 1:65–72.

Livadas, G., and J. Sphangos. 1940–1. *Malaria in Greece.* 2 vols. Athens: Pyrsos.

Lloyd, Seton, and D. Storm Rice. 1958. *Alanya ('Ala'iyya).* London: Institute of Archaeology at Ankara, Occasional Publications, no. 4.

Locuratolo, Lucia. 1937, "Illustrazione grafica delle condizioni demografiche della Lucania dal 1870 al 1937." *Question Meridionali* 6:123–47.

Lombard, Maurice. 1959. "Un problème cartographié: Le bois dans la Méditerranée musulmane (VIIe–XIe siècles)." *Annales: Economies, Sociétés, Civilisations* 14:234–54.

López Casas, José, and Antonio Linde Rodríguez. 1984. *Magnitudes socio-económicas de la provincia de Granada.* Granada: Instituto de Promoción y Desarollo de la Diputación Provincial.

López-Cadenas de Llano, Filiberto. 1982. "La erosión: situación mundial y del mediterráneo andaluz." In Junta de Andalucía, Dirección General de Medio Ambiente, *La erosión de los suelos de Andalucía,* 21–7. Seville.

López-Vera, F., ed. 1986. *Quaternary Climate in Western Mediterranean.* Madrid: Universidad Autónoma de Madrid.

Lopinto, Michele. 1972. "La provincia di Matera tra due alluvioni." *Monti e boschi* 23, no. 6:39–50.

Lopinto, Michele. 1975. "Ancora alluvioni in Provincia di Matera." *Monti e boschi* 26, no. 1:37–43.

Lopreato, Joseph. 1967. *Peasants No More.* San Francisco: Chandler.

Losa Quintana, José María, Joaquín Molero Mesa, and Manuel Casares Porcel. 1984. *El paisaje vegetal de Sierra Nevada: La cuenca alta del Río Genil.* Granada: Universidad de Granada.

Louis, Herbert. 1939. *Das natürliche Pflanzenkleid Anatoliens.* Stuttgart: J. Engelhorns.

Lourido Díaz, Ramón. 1978. *Marruecos en la segunda mitad del siglo XVIII.* Madrid: Instituto Hispano-Arabe de Cultura.

Lucas, Paul. 1714. *Le voyage du Sieur Paul Lucas.* Amsterdam. Aux dépens de la Compagnie.

Luiggi, Luigi. 1916. "Torrenti, paludi, emigrazione e loro rimedi." In *Atti del III congresso forestale italiano e I congresso per l'irrigazione 1914,* 429–41. Portici: Torre.

Luna, Paco. 1984. *Demografía de la Alpujarra: estructura y biodinámica.* Granada: Diputación Provincial, Universidad de Granada.

Luque Baena, Enrique. 1974. *Estudio antropológico social de un pueblo del sur.* Madrid: Tecnos.

M.G.A.M. 1826. *Résumé géographique de la Grèce et de la Turquie d'Europe.* Paris: Dupart.

McCarthy, Justin. 1982. *The Arab World, Turkey and the Balkans. A Handbook of Historical Statistics.* Boston: G. K. Hall.

McCarthy, Justin. 1983. *Muslims and Minorities: The Population of Ottoman Anatolia and the End of the Empire.* New York: New York University Press.

McCay, Bonnie J., and James Acheson, eds. 1987. *The Question of the Commons.* Tucson: University of Arizona Press.

MacDonald, J. S. 1956. "Italy's Rural Social Structure and Emigration." *Occidente* 12:437–53.

MacDonald, J. S. 1963. "Agricultural Organization, Migration and Labor Militancy in Rural Italy." *Economic History Review,* 2d ser., 16:61–75.

McGowan, Bruce. 1981. *Economic Life in Ottoman Europe.* Cambridge: Cambridge University Press.

McGrew, William W. 1985. *Land and Revolution in Modern Greece, 1800–1881.* Kent, Ohio: Kent State University Press.

Madoz, Pascual. 1845–50. *Diccionario geográfico-estadístico-histórico de España y sus posesiones de ultramar.* 16 vols. Madrid: P. Madoz & L. Sagasti.

Magalhaes Godinho, V. 1963. "O milho maiz – Origem e difusão." *Revista de economia* 15, no. 1:33–8.

Maioui, Allal. 1984. "Le développement économique rural du Rif occidental: Un modèle d'administration de mission." Thèse: Université Mohammed V, Rabat.

Malherbe, Raoul de. 1846. *L'Orient, 1718–1845.* 2 vols. Paris: Gide.

Mammopoulos, Alexandros. 1966. *I xenetiea kai to dimotiko tragoudi.* Athens: Society for Epirot Studies.

Manners, Ian R., and Marvin W. Mikesell, eds. 1974. *Perspectives on Environment.* Washington, D.C.: Association of American Geographers.

Mansolas, Alexandros. 1867. *Politeiographikai plerophoriai peri Ellados.* Athens: Ethnikou.

Mansolas, Alexandre. 1872. *Rapport sur l'état de la statistique en Grèce présenté au Congrès International de Statistique de St. Pétersbourg en 1872.* Athens: Perris.

Mansolas, Alexandros. 1876. *Apografikai pliroforiai peri yeorgias kata to eto 1875.* Athens.

Mansolas, Alexandre. 1878. *La Grèce à l'Exposition Universelle de Paris en 1878. Notions statistiques.* Athens: Philocalie.

Marcos de Lanuza, Julio. 1964. *Estudios sobre el corcho de Quercus Suber.* Madrid: Instituto Forestal de Investigaciones y Experiencias.

Margaropoulos, P. 1961. "Reboisement et correction torrentielle des bassins versants de montagne par travaux d'art en Grèce." In Institut Français de Coopération Technique et Faculté d'Agronomie de Karadj (Iran), *Colloque sur la Conservation et la restauration des sols (Teheran, 1960). Compte rendu général,* 400–8. Coutances: Bellée.

Marín, Diego. 1896. "La suiza andaluza." *Boletín de la Sociedad geográfica* 38:177–209.

Marin, Cosme, 1913. "La question d'Epire." *La Vic* [magazine clipping in Gennadeion Library, Athens].

Marino, John. 1988. *Pastoral Economics in the Kingdom of Naples.* Baltimore: Johns Hopkins University Press.

Marlowe, John. 1964. *The Making of the Suez Canal.* London, Cresset.

Marmol Carvajal, Luís del. [1573] 1953. *Descriptión general de Africa.* 4 vols. Madrid: Consejo Superior de Investigaciones Científicas.

Marmol Carvajal, Luís del. 1797. *Historia de la rebelión y castigo de los moriscos del reyno de Granada.* 2 vols. Madrid: Sancha.
Marsh, George Perkins. [1864] 1965. *Man and Nature.* Cambridge: Belknap Press of Harvard University Press.
Martín Bolaños, Manuel. 1943. *Consideraciones sobre los encinares de España.* Madrid: Instituto Forestal de Investigaciones y Experiencias.
Martín Rodríguez, Manuel. 1982. *Historia económica de la Vega de Granada (siglos XV–XX): Una propuesta de interpretación malthusiana.* Granada: Universidad de Granada y Editorial Don Quijote.
Martínez Ruíz, Adolfo. 1977. *El reino de Granada en la guerra de la Independencia (1808–1814).* Granada: Diputación Provincial de Estudios y Promoción Cultural.
Martuscelli, Stefania. 1979. *La popolazione del Mezzogiorno nella statistica di Re Murat.* Naples: Guida.
Mas y Guindal, Joaquín. 1930. "Dos días en los bosques de cedros de Ketama." *Revista de tropas coloniales* 6:103–4.
Masachs Alavedra, V. 1942. "Régimen de los ríos anduluces." *Estudios geográficos* 3:865–80.
Mason, I. L. 1967. *Sheep Breeds of the Mediterranean.* Rome: FAO and Commonwealth Agricultural Bureaux.
Massafra, Angelo, ed. 1981. *Problemi di storia delle campagne meridionali nell'età moderna e contemporanea.* Bari: Dedalo.
Maurer, G. 1960. "Les genres de vie de la vallée de Tarzhout (Rif central)." *Notes morocaines* 15:27–9.
Maurer, G. 1968a. *Les montagnes du Rif central. Etude géomorphologique.* Rabat: Travaux de l'Institut Scientifique Cherifien.
Maurer, G. 1968b. "Les paysans de Haut Rif central." *Revue de géographie du Maroc* 14:3–70.
Maurer, G. 1976. "L'environnement géographique rifaine." In *Abd el-Krim et la République du Rif. Actes du colloque internationale,* 15–24. Paris: Maspero.
Maurer, G. 1979. "Les milieux naturels et leur aménagement dans les montagnes humides du domain rifain et tellien d'Afrique du Nord." *Méditerranée,* 3d ser., 35, no. 1:47–56.
Maurer, M. 1960. "Les pays rifains et prérifains." *Confluent* no. 9:550–6.
Mavrommatis, G. 1973. "Scheseis edafous-vlastiseos." *To dasos* 25, nos. 59–60: 11–20.
Mavrommatis, G. 1980. "To bioklima tis Ellados." *Dasiki erevna* 1, appendix: 9–63.
Meiggs, Russell. 1982. *Trees and Timber in the Ancient World.* Oxford: At the Clarendon Press.
Mekios, Konstantinos. 1909. *Istoria tis Epeirou.* Cairo: Evangelidis.
Melis, Federigo. 1974. "La lana della Spagna mediterranea e della Barberia occidentale nei secoli XIV–XV." In *La lana come materia prima,* 241–51. Florence: Olschki.
Mendras, Henri. 1961. *Six villages d'Epire.* Paris: UNESCO.
Menéndez Amor, J., and F. Florschutz. 1962. "Un aspect de la végétation en

Espagne méridionale durant la dernière glaciation et l'Holocène." *Geologie en Mijnbouw* 41:131–4.

Merchant, Carolyn. 1989. *Ecological Revolutions.* Berkeley: University of California Press.

Merendi, Ariberto. 1952. "Il difesa del suolo." *Il Mezzogiorno* 1, no. 2:7–10.

Merendi, Ariberto. 1957. "Aspetti e problemi forestali del Mezzogiorno." *Il Mezzogiorno* 10–11 (November): 9–12.

Merry del Val, Alfonso. 1920. *Las zonas septentrional y meridional del Protectorado de España en Marruecos.* Madrid: Patronato de Huérfanos de Intendencia e Intervención Militares.

Metaxas, N. 1948. "To nero kai to dasos." *To dasos* 2, no. 5:17–22.

Metaxas, N. 1953. *Oi diavroseis stin Ellada.* Athens: Ipourgeiou Yeorgias.

Metro, André. 1958. *Atlas du Maroc. Notes explicatives. VI. Forêts.* Rabat: Comité de Géographie du Maroc.

Mez, Bruno, and Alfonso Parentini. 1967. "L'opera di conservazione del suolo nel comprensorio del Bradano e Metaponto." In Congresso (XXIII) Nazionale delle Bonifiche, *Atti del XXIII Congresso,* 280–90. Bologna: Il Mulino.

M'Hirit, O. 1982. "Etude écologique et forestière des cedraies du Rif Marocain." *Annales de la recherche forestière au Maroc* 22:1–502.

Michaux-Bellaire, G. 1926. "L'histoire du Rif." *Bulletin de l'Enseignement public au Maroc* 71:35–45.

Michelangeli, Mario. 1940. *Il problemo forestale Albanese.* Rome: Reale Accademia d'Italia.

Miège, Jean-Louis. 1961. *Le Maroc et l'Europe, 1830–1894.* 4 vols. Paris: Presses Universitaires de France.

Mignon, Christian. 1970. "Notes sur l'évolution récente de la population en Andalousie orientale." *Méditerranée,* 2d ser., 1, no. 4:289–319.

Mignon, Christian. 1981. *Campagnes et paysans de l'Andalousie méditerranéenne.* Clermont-Ferrand: Faculté des Lettres et Sciences Humaines.

Mignon, Christian. 1982. *Campos y campesinos de la Andalucía Mediterránea.* Madrid: Ministerio de Agricultura, Pesca y Alimentación.

Mikesell, Marvin W. 1960. "Deforestation in Northern Morocco." *Science* 132: 441–8.

Mikesell, Marvin W. 1961. *Northern Morocco: A Cultural Geography* (Berkeley: University of California Press.

Miller, James A. 1984. *Imlil: A Moroccan Mountain Community in Change.* Boulder: Westview Press.

Milliot, Louis. 1934. "L'éxode saisonnier des rifains vers l'Algérie." *Bulletin économique de Maroc* 1:313–21 and 397–402.

Miñano y Bedoya, Sebastián de. 1826–9. *Diccionario geográfico-estadístico de España y Portugal.* 11 vols. Madrid: Pierart-Peralta.

La mobilité des paysages méditerranéenes: Hommage à Pierre Birot. 1984. Toulouse: Revue Géographique des Pyrénées et du Sud-Ouest.

Molero Mesa, J., and E. García Martínez. 1981. "Resumen fitosociológico de la vegetación de Sierra Nevada." *Cuadernos geográficos* 11:215–66.

Molero Mesa, J., and F. Pérez Raya. 1987. *La flora de Sierra Nevada.* Granada: Universidad de Granada.
Monnier, Marco. [1862] 1986. *Notizie storiche documentate sul brigantaggio nelle provincie napoletane.* [Florence.] N.p.: Adelano Polla Editore.
Montagne, Robert. 1973. *The Berbers: Their Social and Political Organization.* Translated by David Seddon. London: Frank Cass.
Montalembert, René de. 1972. "L'Economie des produits forestiers au Maroc: bilan et perspectives." Thèse, Université de Montpelier.
Montalvo Rodríguez, Manuel. 1975. "La economía granadina y sus comarcas." Tesis, Universidad de Granada.
Monteil, Vincent. 1976. "La guerre révolutionnaire." In *Abd el-Krim et la République du Rif. Actes du colloque internationale* 147–52. Paris: Maspero.
Montoya Oliver, J. M. 1983. *Pastoralismo mediterráneo.* Madrid: Ministerio de Agricultura, Pesca y Alimentación.
Moore, Jack. 1981. *Taming Ancient Rivers of Greece.* London: Faraway Books.
Moral Ruíz, Joaquín del. 1979. *La agricultura española a mediados del siglo XIX (1850–1870). Resultados de una encuesta agraria de la época.* Madrid: Ministerio de Agricultura, Pesca y Alimentación.
Moral Villalobos, Juan G. 1908. "Memorias de la Guerra de Independencia por Juan G. Moral Villalobos, natural del Fondón en la Alpujarra." *Revista de Archivos, Bibliotecas y Museos*: 111–24.
Morales Lezcano, Victor. 1976. *El colonialismo hispano-francés en Marruecos, 1898–1927.* Madrid: Siglo XXI.
Morandini, Ricardo. 1950. "Tra i boschi del Monte Pollino." *Monti e boschi* 1:361–65.
Morano, Michelangelo. 1981. "Tecniche culturali e organizzazione produttiva nelle campagne della Basilicata del secolo XIX." In Angelo Massafra, ed., *Problemi di storia delle campagne meridionali nell'età moderna e contemporanea,* 507–38. Bari: Dedalo.
Moreau de Jonnés, Alexandre. 1826. *Premier mémoire en réponse à la question par l'Académie Royale de Bruxelles: Quels sont les changements que peut occasioner le déboisement* . . . Brussels: De Mat.
Morocco. Direction de la Statistique [Kitabât Al-Dawlah Lil-Takhtit]. 1983 *Population légale du Maroc . . . 1982.* Rabat: Direction de la Statistique.
Morocco. Direction de Statistiques. 1971. *Population légale du Maroc . . . 1971.* Rabat.
Morocco. Gouvernement Cherifien. Secrétariat Général du Protectorat. 1947–9. *Tableaux de statistique graphique marocaine.* 3 vols. Rabat.
Morocco. Ministère de l'Agriculture. 1960. *Le Rif agricole.* Rabat: Ministère de l'Agriculture.
Morocco. Ministère de l'Education Nationale. Division des Programmes et Inspections Specialisées [Wizarat al-Tarbiyah al-Wataniyah]. 1980. *Données écologiques régionales: Taza.* Rabat: C.M.D.
Morocco. Secrétariat d'Etat au Plan et au Développement Régionale [Kitabât al-Dawlah lil-Tahktit . . .]. 1977. *Recueil des données.* Rabat.

Morocco. Service Centrale des Statistiques. 1962. *Recensement démographique (juin 1960). Population légale du Maroc.* 2 vols. Rabat.

Morocco. Wizārat al-Iqtiṣād al-Waṭanī. 1960. *Tableaux économiques du Maroc 1915–1959.* Rabat. Service Central des Statistiques.

Mortara, G. 1910. "Basilicata e Calabria secondo le statistiche demografiche." *Giornale degli economisti e rivista statistica* 1:435–61, 659–76.

Mosso, Angelo. 1906. *Vita moderna degli italiani.* Milan, Fratelli Treves.

Mouliéras, Auguste Jean. 1895–9. *Le Maroc inconnu.* 2 vols. Paris: André.

Mouseio Goulandri Fisikis Istorias. 1989. *Ellinika dasi.* Athens: Mouseio Goulandri, Fisikis Istorias.

Moustaka, Calliope. 1964. *The Internal Migrant: A Comparative Study in Urbanization.* Athens: Social Sciences Centre.

Mouterde, P. 1970. *Nouvelle flore du Liban et de la Syrie.* Beirut: Imprimerie Catholique.

M'Rabet, M. 1986. "Le développement socio-économique de la région rifaine." *Dar el-Niaba,* no. 11:1–11.

Muñoz Torres, Acisclo. 1939. *La patata.* Tangier: Editorial Tanger.

Munson, Henry. 1990. "Slash-and-Burn Cultivation, Charcoal Making, and Emigration from the Highlands of Northwest Morocco." In Muneera Salem-Murdock, Michael M. Horowitz, and Monica Sella, eds., *Anthropology and Development in North Africa and the Middle East.* Boulder: Westview.

Murphey, Rhoads. 1983. "Deforestation in Modern China." In John F. Richards and Richard P. Tucker, eds., *Global Deforestation and the Nineteenth-Century World Economy,* 111–28. Durham, N.C.: Duke University Press.

Nadal, Jordi. 1982. "Andalucía, paraíso de los metales no ferrosos." In *Historia de Andalucía:* 7:179–240. Madrid: CUPSA.

Nadji, Abd-ur-Rahman. 1892. *Empire Ottoman. Géographie botanique. Faits nouveaux relatifs à la Province de Salonique.* Salonica: Muratori.

Navarro Alcalá-Zamora, Pío. 1979. *Mecina: la cambiante estructura social de un pueblo de la Alpujarra.* Madrid: Centro de Investigaciones Sociológicas.

Navarro Alcalá-Zamora, Pío. 1981. *Tratadillo de agricultura popular: el medio, las técnicas, y los personajes en la Alpujarra.* Barcelona: Ariel.

Navarro Garnica, Miguel. 1955. *El pastoreo en los montes: pastizales españoles.* Madrid: Ministerio de Agricultura.

Naveh, Z. 1974. "Effects of Fire in the Mediterranean Region." In T. T. Kozlowski and C. E. Ahlgren, eds., *Fire and Ecosystems,* 401–34. New York: Academic Press.

Naveh, Z., and J. Dan. 1973. "The Human Degradation of Mediterranean Landscapes in Israel." In Francesco Di Castri and Harold Mooney, eds., *Mediterranean Type Ecosystems: Origins and Structure,* 373–90. Berlin: Springer-Verlag.

Naveh, Z., and P. Kutiel. 1990. "Changes in the Mediterranean Vegetation of Israel in Response to Human Habitation and Land Use." In G. M. Woodwell, ed., *The Earth in Transition,* 259–99. Cambridge: Cambridge University Press.

Neboit, R. 1971. "Morphogenèse récente des formations tendres en Lucanie (Italie du Sud)." *Méditerranée,* 2d ser., 2, no. 7:701–19.

Neboit, R. 1977. "Un exemple de morphogenèse accélerée dans l'Antiquité. Les vallées du Basento et du Cavone en Lucanie (Italie)." *Méditerranée,* 2d ser., 28, no. 4:39–50.

Neboit, R. 1979. "Les facteurs naturels et les facteurs humains de la morphogenèse. Essai de mise au point." *Annales de géographie* 88:649–70.

Neboit, R. 1980. "Morphogenèse et occupation humaine dans l'Antiquité." *Bulletin de l'association des géographes francais,* no. 466:21–7.

Neboit, R. 1983. *L'Homme et l'érosion.* Clermont-Ferrand: Faculté des Lettres et Sciences Humaines de l'Université de Clermont-Ferrand II.

Netting, Robert McC. 1981. *Balancing on an Alp: Ecological Change and Continuity in a Swiss Mountain Community.* Cambridge: Cambridge University Press.

Newbigin, Marion. 1932. *Southern Europe: A Regional and Economic Geography of the Mediterranean Lands.* London: Methuen.

Newby, Eric. 1984. *On the Shores of the Mediterranean.* Boston: Little, Brown.

Nicol, Donald M. 1984. *The Despotate of Epirus, 1267–1479.* Cambridge: Cambridge University Press.

Nicolini, Luciano, and Davide Pettener. 1984. "Ricerca antropologica e minoranze etniche: Il caso di San Paolo Albanese." *Monti e boschi* 35, no. 4: 26–9.

Nicotra, Rosario. 1983. *Le crisi demografiche nella Sicilia del Settecento. Cronologia, intensità, fenomenologia.* Catania: Leone.

Nikolaidis, Ioannis. 1939. *Istoriki monografia peri tis en Epeirou choras vezitzis (Vitsis-Monodendriov-Zagoriou).* Ioannina: Ermis.

Nikolaidis, Ioannis. 1982. *Zagori. Dokimio anthropoyeografias.* Ioannina: n.p.

Nir, Dov. 1983. *Man, A Geomorphological Agent: An Introduction to Anthropic Geomorphology.* Jerusalem: Keter.

Nitti, F., ed. 1909–10. *Inchiesta parlamentare sulle condizioni dei contadini nelle provincie meridionali e nella Sicilia. V. Basilicata e Calabria.* Rome: Bertero.

Noé, Luigi, and Manlio Rossi-Doria. 1979. *I problemi della difesa del suolo.* Milan: Angeli.

Noin, Daniel. 1970. *La population rurale du Maroc.* 2 vols. Paris: Presses Universitaires de France.

Nuñez Noguerol, Gregorio. 1964. *Estudio geográfico de la Alpujarra Oriental. (Resumen del tesis).* Madrid: Universidad de Madrid, Departamíento de Geografia.

Nuñez Noguerol, Gregorio. 1969. "La poblición y las formas de aprovechamiento de la Alpujarra oriental." *Estudios geográficos* 30:241–306.

Ocaña Ocaña, María Carmen. 1974. *La Vega de Granada.* Granada: Universidad de Granada.

Oikonomopoulos, Anastaseos. 1966. *I Exelixeis tis dasoponias en ti nea Elladi apo tis apeleftheroseos aftis mechri tou etous 1940.* Athens: Ipourgeion Yeorgias.

Oost, Stewart Irvin. [1954] 1975. *Roman Policy in Epirus and Acarnania in the Age of the Roman Conquest of Greece.* New York: Arno Press.

Orengo, C., and G. Rossi. 1973. "Sur l'évolution des versants denudés par incendie sous climat méditerranéen." *Méditerranée,* 2d ser., 12, no. 1:95–105.

Oriol Catena, F. 1935. "La repoblación de Granada después de la expulsión de los moriscos." *Boletín de la Universidad de Granada* nos. 34–35:305–31; no. 36:499–527.

Orioli, G. 1934. *Moving Along.* London: Chatto.

Ortega Alba, Francisco. 1981. "Granada: Geografía." In Enrique Pareja López, Francisco Ortega Alba, Juan Sanz Sampelayo, Ignacio Henares Cuéllar, Antonio Sánchez Trigueros, *Granada,* 1:18–212. Granada: Excellentíssima Diputación Provincial.

Otero, Santiago S. 1930. *En el corazón del Rif.* Tangier.

Özbayri, Kemal. 1972. *Tahtacılar ve Yörükler.* Paris: Maisonneuve.

Özdönmez, M. 1973. "La rôle des forestiers français dans l'évolution de la foresterie turque." *Revue française forestière* 25, no. 2:170–2.

Özdönmez, M. 1982. "Les droits d'usage dans les forêts de l'état en Turquie." In *Actes du Symposium International d'Histoire Forestière,* 2:213–15. Nancy: Ecole Nationale du Génie Rurale.

Packe, Charles. 1868. "The Sierra Nevada." *Alpine Journal* 4:113–25.

Pagano, Vincenzo. 1892. *Studi sulla Calabria.* 2 vols. Naples: d'Auria.

Pamuk, Sevlet. 1987. *The Ottoman Empire and European Capitalism,* Cambridge: Cambridge University Press.

Panayiotopoulos, Vasilis. 1988. "I 'apochorisi' plithismou apo tin pediada sto vouno sta chronia tis Tourkokratias." In Ethniko Kentro Koinonikon Erevnon and Kentro Neoellinikon Erevnon, *O agrotikos kosmos ston Mesoyeiako choro,* 203–5. Athens.

Panzac, Daniel. 1985. *La peste dans L'Empire Ottoman.* Leuven: Editions Peeters.

Panzac, Daniel. 1988. "L'enjeu du nombre. La population de la Turquie de 1914 à 1927." *Revue du monde Musulman et de la Méditerranée* 50:45–67.

Papageorgiou, Georgios. 1987. "Prospatheies yia tin idrysi Panepistimiou stin Eparchia Zagoriou tis paramones tis Ellinikis Epanastaseis (1813–1820)." *Dodoni* 16:463–81.

Papakostas, N. 1967. *Epeirotika: Istorika, laografika, koinonika. A tomos. Athamanika.* Athens: n.p.

Papanastasis, Vasilios. 1978. "Prota stadia diadakis tis vlastisios meta tin pirkaia stous thamnous aefillon platifillon tis Chalkidikis." *To dasos* 30, nos. 79–80: 19–26.

Papastavrou, Anastasios, and Konstantinos Makris. 1985–6. *Dasiki politiki.* 2 vols. Thessaloniki: n.p.

Papoulias, J. 1973. "To idratikon isozigion eis oreinous stathmous tis Ellados." *To dasos* 25:29–47.

Paranzino, Giuseppe. 1967. "La difesa del suolo e la regolazione delle acque nel territorio dell'Appennino meridionale." In Congresso (XXIII) Nazionale Delle Bonifiche, *Atti del XXIII Congresso,* 298–307. Bologna: Il Mulino.

Pareja López, Enrique. 1981a. "Arqueología medieval." In Enrique Pareja López, Francisco Ortega Alba, Juan Sanz Sampelayo, Ignacio Henares Cuéllar, and Antonio Sánchez Trigueros, *Granada,* 2:377–428. Granada: Excellentíssima Diputación Provincial.

Pareja López, Enrique. 1981c. "Prehistoria, protohistoria y arqueología romana."

In Enrique Pareja López, Francisco Ortega Alba, Juan Sanz Sampelayo, Ignacio Henares Cuéllar, and Antonio Sánchez Trigueros. *Granada*, 1:215–365. Granada: Excellentíssima Diputación Provincial.

Pareja López, Enrique, Francisco Ortega Alba, Juan Sanz Sampelayo, Ignacio Henares Cuéllar, and Antonio Sánchez Trigueros. 1981b. *Granada*. 4 vols. *Granada*: Excellentíssima Diputación Provincial.

Pascon, Paul, and Herman van der Wusten. 1983. *Les Beni Boufrah. Essai d'écologie sociale d'une vallée rifaine (Maroc)*. Rabat: Reproductions Industrielles.

Pastina, F. 1986. "Forest Industry in Italy: Constraints, Solutions and Policy Issues." In *FAO Forestry Paper 68. Appropriate Forest Industries. Selected Papers of an Expert Consultation held in Jakarta, Indonesia 1985*, 145–56. Rome: FAO.

Paulet, J.-P. 1982. "Forêts et pression démographique en Basse Provence." *Méditerranée*, 3d ser., 45, no. 2:31–40.

Pavan, Mario. 1970. *La difesa del suolo nella conservazione della natura*. Rome: Ministerio de Agricultura e Foreste.

Pavari, Aldo. 1925. *Sui rimboschimenti e sistemazioni delle argille eoceniche in Basilicata*. Rome: Pallotta.

Pavari, Aldo. 1951. "I prodotti non legnosi dei boschi italiani." *Monti e boschi* 2:215–19.

Pavari, Aldo. 1955. "Il problema dei boschi cedui in Italia." *Monti e boschi* 6:339–50.

Pecora, Aldo. 1955. "Sullo spopolamento montano negli Abruzzi." *Bollettino della Società geografica italiana* 92, nos. 11–12:508–24.

Pecorano, Antonio. 1974. *La foresta mediterranea*. Rome: Bardi.

Pedio, Tommaso. 1961a. *La Basilicata durante la dominazione barbonica*. Matera: Montemurro.

Pedio, Tommaso. 1961b. "Condizioni della Basilicata in una inchiesta del 1859." In Comitato Provinciale di Potenza dell'Istituto per la Storia del Risorgimento Italiano, *Primo Centenario dello Stato Italiano. Contributi e ricerche storiche*. Matera: Montemurro.

Pedio, Tommaso. 1961c. *Popolazione e prezzi nella Basilicata borbonica*. Matera: Montemurro.

Pedio, Tommaso. 1963–4. *La statistica murattiana del Regno di Napoli*. Potenza: Riviello.

Pedio, Tommaso, ed. 1965. *Le Grandi inchieste sulle regioni meridionali. Vol. 1, La relazione Gaudioso sulla Basilicata (1736)*. Bari: Centro Librario.

Pedio, Tommaso. 1979. *La "questione meridionale" in una provincia del Mezzogiorno: La Basilicata dall'annessione al Piemonte all'inizio del Novecento*. Bari: Levante.

Pedio, Tommaso, ed. 1983. *Inchiesta Massari sul brigantaggio*. Manduria: Lacairta.

Pedio, Tommaso. 1987. *Brigantaggio meridionale (1806–1863)*. Lecce: Capone Editore.

Pedio, Tommaso. 1987–9. *La Basilicata dalla caduta dell'impero romano agli Angioni*. 5 vols. Bari: Levante.

Peña Sánchez de Rivera, Daniel. 1983. *Dependencia dinámica entre precios agrícolas. El trigo en España, 1857–1890*. Madrid: Banco de España.

Pennell, C. R. 1986a. *A Country with a Government and a Flag: the Rif War in*

Morocco, 1921–1926. London: Middle East and North African Studies, Mena Press.

Pennell, C. R. 1986b. "Exito y fracaso de Abd El-Krim." *Historia 16* 126:28–36.

Pennell, C. R. 1987. "Women and Resistance to Colonialism in Morocco: The Rif, 1916–1926." *Journal of African History* 28:107–18.

Pérez de Perceval Verde, Miguel Angel. 1983. "La metalurgía del plomo de la Sierra de Gádor, 1820–1860." *Anuario de Historia Contemporánea* 10:153–81.

Perilla, F. 1932. *A travers la Macédoine.* Athens: F. Perilla.

Perry, Allen. 1981. "Mediterranean Climate: A Synoptic Appraisal." *Progress in Physical Geography* 5:107–33.

Petagna, L., G. Terrone, and M. Tenore. 1827. *Viaggio in alcuni luoghi della Basilicata e della Calabria Citeriore.* Naples: Tipografia francese.

Peters, Robert L., and Thomas E. Lovejoy. 1990. "Terrestrial Fauna." In B. L. Turner, W. C. Clark, R. W. Kates, J. F. Richards, J. T. Matthews, and W. B. Meyer, eds., *The Earth as Transformed by Human Action,* 353–69. Cambridge: Cambridge University Press.

Pezzi, Manuel, and Francisco Pinazo. 1981. "Desarollo de una cartografía temática en base al uso de variables retinianas: la población de Sierra Nevada (Granada)." *Cuadernos geográficos* 11:43–81.

Pezzi Cerreto, M., et al. 1982. "La red hidrográfica de Sierra Nevada (Granada): relación con las pendientes, análisis cuantativo, determinación de caudales y regimenes." *Cuadernos geográficos* 11:183–213.

Philippson, Alfred. 1894. "Northern Greece." *Geographical Journal* 3:323–6.

Philippson, Alfred. 1897. *Thessalien und Epirus.* Berlin: Kühl.

Philippson, Alfred, with Ernst Kirsten. 1956. *Die Griechischen Landschaften. Band I. Der Nordwestern der Greichische Halbinsel. Teil II. Epirus und der Pindos.* Frankfurt: Klostermann.

Pignatti, S. 1983. "Human Impact on the Mediterranean Basin." In W. Holzner, ed., *Man's Impact on Vegetation,* 151–62. The Hague: W. Junk.

Pimentel, D., E. C. Terhune, and R. Dyar-Hudson. 1976. "Land Degradation: Effects on Food and Energy Resources." *Science* 194:149–55.

Pino Artacho, Juan del. 1965, *La familia alpujarreña a finales del siglo XIX.* Granada: Escuela Social de Granada.

Pino Artacho, Juan del. 1978. *Sociología de la Alpujarra. Análisis de un cuestionario aplicado en 1894.* Málaga: Universidad de Málaga.

Pîrî [Reis]. 1988. *Kitab-i Bahriye Pîrî.* 4 vols. Ankara: Ministry of Culture.

Pitcher, Donald Edgar. 1972. *An Historical Geography of the Ottoman Empire.* Leiden: E. J. Brill.

Piussi, Pietro. 1976. *Un inventario forestale del XVIII secolo per i boschi costieri dell'alto adriatico.* Rome: Ministero de Agricoltura e Foreste.

Pizanias, Petros. 1988. "La production agricole en Grèce au XIXe siècle." In Ethniko Kentro Koinonikon Erevnon and Kentro Neoellinikon Erevnon, *O agrotikos kosmos ston Mesoyeiako choro,* 468–89. Athens.

Pizzigallo, V. 1970. "I sessanto anni dell'azienda di stato per le foreste demaniali." *Monti e boschi* 21, no. 3:15–25.

Planhol, X. de. 1950. "Estivage des montagnes en Pisidie." *Bulletin de l'Association de géographes français* 206–7:81–8.
Planhol, X. de. 1952. "Les migrations de travail en Turquie." *Revue de géographie alpine* 40:582–600.
Planhol, X. de. 1956. "Contribution à l'étude géomorphologique du Taurus occidental et de ses plaines bordières." *Revue de géographie alpine* 44:609–85.
Planhol, X. de. 1958. *De la plaine pamphylienne aux lacs pisidiens: Nomadisme et vie paysanne.* Paris: Maisonneuve.
Planhol, X. de. 1959. "Geography, Politics and Nomadism in Anatolia." *International Social Science Journal* 11:525–31.
Planhol, X. de. 1962. "Caractères généraux de la vie montagnarde dans le Proche-Orient et dans l'Afrique du Nord." *Annales de géographie* 71:113–30.
Planhol, X. de. 1965. "Les nomades, la steppe et la forêt en Anatolie." *Geographische Zeitschrift* 53:101–16.
Planhol, X. de. 1966. "Aspects of Mountain Life in Anatolia and Iran." In S. R. Eyre and G. R. J. Jones, eds., *Geography as Human Ecology*, 291–308. London: Edward Arnold.
Planhol, X. de. 1968. *Les fondements géographiques de l'histoire de l'Islam.* Paris: Flammarion.
Plit, Florian. 1983. "La dégradation de la végétation, l'érosion et la lutte pour protéger le milieu natural en Algérie et au Maroc." *Méditerranée*, 3d ser., 49: 79–89.
Polestra, Rocco. 1967. "Rimboschimento e proprietà privata." In Congresso (XXIII) Nazionale delle Bonifiche, *Atti del XXXIII Congresso*, 311–16. Bologna: Il Mulino.
Polunin, Oleg, and Anthony Huxley. 1974. *Flowers of the Mediterranean.* London: Chatto and Windus.
Poncet, J. 1958(?). *Les rapports entre les modes d'exploitation agricole et l'érosion des sols en Tunisie.* Tunis: Publications du Secrétariat d'Etat à l'Agriculture, no. 2.
Poncet, J. 1969. "Sous-développement et facteurs naturels: L'exemple du Mezzogiorno." *Méditerranée*, 1st ser., 10:63–79.
Pons, A. 1984. "Les changements de la végétation de la région méditerranéenne durant le Pliocène et la Quaternaire en relation avec l'histoire du climat et de l'action de l'homme." *Webbia* 38:427–39.
Pons, A., and P. Quézel. 1985. "The History of the Flora and Vegetation and Past and Present Disturbance in the Mediterranean Region." In C. Gómez-Campo, ed., *Plant Conservation in the Mediterranean Area*, 25–43. Dordrecht: W. Junk.
Pons, A., and M. Reille. 1986. "Nouvelles recherches pollenanalytiques à Padul (Granada): La fin du dernier glaciaire et l'Holocène." In F. López-Vera, ed., *Quatenary Climate in Western Mediterranean*, 405–20. Madrid: Universidad Autónoma de Madrid.
Pons, A., and M. Reille. 1988. "The Holocene and Upper Pleistocene Pollen Record from Padul (Granada, Spain): A New Study." *Palaeogeography, Palaeoclimatology, Palaeoecology* 66:243–63.

Poujade, Eugène. 1859. *Chrétiens et Turcs: Scènes et souvenirs de la vie politique, militaire et réligieuse en Orient.* Paris: Didier.

Pouqueville, F. C. H. L. 1805. *Voyage en Morée, à Constantinople, en Albanie . . . pendant les années 1798, 1799, 1800 et 1801.* 3 vols. Paris: Gabon.

Pouqueville, F. C. H. L. 1822. *Travels in Southern Epirus, Acarnania, Aetolia, Atttica, and the Peloponesus, or the Morea in the Years 1814–1816.* London: Phillips.

Pouqueville, F. C. H. L. 1826–7 *Voyage de la Grèce.* 6 vols. Paris: Didot.

Prampolini, Antonio. 1981. *Agricoltura e società rurale nel Mezzogiorno agli inizi del '900.* Milan: Angeli.

Pratesi, Fulco, and Franco Tassi. 1979. *Guida alla natura della Puglia, Basilicata e Calabria.* Milan: Mondadori.

Prentice, A. 1956. "Re-Afforestation in Greece." *Scottish Geographical Magazine* 72:25–31.

Prieto, P. 1971. *Vegetación de Sierra Nevada: La cuenca del Monachil.* Granada: Universidad de Granada Colección Monográfica, no. 11.

Prieto Fernández, Pablo. 1975. *Flora de la tundra de Sierra Nevada.* Granada: Universidad de Granada.

Prieto Fernández, Pablo. 1982. *Flora de Sierra Nevada.* Granada: Servicio de Publicaciones de la Universidad.

Principi, P. 1947. "I terreni agrari della Basilicata." *L'Italia agricola* 84:311–19.

Pryor, John H. 1988. *Geography, Technology and War: Studies in the Maritime History of the Mediterranean, 649–1571.* Cambridge: Cambridge University Press.

Psichoyios, Dimitris, and Yiouli Papapetrou. 1987. "Oi metakiniseis ton nomadon ktinotrofon." In Dimitris Psichoyios, Roxani Kaftantzoglou, Evdokia Manologlou, Ariadne Michalakopoulou, Matina Naoumi, and Yiouli Papapetrou, *O Oikonomikos kai koinonikos metaschimatismos agrotikon koinotiton,* 93–112. Athens: Ethniko Kentro Koinonikon Erevnon.

Puglisi, Salvatore. 1965. "La correzione dei torrenti in Grecia." *Monti e boschi* 16, no. 4:31–43; no. 5:55–64.

Puglisi, Salvatore. 1967. "Criteri per il riordino dei boschi nella provincia di Potenza." *Monti e boschi* 18, no. 5:11–23.

Puglisi, Salvatore. 1968. "Proposta di riordino dei boschi della provincia di Potenza." *Potenza* 2, nos. 1–2:13–28.

Puglisi, Salvatore. 1977. "Il dissesto idrogeologico." In Nino Calice, ed., *Basilicata tra passato e presente,* 90–99. Milan: Teti.

Quaranta, Adele. 1974. "L'evoluzione della popolazione in Basilicata nel decennio 1961–1971." *Nord e Sud* 30:71–5.

Quézel, P. 1953. "Contributions à l'étude phytosociologique et géobotanique de la Sierra Nevada." *Memorias da Sociedade Broteriana* [Coimbra] 9:1–76.

Quézel, P. 1957. *Peuplement végétal des hautes montagnes de l'Afrique du Nord.* Paris: Lechevalier.

Quézel, P. 1967. "Végétation des hauts sommets du Pinde et l'Olympie de Thessalie." *Vegetatio* 14:127–228.

Quézel, P. 1973. "Contribution à l'étude phytosociologique du Massif du Taurus." *Phytocoenologia* 1:131–222.

Quézel, P. 1982. *Bosque y maquis mediterráneos.* Madrid: Serbal.

Quézel, P., and G. Bonin. 1980. "Les forêts feuillues du pourtour méditerranéen." *Revue forestière française* 32:253–68.

Quézel, P., and J. Contandriopoulos. 1965. "A propos de la végétation des forêts de hêtre dans le massif du Pinde." *Bulletin de la société botanique française* 112, nos. 5–6:312–19.

Quézel, P., and A. Pamukçuoğlu. 1973. "Contribution à l'étude phytosociologique et bioclimatique de quelques groupements forestières du Taurus." *Feddes Repertorium* 84:185–229.

Quézel, P., and A. Pamukçuoğlu. 1975. "Contribution à l'étude phytosociologique du Massif du Taurus." *Phytocenologia* 1, no. 2:131–222.

Quilici, Folco. 1967. *Basilicata e Calabria. L'Italia vista del cielo.* Milan: Pizzi.

Quintana, L. de. 1862. "El valle de las Alpujarras." *Revista minera* 13:496–510, 513–23, 545–67.

Quirantes Puertas, José. 1985–6. "Erosión antrópica (Cuenca del Guadalfeo)." *Cuadernos geográficos de la Universidad de Granada* 15:89–98.

Quirantes Puertas, José. 1987. "Guadalfeo el medio y el hombre." In Instituto Nacional para la Conservación de la Naturaleza, *Proyecto Lucdeme III*, 93–174. Madrid.

Racioppi, Giacomo. [1889] 1902. *Storia dei popoli della Lucania e della Basilicata.* Rome: Loescher.

Racioppi, Giacomo. 1910. *Storia dei moti di Basilicata.* Bari: Laterza.

Rackham, Oliver. 1982. "Land-use and the Native Vegetation of Greece." In Martin Bell and Susan Limbrey, eds., *Archaeological Aspects of Woodland Ecology*, 177–98. Symposia of the Association for Environmental Archaeology, no. 2. Oxford: British Archaeological Reports International Series, no. 146.

Rackham, Oliver. 1983. "Observations on the Historical Ecology of Boeotia." *Annual of the British School at Athens* 78:291–351.

Radina, Bruno. 1976. "Geologia e conservazione del suolo in Basilicata." *Sviluppo* 9:98–103.

Ramage, Craufurd Tait. 1868. *Nooks and Byways of Italy.* Liverpool: Howell.

Ramsay, William M. 1890. *The Historical Geography of Asia Minor.* London: John Murray.

Ramsay, William M. 1897. *Impressions of Turkey during Twelve Years' Wandering.* London: Hodder & Stoughton.

Ramsay, William M. 1906. "The Peasant-God: The Destruction and Restoration of Agriculture in Asia Minor." *Contemporary Review* 90:786–800.

Ramsay, William M. 1923. "Geography and History in a Phrygo-Psidian Glen." *Geographical Journal* 61:279–96.

Rangavis, I. R. 1853–4. *Ta hellenika, etoi perigraphi geographiki, historiki, archaiologiki, kai statistiki tis archaias kai neas Hellados.* Athens.

Ranieri, Luigi. 1961. *Basilicata.* Turin: UTET.

Raphael, C. Nicholas. 1978. "The Erosional History of the Plain of Elis in the Peloponnese." In William G. Brice, ed., *Environmental History of the Near and Middle East*, 51–66. London: Academic Press.

Ravenstein, E. G. 1877. "The Populations of Russia and Turkey." *Journal of the Statistical Society* 40:433–67.

Regnault, Felix. 1904. "Déboisement et décadence." *Revue des revues*, 32–45.

Reille, M. 1977. "Contribution pollenanalytique à l'histoire holocène de la végétation des montagnes du Rif (Maroc septentrional)." *Recherches Françaises sur le Quatenaire INQUA 1977*, supplement to *Bulletin de l'Association française pour l'étude du Quatenaire*: 53–76.

Rein, Johannes. 1899. *Beiträge zur Kenntais der Spanischen Sierra Nevada.* Vienna: Lechner.

Reparaz, G. A. de. 1962. "Notes sur les migrations de la population espagnole durant les vingt dernières années." *Méditerranée*, 1st ser., 3, no. 3:67–90.

Rhoades, Robert E. 1979. "Cultural Echoes across the Mountains." *Natural History* 88, no. 1:46–57.

Ricard, Prosper. 1926. "Arts et industries." *Bulletin de l'enseignement public au Maroc* 71:63–71.

Richardson, Bonham C. 1983. *Caribbean Migrants: Environment and Human Survival on St. Kitts and Nevis.* Knoxville: University of Tennessee Press.

Rickard, T. A. 1928. "The Mining of the Romans in Spain." *Journal of Roman Studies* 18:129–43.

Rikli, M. 1943–8. *Das Pflanzenkleid der Mittelmeerländer.* Bern: Hans Huber.

Rivas Goday, S., and S. Martínez Rivas. 1971. "Vegetación potencial de la provincia de Granada." *Trabajos del Deptartamiento de botánica y fisiología vegetal* (Madrid) 4:3–85.

Rivas Goday, S., and M. Mayor López. 1965. "Aspectos de vegetación y flora orófilas del Reino de Granada." *Anales de la real academia de farmacia* 31: 345–400.

Rivera, Vincenzo. 1961. "In Abruzzo la montagna muore." *Monti e boschi* 12: 205–9.

Riviello, Raffaele. 1888. *Cronaca potentina dal 1799 al 1882 . . .* Potenza: n.p.

Roberts, Neil. 1982. "Forest re-advance and the Anatolian Neolithic." In Martin Bell and Susan Limbrey, eds., *Archaeological Aspects of Woodland Ecology*, 231–47. Symposia of the Association for Environmental Archaeology, no. 2. Oxford: British Archaeological Reports International Series, no. 146.

Roberts, Neil. 1989. *The Holocene: An Environmental History.* London: Basil Blackwell.

Roda Jiménez, Rafael de. 1939. *Compendio de sociología marroquí.* Ceuta: Imperio.

Roda Jiménez, Rafael de. 1941. *Economía marroquí. Los problemas del campo.* Ceuta: Impero.

Roda Jiménez, Rafael de. c. 1940s. "Estructura y evolución de la población de Marruecos." *Revista Internacional de Sociologia.*

Roda y Jiménez, Rafael de, and Tomas García Figueras. 1950–5. *Economía social de Marruecos.* 3 vols. Madrid: Consejo Superior de Investigaciones Científicas.

Rodolico, Franco. 1963. *L'esplorazione naturalistica dell'Appennino.* Florence: Le Monnier.

Rodríguez Martínez, Francisco. 1982. "Notas sobre la crisis y las posibilidades de desarollo de la montaña mediterránea andaluza: El caso de Sierra Nevada." *Cuadernos geográficos* 11:267–81.

Rodríguez Martínez, Francisco. 1985. *Granada: medio físico y desarollo.* Granada: Universidad, Instituto de Desarallo Regional.

Rodríguez Martínez, Francisco, et al. 1981. "Bases físicas para la ordenación territorial de la Vertiente Sur de Sierra Nevada (Alpujarra, Granada)." *Cuadernos geográficos* 11:83–125.

Rodríguez Ocaña, Esteban. 1983. *El cólera de 1834 en Granada.* Granada: Universidad de Granada.

Rokou, Vaso. 1988. "Metanastevseis ston oreino choro: paradeigmata apo tin Epeiro." In Ethniko Kentro Koinonikon Erevnon and Kentro Neoellinikon Erevnon, *O agrotikos kosmos ston Mesoyeiako choro,* 252. Athens.

Rolley, Paul, and M. de Visme. 1911–12. "La Macédoine et l'Epire (vilayets de Monastir et de Janina). Etude de géographie physique et d'agrologie." *Annales de l'Institut National Agronomique,* 2nd sér., 10:375–447 and 11:5–75.

Rosenberger, Bernard. 1980. "Cultures complémentaires et nourritures de substitution au Maroc (XVe–XVIIIe siècles)." *Annales: Economies, sociétés, civilisations* 35:477–503.

Rosenberger, Bernard. 1984. "Calamités, securité, pouvoir: Le cas du Maroc." *Peuples méditerrannéens* 27–8:247–71.

Rossi-Doria, Manlio. 1963. *Memoria illustrativa della carta della utilizzazione del suolo della Basilicata.* Naples: Buona Stampa.

Rossi-Doria, Manlio. 1968. "La difesa del suolo nella montagna e nell'alta collina." In Accademia Nazionale Dei Lincei, *Problemi attuali di scienza e di cultura. Atti del convegno sul tema: Le scienze della natura di fronte agli eventi idrogeologici.* Rome.

Roux, Jean-Paul. 1961. "La sédentarisation des nomades Yuruk du Vilayet d'Antalya." *L'ethnographie* 55:64–78.

Roux, Jean-Paul. 1970. *Les traditions des nomades de la Turquie méridionale.* Paris: Adrien-Maisonneuve.

Roux, Jean-Paul. 1987. The Tahtacı of Anatolia. In Aparna Rao, *The Other Nomads,* 229–46. Cologne: Kölner Ethnologische Mitteilungen Bild 8, Böhlau Verlag.

Rowton, M. B. 1967. "The Woodlands of Ancient Western Asia." *Journal of Near Eastern Studies* 26:261–77.

Rubio, Antonio. 1881. *Del mar al cielo. Crónica de un viaje a Sierra Nevada.* Almería: Viuda Cordero.

Rugolo, Carmela M. 1988. "Paesaggio boschivo e insediamenti umani nella Calabria medievale." In Bruno Andreoli and Massimo Montanari, eds., *Il bosco nel medioevo,* 323–48. Bologna: Cooperativa Libraria Universitaria Editrice Bologna.

Ruíz Albeniz, Victor. 1912. *El Rif.* Madrid: Juan Fueyo.

Ruíz Albeniz, Victor. 1930. *Monografía sobre colonización rural en Marruecos español.* Madrid: Saez Hermanos.

Ruíz de la Torre, Juan. 1955. *El matorral en Yebala (Marruecos español). Estudio de la formaciones de matorral.* Madrid: Consejo Superior de Investigaciones Científicas, Instituto de Estudios Africanos.

Ruíz de la Torre, Juan. 1957. "La vegetación natural del norte de Marruecos." *Notes marocaines* 9–10:23–33.

Ruíz de la Torre, Juan. 1979. *Arboles y arbustes de la España peninsular.* Madrid: Escuela Técnica Superior de Ingenieros de Montes.

Russell, Richard. 1954. "Alluvial Morphology of Anatolian Rivers." *Annals of the Association of American Geographers* 44:363–91.

Russo, P. 1928. "Un voyage dans le Rif." *Renseignements coloniaux,* no. 11:683–8.

Ruz Márquez, José Luís. 1981. *Adra. Siglo XIX.* Almería: Cajal.

Saatçioğlu, Fikret. 1952. "A General View on the Importance and Problems of Afforestation in Turkey." *Istanbul Üniversitesi. Orman Fakültesi Dergisi* 2, no. 1:80–3.

Sacchi, Giulio. 1950. "La difesa del suolo della montagna nella evoluzione legislativa italiana." *Monti e boschi* 1:488–91.

Sáenz Lorite, Manuel. 1977. *El valle del Andarax y campo de Níjar. Estudio geográfico.* Granada: Universidad de Granada.

Sagra, R. de la. 1845. *Informe sobre el cultivo de la caña de azúcar en las costas de Andalucía.* Madrid: Imprenta de Colegio de Sordo-Mudos y Ciegos.

Saint-Non, Jean Claude Richard de. 1781–6. *Voyage pittoresque ou description des royaumes de Naples et de Sicilie.* 5 vols. Paris: Clousier.

Sáinz y Gutierrez, Pedro. 1862. *Memoria sobre la utilidad de los montes y necessidad de atender a su conservación.* Granada: Ventura y Sabatel.

Sakellaridis, Spyros. 1968. *Ta orfna edafi tis Kentrikis Pindou.* Thessaloniki: University of Thessaloniki Forestry School.

Sakellaridis, Thanassi. 1984. "Frangadhes, parelthon kai paron." *Deltio kentrou erevnon zagoriou* 4, no. 12:4–10.

Sakellarios, Anastasios. 1888. *To Zagorion kai ai kat'aftou listrikai epidromai.* Athens.

Sale, Kirkpatrick. 1990. *The Conquest of Paradise.* New York: Knopf.

Salomone, Nicola. 1907. *La colonizzazione interna nella Basilicata.* Potenza: Perserveranza.

Salvatore, A. 1938. *La presente crisi foragera.* Potenza: Nucci.

Salvia, G. 1956. "L'emigrazione rurale in un comune montano della Provincia di Potenza. Picerno." *Nuova agricultura lucana,* 138–9.

Samios, K. M. 1900. *Eikones ek ton Ellinikon dason.* Athens: Paliyenesias.

Sánchez Albornoz, Nicolás. 1975. *Los precios agrícolas durante la segunda mitad del siglo XIX.* Madrid: Servicio Estudios del Banco de España.

Sánchez Albornoz, Nicolás, ed. 1985. *La modernización económica de España 1830–1930.* Madrid: Alianza.

Sánchez Cózar, Santiago. 1942. *Explotaciones forestales en el protectorado.* Barcelona: SIEM.

Sánchez Cózar, Santiago. 1948. "Montes." In *Conferencias desarolladas en la Academia de Interventores durante 1947,* 5–25. Tetouan: Imprenta del Majzen.

Sánchez Picón, A. 1983. *La minería del levante almeriense, 1838–1930.* Almería: Cajal.

Sánchez-Montes González, Francisco. 1989. *La población granadina en el siglo XVII.* Granada: Universidad de Granada.

Sánchez Sánchez, J., and V. Rodríguez Rodríguez. 1989. "Politique socio-structurelle pour les zones d'agriculture de montagne." *Méditerrannée,* 3d ser., 67:23–32.

Sángroniz, Jóse Antonio. 1921. *Marruecos, sus condiciones físicas, sus habitantes y las instituciones indígenas.* Madrid: Rivadeneyra.

Sanmarchi, Antonio. 1957. "La montagna e i montanari." *Monti e boschi* 8: 339–52, 387–98.

Sanz Sampelayo, Juan Felix. 1980. *Granada en el siglo XVIII.* Granada: Diputación Provincial de Estudios y Promoción Cultural.

Sanz Sampelayo, Juan Felix. 1981. "Historia." In Enrique Pareja López, Francisco Ortega Alba, Juan Sanz Sampelayo, Ignacio Henares Cuéllar, and Antonio Sánchez Trigueros, *Granada,* 2:429–516. Granada: Excellentíssima Diputación Provincial.

Sari, Djibali. 1970. "Déboisement et reboisement en Algérie." In X. de Planhol, ed., *Maghreb et Sahara; études géographiques offertes à J. Despois,* 351–64. Numéro special de *Acta Geografica.* Paris.

Sari, Djibali. 1977. *L'Homme et l'érosion dans l'Ourensis (Algérie).* Algiers: Société Nationale d'Edition et de Diffusion.

Sari, Djibali. 1979. "Les tentatives de restructuration du monde rural en Algérie." *Méditerranée,* 3d ser., 35:65–72.

Sarre, Friedrich P. T. 1896. *Reise in Kleinasien.* Berlin: D. Reimer.

Sarti, Roland. 1985. *Long Live the Strong: A History of Rural Society in the Apennine Mountains.* Amherst: University of Massachusetts Press.

Sarton, George. 1936. "The Unity and Diversity of the Mediterranean World." *Osiris* 2:406–63.

Sauer, Carl O. 1963. *Land and Life: A Selection from the Writings of Carl Ortwin Sauer.* Edited by John Leighly. Berkeley: University of California Press.

Sauvage, Charles. 1941. "La forêt marocaine." *Bulletin de l'enseignement public du Maroc,* 238–57.

Sauvage, Charles. 1961. *Recherches géobotaniques sur les suberaies marocaines.* Rabat: Institut Scientifique Chérifen.

Schilizzi, Maria S., and Giovanni Viola. 1981. *Terranova di Pollino. Storia e volto di una terra.* Lecce: Adriatica.

Schirò, G., ed. 1975. *Cronaca dei Tocco di Cefalonia.* Vol. 10 of *Corpus Fontium Historiae Byzantinae.* Rome: Accademia Nazionale dei Lincei.

Schläfli, Alexander. 1859. "Reisekizzen aus Epirus." *Das Ausland* 32. Stuttgart. Xerographic copy in the library of the Society for Epirot Studies, Ioannina.

Sciacca, Marisa. 1977. *Le terre del Sud. La formazione del paesaggió agrario meridionale moderno.* Cosenza: Lerici.

Sclafert, Th. 1959. *Cultures en Haute Provence. Déboisement et pâturages au Moyen Age.* Paris: S.E.V.P.E.N.

Scoging, H. 1982. "Spatial Variations in Infiltration, Runoff, and Erosion on Hillslopes in Semi-Arid Spain." In R. B. Bryan and A. Yair, eds., *Badland Geomorphology and Piping,* 89–112. Norwich: Geobooks.

Seddon, David. 1981. *Moroccan Peasants: A Century of Change in the Eastern Rif, 1870–1970.* Folkestone: Dawson.

Segonzac, Marquis de. 1903. *Voyage au Maroc (1899–1901).* Paris: Armand Colin.

Seigue, Alexandre. 1985. *La forêt circumméditerranéenne et ses problèmes.* Paris: Maisonneuve et Larose.

Sempere y Guarinos, Juan. 1821. "Memoria sobre las causas de la decadencia de la seda en el Reino de Granada." In *Biblioteca española económica-política* (Madrid): 4:279–343.

Semple, Ellen Churchill. 1919. "Climatic and Geographic Influences on Ancient Mediterranean Forests and the Lumber Trade." *Annals of the Association of American Geographers* 9:13–40.

Semple, Ellen Churchill. 1931. *The Geography of the Mediterranean Region: Its Relation to Ancient History.* New York: Henry Holt.

Septsault, Abel. 1987 "*Clergé régulier et économie régionale au Royaume de Grenada: Quelques aspects (1700–1800).*" Maîtrise d'Histoire, Université de Paris VII.

Sereléa, G. 1978. "Regards sur la nuptialité et la fécondité en Grèce pendant la deuxième moitié du XIXème siècle." *Epitheorisis koinonikon erevnon* 32:41–50.

Sermet, Jean. 1934. "La Vega de Adra." *Revue de géographie commerciale de Bordeaux* 58:1–30.

Sermet, Jean. 1942. "Sierra Nevada." *Estudios geográficos* 3:727–49.

Sermet, Jean. 1953. *L'Espagne du Sud.* Paris: Arthaud.

Sermet, Jean. 1965. "L'introduction des plantes alimentaires en Andalousie." *Coloquio: Aportación de las investigaciones ecológicas y agrícolas a la lucha del mundo contra el hambre.* Madrid: Consejo Superior de Investigaciones Científicas.

Sermet, Jean. 1973. "Acclimatation: les jardins botaniques espagnols au XVIIIe siècle et la tropicalisation d'Andalousie." In *Histoire économique du monde méditerranéen: Mélanges en l'honneur de F. Braudel,* 555–82. Toulouse: privately published.

Serra, Luigi. 1983. *Sopravvivenza lessicali arabe e berbere in un'area dell'Italia meridionale: la Basilicata.* Naples: Istituto Universitario Orientale.

Sfikas, G. 1985. "To fisiko perivallon tis B. Pindou." *Korfes* 55:24–5.

Shaw, Brent D. 1985. "Climate, Environment, and History: The Case of Roman North Africa." In T. M. L. Wrigley, M. J. Ingram, and G. Farmer, eds., *Climate and History: Studies in Past Climates and Their Impact on Man,* 379–403. Cambridge: Cambridge University Press.

Shaw, Brent D. 1990. "Bandit Highlands and Lowland Peace: The Mountains of Isauria-Cilicia." *Journal of the Social and Economic History of the Orient* 33: 199–233, 237–70.

Siampos, George. 1969. *Dimographikai exilixeis en Elladi, 1950–1980.* Athens: Ministry of Coordination.

Sierra Mola, Luís. 1948. "La industria en el protectorado." In *Conferencias desarolladas en la Academia de Interventores durante el curso de 1947,* 29–50. Tetouan: Majzen.

Siguan Soler, Miguel. 1971. *El medio rural en Andalucía Oriental.* Madrid: Ministerio de Agricultura, Servicio Nacional de Concentración Parcelaria y Ordenación Rural, no. 23.

Simmons, I. G. 1989. *Changing the Face of the Earth: Culture, Environment, History.* Oxford: Basil Blackwell.

Simond, L. 1838. *Voyage en Italie et en Sicilie.* Paris: Raymond-Bocquet.

Sivignon, Michel. 1968. *Les pasteurs du Pinde septentrional.* Lyon: Centre d'Etudes et Recherches sur la Géographie de l'Europe.

Sivignon, Michel. 1975. *La Thessalie. Analyse d'une province grecque.* Lyon: Université de Lyon II.

Sivignon, Michel. 1977. "The Demographic and Economic Evolution of Thessaly (1881–1940)." In Francis Carter, ed., *An Historical Geography of the Balkans,* 379–407. London: Academic Press.

Skafidas, Basileos K. 1952. "Pronomia tou Metsovou." *Epeirotiki Estia* 1:657–60.

Soler y Pérez, Eduardo. 1906. *La Alpujarra y Sierra Nevada.* Madrid: Arias.

Soria Marco, B. 1948. *Al través de Marruecos español.* Barcelona: Myria.

Sorre, M. 1932. "Nomadisme agricole et transhumance dans la Sierra Nevada." *Annales de géographie* 41:301–8.

Spain. 1801. *Censo de la población de España de el año de 1797.* Madrid.

Spain. Alta Comisaria de España en Marruecos. 1955. *Marruecos: Zona Jalifiana.* Madrid.

Spain. Alta Comisaria de España en Marruecos. Delegación de Asuntos Indígenas. 1929. *Campaña profiláctica contra el paludismo.* Tetouan.

Spain. Alta Comisaria de España en Marruecos. Delegación de Asuntos Indígenas. 1948. *Conferencias desarolladas en la Academia de Interventores durante el curso de 1947.* Tetouan: Majzen.

Spain. Alta Comisaria de España en Marruecos. Delegación de Asuntos Indígenas. n.d. *Datos estadísticos de la superficie cultivada, producción agricola obtenida y arbolado correspondiente a los años 1935–1938.* N.p.

Spain. Alta Comisaria de España en Marruecos. Delegación de Asuntos Indígenas. 1935a. *Datos estadísticos de la superficie cultivada, producción agricola obtenida, arbolado e industria apícola.* N.p.

Spain. Alta Comisaria de España en Marruecos. Delegación de Asuntos Indígenas. 1935b. *Régimen de le propriedad en Marruecos.* Ceuta: M. Alcala.

Spain. Alta Comisaria de España en Marruecos. Delegación de Economía. 1944. *Datos utiles para los comerciantes y agricultores.* Tetouan: Majzen.

Spain. Alta Comisaria de España en Marruecos. Delegación de Economía. 1945–7. *Memoria-estadística 1944, 1945, 1946.* Tetouan.

Spain. Alta Comisaria de España en Marruecos. Delegación de Economía. Servicio de Montes. 1943. *Memoria de trabajos realizados en el año 1943.* Tetouan.

Spain. Alta Comisaria de España en Marruecos. Delegación de la Hacienda. 1934. *Estadísticas de importación y exportación por las Aduanas de la Zona de Protectorado español en Marruecos: Año 1933.* Madrid: Rivadeneyra.

Spain. Alta Comisaria de España en Marruecos. Inspección de los Servicios de Ganadería. 1945. *Memoria-estadística 1944.* Tetouan.

Spain. Alta Comisaria de España en Marruecos. Intervenciones Militares. 1929–41. *Vademecum.* Tetouan.

Spain. Comisión Histórica de la Campañas de Marruecos. 1935–6. *Geografía de Marruecos.* 3 vols. Madrid: Ministerio de la Guerra.

Spain. Dirección General de Agricultura y Montes. 1922. *Memorial sobre la estadística general de la producción de los montes de utilidad pública.* Madrid: Hijos de M. G. Hernández.

Spain. Dirección General de Estadística. 1941–55. *Zona de Protectorado . . . en el Norte de Africa. Annuario Estadístico.* Madrid: Dirección General de Estadística.

Spain. Dirección General de Marruecos. 1931. *Datos estadísticos relativos a la Zona de Protectorado español y a las colonias españoles de Africa Occidental.* Madrid: Rico.

Spain. Instituto Geográfico, Catastral y de Estadística. 1932. *Censo de la población de España 1930.* Madrid.

Spain. Instituto Geográfico y Estadístico. 1858. *Censo de la población de España . . . 1857.* Madrid: Imprenta Nacional.

Spain. Instituto Geográfico y Estadístico. 1876. *Nuevo compendio de los ciudades, villas, aldeas, y lugares de España.* Madrid: Imprenta Nacional.

Spain. Instituto para la Conservación de la Naturaleza. 1982. *Paisajes erosivos en el sureste español.* Madrid: Ministerio de Agricultura, Pesca y Alimentación. Monografía, no. 26.

Spain. Instituto para la Conservación de la Naturaleza. 1987a. *Mapa de estados erosivos: cuenca hidrográfica del Guadalquivir.* Madrid.

Spain. Instituto para la Conservación de la Naturaleza. 1987b. *Panorama de la actualidad forestal en España.* Madrid: Ministerio de Agricultura, Pesca y Alimentación.

Spain. Instituto para la Conservación de la Naturaleza. 1987c. *Proyecto Lucdeme III.* Madrid.

Spain. Instituto para la Conservación de la Naturaleza and Ministerio de Agricultura. 1977. *Inventario forestal nacional, region Andalucía Oriental.* Madrid.

Spain. Instituto Nacional de Estadística. 1952–9. *Censo de la población de España . . . 1950.* Madrid.

Spain. Instituto Nacional de Estadística. 1956. *Reseña estadística de la Provincia de Granada.* Madrid.

Spain. Instituto Nacional de Estadística. 1957. *Centenario de la estadística española.* Madrid.

Spain. Instituto Nacional de Estadística. 1977. *Reseña estadística de la provincia de Granada.* Madrid.

Spain. Junta de Andalucía. 1987. *Informe General del medio ambiente en Andalucía 1987.* Seville.

Spain. Junta de la Cría Caballar del Reino. 1904. *Censo del ganado caballar y mular de España e islas adyacentes.* Madrid: Imprenta de Administración Militar.

Spain. Ministerio de Agricultura. 1969. *Censo de la ganadería española 1969.* Madrid.

Spain. Ministerio de Agricultura, Industria, Comercio y Obras Públicas, Dirección General de Agricultura. 1905. *Prados y Pastos.* Madrid: Hijos de M. G. Hernández.

Spain. Ministerio de Agricultura, Pesca y Alimentación. 1986. *Mapa de cultivos y aprovechamientos de la Provincia de Granada. Memoria.* Madrid.

Spain. Ministerio de Fomento. Dirección General de Agricultura, Industria y Comercio. 1892. *La ganadería en España. Avance sobre la riqueza pecuaria en 1891.* 5 vols. Madrid: L. Péant e Hijos.

Spain. Ministerio de Fomento. Dirección General de Agricultura, Minas y Montes. 1914. *Avance estadístico de la riqueza que en España representa la producción media anual de pastos, prados . . .* Madrid: Hijos de M. G. Hernández.

Spain. Ministerio de Fomento. Dirección General de Agricultura, Minas y Montes. 1921. *Estudio de la ganadería en España.* 2 vols. Madrid: Hijos de M. G. Hernández.

Spain. Ministerio de Instrucción Pública y Bellas Artes. Dirección General del Instituto Geográfico y Estadístico. 1907. *Censo de la población de España . . . 1900.* Madrid.

Spain. Ministerio de Instrucción Pública y Bellas Artes. Dirección General del Instituto Geográfico y Estadístico. 1916. *Censo de la población de España . . . 1910.* Madrid.

Spencer, Edmond. 1851. *Travels in European Turkey in 1850.* 2 vols. London: Colburn.

Spera, G. 1903. *La Basilicata. Studi e proposte per la sua rigenerazione economica.* Rome: Cooperativo Sociale.

Spon, Jacob, and G. Wheler, 1724. *Voyage d'Italie, de Dalmatie, de Grèce et du Levant fait aux années 1675 et 1676.* The Hague: Alberts.

Sprengel, U. 1975. "La pastorizia transumante nell'Italia centro-meridionale." *Annali del Mezzogiorno* 15:271–327.

Stanford, Edward. 1877. *Carte ethnologique de la Turquie d'Europe et de la Grèce.* Paris: Dentu.

Stanley, Daniel Jean. 1990. "Med Desert Theory Is Drying Up." *Oceanus* 33: 14–23.

Stéphanos, Clon. 1884. *La Grèce au point de vue naturel, ethnologique, anthropologique, démographique et médical.* Paris: Masson.

Stergiopoulos, Konstantinos. 1937. *Paratiriseis eis tin neoteron yeografian tis Epeirou.* Athens: Dimitrakou.

Sterrett, J. R. S. 1888. "The Wolfe Expedition to Asia Minor." *Papers of the American School of Classical Studies at Athens* 3 (1884–85): 3–448. Boston: Damrell and Upham.

Stewart, Charles F. 1964. *The Economy of Morocco.* Cambridge: Cambridge University Press.

Stoianovich, Traian. 1960. "The Conquering Balkan Orthodox Merchant." *Journal of Economic History* 20:234–313.

Stoianovich, Traian. 1962. "Le maïs arrive dans les Balkans." *Annales: Economies, Sociétés, Civilisations* 17:84–7.

Stoianovich, Traian. 1966. "Le maïs dans les Balkans." *Annales: Economies, Sociétés, Civilisations* 21:1026–40.

Stoianovich, Traian. 1967. *A Study in Balkan Civilization.* New York: Knopf.

Straforello, Gustavo, et al. 1899. *Provincie di Bari, Foggia, Lecce e Potenza.* Turin: UTET.

Strid, Arne. 1986. *Mountain Flora of Greece.* Cambridge: Cambridge University Press.

Strid, Arne, and K. Papanicolau. 1985. "The Greek Mountains." In C. Gómez-Campo, ed., *Plant Conservation in the Mediterranean Area,* 89–111. The Hague: W. Junk.

Stuart, Robert. 1868. "The Vlakhs of Mount Pindos." *Transactions of the Ethnological Society of London* 6:311–27.

Stuart, Robert. 1869. "On the Physical Geography and Natural Resources of Epirus." *Journal of the Royal Geographical Society* 39:276–96.

Suc, J.-P. 1984. "Origin and Evolution of the Mediterranean Vegetation and Climate in Europe." *Nature* 307:429–32.

SVIMEZ [Associazione per lo Sviluppo dell'Industria nel Mezzogiorno]. 1954. *Statistiche sul Mezzogiorno d'Italia 1861–1953*. Rome: Failli.

SVIMEZ. 1978. *Un quarto di secolo nelle statistiche nord–sud 1951–1976*. Rome: Giuffré.

Syndesmos Koinotitos Syrrakou. 1969. *Syrrakon*. Ioannina: Edra.

Synvet, A. 1877. *Carte ethnographique de la Turquie d'Europe et dénombrement de la population Grecque de l'empire Ottoman*. Paris: A. Lassailly.

Synvet, A. 1878a. *Les Grecs de l'empire Ottoman. Etude Statistique et Ethnographe*. Constantinople: Orient Illustré.

Synvet, A. 1878b. *The Greeks of the Ottoman Empire: Statistical and Ethnographical*. London: Hellenic Committee.

Taha, Abdulwahid Dhanun. 1988. *Muslim Conquest and Settlement of North Africa and Spain*. London: Routledge.

Tapia Garrido, José Ángel. 1965. *Historia de la baja Alpujarra*. Almería: Artes Gráficas.

Tapia Garrido, José Ángel. 1981. *Historia General de Almería y su Provincia*. 4 vols. Almería: Cajal.

Targioni, L. 1812–14. *Scelta di notizie interassanti l'agricoltura, la pastorizia . . .* 5 vols. Naples: Masi.

Tassinari, Franco. 1967. *Ricerche sul movimento naturale della popolazione italiana secondo le zone altimetriche*. Rocca San Casciano: Cappelli.

Tayaa, M'hammad. 1985. "Modeling Storm Induced Erosion and Sedimentation in Northern Morocco." Thesis, Institut Agronomique et Vétérinaire Hassan II, Rabat.

Tayaa, M'Hammed, and Kenneth N. Brooks. 1984. "Erosion and Sedimentation in the Rif Mountains of Northern Morocco." Unpublished paper presented at the International Union of Forest Research Organizations Symposium on Effects of Forest Land Use on Erosion and Slope Stability, May 7–11, Honolulu.

Tenore, Michele. 1827. *Essai sur la géographie physique et botanique du royaume de Naples*. Naples: L'Imprimerie française.

Teran, Manuel de. 1943. *Geografía histórica de España, Marruecos y colonias*. Madrid: E. Prieto.

Thalassinos, Dimitrios. 1948. "Iperkarposis." *To dasos* 2, no. 5:18–25.

Thalassinos, Dimitrios. 1953. "Apo tis Aitias tis katastrofis tou Ellinikou dasous." *To dasos* 7, nos. 23–4:41–6.

Théophilou, Meletios D. 1983. *La vie agro-pastorale dans un village montagnard d'Epire*. Salonika: Grafikis Technis.

Théry, Edouard. 1905. *La Grèce actuelle au point de vue économique et financière*. Paris: Economiste Européen.

Thirgood, J. V. 1981. *Man and the Mediterranean Forest*. London Academic Press.

Thomas, William L., ed. 1956. *Man's Role in Changing the Face of the Earth.* 2 vols. Chicago: University of Chicago Press.

Thornes, J. B. 1976. *Semi-Arid Erosional Systems.* London School of Economics Geographical Paper No. 7. London: London School of Economics.

Thornes, J. B. 1987. "The Palaeoecology of Erosion." In J. Wagstaff, ed., *Landscape and Culture: Geographical and Archaeological Perspectives,* 37–55. New York: Basil Blackwell.

Thornes, J. B. 1988. "Erosional Equilibria under Grazing." In John L. Bintliff, Donald A. Davidson, and Eric G. Grant, eds., *Conceptual Issues in Environmental Archaeology,* 193–210. Edinburgh: Edinburgh University Press.

Tichy, Franz. 1957. "Die entwaldungsvorgänge des 19 jahrhunderts in der Basilicata." *Erdkunde* 11:288–96.

Tichy, Franz. 1962. *Die Wälder der Basilicata und die Entwaldung im 19 Jahrhundert.* Heidelberg: Keysersche Verlagsbuchhandlung.

Todd, I. A. 1980. *The Prehistory of Central Anatolia. I. The Neolithic Period.* Göteborg: Paul Astroms Förlag.

Tomaselli, R., ed. 1977. *Mediterranean Forest and Maquis.* Paris: UNESCO.

Tonar, Giancarlo. 1957. "Indagine storico-economica sulla consistenza delle superfici boschive in Toscana sul finire del 1800." *Monti e boschi* 8:255–64.

Tortella Casares, Gabriel. 1985. "Producción y productividad agraria, 1830–1930." In N. Sánchez Albornoz, ed., *La modernización económica de España 1830–1930,* 63–88. Madrid: Alianza.

Touring Club Italiano. Commissione di Propaganda per il Bosco e per il Pascolo. 1913 (?). *Il bosco, il pascolo, il monte.* Milan.

Touring Club Italiano. 1915 (?). *Il bosco contro il torrente. La redenzione delle terre povere.* Milan.

Touring Club Italiano. 1950. "La conservazione del suolo." *Monti e boschi* 1, nos. 10–11. Special issue.

Toynbee, A. J. 1965. *Hannibal's Legacy.* 2 vols. London: Oxford University Press.

Toynbee, A. J. 1969. *Experiences.* London: Oxford University Press.

Tozer, Henry F. 1869. *Researches in the Highlands of Turkey.* London: John Murray.

Trapman, A. 1913. *Greek Operations in Epirus.* London: Royal United Service Institution.

Trifone, Romualdo. 1957. *Storia del diritto forestale in Italia.* Florence: B. Coppini.

Troin, J.-F. 1975. *Les souks marocains.* 2 vols. Aix-en-Provence: EDISUD.

Trolard, Paulin. 1891. *La colonisation et la question forestière.* Algiers: Ligue du reboisement.

Trotta Treyden, H. von. 1916. "Die Entwaldung in den Mittelmeerländern." *Petermanns Geographische Mitteilungen* 62:248–53, 286–92.

Trotter, Alessandro. 1952. "L'Aquifolio nella Lucania." *Monti e boschi* 3:104–7.

Trump, D. H. 1980. *The Prehistory of the Mediterranean.* New Haven: Yale University Press.

Tsoumis, G. 1964. "Forestry in Greece." *Yale Forest School News* 52, no. 3:35–7.

Tuckerman, C. K. 1871. *Brigandage in Greece.* London: Clayton.

Tunçdilek, Necdet. 1951. "The Observation and Some Ideas about Soil Erosion in Turkey." *Istanbul Üniversitesi. Cografaya Enstitüsü Dergisi* 1:91–104.

Tunçdilek, Necdet. 1955. "On the Fuel Problem of the Rural Population in Turkey." *Review of the Geographical Institute of the University of Istanbul (International Edition)* 2:159–67.

Tunçdilek, Necdet. 1963–4. "'Yayla' Settlements and Related Activities in Turkey." *Review of the Geographical Institute of the University of Istanbul (International Edition)* 9–10:58–71.

Turkey. Başbakanlık Devlet İstatistik Enstitüsü. 1957. *Genel Nufüs Sayımı.* Ankara.

Turkey. Başbakanlık Devlet İstatistik Enstitüsü. 1963. *Genel Nüfus Sayımı 23 Ekim 1965.* Ankara.

Turkey. Başbakanlık Devlet İstatistik Enstitüsü. 1968. *Genel Nüfus Sayımı 24.10.1965.* Ankara.

Turkey. Başbakanlık Devlet İstatistik Enstitüsü. 1977. *Genel Nüfus Sayımı 26.10.1975.* Ankara.

Turkey. Başbakanlık Devlet İstatistik Enstitüsü. 1986. *Genel Nüfus Sayımı 20.10.1985.* Ankara.

Turkey. Başbakanlık Devlet İstatistik Enstitüsü. 1988. *Türkiye Istatistik Yıllığı 1987.* Ankara.

Turkey. Başbakanlık İstatistik Genel Direktörlüğü. 1937. *Genel Nüfus Sayımı. 20 Ilk Teşrin 1935.* Istanbul.

Turkey. Başbakanlık İstatistik Genel Müdürlüğü. 1948. *Genel Nüfus Sayımı 21 Ekim 1945.* Ankara.

Turkey. İstatistik Umum Müdürlüğü. 1929. *Umumî Nüfus Tahriri (1927).* Ankara.

Turkey. İstatistik Umum Müdürlüğü. 1954. *Umumî Nüfus Sayımı 22 Ekim 1950.* Ankara.

Turkey. Orman Genel Müdürlüğü. 1967. *Turkey's Forests.* Ankara.

Turkey. Orman Genel Müdürlüğü. 1969. *Antalya Orman Başmüdürlüğü Envanteri 1968.* Ankara: Alkan.

Turrill, W. B. 1929. *Plant Life in the Balkan Peninsula.* Oxford: At the Clarendon Press.

Typaldos Forestis, Angelos. 1912. *Ekthesis peri yeorgias, emborias kai viomichanias en Epeirou (1910–1911) [Deltion tou epi ton Exoterikon Ipourgeion].* Athens: Ethnikou.

Ubicini, M. A. 1853. *Lettres sur la Turquie.* 2 vols. Paris: Dumaine.

Ubicini, M. A., and P. de Courteille. 1876. *Etat présent de l'empire Ottoman.* Paris: Dumaine.

Urquhart, David. 1833. *Turkey and Its Resources.* London: Saunders & Otley.

Urquhart, David. 1838. *The Spirit of the East.* 2 vols. London: Colburn.

Uslu, Selman. 1971. "Erosion Control and Vegetative Cover under Dryland Conditions in Turkey." *Istanbul Üniversitesi Orman Fakültesi Dergisi* 21, no. 2:28–33.

Vacalopoulos, A. E. 1963. "La retraite des populations grecques vers les régions eloignées et montagneuses pendant la domination turque." *Balkan Studies* 4:265–76.

Vacalopoulos, A. E. 1976. *The Greek Nation, 1435–1667.* New Brunswick, N.J.: Rutgers University Press.
Vaccari, Lino. 1932. *Difendiamo i nostri boschi.* Milan: Popolo d'Italia.
Van Andel, Tjeerd H., and Curtis Runnels. 1987. *Beyond the Acropolis: A Rural Greek Past.* Stanford: Stanford University Press.
Van Slyke, Lyman. 1988. *Yangtze.* Reading, Mass.: Addison-Wesley.
Van Zeist, W., and S. Bottema. 1982. "Vegetational History of the Mediterranean During the Last 20,000 years." In J. L. Bintliff and W. van Zeist, eds., *Palaeoclimates, Palaeoenvironments and Human Communities in the Eastern Mediterranean Region in Later Prehistory,* 277–321. Oxford: British Archaeological Reports, International Series, no. 133.
Van Zeist, W., and H. Woldring. 1978. "A Postglacial Pollen Diagram from Lake Van in East Anatolia." *Review of Palaeobotany and Palynology* 26:249–76.
Van Zeist, W., H. Woldring, and D. Stapert. 1975. "Late Quatenary Vegetation and Climate of Southwestern Turkey." *Palaeohistoria* 17:53–143.
Vavaretos, G. 1929. *I Epeiros allote kai tora.* Athens: Dedi & Tsipi.
Vázquez del Río, Manuel. 1935. *Arbolado y bosque in Marruecos.* Ceuta: M. Alcalá.
Vergopoulos, Kostas. 1975. *To Agrotiko zitima stin Ellada.* Athens: Exantas.
Versace, Pasquale. 1986. "Aspetti idrologici nei problema di difesa del suolo in Basilicata." In Basilicata (Regione), *Programmi ed interventi nel settore della difesa del suolo,* 123–9. Matera: EDAD.
Viazzo, Pier Paolo. 1989. *Upland Communities: Environment, Population and Social Structure in the Alps since the Sixteenth Century.* Cambridge: Cambridge University Press.
Viggiani, G. 1931. "Lo sviluppo dell'ecologia agraria in Provincia di Potenza." Extract from *Atti del Raduno dei Tecnici agricoli del Mezzogiorno e delle Isole d'Italia.* Naples: SIEM.
Viggiani, G. 1935. "Il clima della Lucania." *Bollettino del naturalista in Napoli* 47:45–68.
Viggiani, G. 1946. *I problemi agrari della provincia di Potenza.* Potenza: Armento.
Viggiani, G. 1953. "Agricoltura montana in Lucania." *L'Italia agricola* 90, no. 3.
Villar, Sergio. 1984. *Historia di antifranquismo 1939–75.* Madrid: Plaza & Jones.
Villar Amador, Pablo. 1984. *Bibliotecas de Granada.* Granada: Universidad de Granada.
Vincent, B. 1969. "Les pestes dans le royaume du Grenade aux XVIe et XVIIe siècles." *Annales: Economies, Sociétés, civilisations* 74:1511–13.
Vincent, B. 1982a. "Economía y sociedad en el reino de Granada en el siglo XVI." In *Historia de Andalucía,* 4:161–224. Madrid: CUPSA Editorial.
Vincent, B. 1982b. "Economía y sociedad en el reino de Granada (siglo XVIII)." In *Historia de Andalucía,* 5:377–405. Madrid: CUPSA Editorial.
Vita-Finzi, Claudio. 1969. *The Mediterranean Valleys: Geological Changes in Historical Times.* Cambridge: Cambridge University Press.
Vita-Finzi, Claudio. 1975. "Chronology and Implications of Holocene Alluvial History of the Mediterranean Basin." *Biuletyn Geologiczny* 19:137–47.
Vogel, Horst. 1988. "Deterioration of a Mountainous Agro-Ecosystem in the

Third World due to Emigration of Rural Labor." *Mountain Research and Development* 8:321–9.

Vogiatzoglou, V. H. 1986. *I Sparti tis M. Asias.* Athens: Ekdoseis Enosi Spartis M. Asias.

Vogt, Jean. 1956. "Culture sur brûlis et érosion des sols." *Bulletin de la Section de Géographie du Comité des Travaux historiques et scientifique du Ministère de l'Education nationale* 49:337–42.

Vogt, Jean. 1957. "La dégradation des terroirs Lorrains au milieu du XVIIIe siècle." *Bulletin de la Section de Géographie du Comité des travaux historiques et scientifique du Ministère de l'Education nationale* 70:111–16.

Voigt, Paul. 1937. *Die Sierra Nevada.* Hamburg: Evert.

Volkan, Vamik, and Norman Itzkowitz. 1984. *The Immortal Atatürk.* Chicago: University of Chicago Press.

Volpini, Cesare. 1979. "Il legno dalla foreste all'industria." In *Atti del Convegno sulla forestazione produttiva per una industria del legno nel sud,* 85–93. Salerno: Federlegno-Arredo.

Von Haláczy, E. 1901–4. *Conspectus Florae Grecae.* Leipzig: Engelmann.

Von Luschan, F. 1891. "Die Tachtadschy und andere Uberreste der alten Bevölkerung Lykiens." *Archiv für Anthropologie* 19:31–53.

Vouzaris, A., and E. Karamitros. 1981. "Metavoli idrologikon charaktiristikon meta apo pyrkougia stin periochi ton lekanon aporrois seifyllon platyfillon Ossa." *Dasiki erevna* 2, no. 2:135–72.

Vranousis, L. I., ed. 1946. *Mia emmetri istoria tou Zagoriou.* Athens: Mavridis.

Vranousis, L. I. 1962. *Chronika tis mesaionikis kai Tourkokratoumenis Epeirou.* Ioannina.

Vroucha, Panayiota. 1988. "Oi idromiloi tis Epeirou kai ena singrotima idromilou sta Doliana." 1988. In Ethniko Kentro Koinonikon Erevnon and Kentro Neoellinikon Erevnon, *O agrotikos kosmos ston Mesoyeiako choro,* 361–82. Athens.

Wace, A. J. B., and M. S. Thompson. [1913] 1972. *The Nomads of the Balkans: An Account of Life and Customs among the Vlachs of Northern Pindus.* London: Methuen.

Wagstaff, J. M. 1981. "Buried Assumptions: Some Problems in the Interpretation of the 'Younger Fill' Raised by Recent Data in Greece." *Journal of Archaeological Science* 8:247–64.

Wagstaff, J. M. 1985. *The Evolution of Middle Eastern Landscapes: An Outline to 1840.* Totowa, N.J.: Barnes & Noble Books.

Wagstaff, J. M., ed. 1987. *Landscape and Culture: Geographical and Archaeological Perspectives.* New York: Basil Blackwell.

Watson, Andrew M. 1983. *Agricultural Innovation in the Early Islamic World: The Diffusion of Crops and Farming Techniques.* Cambridge: Cambridge University Press.

Weigand, Gustav. 1894–5. *Die Aromunen. Ethnographisch-philologisch-historisch Untersuchungen über das Volk der sogenannten Makedo-Romanen oder Zinzaren.* 2 vols. Leipzig: Barth.

Westermarck, Eduard Alexander. 1926. *Ritual and Belief in Morocco.* 2 vols. London: Macmillan.

Wickham, C. J. 1988. *The Mountains and the City: The Tuscan Apennines in the Early Middle Ages.* Oxford: At the Clarendon Press.

Widdrington, Samuel Edward. 1834. *Sketches in Spain during the Years 1829, 30, 31 & 32; Containing Notices of Some Districts Very Little Known.* 2 vols. London: T. and W. Boone.

Willcox, G. H. 1974. "A History of Deforestation as Indicated by Charcoal Analysis of Four Sites in Eastern Anatolia." *Anatolian Studies* 24:117–33.

Willkomm, Heinrich M. 1882. *Aus den Hochgebirgen von Granada. Naturschilderungen, Erlebnisse, und Erinnerungen. Nebst granadinischen Volkssarungen und Märchen.* Vienna: C. Gerold's sohn.

Wilson, Stephen. 1988. *Feuding, Conflict, and Banditry in Nineteenth-Century Corsica.* Cambridge: Cambridge University Press.

Windus, John. 1725. *A Journey to Mequinez*. London: Jonson.

Winnifrith, T. J. 1987. *The Vlachs: The History of a Balkan People.* New York: St. Martin's.

Winters, Robert K. 1974. *The Forest and Man.* New York: Vantage Press.

Wise, S. M., et al. 1982. "How Old Are the Badlands? A Case-Study from South-East Spain." In R. Bryan and A. Yair, eds., *Badland Geomorphology and Piping,* 259–78. Norwich: Geobooks.

Woolman, David S. 1968. *Rebels in the Rif.* Stanford: Stanford University Press.

Yannoulopoulos, George. 1981. "State and Society in the Ionian Islands 1800–1835." In Richard Clogg, ed., *Balkan Society in the Age of Greek Independence,* 40–62. Totowa, N.J.: Barnes & Noble.

Zamir, Meir. 1981. "Population Statistics of the Ottoman Empire in 1914 and 1919." *Middle Eastern Studies* 17:85–106.

Zervas, P. 1932. *Ta Elleniki dasi.* Athens.

Ziangos, N. G. 1974. *Tourkokratoumeni Epeirou.* Athens: n.p.

Zohary, Michael. 1973. *Geobotanical Foundations of the Middle East.* 2 vols. Stuttgart: Gustav Fischer.

Zohary, Michael. 1983. "Man and Vegetation in the Middle East." In W. Holzner, M. J. A. Werger, and I. Ikusima, eds., *Man's Impact on Vegetation,* 287–96. London: Junk.

Zorzi, L., and C. Reina. 1964. "Aspetti della circolazione idrica sotterranee della pianura di Arta in Epiro e rapporti con le sorgenti, i fiumi ed i complessi rocciosi del bacino." *Tempo tecnico* 1, no. 1:15–44.

Zotos, Dimitrios. 1935. *I Xeniteia ton Epeiroton.* Athens: Ta Chronika.

Zugasti y Sáenz, Julian de. [1876–80] 1982. *El bandolerismo. Estudio social.* Madrid: Alianza.

INDEX

Abdel Krim, 134, 152, 208
Adana plain, *see* Çukurova
agriculture
  in the Alpujarra, 40–1, 252–7
  commercial, 249–60
  environmental effects of, 272–6
  feminization of, 274–5
  lowland, 230–2, 234–6
  in Lucania, 34, 332–4
  mechanization of, 230–1
  medieval expansion of, 87–91
  mountain, 87–91, 102–3
  in the Pindus, 29, 249–52
  in the Rif, 47–9
  and seasons, 108–12
  in the Taurus, 24–6, 74–5
  and yields, 331–6
Albania, 26, 120
Albanians, 32, 119
Algeria, 110, 213, 216, 218, 355
Ali Pasha, 163, 169–72, 267, 292, 293, 294, 299–300
Alpes Maritimes, 13, 70
Alps, 355
Alpujarra
  climate, 36–7
  economy, 40–1
  landscape change, 96–9
  physical geography, 35–6
  population, 189–202
  vegetation, 37
Altınkaya, *see* Zerk
Antalya
  city of, 22, 51, 159
  Plain 20, 22
  timber exports from, 25, 93–4
Apennines, Lucanian
  climate, 31
  economy, 34
  landscape change, 95–6
  physical geography, 30–1
  population, 176–89
  vegetation, 31–2
Arachthos River, 79, 343
Arta, 29
  plain of, 343–4

Basilicata, *see* Apennines, Lucanian
Berbers, 46–7, 88, 91
blood feud, 205–7
brigandage, 91, 117–19, 172, 228–9, 266–70
Büyük Menderes, 77

Calabria (*see also* Sila), 82, 128, 130, 302
California, 68, 255
Calvello, 59, 186
Capileira, 64, 305
caravan trade, 22–4, 224, 232–4
carrying capacity, 4–6
Carthaginians, 82, 84
Castelmezzano, 30
charcoal, 72, 90, 109, 122–3, 137, 243, 245–6, 248, 283–4
Chechouan, 16, 42, 47, 65–6, 309
chestnuts, 130, 330
climate (*see also* Pindus mountains, climate; Taurus mountains, climate, etc.)
  mediterranean, 14–16
commons, problems of, 108, 265–6, 275–6

Corfu, *see* Ionian Islands
Çukurova, 20, 290

Dalías, Campo de, 235
deforestation (*see also* forests), 349–50
  in the Alpujarra, 304–6
  ancient, 72–4
  in ancient Lucania, 81–2
  in the Apennines, 301–4, 349
  in the Pindus, 292–300
  in the Rif, 100–1, 306–10
  in the Taurus, 286–92
DERRO, 322–3
Dinaric Alps, 13, 26, 104, 232
disease (*see also* malaria), 85
  in the Alpujarra, 190, 195, 197
  in Lucania, 177–9, 183
  in the Pindus, 163, 167, 171, 172
  in the Rif, 204
  in the Taurus, 75
drought, 209

earthquakes, 30, 183
Eğridir
  district population, 152–3
  Lake, 22
  town of, 22, 25–6
Egypt, 88, 246
emigration
  from the Alpujarra, 196, 197–8, 200
  from Lucania, 184–5, 186–7
  from the Pindus, 172–3, 175
  from the Rif, 209, 211–18
  from the Taurus, 53, 161
Epirus (*see also* Pindus mountains)
  ancient, 78–9
  medieval, 94–5
erosion, *see* soil erosion
European Community, 55, 161, 187, 239

fauna, 328–30
firewood, *see* fuelwood
fishing, 125
  Rifian, 48–9
floods, *see* watersheds
food, 126–33
forests (*see also* deforestation)
  Alpujarra, 37
  ancient, 72–4
  Apennine, 31
  medieval, 93–101
  Mediterranean, 18–19
  pastoralism and, 277, 281–2
  Pindus, 26–9
  privatization of, 261–2, 263–4
  Rif, 43–4
  Taurus, 21–2, 75–6
fuelwood, 135–8, 283–4
  at Calvello, 59
  in Egypt, 248
  in the Pindus, 95, 243, 250, 251t
  prices of, 326–7
  in the Rif, 246
  shortages of, 326–8
  smelting and, 238–40
  at Trevélez, 64

garrigue, 17–18
goats, *see* pastoralism
Granada
  city of, 35, 40, 190
  Vega de, 35, 200–1, 256
Guadalfeo River (*see also* watersheds), 35, 97, 143

honey, 18, 131
housing, 133–5
hunting, 125, 328–30
  prehistoric, 70
  in the Rif, 48
hydrology, *see* watersheds

Ioannina, 26, 29–30, 293, 300
  artisans of, 121
  medieval, 95
  population, 168–70, 172
Ionian Islands, 10, 249–52
irrigation
  in the Alpujarra, 39–40
  in Antalya Plain, 20
  Arab diffusion of, 87–8
  Campo de Dalías, 235

in the Rif, 48
at Trevélez, 62
in Yılanlı, 51

Kallarites, 57–8, 120, 168–9, 171, 223, 297–8
Ketama, 42, 43, 64–5
kif, 48, 64–5, 114, 257–60
Koçular, 51–3
Kónitsa, 30, 121, 142, 166, 296–7, 315

labor migration, 88–9, 109–10, 227–32
artisan, 119–20
land alienations, 260–2
in Italy, 262–6
in Spain, 261–2
land reform, 185, 304
land tenure, 106–8, 114
landslides, 30–1
Leake, William, 150, 285, 289, 294, 295
Lebanon mountains, 14, 104, 248, 269
Leo Africanus, 100, 152, 203, 204
Levi, Carlo, 117, 120
logging, *see* timber trade
Louros River, 79, 343
Lucania, *see* Apennines, Lucanian

Madeira, 68–9
maize, 5, 89–90, 111, 112, 114, 127
in the Alpujarra, 39, 98, 191–3
in the Apennines, 34
in the Pindus, 29
in the Rif, 48, 101
in the Taurus, 25
yields of, 332–4
at Zerk, 50
malaria, 86–7, 312, 350
in ancient times, 74
and hydrology, 343–5
in Lucania, 176–7, 183, 185
maquis, 17
market integration, 8–11, 221–2
Marquesado del Cenete, 35, 234
Metaponto, 81, 236
Metsovo, 26, 29, 30, 166, 295

mining
in the Alpujarra, 9, 40, 198, 236–40
ancient, 72
in the Rif, 240–1
Montenegro, 206–7
Monticchio, Lake, 267, 303–4
muleteers, 124, 232–4
Mut, 20, 22, 24

Naples, 34, 106, 177, 189
navies (and naval timber), 72, 98–9, 241–2, 247
neolithic revolution, 69–71
New England, 9, 66
nomadism (*see also* pastoralism), 91–2, 157, 219, 281–2, 292
decline of, in Anatolia, 158–60

overshoot, 2–7, 220
in the Alpujarra, 195–7
in the Apennines, 181–4
in the Pindus, 171–2
in the Rif, 211–12, 218

Palestine, 248
Paramythia, 297
pastoralism (*see also* nomadism; wool trade), 91–2, 97, 112–14
in the Apennines, 95–6
decline of, 234–6
environmental effects of, 276–82
instability in, 279
overgrazing and, 279–81
productivity of, 336–8
phylloxera, 195, 253, 254
Pindus mountains
climate, 26
economy, 29–30
landscape change, 94–5
peasant crafts, 223–5
physical geography, 26
population, 162–76
vegetation, 26–9
Pisidia, 74, 75, 93, 119
population (*see also* Pindus mountains, population; Taurus mountains, population, etc.)

instability and landscapes, 7, 147–8
pressure, 2–7, 147–8, 208–9
shortage, 6–7
stability and landscapes, 188–9
potatoes, 5, 90–1, 129–30, 251
in the Alpujarra, 39, 98
in Lucania, 34
Potosí, 9
Pramanda, 58, 120
Pyrenees, 355

rainfall, *see* climate
reforestation, 346–7
Rif mountains, 13
climate, 42–3
economy, 47–9
landscape change, 99–101
physical geography, 41–2
population, 202–18
vegetation, 43–4
Rif War, 208–9, 268–9
roads, 143–4
Alpujarra, 39, 99
Apennines, 32
Pindus, 27–9, 55
Rif, 44
Taurus, 22–4

Samarina, 29, 53–4, 113, 119, 121, 128, 140
scirocco, 15–16, 121
Sebou River, 42, 323
Selge (*see also* Zerk), 50, 74–5
Sierra Contraviesa, 83, 126, 253, 321
Sierra de Gádor, 9, 236–40
Sierra Nevada, *see* Alpujarra
Sila, 10, 82, 244
silk culture, 97, 123, 172, 225–7
Sirako, 56–7
snow trade, 124
social conditions, 116–17
gender and, 109, 144, 330–1
life cycle and, 138–42
literacy and, 144
soil conservation, 174–5
soil erosion, 311–25
Alpujarra, 319–22
ancient, 72–3, 85
Apennines, 80–1, 316–19
and coastal deposition, 312, 318, 320, 342–6
and ethnic differences, 313
neolithic, 71
pastoralism and, 313
Pindus, 314–16
Rif, 322–5
Taurus, 313–14
vegetation and, 311–12
Suez Canal, 158, 247
sugar cultivation, 88, 99, 110, 131, 230, 255–6
sustainability, ecological, 105–6, 114–17, 145–6, 352–4
in the Alpujarra, 201–2
Sütçüler, 22, 124, 154–5

tahtaçı, 122, 247, 286, 291
Tarzhoute, 121, 209
Taurus mountains
climate, 20–1
economy, 24–6
landscape change, 93–4
physical geography, 19–20
population, 152–61
vegetation, 21–2
Termessos, 75
Terranova di Pollino, 31, 59–60, 106, 179
timber, 73, 79, 93–4, 95, 99–100, 122
timber trade, 241–9, 282–3
Apennines, 244
Pindus, 242–3, 252
Rif, 245–6
Taurus, 246–8
tourism, 25, 40–1, 51, 64
Trevélez, 36, 60–4, 128, 197, 305
Türkmen, 91, 122, 270

undershoot, 6–7 173–5

vineyards, 252–5, 336
Vlachs, 29, 54, 56–9, 120, 123, 124, 234, 270
Vovousa, 29, 55–6

warfare, 105, 117, 119, 219–20, 228, 266–70
guerilla, 266–8
in the Pindus, 267–8
and population (Alpujarra), 190–1, 193
and population (Apennines), 185
and population (Pindus), 169–70, 173, 174
and population (Rif), 208–9
water, 131–2
watersheds
Alpujarra (Guadalfeo), 339–43
Apennines, 339, 343
and malaria, 343–5
Pindus, 339
Taurus, 340, 344–5
wheat, 34, 39, 230, 332–4, 335
wine, 131, 252–3
wool trade, 92, 121, 223–4, 280

Yılanlı, 24, 51–3, 152–5

Zagori (*see also* Pindus mountains), 26, 29, 54–6, 106, 115, 120, 162–3, 243, 298
Zerk, 49–51, 155
population (Taurus), 160–1

For EU product safety concerns, contact us at Calle de José Abascal, 56–1°,
28003 Madrid, Spain or eugpsr@cambridge.org.

www.ingramcontent.com/pod-product-compliance
Lightning Source LLC
LaVergne TN
LVHW100133061126
851857LV00010B/71
* 9 7 8 0 5 2 1 5 2 2 8 8 5 *